Lecture Notes in Chemistry

Edited by G. Berthier M. J. S. Dewar H. Fischer
K. Fukui G. G. Hall H. Hartmann H. H. Jaffé J. Jortner
W. Kutzelnigg K. Ruedenberg E. Scrocco

31

H. Hartmann
K.-P. Wanczek

Ion Cyclotron Resonance Spectrometry II

Springer-Verlag
Berlin Heidelberg New York 1982

Authors

H. Hartmann
Dept. of Physical Chemistry, University of Frankfurt
Robert-Mayer-Str. 11, D-6000 Frankfurt

K.-P. Wanczek
Dept. of Inorganic Chemistry
University of Bremen
D-2800 Bremen

ISBN 978-3-540-11957-9 ISBN 978-3-642-50207-1 (eBook)
DOI 10.1007/978-3-642-50207-1

2152/3140-543210

PREFACE

In this volume 28 papers are presented. They cover all the fields studied with ion cyclotron resonance today, including reviews on important fields as well as short contributions on special topics.

This report is devided into four parts:

1. Detailed studies on simple molecules,
2. Systematic studies of the ion chemistry,
3. Spectrometer development and
4. Theory.

The plan to edit a progress report of the complete field was projected at the 2nd International Symposium on Ion Cyclotron Resonance Spectrometry, held at the Akademie der Wissenschaften und der Literatur, Mainz, March, 1981. Most of the contributions were written in late 1981 or in 1982.

Hermann Hartmann
Akademie der Wissenschaften und der Literatur
Mainz

Karl-Peter Wanczek
Institut für physikalische und theoretische Chemie
Universität, Frankfurt

Acknowledgements

The editors gratefully acknowledge support of the Symposium by the Stiftung Volkswagenwerk and the Fonds der Chemischen Industrie.

The editors want to take the opportunity to thank H. Otten, President, Akademie der Wissenschaften und der Literatur, for the hospitality all the participants of the Symposium enjoyed.

Delegates attending the 2nd International Symposium on Ion Cyclotron Resonance Spectrometry, Akademie der Wissenschaften und der Literatur, Mainz (numbers refer to the photograph)

1. M. B. Comisarow
2. K.-P. Wanczek
3. H. Hartmann
4. D. P. Ridge
5. F. S. Klein
6. J. Rasch
7. G. Mauclaire
8. V. G. Anicich
9. J. H. Futrell
10. J. B. Laudenslager
11. B. S. Freiser
12. J. R. Eyler
13. G. Baykut
14. K.-M. Chung
15. W. Ilse
16. D. Schuch
17. J. H. J. Dawson
18. M. Böttger
19. R. C. Dunbar
20. R. Houriet
21. M. M. Kappes
22. J. L. Beauchamp
23. G. Boand
24. R. H. Staley
25. M. T. Bowers
26. D. Parent
27. R. T. McIver, Jr.
28. P. R. Kemper
29. M. L. Gross
30. C. L. Wilkins
31. P. Kemper
32. T. Francl
33. R. L. Hunter
34. M. Inoue
35. J. C. Kleingeld
36. Hp. Kellerhals
37. A. J. Noest
38. M. Allemann
39. S. G. Lias
40. T. Gäumann
41. W. J. van der Hart
42. N. M. M. Nibbering
43. M. Wanczek

Not on the photograph:

F. Becker

R. Derai

G. Gräff

K. Hensen

G. Thews

P. N. T. van Velzen

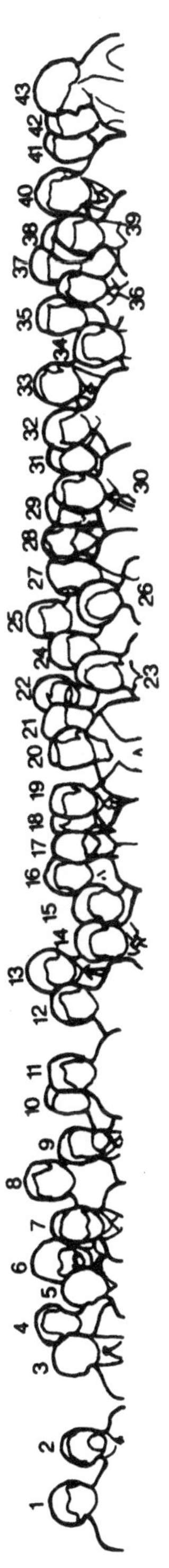
1
2
3
4
5
6
7
8
9
10
11
12
13
14
15
16
17
18
19
20
21
22
23
24
25
26
27
28
29
30
31
32
33
34
35
36
37
38
39
40
41
42
43

TOPICS IN ION PHOTODISSOCIATION

Robert C. Dunbar
Chemistry Department
Case Western Reserve University, Cleveland, OH 44106

1. Introduction

The ICR spectrometer has come to be recognized as one of the best ways of looking at the photochemistry of gas-phase ions, and by the same token, ion photochemistry is an important, lively and (to take a biased view) exceptionally interesting area of ICR research. There are at least five ICR laboratories following such studies, all of them represented at the 1981 ICR Conference at Mainz, and pursuing a variety of aspects of ICR photochemistry. Here, the aim will not be to survey this field, but rather to note some of the different ideas and perspectives that have been current in our laboratory recently. (Recent reviews [1-4] are more comprehensive).

It might be interesting to note several rather different points of view on this general field, each which has generated distinctive types of experiments and bodies of results in the decade or so of research involving shining lights into ICR cells:

- Ion photodissociation as spectroscopy. The photodissociation spectrum provides a reflection of the optical absorption spectrum of an ion (For example, Benz & Dunbar [5]).
- Photodissociation spectra as structural fingerprints for distinguishing and identifying isomeric ions and characterizing rearrangements (Hays and Dunbar [6]).
- The ICR trap for photofragmentation dynamics. Kinetic energy release and angular distribution of fragments can be observed (Orth & Dunbar [7]).
- The ICR cell as a photochemical reactor. Complex one-photon and many-photon photochemical sequences can be untangled (Kim & Dunbar [8]).

Aspects of all four of these points of view will appear in the following sections. The first section describing spectra of "proton-labelled" molecules has a spectroscopic point of view. The next three sections concern various ideas in characterizing isomers and rearrangements. A section on one-photon IR photodissociation illustrates some aspects of ion fragmentation dynamics, although the diligent reader will notice that this work is not ICR work at all, but is included for its interest without apology. The last two sections are concerned with the complex multiphoton photochemistry of ions under laser illumination.

2. Experimental Aspects

Little description of the experimental details of work done in the CWRU laboratory seems necessary, since the ICR experiments use well known, conventional methods, which have frequently been described in detail. (Although a new Fourier-transform, superconducting-magnet instrument has now been built, no results using it will be described here.) In our pulsed ICR instrument, ions are trapped typically for 1 to 5 seconds at pressures between 1×10^{-8} and 1×10^{-6} torr, so that the number of ion-neutral collisions during the trapping period varies from perhaps 0.3 to 200, as desired. The cell is illuminated by one (or more) of several light sources, including: wavelength - selected arc lamp (1000 - 200 nm), Argon-ion laser (458 - 514 nm), pulsed or cw dye laser (700 - 500 nm), and CO_2 laser (lines between 900 and 1100 cm^{-1}).

The fast ion beam instrument at SRI has been described in detail (Huber et al [9]). It consists essentially of a low-pressure electron-impact ion source, a three-section flight tube, and a mass-selecting ion detector. The three flight regions are about 200, 60 and 130 cm in length (going from source to detector) and the ion energy is 3 keV, giving flight times of the order of 100 μsec; laser illumination is in the central region. As is well known, the fast ion beam permits the measurement of kinetic energy release in ion fragmentations with essentially no thermal spread, allowing energy resolutions of fractions of a meV in favorable cases. The possibility of observing fragmentation in both the central region and the third region allows determination of an average time constant for dissociation, both for metastable fragmentation and for photodissociation. The CO_2 laser was directed along the beam axis in the central flight region, with a power of several W/cm^2.

3. Ion Spectroscopy

"Proton-Labelling" Spectroscopy

It was pointed out some time ago by Freiser and Beauchamp [10] that the photodissociation spectra of some protonated molecules closely resemble the UV spectra of the neutral molecules, with a shift in the wavelength region of the spectra. The similarity in appearance of the spectrum or neutral and protonated forms of the molecules is understood as reflecting the minor extent of disturbance of the spectroscopically dominant π electron structure caused by the peripheral attachment of a proton. The energy shift of the spectral features is understood as arising from the differential stabilization of the ground and excited π states due to attachment of the proton. This differential stabilization can equally well be seen as a difference in stabilization of the bound proton, and Freiser and Beauchamp formulated these ideas in terms of a difference in proton affinity of the ground and excited states of the molecule, using a Förster cycle diagram to help clarify the energy relationships. The concept that attachment of a proton has little effect on the nature of the π states is particularly appealing when the proton attachment site involves interaction with non-bonding electrons of an n-donor basic site, while lying in a nodal region of the π system. This is presumably the situation for protonation at the oxygen of acrolein and its derivatives. Since the π spectroscopy of the neutral α,β-unsaturated ketone system has been extensively worked out, it seemed of interest to look systematically at the photodissociation spectroscopy of a series of corresponding protonated molecules.

It was indeed the case for the protonated α,β-unsaturated ketones that the photodissociation spectra closely resembled the optical spectra of the neutrals (Honovich & Dunbar [11]). Figure 1 shows a typical example of this correspondence. The spectra of the protonated molecules were all red-shifted relative to the corresponding neutrals by amounts ranging from 1.1 to 1.7 eV. The amount of red shift (expressed in energy units) seems to group the molecules into three classes: the simple α,β- unsaturated ketones (1.2 eV of red shift); those with an additional conjugated double bond, (1.4 eV of red shift); and those with a conjugated benzene ring (1.6 ev of red shift).

In the spirit of looking at the spectroscopy of the protonated molecules as a "proton-labelled" reflection of the neutral-molecule spectroscopy, it is interesting to organize these results in terms of a set of "Woodward-Fieser" parameters, expressing the position of the spectral peaks in terms

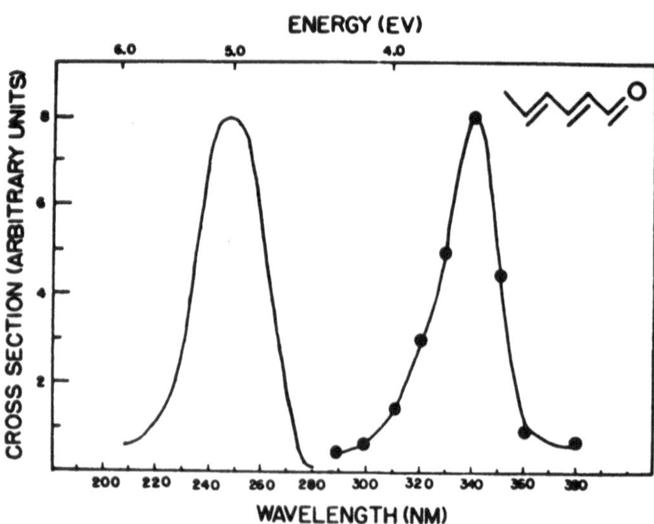

Fig. 1. Gas-phase optical spectrum of neutral molecule (solid curve) and photodissociation spectrum of protonated molecule (curve with experimental points) for the indicated compound. (Reprinted by permission from J. P. Honovich and R. C. Dunbar, J. Phys. Chem. 85(1981)1558).

of a basic $\pi \rightarrow \pi$ transition energy modified by empirical corrections for various substituent and structural features in the molecule. A set of such parameters which gives a satisfactory fit to our results on seven ions is given in Table 1. The "Woodward-Fieser" parameters for the ions are qualitatively similar to those for the neutral molecules: the similarity is most apparent when the parameters are expressed in terms of energy (units of eV in Table 1), in order to compensate for the difference in spectrals regions (Some of the quantitative differences between ions and neutrals may be interesting to think further about, but of course the construction of Table 1 from only seven ion spectra implies a

Table 1. Proposed "Woodward-Fieser" Parameters for Protonated Acroleins

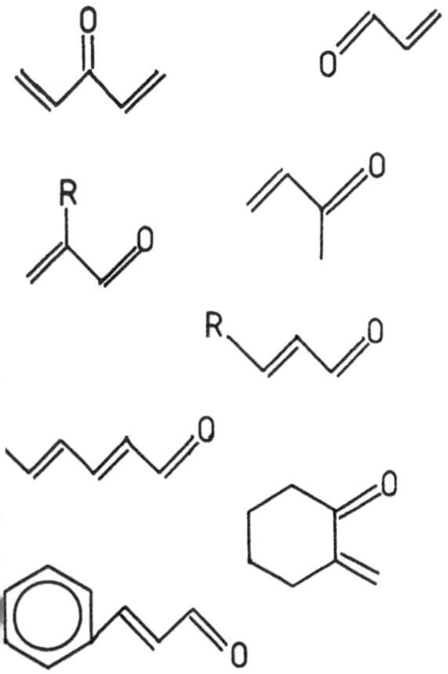

Structural Features or Substituent	Neutral* (Gas-phase)	Ion*
(Base Chromophore)	193 (6.42)	240 (5.17)
(Cross-conjugated chromophore)	200 (6.21)	280 (4.43)
(ketone)	+5 (-.15)	+10 (-.19)
(α-alkyl)	+10 (-.30)	+20 (-.37)
(β-alkyl)	+12 (-.36)	+15 (-.28)
(β-vinyl R)	+53 (-1.37)	+100 (-1.52)
(Exocyclic double bond)	+5 (-.15)	+10 (-.19)
(β-phenyl)	+70 (-1.68)	+140 (-1.88)

*Wavelength or wavelength increment in nm (corresponding energy (eV) in parentheses).

good deal of uncertainty in the ion values.). It seems that the successful ideas developed about the systematic spectroscopy of the neutral acrolein chromophore carry over to the protonated systems. The ultimate attraction of this is the possibility of doing systematic analytical spectroscopy on a mass-spectrometer sample, using chemical ionization methods to attach protons to the sample molecules, and then using photodissociation spectroscopy as a systematic structure probe in ways already well established in analytical organic spectroscopy.

4. Ion Structure Studies

The use of photodissociation spectroscopy to characterize ion structures is by now a familiar approach, and it will be sufficient here to illustrate these possibilities with just a few diverse examples.

4.1 Chloropropene Ions

The capability of laser photodissociation at reasonable resolution for distinguishing isomeric ion structures is illustrated by results for chloropropene ions $C_3H_5Cl^{+\cdot}$ (Orth & Dunbar [12]). Of the four isomers studied (I - IV), only allyl chloride (IV) would be readily distinguishable by

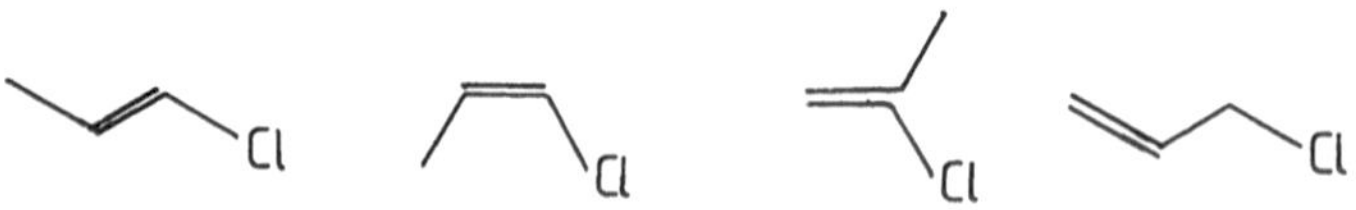

Cl Cl Cl Cl

I II III IV

its low resolution photodissociation spectrum. The laser spectra of the four isomers at 1 nm resolution are shown in Figure 2. It is evident that the spectra of the other three isomers (I, II and III) are similar in overall features, but at this resolution they are readily distinguished by their characteristic vibrational structure, and by the difference in onset wavelength. The overall position of the photodissociation peak in 1-chloropropene (I or II) corresponds excellently to the peak in the photoelectron spectrum of the neutral lying 2.0 eV above the ground state (Katrib et al [13]). This excited state has been confidently assigned as corresponding to the a' out-of-plane chlorine lone-pair orbital; optical excitation from the ion ground state to this is weakly allowed, with poor orbital overlap accounting for the low observed photodissociation cross section of 5×10^{-20} cm^2 (at 579 nm). The UV photodissociation spectra of I, II and III are all similar, showing peaks at 3.65 and 4.7 eV which would be wholly useless in attempting structural distinction of the isomers.

The usefulness of high-resolution laser photodissociation as a structure probe was pointed up by applying the spectroscopy of the previous paragraph to a study of the ion-molecule reaction products of the reaction

$$C_3H_6^+ + C_3H_7Cl \longrightarrow C_3H_8 + C_3H_5Cl^+ \qquad (1)$$

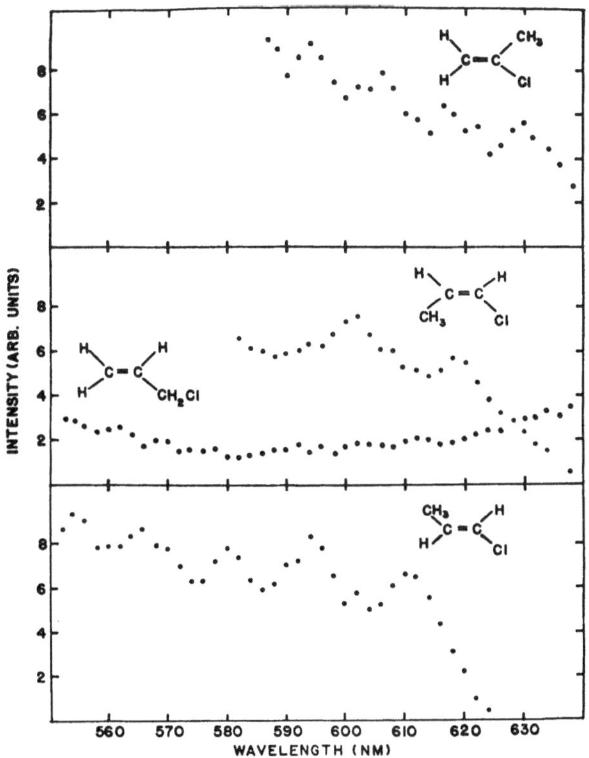

Fig. 2. Laser photodissociation spectra of parent cations of four $C_3H_5Cl^+$ isomers. (Reprinted by permission from R. G. Orth and R. C. Dunbar, J. Am. Chem. Soc. 100(1978)5949).

When the neutral reactant was 2-chloropropane, the $C_3H_5Cl^+$ product ion gave a photodissociation spectrum like that of authentic 2-chloropropane ion (III). On the other hand, the product ion from reaction of 1-chloropropane gave a photodissociation spectrum like that expected for a mixture of a cis and trans 1-chloropropene ions (I and II). The conclusion that reaction (1) proceeds, formally, by abstraction of H_2^- from the chloroalkane and retention of the C_3Cl skeletal structure in the $C_3H_5Cl^{+\cdot}$ product is not surprising, and has precedent in olefin ion-molecule reactions (Sieck & Futrell [14], Henis [15]).

4.2 Benzyl Chloride Ion

The cation obtained from benzyl chloride has provided an interesting story in the study of isomeric ions. First results, (Fu et. al. [16]) based on photodissociation in a continuous - mode ICR spectrometer, showed the presence of two spectroscopically distinct ions of m/e 126, one predominantly absorbing red light, the other blue light. It was believed that these two ions were formed in comparable abundance by the electron impact ionization process, and it was proposed that the red-absorbing ion had the benzyl chloride structure, the blue-absorbing ion some other structure, (possibly chlorotoluene).

This picture was upset by Morgenthaler and Eyler's [17] finding that at the pressures and trapping times of their pulsed ICR, pulsed-dye-laser instrument, the ions were entirely of the red-absorbing form, with no suggestion of formation of another structure.

Most recently we have reinvestigated this puzzling problem with pulsed ICR and continuous (but time-gated) lasers of both red and blue (or green) wavelengths. At low pressures and short delays before the beginning of irradiation, all of the ions photodissociate readily with red light, and the ion population appears homogeneous in red-absorbing ions. Part of the earlier confusion is accounted for by the observation that these ions apparently have modest but significant photodissociation cross section in the blue as well as around 420 nm.

If the benzyl chloride pressure is raised, and/or if a sufficient delay time is allowed before irradiation, a collisional rearrangement occurs, yielding a new ion structure with very intense photodissociation in the blue (see Figure 3). We have characterized this rearrangement carefully: it has a collisional rate constant k_R of 1.7×10^{-11} cc sec^{-1} molecule^{-1} in benzyl chloride neutral. On the other hand, collisions with N_2 neutrals seem to be ineffective, and N_2 at the highest usable pressures gave no yield of rearranged, blue-absorbing ions. No evidence is available on whether the rearrangement process is a true collision-induced rearrangement, or alternatively an atom-exchanging reactive process, although the ineffectiveness of N_2 gives some weight to the latter possibility. In any event, this seems to be the first well characterized instance of a collisonal ion-rearrangement process yielding an isomerized but equal-mass product ion. In summary, our current understanding of this system is described by the reaction scheme.

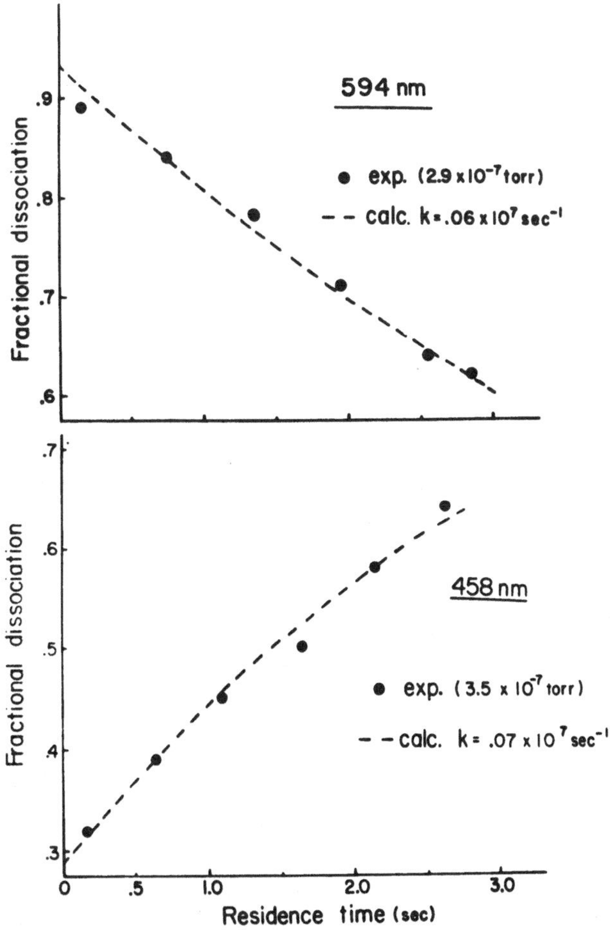

Fig. 3. Comparison of $C_7H_7Cl^+$ photodissociation at red and blue wavelengths, as a function of residence time. Following ion formation at time zero, ions are stored and undergo collisions for the indicated residence time, and are then sampled by a brief (300 msec) pulse of laser irradiation at either 458 or 594 nm. The increase in blue-light dissociation and decrease in red-light dissociation match the expectation for a simple collisional rearrangement process (solid lines).

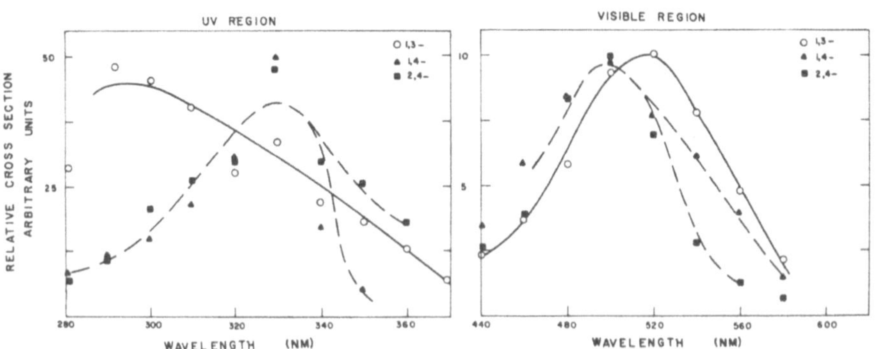

Fig. 4. Photodissociation spectra of three isomers of octadiene parent ions.

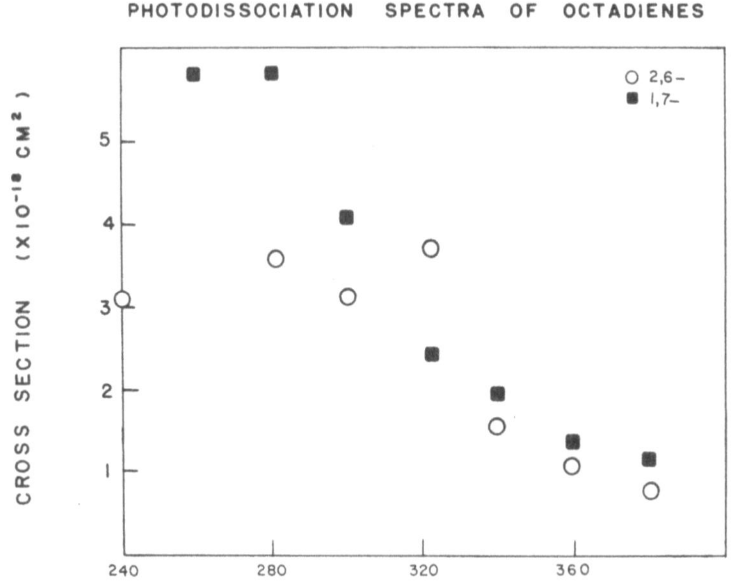

Fig. 5. Photodissociation spectra of two unconjugated octadiene-ion isomers.

$$C_6H_5CH_2Cl \xrightarrow{EI} C_6H_5CH_2Cl^{+\cdot} \xrightarrow{600\ nm} \text{dissociation}$$

$$C_6H_5CH_2Cl^{+\cdot} + C_6H_5CH_2Cl \xrightarrow{k_R} C_6H_4ClCH_3^{+} \xrightarrow{400\ nm} \text{dissociation}$$

4.3 Diene-Ion Rearrangements

Photodissociation spectroscopy has given an effective means of investigating the rearrangements of ions following electron impact ionization. A recently reported series of experiments on hexadienes (Benz et al. [18]) explored two types of rearrangement process, migration of the double bonds into conjugation, and cis-trans isomerization around a double bond. It was found that the terminal double bond of 1.4-hexadiene ion moved into conjugation, but that 1.5-hexadiene ion retained its unconjugated structure. The three cis-trans isomers of 2,4-hexadiene gave distinct photodissociation spectra, indicating lack of rotation about the double bonds.

A recent investigation of the octadiene ions adds to our increasing understanding of the patterns of hydrocarbon cation rearrangements. As shown in Figure 4, the conjugated 1.3 and 2.4-octadiene ions give spectra with general similarity, but they are clearly distinguishable, with the 1.3-ion having its visible peak farther to the red and its UV peak farther to the blue. The ion obtained from 1.4 octadiene clearly rearranges to a conjugated structure, and almost certainly has the 2.4-octadiene structure. 2.6- and 1.7-octadienes, in constrast, give spectra, shown in Figure 5, which are clearly characteristic of unconjugated olefin ions, and show no evidence for any rearrangement. These results confirm a pattern observed in several series of radical ions that shift of a double bond by one position into conjugation upon ionization is rapid and complete, but a shift of two or more positions is difficult.

The spectra shown in Figure 6 for the cis and trans isomers of 1.3-pentadiene add more evidence concerning the cis-trans isomerization of diene ions. Although the low resolution and the scatter of the data make precise comparison hard, it seems quite certain that the UV absorption peak of the trans ion (near 295 nm) lies at significantly shorter wavelength than that of the cis ion (near 310 nm). These results are evidence against complete cis-trans equilibration in the pentadiene ion case, in accord with the similar result for hexadiene ions. It is interesting that

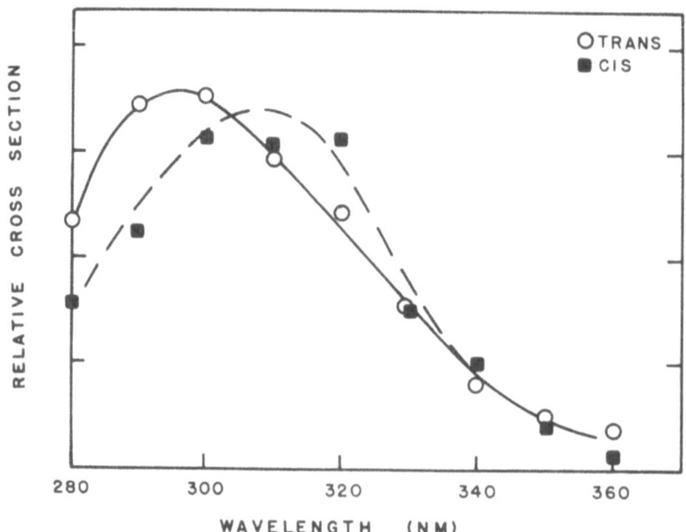

Fig. 6. Photodissociation spectra of cis- and trans-1,3-pentadiene parent ions.

the difference in the UV peak position is substantial for the two pentadienes: in the hexadiene case, the UV peak of trans-trans hexadiene ion is probably about 5 nm to the blue of the other two isomers, (which are similar). For the pentadienes, visible spectra at sufficient resolution to distinguish the isomers still need to be obtained.

5. Photofragmentation

IR Photodissociation of Ions in a Fast Ion Beam

ICR experiments with infrared excitation, such as those described below or those of other laboratories involving IR photodissociation (for instance, Bomse et al. [19]), extract information about the lower part of the IR excitation ladder: most of the infrared photons are absorbed by ions with internal energy much less than dissociation threshold. A complementary experiment yielding information about the very top of the excitation ladder is possible with ion beam instruments, and was carried out at Stanford Research Institute in collaboration with the Molecular Physics Group there (Coggiola et al. [20], report earlier results of this experiment). Ions are produced in a reasonably conventional electron impact source, and take on the order of 50 μsec to traverse the distance to the interaction region. There they interact with a cw CO_2 laser for a few μsec, long enough for a small fraction of the ions to absorb one IR photon. Photoexcited parent ions and dissociation products traverse an additional free-flight path, and then are mass-analysed and kinetic-energy analyzed to give the relative cross section, the time constant, and the kinetic energy release for photodissociation. Since the photodissociation process observed is definitely a one-photon process, the interesting feature of this experiment is that it observes the photochemistry only of that small fraction of the primary ion beam consisting of ions whose internal energy places them within one IR photon of dissociation (or even metastables above the photodissociation threshold). In other words, this experiment corresponds specifically to the _last_ photon of a multiphoton IR photodissociation experiment.

Over 30 ions have been examined in this way, and some interesting observations have resulted. Figure 7 illustrates a typical spectrum of photodissociation cross section as a function of IR frequency: Many ions show well defined peaks in their spectra. and it is certainly true that

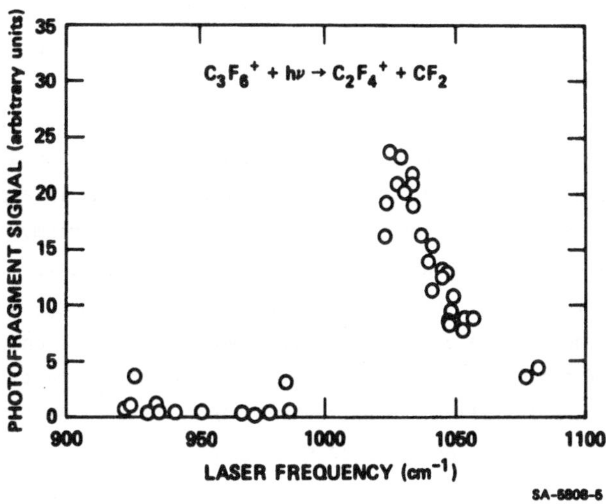

Fig. 7. Photodissociation spectrum of $C_3F_6^+$ irradiated by CO_2 laser in a fast ion beam experiment.

the presence of several eV in internal energy in the ion does not destroy the IR spectral peaks which would be expected if the ions were vibrationally cool. This is easily understood in a nearly-harmonic picture of the poly-atomic molecule, because in molecules this large the individual vibrational modes are not highly excited (many are in v=0) even at rather high levels of internal energy. Thus v=0 to v=1 transitions in the IR-active mode are still well-defined, although possibly broadened or shifted by anharmonic perturbations, and give normal-looking IR spectra. The only ion for which a comparison has yet been made between the one-photon hot-ion spectra of the present experiments and a multiphoton IR dissociation spectrum is $C_3F_6^+$: The multiphoton spectrum peaks at 1043 cm^{-1} (Bomse et al.[21]), while the present one-photon results show the peak red-shifted, presumably by hot-ion anharmonicity, to 1030 cm^{-1}.

These ion-beam experiments also give a unique opportunity to look at the details of the IR multiphoton fragmentation process. Table 2 shows the kinetic energy release and average fragmentation time for photodissociation of four ions which have been carefully analyzed, along with the corresponding quantities for the metastable (laser independent) fragmentation also observed. Also shown in the table are results of calculations of these quantities based on a quasiequilibrium (RRKM) theory model of the fragmentation process (implying complete statistical randomization of the internal energy prior to fragmentation.) for $C_2F_5I^+$, $C_3F_7I^+$ and $C_6H_5I^+$. The RRKM model evidently gives a very satisfactory description of the dissociation process; but for CF_3I^+, the kinetic energy released is far smaller than can be reconciled with RRKM calculations, and it appears that this dissociation must occur by some non-statistical mechanism, perhaps involving direct dissociation from an excited electronic state. The counter-intuitive observation that $C_3F_7I^+$ releases more kinetic energy than $C_2F_5I^+$ is an interesting feature of the ion beam experiment itself, and arises because many of the $C_3F_7I^+$ ions arriving at the photon-interaction region still possess substantially more energy than the dissociation threshold energy, while such super-threshold ions are largely depleted in the $C_2F_5I^+$ case.

Table 2. Observed and RRKM Quantities for IR One-Photon Fragmentation

	Metastable Fragmentation		IR-Induced Fragementation	
	Energy Release (meV)	Average Lifetime (μsec)	Energy Release (meV)	Average Lifetime (μsec)
CF_3^+				
Obs.	--	--	4.4	<<1
Calc.	--	--	24	1
$C_2F_5I^+$				
Obs.	2.8	26	10	2.2
Calc.	2.5	25	12	2.8
$C_3F_7I^+$				
Obs.	10	>100	14	5.5
Calc.	10	170	21	5.5
$C_6H_5I^+$				
Obs.	29	>100	28	N.A.
Calc.	39	250	50	30

6. Multiphoton Photochemistry

An ion trapped in an ICR cell provides a uniquely attractive opportunity for photochemistry, in that it is a molecule isolated from uncontrolled collisions and wall interactions on a time scale of seconds, yet its photochemical behavior is readily monitored by the various techniques of ICR spectrometry. Some of the various processes whose rates and competition can be studied are suggested pictorially in Figure 8. A few currently interesting aspects of these possibilities will be discussed.

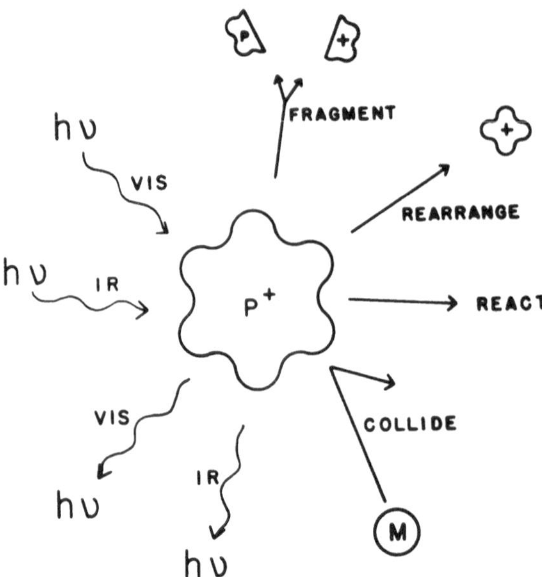

Fig. 8. Events in the life of a trapped ion.

6.1 Two-Photon Dissociation and Collisional Relaxation

Photodissociation involving one, two (Freiser & Beauchamp [22], Dunbar & Fu [23]) or four (Kim & Dunbar [8]) visible photons is well known, but the possibilities of exploiting the kinetics of two-photon dissociations to study collisional energy transfer processes have not been so widely appreciated. Looking at the accepted Freiser-Beauchamp [22] sequential two-photon mechanism,

$$A^+ \xrightarrow[\sigma_1]{h\nu} A^{+*} \xrightarrow[\sigma_2]{h\nu} \text{products} \qquad A^{+*} \xrightarrow[k_3]{M} A^+ \tag{2}$$

analysis of the competition between σ_2 photodissociation and k_3 collisional relaxation by neutral collision partner [M] readily yields the yield k_3 for collisional quenching of A^+ . One ion which was systematically studied was brombenzene parent ion (Kim & Dunbar [24]),

$$C_6H_5Br^+ \xrightarrow{h\nu} C_6H_5^+ + Br \tag{3}$$

for which the photodissociation threshold is 2.75 eV. 514.5 nm photons deposit 2.41 eV of energy, so that the internal energy needed in A^{+*} is 0.34 eV in order to lie above the one-photon threshold for the σ process. The collisional quenching rate k_3 determined in this way thus corresponds to the rate at which ions are lowered from 2.4 eV of internal energy to below 0.34 eV. Since the average thermal internal energy of an ion at 300 K is on the order of 0.1 eV, the "relaxation" process as defined for this experiment corresponds to removal of nearly all the deposited excitation, and cooling to vibrational temperatures not far above thermal. (In fact, 0.34 eV corresponds to about 550 K.) k_3 rates are listed in Table 3 for several collision partners: the rates have been reduced by dividing by the Langevin orbiting rate in order to express them as the number of orbiting collisions required for quenching. This does not imply that quenching actually takes place by orbiting collisions, nor does it imply anything about whether quenching occurs by few strong collisions or many weak ones - this method does not give useful information about these questions.

Table 3. Collisional Quenching of Internal Excitation

	$C_6H_5Br^+$		$C_6H_5I^+$	
	Rate*	Number of Collisions	Rate*	Number of Collisions
CH_4	$\leq$.02	$\geq$50	0.05	20
C_3H_8	$\leq$.05	$\leq$20		
Cyclo-C_6H_{12}			0.5	3
C_6H_5F	.46	3		
Parent	1.5	1	1.5	1

*10^{-9} cm^3/molecule-s

Some similar results have been obtained for iodobenzene ion.

$$C_6H_5I^+ \xrightarrow{h\nu} C_6H_5^+ + I \tag{4}$$

The two-photon threshold in this ion is 2.4 eV, and at 600 nm the one-photon threshold is at an internal energy of only 0.3 eV. Since the average thermal energy of iodobenzene is 0.1 eV at 300 K, it is seen that "quenching" in this case is nearly equivalent to thermalization. Quenching efficiencies are shown in Table 3.

For ions in their parent gases, symmetric charge transfer provides a plausible mechanism for efficient collisional quenching. It seems necessary to accept this mechanism for both iodobenzene and bromobenzene ions, because, if charge transfer is not allowed, it would require at least three collisions even with complete energy equipartitioning to quench A^{+*}, and

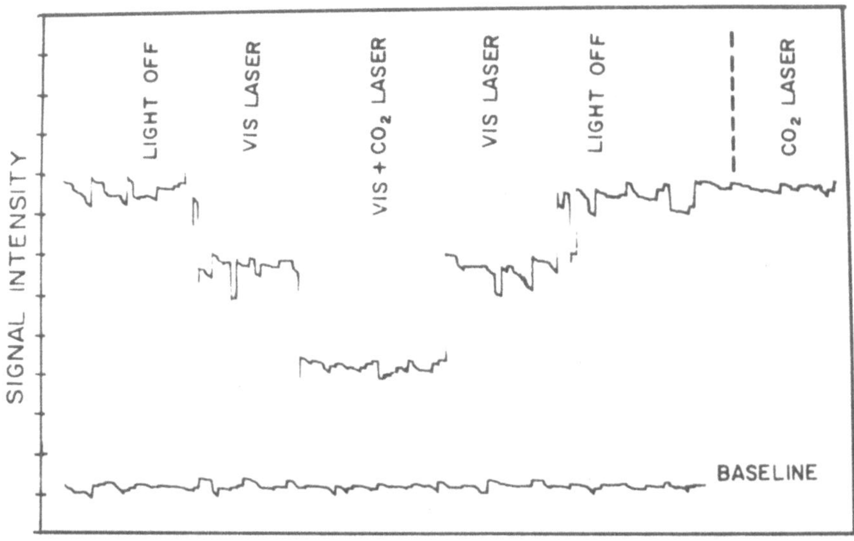

Fig. 9. Photodissociation of $C_6H_5I^+$ with visible radiation (600nm) alone, with visible radiation plus IR radiation, and with IR radiation alone. The drop of the $C_6H_5I^+$ abundance trace below the no-light level indicates the extent of photodissociation . "Baseline" indicates the signal observed with no ions in the cell.

the observed quenching is more efficient than this. By the same token, these results indicate that charge transfer occurs with relatively little energy sharing (since extensive energy sharing sharply reduces the quenching efficiency) .

It is striking that some non-charge-transfer quenching partners, e.g. fluorobenzene in the case of bromobenzene ions, cyclohexane in the case of iodobenzene ions, have such high quenching efficiencies. For iodobenzene ions, the finding that three collisions with cyclohexane result in quenching. indicates very extensive energy sharing on each collision, such that the photoexcited A^{+*} ion loses 1/3 to 1/2 of its internal excitation on each collision.

---

* However, recent evidence (Derai et al. [25]) suggests that charge transfer in polyatomics at low kinetic energy is a strong-collision process with substantial internal-to-translational energy conversion, and it would seem likely that strong collisions with some energy sharing occur in the cases of interest here. This is a very interesting question for further study.

6.2 Two-Laser Photodissociation

We have found a novel photodissociation process involving the concerted effects of a visible laser and a CO_2 laser, which is most prominent in iodobenzene ions (Dunbar et al.[26]), and less so in bromobenzene ions, among the ions we have investigated. As shown in Figure 9, the extent of photodissociation of iodobenzene ion by modest power from a cw visible laser at 600 nm can be nearly doubled when the ions are simultaneously irradiated by a cw CO_2 laser at about 10 Watts (giving a maximun IR flux of several tens of Watts/cm^2). This enhancement shows a strong dependence on IR frequency, as shown in Figure 10 (This wavelength dependence is evidence that the observed effect is not due to an unrecognized cell-heating artifact). The dependence of photodissociation rate on the intensity of both visible and IR irradiation is approximately linear (or a bit stronger).

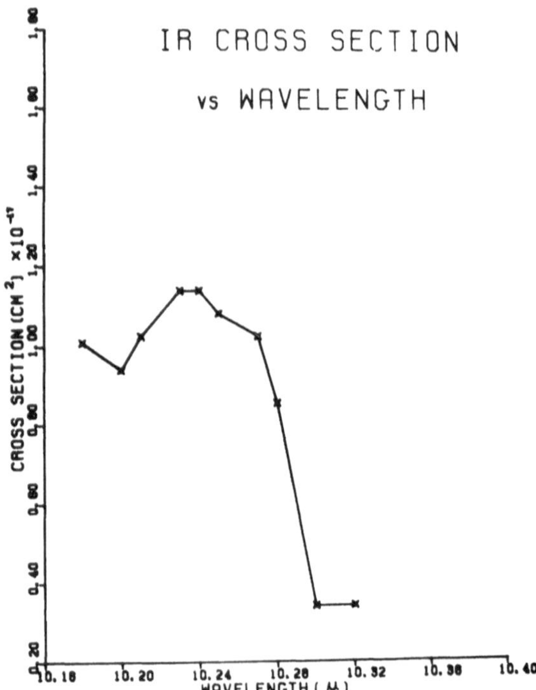

Fig. 10. Dependence of the IR enhancement effect on the wave length of the IR light, with visible light irradiation constant at 600 nm.

Qualitatively we understand this effect as arising from the interplay of the various excitation and relaxation processes in this dissociation system. The one-photon* lies at $\sim$ 0.3 eV above the ground state, which is significantly higher than the average thermal energy of 0.08 eV, and corresponds to the energy of about three IR photons. IR pumping is presumed to be relatively intensitive to the internal energy content of the ion, but IR radiative relaxation certainly increases very fast with increasing internal energy. Accordingly IR irradiation is effective in lifting some steady-state fraction of the ions above the one-photon threshold, where they are efficiently photodissociated by the visible light, but at the same time is entirely unable to drive ions above the dissociation threshold to give pure IR multiphoton dissociation (which is not observed). Collisional relaxation is an important additional relaxation process, and the IR enhancement is strongly pressure sensitive.

The suggested mechanism of IR enhancement, involving the accumulation of substantial internal energy in the ion by IR pumping, followed by visible-photon dissociation, implies the possibility that the IR pumping part of

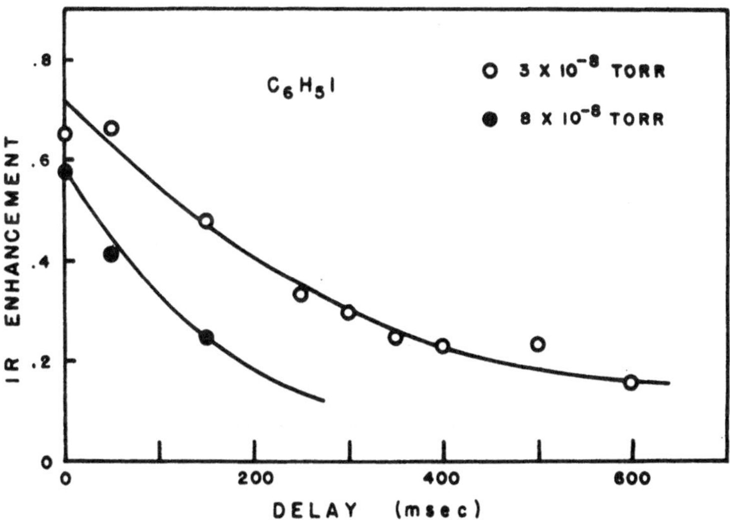

Fig. 11. Extent of IR enhancement of $C_6H_5I^+$ photodissociation when IR irradiation precedes visible irradiation by a delay time between zero and 0.6 sec.

---

*That is the threshold value of internal energy above which one visible photon can raise the ion above the photodissociation threshold.

the experiment might be equally well gated to precede the visible-laser photodissociation. In fact this is the case, as Figure 11 indicates. IR enhancement of photodissociation is observed in an experiment in which the IR laser (alone) illuminates the cell for a period, is then gated off for a dark period, followed by turning on the visible laser (alone) for a photodissociation period. As expected, the extent of IR enhancement in this sequential two-laser declines rapidly as the length of the dark period between the two lasers increases: the persistent effect of the initial CO_2 laser illumination dies away with a time constant of 400 msec at low pressure. As the pressure is raised this fall-off becomes faster, so that at 9×10^{-8} torr, the IR-laser effect is relaxed in 150 msec.

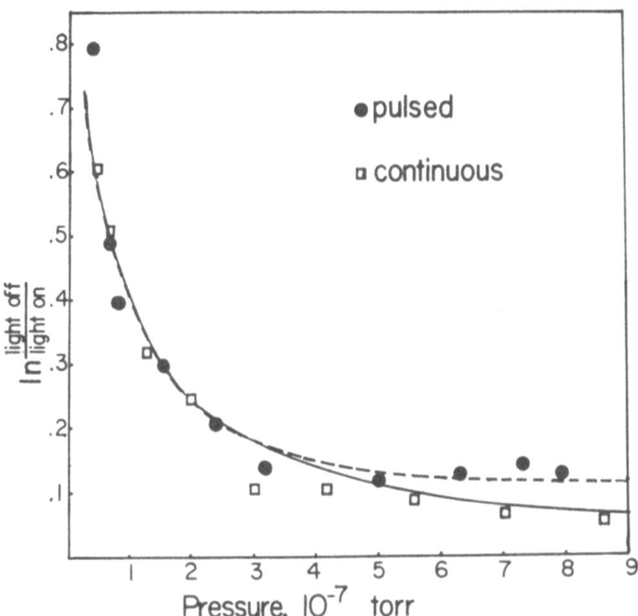

Fig. 12. Comparison of $C_6H_5I^+$ photodissociation at 600 nm with continous irradiation and with pulsed irradiation with a pulse repetition rate of 10 Hz.

This pressure dependence is consistent with a collisional relaxation of 2×10^{-9} cm^3 $molecule^{-1}$ sec^{-1} which is of the order of the charge-transfer rate.

The picture developed above of iodobenzene ion multiphoton chemistry is readily adapted to numerical simulation using a computer routine which includes the processes of visible-laser photoexcitation, IR laser photo-excitation, IR-radiative relaxation, collisional relaxation, collisional excitation, and photodissociation. The rates of many of these individual processes are not well characterized so that numerical simulation involves fixing several adjustable parameters. It is encouraging to find that there is some latitude in the assignments of these: the computer experiment readily reproduces the broad features of the observed results, specifically giving good agreement with the observed pressure and intensity dependences, for both the visible laser experiment and the two-laser experiment. Granted that this agreement cannot be taken as proving the assumed mechanism, still it shows that the picture is a reasonable one.

Similar results are obtained in this two-laser experiment if a Chromatix CMX-4 pulsed dye laser is used in place of the cw dye laser described above. In fact, repetitively pulsed irradiation is entirely equivalent to cw irradiation in sequential two-photon photodissociation, as long as the relaxation rate (whether collisional or IR-radiative) is much slower than the repetition rate of pulsed irradiation. However, as Figure 12 illustrates, when the relaxation rate is increased (by raising pressure) to the point that significant relaxation takes place between each laser pulse, the pulsed laser becomes more effective, for equal average light flux, than the cw laser. This is most readily seen as reflecting the fact that relaxation <u>during</u> the laser pulse cannot occur so that increasing pressure cannot decrease the pulsed-laser photodissociation rate below an irreducible minimum.

In addition to its inherent interest in providing various kinds of new evidence about the various photophysical processes occuring in gas-phase ions, the two-laser experiment, which we can characterize as IR-enhanced two-photon photodissociation of the ion, is also potentially interesting in providing a new way of doing infrared spectroscopy of ions. Since the ions need only absorb a few IR photons for the effect to appear, this experiment does not share with pure IR multiphoton spectroscopy the drawback of dealing with very highly excited ions, but instead looks hopeful as a way of looking at IR absorption of ions not far above thermal excitation. The development of means of studying IR spectroscopy of ions seems likely to be one of the exciting areas of future ICR photochemical and spectroscopic research, and it is becoming clear that there are many useful ways of attacking this problem.

7. Acknowledgements

Although this chapter has only one author, the results described owe their existence to the major efforts of a number of people, who should rightfully have the status of coauthors: Jeffrey Honovich brought about the "proton-labelled" spectroscopic experiments, and substantial parts of the IR-laser ICR experiments; Naomi Lev was involved in much of the two-photon and two-laser work (and earlier, Myung Kim); Robert Orth and George Fitzgerald are largely responsible for the ion-structure results described; and the author formed only a modest part of the effort at SRI on the fast beam experiments, for which Philip Crosby, Michael Coggiola, Peter Helm and James Peterson were the largest contributors. Various aspects of the work described were supported by the donors of the Petroleum Research Fund, administered by the American Chemical Society; the National Science Foundation; and the U.S. Air Force Geophysics Laboratory; whose support is gratefully acknowledged.

8. References

1. R.C. Dunbar, "Gas Phase Ion Chemistry", Vol. II, M.T. Bowers, Ed., Academic Press, 1979.

2. R.C. Dunbar, "Ion Photodissociation" in "Kinetics of Ion-Molecule Reactions", P. Ausloos, Ed., Plenum Press, 1979

3. R.C. Dunbar, "Physical Methods of Modern Chemical Analysis", Vol. 2, T. Kuwana, Ed., Academic Press, 1980

4. R.C. Dunbar, Specialist Periodical Reports, Mass Spectrometry, Volume 6, R.A.W. Johnstone, Ed., 1981

5. R.C. Benz and R.C. Dunbar, J. Am. Chem. Soc., 101 (1979) 6363

6. J.D. Hays and R.C. Dunbar, J. Phys. Chem. 83 (1979) 3183

7. R.G. Orth and R.C. Dunbar, J. Chem. Phys. 68 (1978) 3254

8. M.S. Kim and R.C. Dunbar, J. Chem. Phys. 72 (1980) 4405

9. B.A. Huber, T.M. Miller, P.C. Cosby, H.D. Zeman, R.L. Leon, J.T. Moseley and J.P. Peterson (1977), Rev. Sci. Instrum. 48 (1977) 1306

10. B.S. Freiser and J.L. Beauchamp, J. Am. Chem. Soc., 99 (1977) 3214

11. J.P. Honovich and R.C. Dunbar, J. Phys. Chem. 85 (1981) 1558

12. R.G. Orth and R.C. Dunbar, J. Am. Chem. Soc. 100 (1978) 5949

13. A. Katrib, T.P. Debies, R.J. Colton, T.-H. Lee and J.W. Rabalais, J. Chem. Phys. Lett. 23 (1973) 196

14. L.W. Sieck and J.H. Futrell, J. Chem. Phys. 45 (1966) 560

15. J.M.S. Henis, J. Chem. Phys. 52 (1970) 282

16. E. Fu, P.P. Dymerski and R.C. Dunbar, J. Am. Chem. Soc. 98 (1976) 337

17. L.N. Morgenthaler and J.R. Eyler, Int. J. Mass. Spectrom. Ion Phys. 37 (1981) 153

18. R.C. Benz. P.C. Claspy and R.C. Dunbar, J. Am. Chem. Soc. 103 (1981) 1799

19. D.S. Bomse, R.L. Woodin and J.L. Beauchamp, J. Am. Chem. Soc., 101 (1979) 5503

20. M.J. Coggiola, P.C. Cosby and J.R. Peterson, J. Chem. Phys. 72 (1980) 6507

21. D.S. Bomse, R.L. Woodin and J.L. Beauchamp, "Advances in Laser Chemistry", A.H. Zewail, Ed., Springer Series in Chemical Physics, Springer, New York, 1978

22. B.S. Freiser and J.L. Beauchamp, Chem. Phys. Lett., 35 (1975) 35

23. R.C. Dunbar and E.W. Fu, J. Phys. Chem., 81 (1977) 1531

24. M.S. Kim and R.C. Dunbar , Chem. Phys. Lett. 60 (1979) 247

25. R. Derai, R. Marx and G. Mauclaire, results presented at the Mainz Conference, 1981

26. R.C. Dunbar, J.D. Hays, J. Honovich and N.B. Lev, J. Am. Chem. Soc., 102 (1980) 3950

ION STRUCTURES AND RELAXATION OF VIBRATIONALLY EXCITED IONS AS STUDIED BY PHOTODISSOCIATION

P.N.T. van Velzen and W.J. van der Hart
Gorlaeus Laboratories, University of Leiden
P.O. Box 9502, 2300 RA Leiden. The Netherlands.

1. Introduction

The rapid development of experiments on the photodissociation of ions in beams or trapped in an ICR cell has been described in several review papers [1]. Therefore, in this paper we will only give a survey of some experimental results recently obtained in this laboratory.

Irradiation of the ion mixture in an ICR cell at low pressures may result in a decrease of the intensity of ions of a certain mass by light-induced fragmentation processes. Simultaneously, the intensity of the corresponding fragment ions will increase. It follows that photodissociation can be studied by measurement of both the dissociating ions and of their fragments. The latter type of experiment is especially interesting in cases where the fragmentation processes depend on the wavelength of the irradiating light beam. This is observed e.g. in the photodissociation of the molecular ions of (cyclo)alkanes [2]. In the present paper, however, only experiments on the photodissociating ions will be discussed. It will be shown that in many cases the observation of a wavelength dependent photodissociation alone is not sufficient to obtain reliable conclusions about the ions involved.

As an introduction to the experimental results we will first give a general description of some important cases.

1.1. Photodissociatiion of Ions of one Structural Form

For this most simple case the general reaction scheme is:

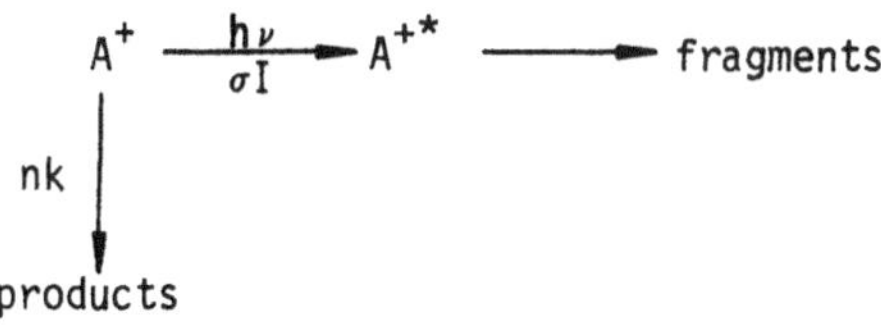

Scheme 1

with k being the reaction rate constant for reactions with the neutral background, n the density of the neutral molecules, σ the cross section for photodissociation, and I the light intensity. It is easy to show that for both reactive and non-reactive ions the ratio of the ion intensities after f seconds with and without irradiation should decay with time according to a single exponential:

$$1-x(t) = \frac{A(t,I)}{A(t,I=0)} = e^{-I\sigma t} \qquad (1)$$

In this equation x(t) is the fraction of dissociated ions and A(t,I) the ion intensity after t seconds. Since in equation 1 the reaction rate constant k cancels, photodissociation of ions of a single structure is indenpendent of the pressure of the neutral molecules.

1.2. Photodissociation of a Mixture of Different Ions of Equal Mass

Even for a mixture of two ions having different structures the general reaction scheme is quite complicated:

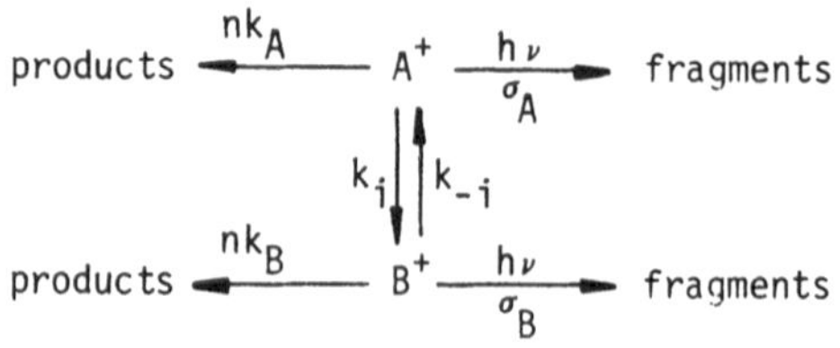

$$\begin{array}{ccccc} \text{products} & \xleftarrow{nk_A} & A^+ & \xrightarrow[\sigma_A]{h\nu} & \text{fragments} \\ & & k_i \downarrow \uparrow k_{-i} & & \\ \text{products} & \xleftarrow{nk_B} & B^+ & \xrightarrow[\sigma_B]{h\nu} & \text{fragments} \end{array}$$

Scheme 2

Both A^+ and B^+ can photodissociate with different cross sections σ_A and σ_B, might react with the neutral background with rate constants k_A and k_B, and possibly interconvert with or without collisions, k_i and k_{-i}. The single exponential decay with irradiation time, mentioned above for a single ion structure, will not apply to ion mixtures. Besides, if k_A and k_B are largely different, the photodissociation spectrum (σ as a function of wavelength) will be strongly dependent on the pressure of the neutral background.

Studies on ion mixtures are relatively easy if the photofragments from A^+ and B^+ have different masses. However, for the ion mixtures studied in the present work we always found that the fragments from isomeric ions are identical.

1.3. Relaxation of Ions Produced with a Large Amount of Internal Energy

Ions produced by electron impact might have an appreciable amount of internal energy, especially if the ion rearranges to a more stable structure or is produced by a fragmentation process. Relaxation of these ions might change the reactivity as can be observed in trapped ICR measurements of the ion intensities. It is also possible that the cross section for photodissociation changes, either by a change in Franck-Condon factors or by differences in the competition between fluorescence and fragmentation. The overall process is described in Reaction Scheme 3.

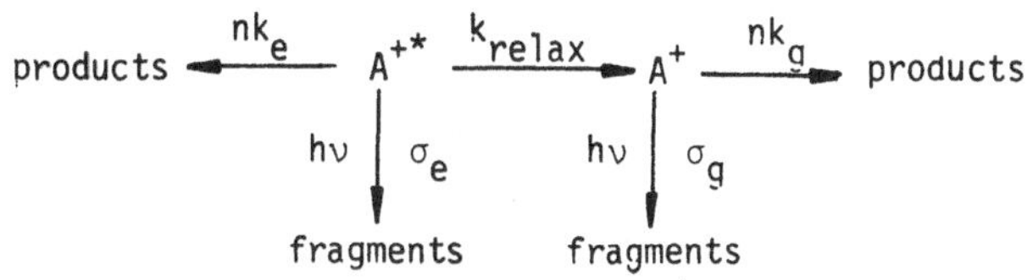

$$\text{products} \xleftarrow{nk_e} A^{+*} \xrightarrow{k_{relax}} A^{+} \xrightarrow{nk_g} \text{products}$$
$$A^{+*} \xrightarrow[h\nu]{\sigma_e} \text{fragments} \qquad A^{+} \xrightarrow[h\nu]{\sigma_g} \text{fragments}$$

Scheme 3

In this case the observed photodissociation rate should depend on the radiative lifetime of the excited ions (collisionless relaxation) and on the pressure of the neutral background (collision-induced relaxation). As discussed later in this paper we believe that the photodissociation of some C_2H_4O radical cations is the first example of such a process.

1.4. Multiphoton Processes

For several ions photodissociation has been observed despite the fact that the photon energy is well below the threshold energy for fragmentation [3,4]. Photodissociation is then caused by the successive absorption of two or more photons. For two-photon processes this results in Reaction Scheme 4.

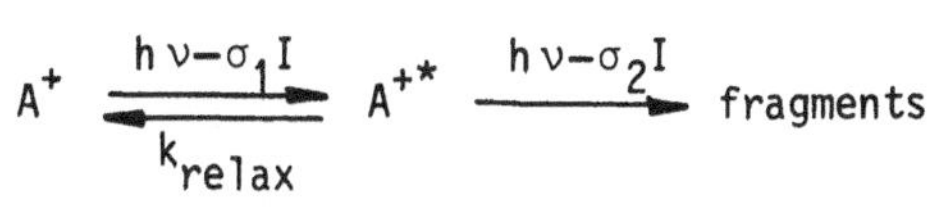

$$A^{+} \underset{k_{relax}}{\overset{h\nu-\sigma_1 I}{\rightleftarrows}} A^{+*} \xrightarrow{h\nu-\sigma_2 I} \text{fragments}$$

Scheme 4

Exact solution of the kinetic equation leads to a double exponential decay of the ion intensity. In practical cases, however, it appears that the relaxation rates are such that in very good approximation the ion intensity satisfies equation 2 [4].

$$1-x(t) = \exp \frac{-\sigma_2\sigma_1 I^2 t}{(\sigma_1 + \sigma_2) + k_3 + nk_3'} \qquad (2)$$

In equation 2 k_3 is the rate constant for collisionless relaxation (by IR emission) and k_3' the rate constant for collision-induced relaxation. In order to test the occurence of multi-photon processes, it is necessary to measure the photodissociation rate as a function of pressure and light intensity. Multi-photon processes offer an interesting possibility to study the relaxation, with and without collisions, of highly vibrationally excited ions.

The different cases described above appeared to be important for an understanding of the photodissociation experiments described in this paper. It follows that it is in general not sufficient to measure a photodissociation spectrum at arbitrary pressures or irradiation times. Studies of photodissociation as a function of pressure, irradiation time and in some cases, light intensity add essential information. This will be illustrated for investigations on the structures of C_6H_6O ions from various sources, the structures of alkene molecular ions and for relaxation processes in $C_2H_4O^{+\cdot}$ ions. In addition, the results for two-photon photodissociation of some (substituted) benzene ions are discussed and at some points appear to be essentially different from those published in previous papers [3,4].

2. Ion Structures

2.1. $C_6H_6O^{+\cdot}$ Ions [5]

From many different mass spectrometric studies on $C_6H_6O^{+\cdot}$ ions from various sources it has been concluded that two stable ion structures exist: the phenol structure A and its keto tautomer, the cyclohexadienone structure B. The cyclohexadienone structure seems to be formed in the decomposition of bicyclo-[2,2,2]-oct-2-en-5,7-dione (equation 3) and of certain phenoxyethyl halides.

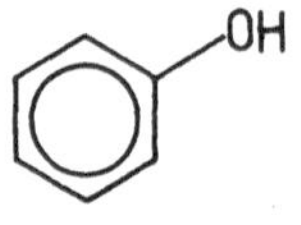

A 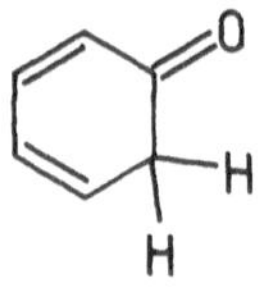 B

$$\longrightarrow \quad + \; CH_2CO \qquad (3)$$

However, it is difficult to establish the existence of mixtures of ions with different structures by the usual mass spectrometric methods. Therefore, we studied this problem by measurement of the spectroscopic properties of C_6H_6O ions from phenol, phenylethyl ether, bicyclo-[2,2,2]-oct-2-en-5,7-dione and 2-phenoxyethyl chloride. The photodissociation spectra of these ions are presented in Figure 1. From this figure it is immediately clear that the spectra of the ions from the bicyclo compound and from 2-phenoxyethyl chloride are essentially different from those of the ions from phenol and phenoxy ethyl ether (note that the spectra of the latter two ions are multiplied by a factor of 5). In view of previous results, the two different types of spectra can be ascribed to structures B and A respectively.

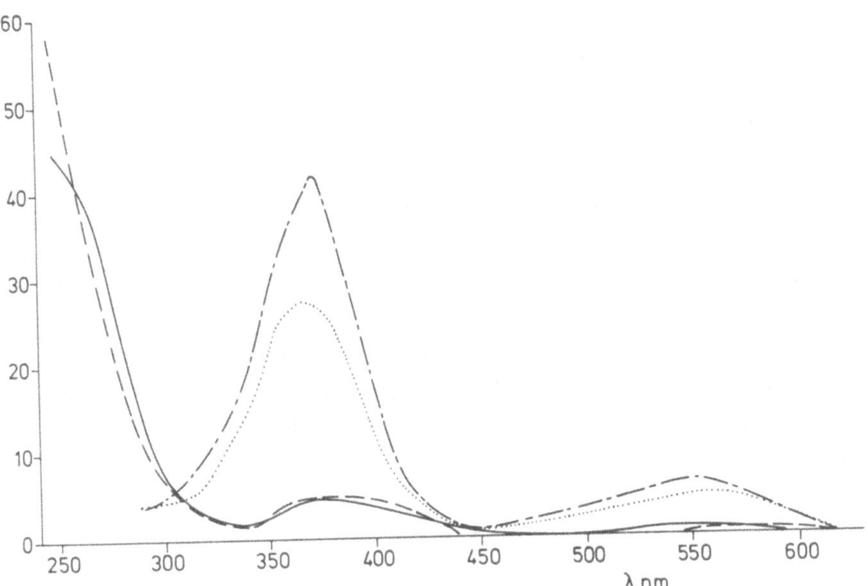

Fig. 1. Photodissociation spectra (relative units) of the m/z 94 ions from phenol (———), phenetole (----), bicyclo-[2.2.2]-oct-2-en-5,7-dione (....) and 2-phenoxyethyl chloride (— - — -). The spectra of ions from phenol and phenetole have been multiplied by a factor of 5.

However, it is still possible that mixtures of keto and enol tautomeric ions are observed. This is due to the orders of magnitude of the photodissociation rates of structures A and B, which are so different that the shape of the spectrum of an ion mixture will almost completely be determined by structure B. Therefore, to answer two problems immediately arising from Figure 1, we performed some additional measurements, especially of the decay of the ion intensity as a function of irradiation time:

1. The first question concerns the weak bands in the visible part of the spectrum of the ions from phenol and phenyl ethyl ether: are these due to a small admixture of ions with structure B ?
   Single exponential decay of the intensity of the C_6H_6O ions from phenol was observed upon variation of the irradiation time up to 13% dissociation after 10 seconds (wavelength 390 nm). Given the accuracy of the experiment and the ratio of the photodissociation rates in Figure 1 it follows that at most a very small fraction of the ions, say 2%, can have structure B. The major part of the weak photodissociation bands must be ascribed to ions having structure A.
2. The photodissociation rates of the ions from 2-phenoxyethyl chloride and the dione differ by about 30%. This is substantially larger than the experimental error in the determination of the relative rates of photodissociation if all ions were assumed to have structure B. The ions from the bicyclo compound react at most very slowly with the parent neutral. Time dependent measurements are then easy to interpret provided the ions do not isomerize on the ICR time scale. From trapped ICR measurements at considerably higher pressures it followed that this latter condition is fulfilled. The photodissociation time plot (Figure 2) then clearly shows that about 75% of the ions from the dione have structure B.

For the C_6H_6O radical cations from 2-phenoxyethyl ether the interpretation of the experiments is more complicated, because part of the ions appear to react rapidly with neutral 2-phenoxyethyl chloride by charge transfer. This results in a strong pressure dependence of the photodissociation spectrum. By comparison of photodissociation time plots with the results of trapped ICR measurements of m/z 94 ions in 2-phenoxyethyl chloride and of m/z 94 and m/z 99 ions in a mixture of 2-phenoxyethyl chloride and pentadeuterated phenol it followed that about 50% of the ions are reactive ($k=0.7\times10^{-9}cm^3mol^{-1}s^{-1}$) and that these ions have the phenol structure A. The remaining ions have structure B and are responsible for the photodissociation spectrum.

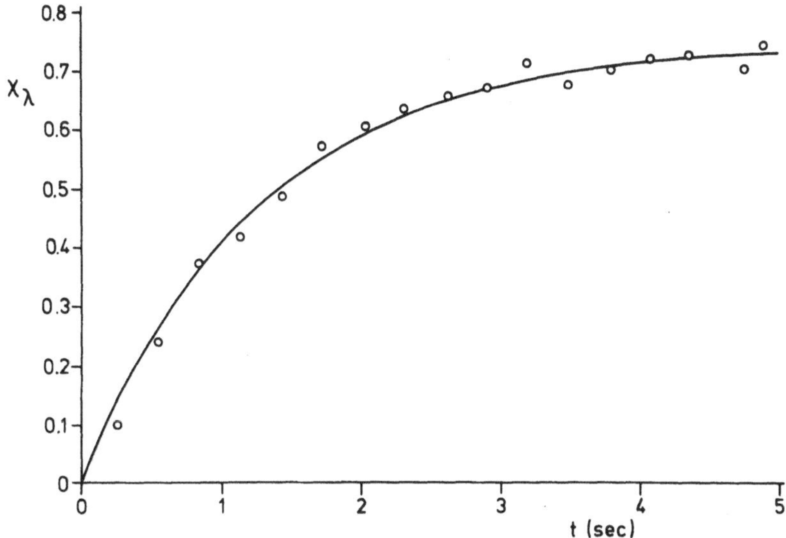

Fig. 2. Time dependence of the photodissociation of m/z 94 ions from bicyclo-[2.2.2]-oct-2-en-5,7-dione at 380 nm and at a pressure of 4 x 10^{-8} torr.

At higher pressures the ions with structure A disappear by reactions. Consequently, the photodissociation rate increases. This is the reason why in Figure 1 the apparent photodissociation rate form $C_6H_6O^{+\cdot}$ from 2-phenoxyethyl chloride is higher than the rate observed for the ions from the dione, despite the fact that the fraction of the ions B is lower.

2.2. Alkene Ions [7]

In previous papers [2,6] some aspects of the photodissociation spectra of simple alkenes have been discussed. The spectra, published sofar, are characterized by a relatively strong photodissociation in the UV region and a very weak band in the visible region. An example is the spectrum observed at 10^{-7} Torr for the molecular ions of 1-hexene (Figure 3). On closer inspection of the spectra of the molecular ions of C_5- and C_6- alkenes we found that in general ionization leads to a mixture of at least two ions with different structures. These ions have both different photodissociation spectra and different reactivities with respect to the parent neutral alkene. We will discuss the experiments for 1-hexene in more detail. The results for the isomeric C_5- and C_6-alkenes studied are summarized in Table 1.

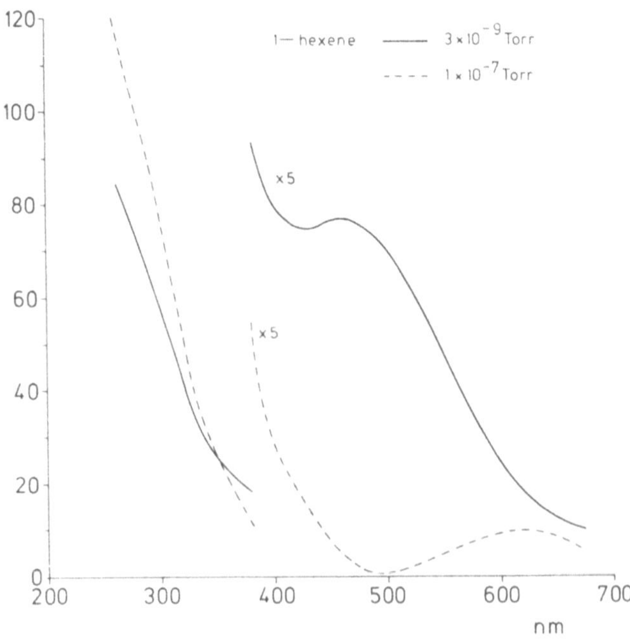

Fig. 3. The photodissociation spectrum (in apparent relative units of the molecular ions of 1-hexene at two different pressures of neutral 1-hexene and 16 eV ionizing energy.

Table 1. Ion mixtures in C_5- and C_6-alkenes

alkene	percentage	rate constant $(x10^{11})$	percentage	rate constant $(x10^{11})$
2-methyl-1-butene 16 eV	60	40	40	<1
2-methyl-1-butene 9.0 eV	<10	-	>90	<1
3-methyl-1-butene	-	-	100	<1
2-methyl-2-butene	-	-	100	<1
1-pentene	40	20	60	~1
trans-2-pentene	87	15	13	2
cis-2-pentene	55	20	45	5
1-hexene	65	19	35	<1
3,3-dimethyl-1-butene	-	-	100	<1
2,3-dimethyl-2-butene	-	-	100	<1
2,3-dimethyl-1-butene	-	-	100	<1

As shown in Figure 3, the photodissociation spectrum of 1-hexene molecular ions is strongly dependent on pressure. As mentioned above this behaviour suggests the existence of an ion mixture (see also Scheme 2 in the introduction). One of the isomeric ions has a low reactivity and a spectrum that resembles the alkene ion spectra published before. The other ion structure has a high reactivity and a strongly enhanced photodissociation cross section for irradiation in the region around 480 nm. This ion mixture could again be studied in more detail by measurements of the reaction kinetics at pressures around 10^{-6} Torr (Figure 4) and time plots of the fraction of photodissociated ions at 480 nm and different pressures (Figure 5). The trapped ICR plots show that about 70% of the ions have a high reactivity leading to a fast decay during the first 250 ms. The remaining ions are nearly non-reactive. The time dependent photodissociation plots show that at higher pressures and long irradiation time the photodissociation yield goes to zero. This is due to the fact that the fraction of photodissociating and reactive ions after 10 seconds at 2.5×10^{-8} Torr is very small. Consequently, there is no difference between the intensities after 10 seconds with and without irradiation. This result furthermore shows that there is no conversion from reactive to non-reactive ions on the time scale of the photodissociation experiments.

As mentioned above, the production of two different ions is not observed for 1-hexene only, but is presumably typical for alkene ions with more than 4 C-atoms. As shown in Table 1 the fraction of ions which are both reactive and have an enhanced photodissociation in the visible region varies from nearly 100% for trans-2-pentene to practically zero for 2 of the branched C_5-alkenes. It is also noteworthy that the fraction of reactive ions from 2-methyl-1butene decreases when the energy of the ionizing electrons is lowered to a value close to the ionization threshold.

Although the results are quite clear, it appears that a general interpretation in terms of ion structures is rather difficult. It is well-known that shifts of double bonds in alkene ions are easy and fast. However, they cannot explain the production of reactive ions from 2-methyl-1-butene at 16 eV. Besides, from CNDO/S calculations there seems to be no simple explanation for differences in the photodissociation rate in the visible spectrum between isomeric alkene structures. On the other hand these calculations also indicate that the oscillator strength for the lowest transitions is strongly dependent on the sterical structure of the ions. We thus believe that the effects observed are presumably determined both by shifts in the position of the double

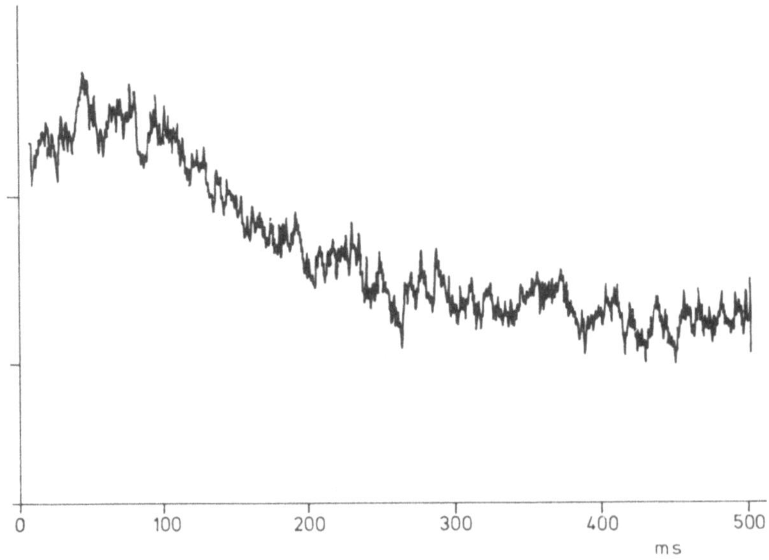

Fig. 4. Decay of the ion intensity (in relative units) of the $[C_6H_{12}]^{+\cdot}$ ions from 1-hexene at a pressure of 1.3 x 10^{-6} torr and 16 eV ionizing energy.

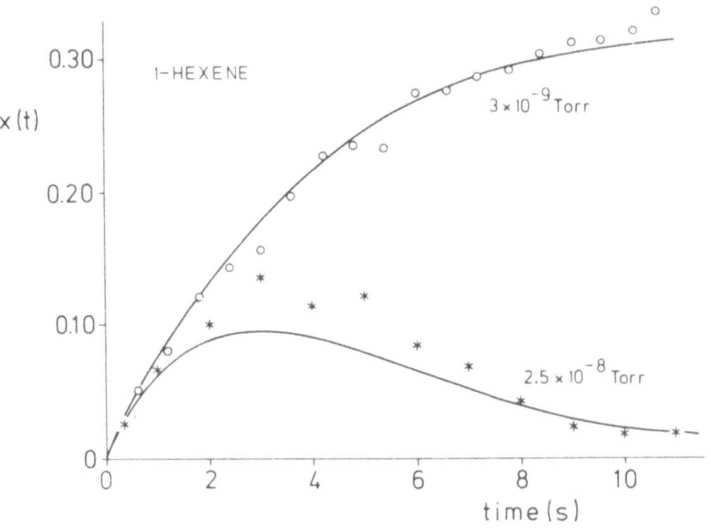

Fig. 5. The percentage of photodissociation, x(t), of the $[C_6H_{12}]^{+\cdot}$ ions from 1-hexene at two different pressures and irradiation at a wavelength of 480 nm.

bond and to differences in the sterical structure. If this latter assumption is correct, the experiments predict a strong correlation between configuration, ion reactivity and photodissociation.

3. Relaxation in Single Photon Processes, $C_2H_4O^{+\cdot}$ Ions from Ethylene Oxide and 1,3-Dioxolan [8]

According to Corderman et al. [9] ethylene oxide ions prepared with photoionization in the electronic and vibrational ground state remain cyclic, whereas vibrationally excited ions ring open to the $CH_2OCH_2^{+\cdot}$ structure. Relaxation of the resulting vibrationally excited ions e.g. by collisions with SF_6 can be observed thanks to differences in the reactivities of ground and excited state ions.

The photodissociation spectrum of $C_2H_4O^{+\cdot}$ ions from ethylene oxide by electron impact is presented in Figure 6. This spectrum is also observed for the $C_2H_4O^{+\cdot}$ fragment ions from 1,3-dioxolan. In both cases it is found that, by using long irradiation times, more than 90% of the ions can be photodissociated.

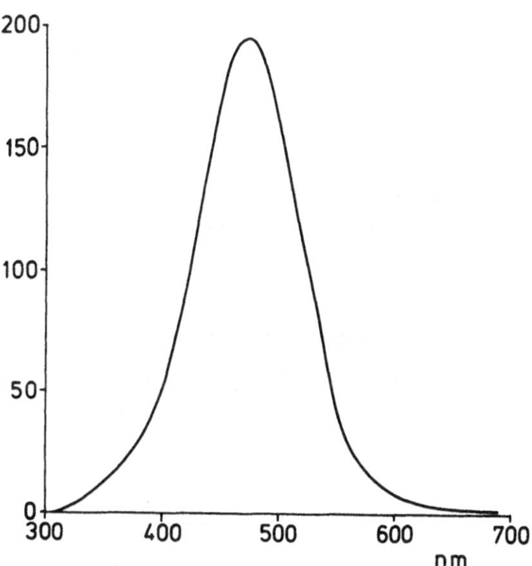

Fig. 6. The photodissociation spectrum of ethylene oxide molecular ions at 1.0×10^{-8} torr and 15 eV. Rate constants are given as apparent relative units.

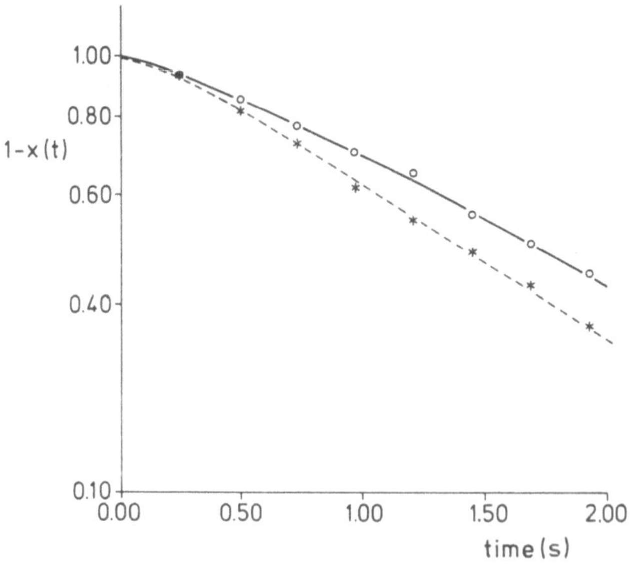

Fig. 7.
Time resolved photodissociation decay of ethylene oxide molecular ions at 3.5×10^{-9} torr (0) and at 3.5×10^{-9} torr with addition of 6×10^{-7} torr SF_6 (*) for irradiation at 480 nm.

For the $C_2H_4O^{+\cdot}$ ions from acetaldehyde and ethyl vinyl ether, which are assumed to have the CH_3CHO and CH_2CHOH structures respectively [10], no photodissociation could be observed. One might therefore conclude that $C_2H_4O^{+\cdot}$ ions from ethylene oxide and 1,3-dioxolan have one single structure: $CH_2OCH_2^{+\cdot}$. However, on closer inspection, the photodissociation of the ions from ethylene oxide appears to be more complicated than in other cases where the ions have one single structure. Time resolved experiments at 3.5×10^{-9} Torr and 480 nm show that for irradiation times up to 1.0 s a gradual increase in the photodissociation rate is observed. At longer times single exponential decay is satisfactorily obeyed (Figure 7). Clearly collisionless conversion to a state with a higher photodissociation rate occurs. This process is complete at much shorter times when the pressure of the parent neutral is raised or SF_6 is added (Figure 7), (collision-induced conversion). For $C_2H_4O^{+\cdot}$ ions from 1,3-dioxolan the same effects are observed but in this case they are less pronounced.

The results suggest that in the photodissociation of vibrationally excited $CH_2OCH_2^{+\cdot}$ ions produced by ring opening of the ethylene oxide, relaxation is important (Reaction Scheme 3). From the curves in Figure 7 it is calculated that in the absence of collisions the life time of the excited ions is 350 ms. For relaxation by collsions with SF_6 a rate constant of 7×10^{-10} $cm^3 mol^{-1} s^{-1}$ is obtained.

4. Two Photon Processes [11]

Two-photon photodissociation (Scheme 4) has been shown to occur upon irradiation of a number of (substituted) benzene ions with visible light [3,4]. For pure two-photon processes it follows from equation 2 that by variation of pressure and light intensity the quantities $\sigma_1\sigma_2/(\sigma_1+\sigma_2)$, $\sigma_1\sigma_2/k_3$ and $\sigma_1\sigma_2/k_3'$ can be determined. The value of σ_1,σ_2 and k_3 are then obtained from the assumption that the rate constant for charge transfer between the molecular ion and its parent neutral. (The values for σ_1 and σ_2 can be interchanged. Beauchamp and coworkers [3] showed that this relation explains the photodissociation kinetics of the benzene and cyanobenzene molecular ions. However, in their work the irradiation times were too short and the pressure too high to observe collisionless relaxation. From our measurements on these two ions it followed that the life time of the intermediate state for collisionless relaxation (IR emission) $\tau = 1/k_3$ is 125±25 ms for the benzene ion and 70±20 ms for the molecular ion of cyanobenzene.

The bromobenzene ion was studied by Dunbar and Fu [4]. From the value of 3.5 eV for the appearance energy of the $C_6H_5{}^+$ fragment, used by these authors, it was concluded that bromobenzene ions photodissociate by a two-photon process. However, according to recent measurements [12] the appearance energy is 2.7 eV, thus being close to the photon energies of 500 nm light. From a reinvestigation of the bromobenzene ion we concluded that in the visible band of the photodissociation spectrum both single and double photon photodissociation processes are involved. This is nicely illustrated by the spectra at different pressures presented in Figure 8. As expected for single-photon processes the UV band is independent of pressure. The visible band, however, is strongly pressure-dependent. It is furthermore seen that the effect of pressure on the photodissociation rate gradually increases with increasing wavelength through the visible band. This results in a shift of the maximum to lower wavelength at higher pressures. It follows that with increasing wavelength the photodissociation mechanism changes from a pure single-photon process to a mixture of single- and double-photon dissociation. The excistance of a mixture of different processes also follows from measurements of the fraction of photodissociated ions as a function of pressure and light intensity at 488 nm. According to equation 2 for two-photon processes a plot of $-I/\ln(1-x)$ at constant irradiation time versus I^{-1} should be linear and the slope of these curves should depend linearly on pressure. As shown in Figure 9, these conditions are not fulfilled for irradiation at 488 nm. Similar results were obtained at a wavelength of 514 nm.

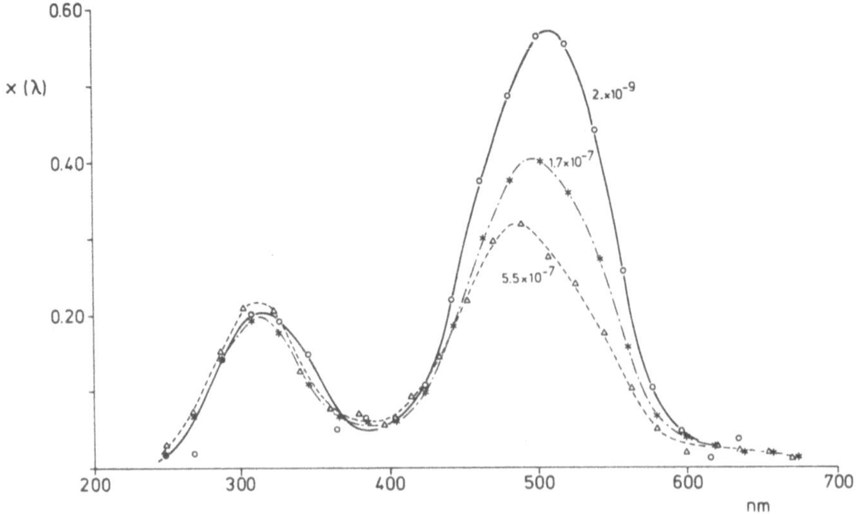

Fig. 8. The photodissociation spectrum of the molecular ions of bromobenzene at three different pressures and a constant irradiation time of 5.5 s.

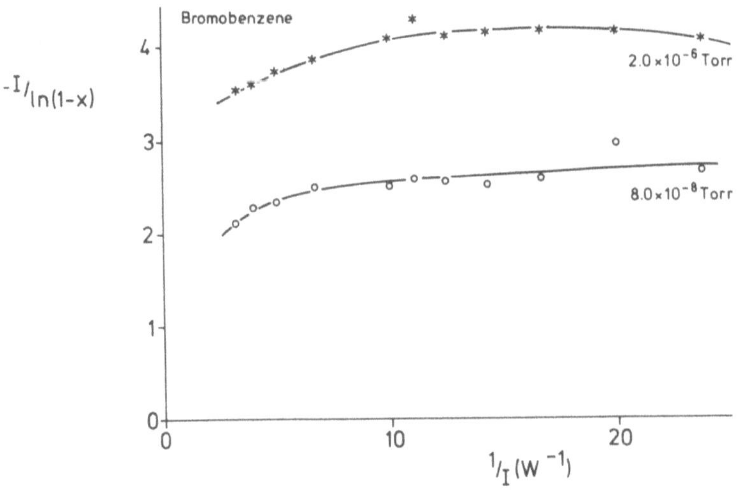

Fig. 9. $-I/\ln(1-x)$ versus $1/I$ for the molecular ions of bromobenzene at two different pressures and a constant irradiation time of 1.0 s.

The measurements suggest that the transition from single to double-photon processes takes place over a rather broad energy range. Therefore, we propose the mechanism in Scheme 5:

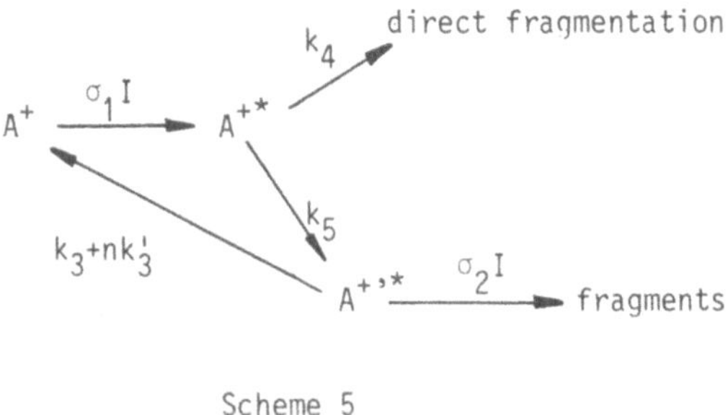

Scheme 5

After absorption of the first photon the excited ion A^{+*} can, as a result of the coupling with two di-ferent potential energy surfaces, either dissociate directly or convert to a mestastable state $A^{+,*}$ that requires another photon to dissociate. The curves drawn in Figure 10 are calculated from this mechanism.

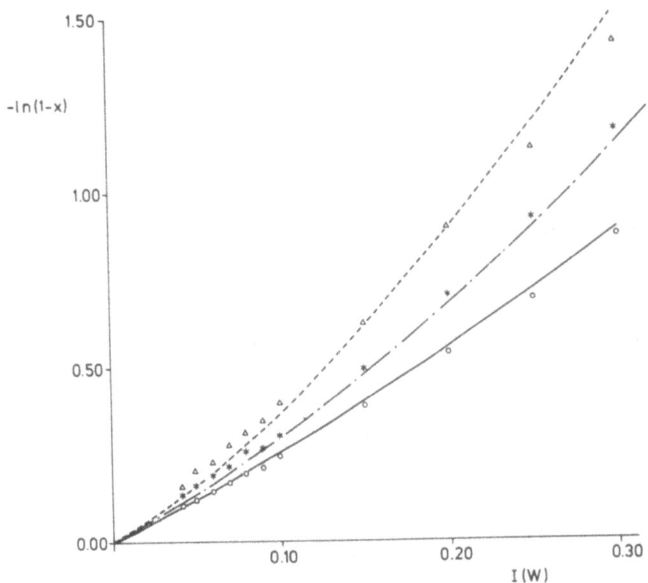

Fig. 10. -ln(1-x) versus I for the molecular ions of bromobenzene at 8 x 10^{-8} (Δ), 6. x 10^{-7} (*), and 2 x 10^{-6} (o). The drawn curves are calculated according to scheme 5.

At low pressures the major part of the ions (80 or 90%) dissociate according to the two-photon route. Just as observed for the benzene and cyanobenzene ions the collisionless lifetime of the intermediate state $A^{+,*}$ is of the order of o.1 s: $\tau=1/k_3= 150 \pm 50$ ms.

5. Acknowledgement

We thank the Netherlands Organization for the Advancement of Pure Research (SON/ZWO) for financial support.

6. References

1. R.C. Dunbar in Kinetics of Ion-Molecule Reactions, P. Ausloos editor, Plenum Press, (1979); R.C. Dunbar in Gas-Phase Ion Chemistry, M.T. Bowers editor, Academic Press (1979).

2. E.F. van Dishoeck P.N.T. van Velzen and W.J. van der Hart, Chem. Phys. Letters, 62 (1979) 135; R.C. Benz and R.C. Dunbar, J. Am. Chem. Soc., 101 (1979) 6363.

3. B.S. Freiser and J.L. Beauchamp, Chem. Phys. Letters, 35 (1975) 35; T.E. Orlowski, B.S. Freiser and J.L. Beauchamp, Chem. Phys. 16 (1976) 439.

4. R.C. Dunbar and E.W. Fu, J. Phys. Chem., 81 (1977) 1531.

5. P.N.T. van Velzen, W.J. van der Hart, J. van der Greef, N.M.M. Nibbering and M.L. Gross, submitted for publication in J. Am. Chem. Soc.

6. R.C. Dunbar, Anal. Chem. 48 (1976) 723.

7. P.N.T. van Velzen and W.J. van der Hart, to be published.

8. P.N.T. van Velzen and W.J. van der Hart, submitted for publication in Chem. Phys. Letters.

9. R.R. Corderman, P.R. LeBreton, S.E. Buttrill, A.D. Williamson and J.L. Beauchamp, J. Chem. Phys., 65 (1976) 4929.

10. J.L. Holmes and J.K. Terlouw, Can. J. Chem., 53 (1975) 2076 ; C.C. Van der Sande and F.W. McLafferty, J. Am. Chem. Soc., 97 (1975) 4613.

11. P.N.T. van Velzen and W.J. van der Hart, to be published.

12. H.M. Rosenstock, K. Draxl, B.W. Steiner and J.T. Herron, J. Phys. Chem. Ref. Data, Vol. 6, 1977 supp. 1; H.M. Rosenstock and R. Stockbauer, J. Chem. Phys., 73 (1980) 773.

INFRARED PHOTOCHEMISTRY OF GAS PHASE IONS

L. R. Thorne*, C. A. Wright and J. L. Beauchamp
Arthur Ames Noyes Laboratory of Chemical Physics
California Institute of Technology, Pasadena, CA 91125[a]

1. Introduction

Experiments involving molecules which are truly isolated for time periods exceeding several milliseconds require special techniques for particle storage at low pressure. In the case of charged species, crossed electric and magnetic fields can be used to restrain particle motion for time periods exceeding several hours, establishing conditions in the laboratory which exist in nature only in the interstellar medium. The phenomenon of ion cyclotron resonance provides a sensitive and selective means to detect charged particles stored in a magnetic field. At pressure below 10^{-5} torr stored ions are forced to maintain equilibrium with their environment by the absorption and emission of infrared radiation rather than in collisions with other molecules. Infrared lasers offer the possibility of upsetting this equilibrium by exposing molecules to an enormous photonflux at specific wavelengths. What fraction of the ion population will absorb infrared radiation at a specific wavelength? Can more than one photon be absorbed?

A photodissociation "spectrum" can be recorded by measuring dissociation yields as a function of wavelength. How does this spectrum compare to the small signal absorption spectrum? Is the spectrum homogeneously broadened (that is, does each molecule absorb at all wavelengths within the photodissociation band)? Or is the spectrum comprised of many, overlapping, narrow bands resulting from transitions between different states of the system (heterogeneously broadened)? These intriguing questions have recently been answered using the techniques of ion cyclotron resonance spectroscopy to monitor ion populations irradiated with both low

* NRC NASA Resident Research Associate

[a] Contribution No. 6516

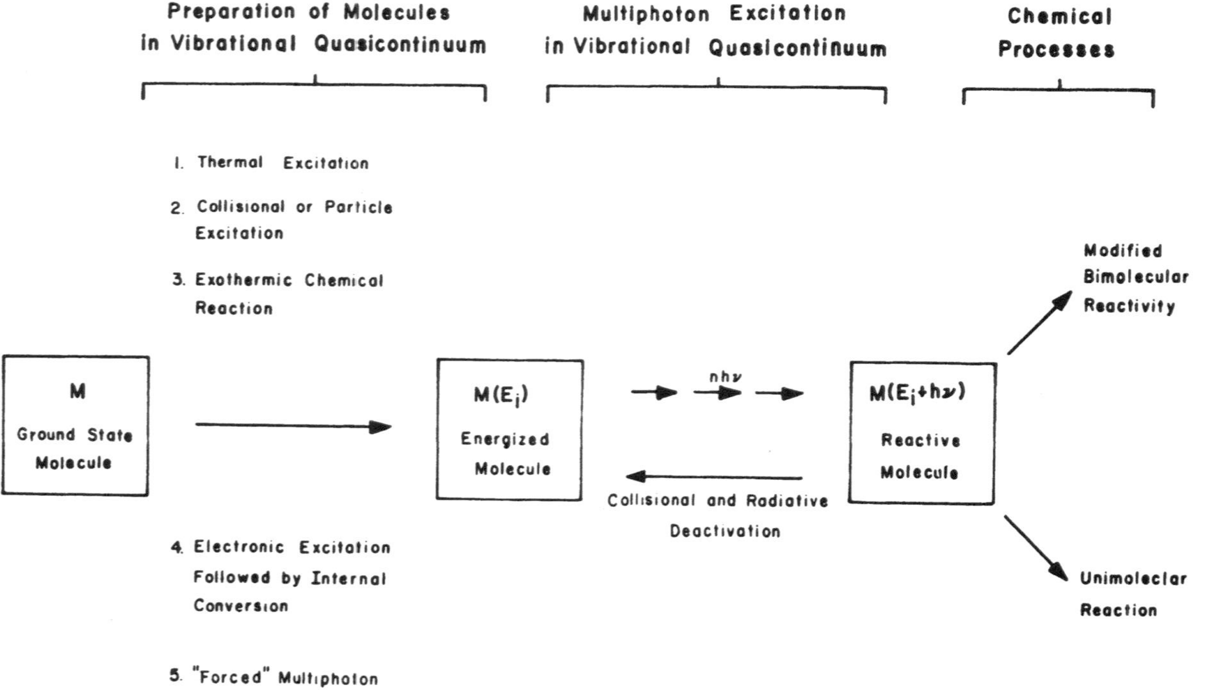

Fig. 1. Schematic representation of laser driven photochemistry indicating processes which promote molecules into the vibrational continuum, cause deactivation and lead to chemical reactivity.

power cw and high power pulsed CO_2 lasers. These experiments represent a unique application of ion cyclotron resonance spectroscopy and lay the groundwork for many interesting applications which are discussed at the end of this article.

The important facets of infrared multiphoton absorption are illustrated in Figure 1. We let M represent a molecular species (neutral or ion) in its ground state. Infrared absorption will occur only at discrete wavelengths if intramolecular vibrational relaxation (IVR) is slow. IVR rates increase rapidly with increasing internal energy due to rapidly increasing density of vibrational energy levels and increasing anharmonic coupling between different vibrational modes. A sufficiently energized molecule can be in what is termed the vibrational quasicontinuum. Onset of the vibrational quasicontinuum can be considered to occur when the density of vibrational levels of a molecule exceeds the inverse of the transition bandwidth (or for a homogeneously broadenend transition exceeds the inverse of the laser bandwidth). Under these circumstances it is highly probable that at least one vibrational level matches the energy level pumped by the laser. This "resonance" enhances IVR rates and energy can be transferred rapidly out of the pumped mode into other vibrational modes of the molecule. Perturbations which occur during collisions can also increase IVR rates.

There are several ways in which a molecule can be energized to the point where it is in the vibrational quasicontinuum. A sufficiently large molecule may already qualify as a result of its thermal energy content at ambient temperature. Other methods of excitation, listed in Figure 1, include collisional or particle impact processes, highly exothermic chemical reactions which yield vibrationally excited products, and electronic excitation followed by radiationless relaxation to the ground electronic state (internal conversion). Molecules can be "forced" into the vibrational quasicontinuum by the resonant or near resonant absorption of several photons through a ladder of discrete levels, with IVR occuring initially only from a high level of the pumped mode. A characteristic of "forced" multiphoton excitation is that measurements of product yield as a function wavelength give photodissociation spectra which are shifted to the red of the fundamental absorption due to anharmonicity. Although very few such photodissociation spectra have been reported, this appears to be a general characteristic of multiphoton dissociation of small molecules effected by high power pulsed infrared lasers.

The total number of photons to which a molecular system is exposed during excitation is measured by laser fluence, usually expressed in Joules cm^{-2}. Fluences in excess of 1 Joule cm^{-2} (e.g. 1 megawatt cm^{-2} in 1 sec with a pulsed laser) are sufficient to effect dissociation of adsorbing species. Although non-linear effects are now well recognized, early experiments with SF_6 suggested that at low conversion, the probability of dissociation is independent of peak power at constant fluence. This lead us to consider the use of cw infrared lasers for such experiments, since 50 watts cm^{-2} which is easily obtained from commercial line-tunable cw lasers, delivers 1 Joule cm^{-2} in only 20 msec. The success of such experiments depends on the radiative and collisional relaxation rates being low because of the comparatively long time scale of the experiment. Estimates of infrared cooling rates of ions suggested that the former would not be a problem; the effects of collision can be avoided by lowering the pressure and using trapped ion techniques. Using cw infrared lasers we have observed that infrared multiphoton dissociation is indeed a facile process with (1) "large" molecules which are in the vibrational quasicontinuum at room temperature, (2) "medium" size molecules, energized by one of the processes indicated in Figure 1.

The majority of our observations suggest that to effect infrared multiphoton dissociation with cw lasers it is necessary for the molecule to be in the vibrational quasicontinuum before irradiation or there must be some mechanism to achieve this condition other than the "forced" excitation which characterizes the dissociation of small molecules with high power pulsed lasers.

As illustrated in Figure 1, multiphoton excitation can lead to modified reactions. At intermediate levels of excitation the energized molecule can be relaxed by spontaneous emission or collisional deactivation. Appropriate experimental procedures allow for the study of these important processes. In the present paper we summarize our observations of multiphoton excitation and dissociation processes using both cw and pulsed lasers to excite ions which are formed, stored, and detected using the techniques of ion cyclotron resonance spectroscopy.

2. Experimental Section

The theory, techniques, and instrumentation of trapped ion ICR spectroscopy have been previously described in detail [1]. The spectrometer used in

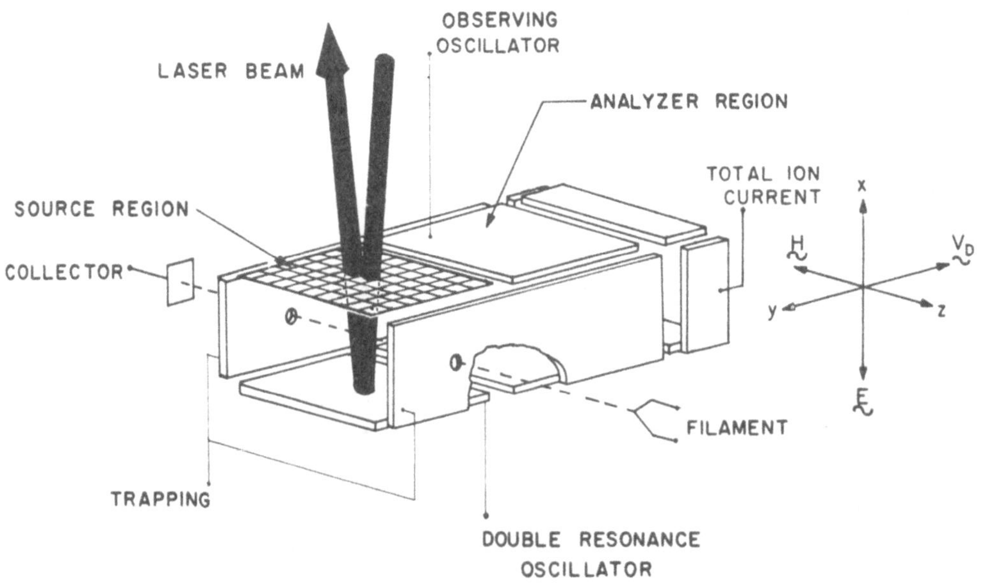

Fig. 2. Cutaway view of cyclotron resonance cell. The electron beam is collinear with the magnetic field. The laser beam enters through and open mesh grid comprising the top plate of the source region and is reflected by the lower plate.

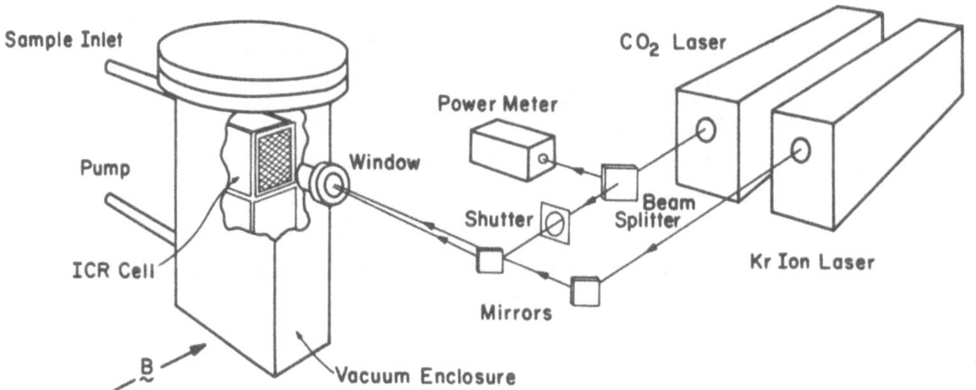

Fig. 3. Schematic view of the ICR cell and optical configuration. The vector B indicates the orientation of the magnetic field.

our studies was built at Caltech and incorporates a 15-in. electromagnet capable of 23.4 kG.

All ICR experiments were carried out in the range 10^{-7}-10^{-5} torr, corresponding to neutral particle densities of 3 x 10^9 to 3 x 10^{11} molecules cm^{-3}. Pressure is measured with a Schulz-Phelps type ionization gauge calibrated against an MKS Instruments Baratron Model 90H1-E capacitance manometer. It is expected that absolute pressure determinations are within ± 20% using this method, with pressure ratios being somewhat more accurate. Sample mixtures are prepared directly in the instrument using three sample inlets and the Schulz-Phelps gauge.

Replacement of one of the source plates of the ICR cell with a screen grid, as shown in Figure 2, permits irradiation of trapped ions. The laser beam is directed through the grid into the cell and is reflected back out by a mirror finish on the back source plate. Assuming the source plate to be 100% reflective, the beam irradiance inside the cell is 1.84 times the incident irradiance.

Figure 3 shows a schematic view of the experimental apparatus and optical configuration used for two laser photodissociation studies. Either an Apollo 550A line-tunable cw CO_2 laser or a Tachisto 215 G line-tunable CO_2 TEA laser is used as the infrared radiation source. Both are operated in TEM_{00} and have beam intensity profiles which are nearly Gaussian (FWHM = 6 mm). For the cw laser, reported irradiances are calculated by dividing the total beam power in the cell by the area of the 6 mm diameter beam. Thus, the irradiance (W cm^{-2}) to which the ions are exposed is 6.51 times the total power (W) of the incident laser beam. The fluences (J cm^{-2}) to which the ions are exposed by the pulsed laser are calculated in a similar way. Both laser beams are used unfocused so that ions are uniformly irradiated. The ions can be exposed to irradiances up to 120 W cm^{-2} or fluences up to 1.6 J cm^{-2} TEM_{00}. Irradiation of the ions with the cw laser is controlled by an Uniblitz Model 255L0A14x5 mechanical shutter having a 5 msec opening time. The frequency of the laser radiation is measured with an Optical Engineering Model 16A spectrum analyzer. Additional experimental details are given elsewhere [2]. For the experiments requiring visible laser radiation, a coherent radiation CR-2000K cw Krypton ion laser was used. The beam diameter is 2 mm, and the effective irradiance was calculated in the same manner as for the CO_2 laser.

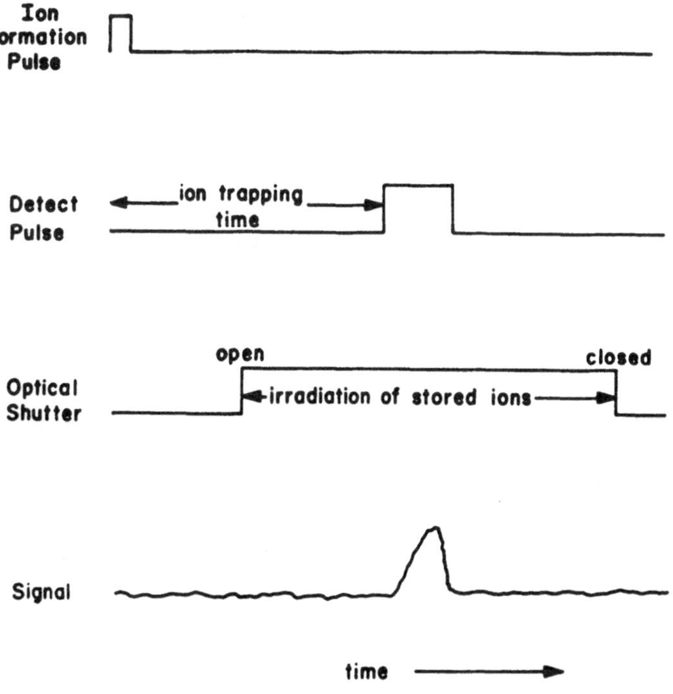

Fig. 4. Timing sequence for trapped ion ICR photodissociation experiments. Ions may be irradiated during any portion of the trapping period.

The timing sequence for a typical experiment is illustrated in Figure 4. A 10-ms electron beam pulse generates ions (typically 10^5 ions cm^{-3}) which can be stored for several seconds during which time reactions may occur. The ions are then mass analyzed to determine concentrations of the various species present. Ions of particular charge to mass ratio can be selectively ejected by ICR double resonance allowing positive identification of reaction pathways. Electronic control of the optical shutter allows the infrared beam to irradiate on alternate trapping cycles and corresponding ion intensities (laser on and laser off) are monitored by a two-channel boxcar integrator. These signals are then processed in a straightforward fashion to yield photodissociation rate constants, even in the presence of ion loss due to diffusion and reaction.

3. Results

3.1 Multiphoton Dissociation of Ions with Low Intensity cw Infrared Radiation

3.1.1 Summary of Multiphoton Dissociation Results

ICR techniques have been used successfully to study infrared multiphoton dissociation of ions using laser irradiances of 1-100 W cm^{-2}. Table 1 lists those ions studied in our laboratory. In each case the decomposition pathway was confirmed by direct observation of the product ion or by using double resonance techniques. The identity of the neutral product was inferred from mass balance considerations. All observed dissociation reactions are identical to lowest energy thermal decomposition pathway, where known. Proton bound dimers of some alcohols show exception to this, giving multiple products. However, the lowest energy thermal process is always observed (see section 3.2). Enthalpy changes are calculated from heats of formation. Activation energies are not known except where noted and may be in excess of the reaction endothermicity. The minimum number of photons needed to reach thermodynamic threshold for dissociation, n, is given by the enthalpy change divided by the photon energy.

For all ions studied, the observed decomposition pathway is not dependent on laser wavelength. However, the photodissociation yield, P_D, is generally dependent on laser wavelength. P_D is definded as the fraction dissociated during a specified exposure time and with a specified laser irradiance.

$$P_D = 1 - I/I_o \tag{1}$$

Here, I is the signal intensity of the dissociating ion measured at the end of the exposure period and I_o is the signal without irradiation at the end of the same period. In almost every case studied (the exceptions are discussed in section 3.1.5) the photodissociation process is characterized by first order decay kinetics given by eq 2 where k_D is the dissociation rate constant.

$$I/I_o = \exp(-k_D t) \tag{2}$$

Table 1: Observed Low Intensity IR Multiphoton Dissociation Reactions

Reactants	Products	ΔH(kcal/mole)	n[a]
$(C_2H_5)_2O\ _2H^+$ [b,c]	$(C_2H_5)_2OH^+ + (C_2H_5)_2O$	31	12
$(C_2H_5)_2OH^+$ [b,c]	$C_2H_5OH_2^+ + C_2H_4$	27	10
$(C_2D_5)_2OD^+$ [b,c]	$C_2D_5OD_2^+ + C_2D_4$	27	10
$(C_2D_5)(C_2H_5)OH^+$ [b,c]	$C_2D_5OH_2^+ + C_2D_4$	27	10
$CH_3CHOC_2H_5^+$ [b]	$CH_3CHOH^+ + C_2H_4$	34	13
$(CH_2CH_2CH_2CH_2O)_2H^+$	$CH_2CH_2CH_2CH_2OH^+ + CH_2CH_2CH_2CH_2O$	31	12
$C_3F_6^+$ [d,e]	$C_2F_4^+ + CF_2$	56[f]	20
$C_6H_{12}^+$ [d]	$C_5H_9^+ + CH_3$	30	11
$CpRh(CO)_2H^+$ [b,g]	$CpRhH^+ + 2CO$	--[h]	--[h]
CF_3I^+ [i]	$CF_3^+ + I$	10	3
$(CH_3OH)_2H^+$ [j]	$(CH_3)_2OH^+ + H_2O$	17	7
$(C_2H_5OH)_2H^+$ [j]	$(CH_2H_5)_2OH^+ + H_2O$	15	6
$(n\text{-}C_3H_7OH)_2H^+$ [j]	$n\text{-}C_3H_7OH_2^+ + C_3H_7OH$(71%)	33	12
	$(n\text{-}C_3H_7)_2OH^+ + H_2O$(17%)	17	6
	$(n\text{-}C_3H_7OH)H^+(H_2O) + C_3H_6$(12%)	19	7
$(i\text{-}C_3H_7OH)_2H^+$ [j]	$(i\text{-}C_3H_7)_2OH^+ + H_2O$	19	7
$(s\text{-}C_4H_9OH)_2H^+$ [j]	$(s\text{-}C_4H_9)_2OH^+ + H_2O$(38%)	15	6
	$(s\text{-}C_4H_9)H^+(H_2O) + C_4H_8$(62%)	20	7
$(i\text{-}C_3H_7OH)H^+(s\text{-}C_4H_9OH)$ [j]	$(i\text{-}C_3H_7)(C_4H_9)OH^+ + H_2O$(37%)	19	7
	$(s\text{-}C_4H_9)H^+(H_2O) + C_4H_8$(63%)	20	7
$(t\text{-}C_4H_9OH)_2H^+$ [j]	$(t\text{-}C_4H_9)_2OH^+ + H_2O$	20	7
$(t\text{-}C_5H_{11}OH)_2H^+$ [j]	$(t\text{-}C_5H_{11}OH)H^+(H_2O) + C_5H_{10}$	22	8
$(i\text{-}C_3H_7OH)H^+(t\text{-}C_4H_9OH)$ [j]	$(i\text{-}C_3H_7)(t\text{-}C_4H_9)OH^+ + H_2O$	13	5

$(CD_3)_2{}^{35}Cl^+$ [k]	$CD_2{}^{35}Cl^+ + CD_4$	18	7
$(CD_3)_2{}^{37}Cl^+$ [k]	$CD_2{}^{37}Cl^+ + CD_4$	18	7
$C_6H_5CN^+$ [l]	$C_6H_4{}^+ + HCN$	--	--
$i\text{-}C_3H_7ClLi^+$	$C_3H_6Li^+ + HCl$		
(p-methoxybenzaldehyde)Li^+	p-methoxybenzaldehyde + Li^+		

[a] Calculated for 944 cm^{-1} except for $C_3F_6{}^+$ which is given for 1047 cm^{-1} (maximum of photodissociation probability curve). [b] Ion-molecule reaction product. [c] Ref. 2-4. [d] Formed directly by electron impact ionization. [e] Ref. 4-6. [f] Ref. 34. [g] Cp = (η^5-C_5H_5). [h] Value not known. [i] Dissociation from an electronic excited state, ref. 7. [j] Ref. 8. [k] Ref. 9. [l] Ref. 10.

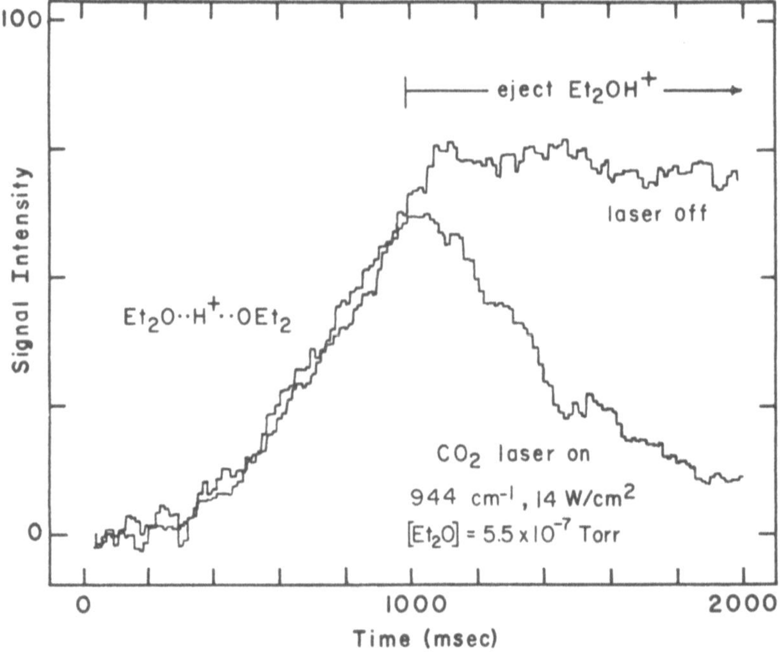

Fig. 5. Ion intensity versus trapping time for a typical multiphoton dissociation experiment . At a diethylether pressure of 5.5×10^{-7} Torr ions are formed by a 10 msec 70 eV electron beam pulse. The upper trace is the proton bound dimer signal with the laser off. Ejection of $(C_2H_5)_2OH^+$ beginning at 1 second trapping time halts further dimer formation. CW irradiation by the infrared laser coincident with ejection of $(C_2H_5)_2OH^+$ (lower trace) results in photodissociation of the dimer. At this pressure the time between collisions is approximately 50 msec [2].

For all ions studied, the dissociation rate constant, typically 2-30 sec^{-1}, is dependent on laser irradiance. If the dependence is linear, k_D can be expressed in terms of a phenomenological dissociation cross section, σ_D, where Φ is the photon flux.

$$k_D = \sigma_D \Phi \tag{3}$$

At a given laser wavelength the photon flux is proportional to the irradiance (1 W cm^{-2} = 5.34 x 10^{19} photons $cm^{-2}sec^{-1}$ for 947 cm^{-1} photons). Typical values of σ_D are in the range of 0.5 - 2 x 10^{-20} cm^2.

Diethyl Ether. Major ions in the electron impact mass spectrum of diethyl ether at long trapping times and low pressure are $(C_2H_5)_2OH^+$ and $CH_3CHOC_2H_5{}^+$. At higher diethyl ether pressures (> 5x10^{-7} Torr) and long trapping times (> 500 msec) appreciable amounts of proton bound dimer are formed according to eq 4 having a

$$(C_2H_5)_2OH^+ + (C_2H_5)O \longrightarrow [(C_2H_5)_2O]H^+ \tag{4}$$

bimolecular reaction rate constant of k = 1.9(2) x 10^{-11} $cm^3molecule^{-1}sec^{-1}$. Details of the diethyl ether ion-molecule chemistry are given in ref. 2. A typical experiment monitoring $[(C_2H_5)_2O]_2H^+$ is shown in Figure 5. Ions are produced by a 10 msec electron beam pulse and stored for up to 2 sec. At 1 sec of trapping time the remaining $(C_2H_5)_2OH^+$ is rapidly ejected by double resonance, thus preventing further formation of the dimer. This is evidenced (Figure 5, upper trace) by the constant abundance of $[(C_2H_5)_2O]_2H^+$ after 1 sec. The laser, tuned to 944 cm^{-1} and 14 W cm^{-2} irradiance, gated on at 1 sec of trapping time, coincident with ejection of $(C_2H_5)_2OH^+$ and effects an exponential decay of the dimer (Figure 5, lower trace). At this irradiance no appreciable photodissociation of $(C_2H_5)_2OH^+$ is observed, and the increase in abundance of this species exactly matches the decrease in abundance of the proton bound dimer verifying that there is only one decay channel open, eq 5.

$$[(C_2H_5)_2O]_2H^+ + nh\nu \longrightarrow (C_2H_5)_2OH^+ + (C_2H_5)_2O \tag{5}$$

At low pressures where proton bound dimer formation is not significant, multiphoton dissociation of the protonated ether, $(C_2H_5)_2OH^+$, can be studied. The laser induced process and postulated four-center intermediate (11) are shown in eq 6. The other major ion present at long times,

$$(C_2H_5)_2OH^+ + nh\nu \longrightarrow \left[\begin{array}{c} H^+ \\ | \\ C_2H_5O\!-\!CH_2 \\ \vdots \quad \vdots \\ H\!-\!CH_2 \end{array}\right]^* \longrightarrow C_2H_5OH_2^+ + CH_2CH_2 \qquad (6)$$

$$\Delta H = 27 \text{ kcal/mole}$$

$CH_3CHOC_2H_5{}^+$, also undergoes infrared multiphoton dissociation, eq 7.

$$CH_3CHOC_2H_5{}^+ + nh\nu \longrightarrow CH_3CHOH^+ + CH_2CH_2 \qquad (7)$$

Ions derived from electron impact ionization of $(C_2D_5)_2O$ undergo the identical reactions as the corresponding unlabelled species. However, with the partially deuterated ether, $C_2H_5OC_2D_5$ an interesting isotope effect is observed in the decomposition of the protonated molecular ion. Chemical ionization of $C_2D_5OC_2H_5$ at low (12 eV) electron energies using cyclohexane as protonating agent allows for selective formation of $(C_2H_5)(C_2D_5)OH^+$ with only trace amounts of $(C_2H_5)(C_2D_5)OD^+$. By analogy with eq 6 there are two possible product ions from the decomposition of $(C_2H_5)(C_2D_5)OH^+$, eq 8. Yet during laser irradiation,

$$C_2H_5\overset{H}{\underset{+}{O}}C_2D_5 + nh\nu \longrightarrow C_2D_5OH_2{}^+ + CD_2CD_2 \qquad (8a)$$

$$C_2H_5\overset{H}{\underset{+}{O}}C_2D_5 + nh\nu \not\longrightarrow C_3H_5OHD^+ + CH_2CH_2 \qquad (8b)$$

$C_2D_5OH_2{}^+$ is the only product ion detected, eq 8a. Thus hydrogen transfer in the four center intermediate is more facile than deuterium transfer. Consideration of ion detection limits in this experiment provides a lower limit for the combined primary and secondary isotope effects (defined as the ratio of rates of product ion formation) as ≥ 6. It is assumed that the observed specificity arises from energetics of decomposition and is in no way attributable to selective laser pumping of only one half of the ion. In comparison, when $(C_2D_5)(C_2D_5)OH^+$ is formed by highly exothermic proton transfer such that the protonated ether internal energy greatly exceeds the threshold for decomposition in accordance with eq 8 the observed isotope effect is ~ 2.

These results imply that multiphoton dissociation occurs at an energy only slightly in excess of thermodynamic threshold. Large primary isotope effects have also been reported for metastable ion decompositions at threshold energies.

Perfluoropropylene Cation. Electron impact ionization (14-70 eV) of perfluoropropylene, $CF_3CF=CF_2$, produces only four major ions: $C_3F_6^+$, $C_3F_5^+$, $C_2F_4^+$, and CF_3^+ [6]. Fluoride abstraction by CF_3^+ occurs, reaction 9, leaving three

$$CF_3^+ + C_3F_6 \longrightarrow C_3F_5^+ + CF_4 \qquad (9)$$

ions at long trapping times. Dissociation of the parent cation $C_3F_6^+$, reaction 10, is the only observed infrared laser

$$C_3F_6^+ + nh\nu \longrightarrow C_2F_4^+ + CF_2 \qquad H = 56 \text{ kcal/mole} \qquad (10)$$

induced process [12]. The product ion is both stable to laser irradiation and chemically unreactive.

$C_3F_5^+$ is totally unaffected by laser irradiation despite the availability of a decay channel, reaction 11, which has an energy requirement comparable to that of the observed photodissociation reaction [12]. The inertness of $C_3F_4^+$

$$C_3F_5^+ + nh\nu \not\longrightarrow CF_3^+ + C_2F_2 \qquad H = 53 \text{ kcal/mole} \qquad (11)$$

toward the laser field proves that those photodissociation reactions which are observed result from absorption by the ions and are not due to some general non-specific heating of the ICR cell contents.

Dimethyl Chloronium Cation. Ion-molecule chemistry of CH_3Cl is described in detail elsewhere [13]. Briefly, the dimethyl chloronium ion $(CH_3)_2Cl^+$ is formed in a two-step sequence, $(CH_3)_2Cl^+$ does not react further

$$CH_3Cl^+ \; CH_3Cl \longrightarrow CH_3ClH^+ + CH_2Cl \qquad (12)$$

$$CH_3ClH^+ + CH_3Cl \longrightarrow (CF_3)_2Cl^+ + HCl \qquad (13)$$

with methyl chloride. In a mixture of CD_3Cl and CH_3Cl the isotopically mixed

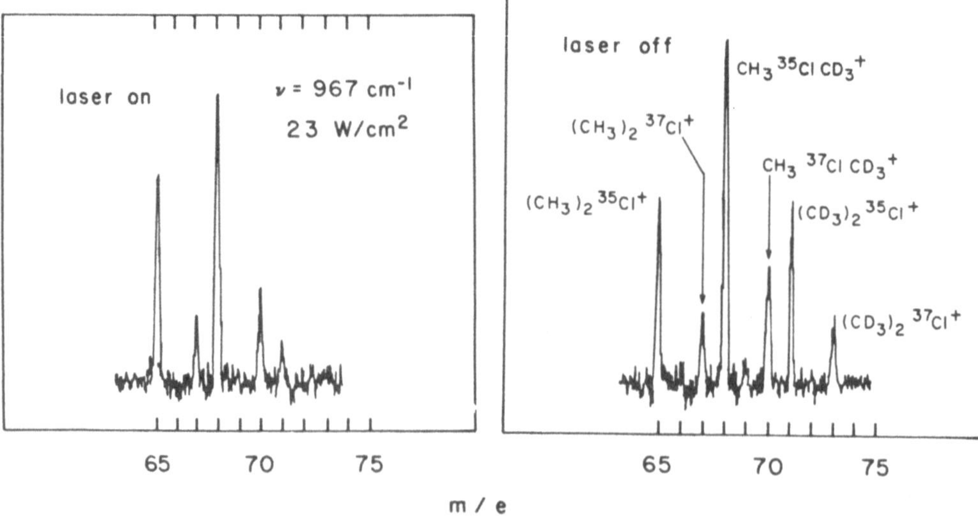

Fig. 6. Right trace is 2 s time delay mass spectrum of 1:1 $CH_3Cl:CD_3Cl$. Ionization is by 50 ms pulse of 14 eV electrons. Total pressure is 1.9×10^{-7} Torr. Left trace shows the same mass region for experiment in which ions undergo 2 s continous infrared laser irradiation 23 W cm^{-2} at 967 cm^{-1}).

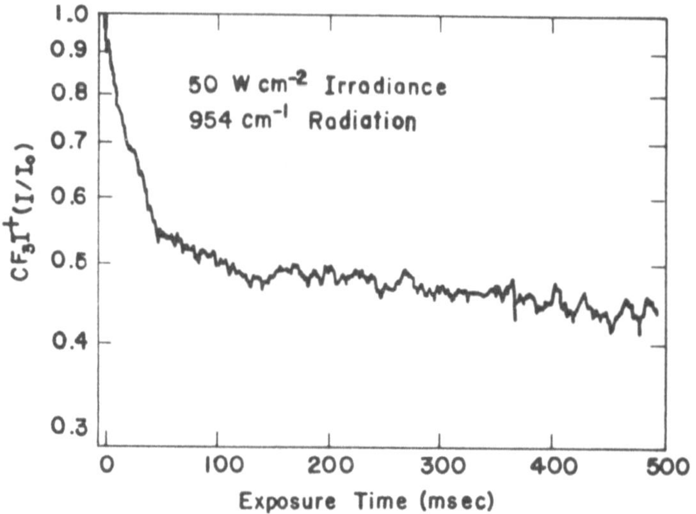

Fig. 7. Photodissociation kinetics of CF_3I^+ at 70 eV electron impact energy, 1.5×10^{-7} Torr CF_3I pressure and 50 W cm^{-2} laser irradiance at 954.6 cm^{-1}. The slope of the line at a given point fives the apparent photodissociation rate constant.

species $(CH_3)_2Cl^+$, $(CD_3)_2Cl^+$ and $CH_3CD_3Cl^+$ are formed as indicated in Figure 6. Only $(CD_3)_2Cl^+$ photodissociates with laser irradiance ≤ 100 Wcm^{-2} according to eq 14.

$$(CD_3)_2Cl^+ + nh\nu \longrightarrow CD_2Cl^+ + CD_4 \quad H = 19\ kcal/mole^{-1} \quad (14)$$

No chlorine isotope effect is observed.

Trifluoroiodomethane. The major ions formed by 70 eV electron impact on CF_3I are CF_3^+, I^+, CF_2I^+ and CF_3I^+ [14]. Only the concentration of CF_3I^+, CF_2I^+ and CF_3^+ are affected by CO_2 laser radiation at CF_3I pressures below 10^{-6} Torr. CF_3I^+ decreases, with a concomitant increase in CF_2I^+ and CF_3^+. Ion cyclotron double resonance results show the increase in CF_2I^+ is due entirely to the increase in CF_3^+ via its reaction with neutral CF_3I. This establishes CF_3^+ as the exclusive photoproduct, reaction 15. Unlike the ions

$$CF_3I^+ + nh\nu \longrightarrow CF_3^+ + I \qquad \Delta H = 24\ kcal/mole \qquad (15)$$

mentioned previously, the photodissociation kinetics of CF_3I^+ cannot be characterized by a simple exponential decay, as indicated in Figure 7, where a single exponential decay would give a straight line. Rather, the CF_3I^+ photodissociation kinetics show evidence of at least two populations of CF_3I^+, one which dissociates rapidly and another which dissociates slowly or not at all. These two populations are made in roughly equal amounts by electron impact ionization of CF_3I. Experimental evidence [7] supports the conclusion that the rapidly dissociating fraction is CF_3I^+ in the higher energy level of the ground state spin-orbit doublet, $\tilde{X}^2E_{1/2}$. The slowly dissociating fraction is in the lower doublet level, $\tilde{X}^2E_{3/2}$. Kinetics of the rapidly dissociating fraction follow approximately a simple exponential decay which is proportional to laser irradiance, $k_D = 0.80(6)\ sec^{-1}W^{-1}cm^2$ or $\sigma_D = 1.5 \times 10^{-20}\ cm^2$.

Proton Bound Alcohol Dimers. Proton bound alcohol dimers of CH_3OH, C_2H_5OH, $i\text{-}C_3H_7OH$, $n\text{-}C_3H_7OH$, $s\text{-}C_4H_9OH$, $t\text{-}C_4H_9OH$, $t\text{-}C_5H_{11}OH$ have been investigated. Full details of alcohol ion-molecule chemistry including reaction sequences leading to $(ROH)_2H^+$ formation [15-21] and pertinent thermochemistry are given in ref 22. Infrared laser irradiation leads to multiphoton dissociation of each of the proton bound dimers studied. The three lowest energy routes to decomposition are given by eqs 16-18. Temporal variation of the ion population during laser irradiation is first order in all cases and it is possible to dissociate 100 % of the dimers

$$ROH_2^+ + ROH \longleftrightarrow [(ROH)_2H^+]^* \longrightarrow \begin{cases} ROH_2^+ + ROH & (16) \\ R_2OH^+ + H_2O & (17) \\ (ROH)H^+(H_2O) + \text{olefin} & (18) \end{cases}$$

Cyanobenzene Molecular Ion. Cyanobenzene cation undergoes a sequential two-photon dissociation, in the presence of visible laser irradiation between

$$C_6H_5CN^+ \xrightarrow{h\nu} [C_6H_5CN^+]^* \xrightarrow{h\nu} C_6H_5^+ + HCN \qquad (19)$$

450 and 600 nm (2.1-2.8 eV). The overall endothermicity of the reaction is 3.2 eV. Kinetic modelling of the dissociation reveals that the intermediate state is long lived and that the principal mode of deactivation is charge transfer to neutral cyanobenzene [23]. Based on a comparison of photodissociation and photoeletron spectra [24,25] the initial excitation is thought to involve promotion of a π electron on the nitrile group to the highest occupied ring orbital of cyanobenzene. Internal conversion to excited vibrational levels of the ground state is followed by absorption of a second photon which raises the energy of the molecule above the threshold for the lowest energy dissociation channel.

When the ions are irradiated with up to 60 W cm^{-2} from a cw CO_2 laser for 1 s, no dissociation is observed at any of the available laser wavelengths. Irradiation by both visible and infrared lasers, however, results in a significant increase in the overall dissociation rate compared with visible irradiation alone under the same conditions. This enhancement of the dissociation rate is due to infrared multiphoton dissociation of vibrationally excited cyanobenzene cations which have undergone internal conversion. Since the infrared laser is effective in dissociating only those molecules which have been "activated" by absorption of a visible photon, this technique represents a convenient photochemical probe of molecules with a large degree of internal excitation.

3.1.2 Variation of Photodissociation Yield with Laser Wavelength

For all of the ions studied, the decomposition pathway does not depend on laser wavelength, but the photodissociation yield does. Of the ions listed in Table 1, only the larger proton-bound alcohol dimers show little wavelength dependence.

Analysis of photodissociation spectra is not always straightforward, partly because our studies have been limited to the CO_2 laser wave number range, 925-1090 cm^{-1}. However, comparisons with gas phase spectra of related neutrals and consideration of changes in bonding due to ionization allows a reasonable interpretation of the features present in the infrared dissociation spectra. The analysis rests on the assumption that multiphoton dissociation spectra somewhat mimic the small signal absorption spectrum of closely related neutral molecules(see section 3.1.5). The close relationship between small signal absorption spectra and the wavelength dependence of dissociation yield using pulsed megawatt lasers is well established [26,27]. The primary differences between small signal absorption spectra of neutrals and the photodissociation spectra of the corresponding ions are that the latter can be broadened by up to 50 % and shifted by up to 10 cm^{-1}. Further, photodissociation bands are generally featureless without any P(Q)R type band contours. Despite these differences multiphoton dissociation represents one of the few techniques available for obtaining information on gas phase ions.

Perfluoropropylene. Figure 8 shows the variation of $C_3F_6^+$ photodissociation yield, P_D, with laser wavelength. P_D is defined, in this case, as the fraction of ions dissociated during 2 seconds of irradiation at 34 W cm^{-2} in 4.8 x 10^{-7} Torr C_3F_6. The two sets of data are for electron impact ionization at 70 and 20 eV. Also shown in Figure 8 is the gas phase perfluoropropylene absorption spectrum over the same wavelength range. The infrared absorption band of the neutral molecule (ν_{max} = 1037 cm^{-1}) has been assigned to a C-F stretching mode of A' symmetry [28]. Comparison with other fluorinated species suggests that this vibrational mode involves primarily motion of the CF_3 group. Therefore the photodissociation spectrum and the neutral absorption band are nearly superimposed. Although the infrared absorption spectrum of the neutral shows a combination band at 978 cm^{-1}, its intensity relative to the major peak at 1037 cm^{-1} is considerably smaller than the feature at 985 cm^{-1} in the $C_3F_6^+$ photodissociation spectrum. The small peak in the photodissociation spectrum occurs at an

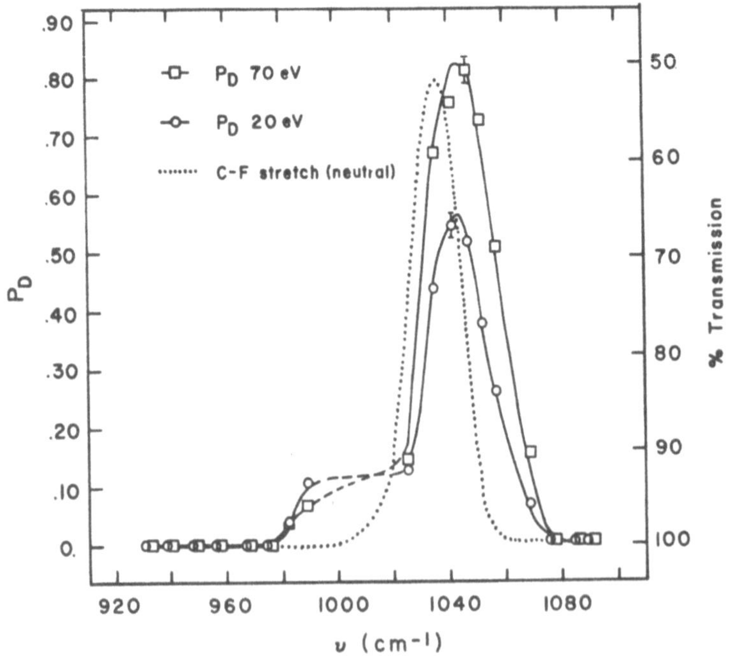

Fig. 8. Photodissociation spectrum of $C_3F_6^+$ over the CO_2 laser spectral range [4] Left ordinate is fraction of $C_3F_6^+$ dissociated after 2 seconds of irradiation at 34 W/cm^2. The two solid curves are for ionization energies of 70 eV (□) and 20 eV (o). Perfluoropropylene pressure is 4.8 x 10^{-7} Torr. Dotted line is infrared absorption spectrum of perfluoropropylene at 0.8 Torr in a 10 cm length cell.

energy with is too low to attribute it to a v = 1→v = 2 transition. Thus, it is tentatively assigned as a combination band. Attempts to probe in detail the characteristics of $C_3F_6^+$ photodissociation at 985 cm^{-1} were thwarted by a lack of laser intensity in that special region.

Energy deposition into internal degrees of freedom of ions formed by electron impact ionization increases with increasing electron energy. Hence the population of vibrationally excited $C_3F_6^+$ will be greater when formed with 70 eV electrons than with 20 eV electrons. The increase in photodissociation yield a 70 eV compared to 20 eV (Figure 8) is attributed to the presence of "hot" ions formed at 70 eV.

Diethyl Ether. The wavelength dependences for multiphoton dissociation of $(C_2H_5)_2OH^+$ and $(C_2H_4)_2O\ _2H^+$ are shown in Figure 9a and data for $(C_2D_5)_2OD^+$ are shown in Figure 9b. Also shown are the gas phase absorption spectra of

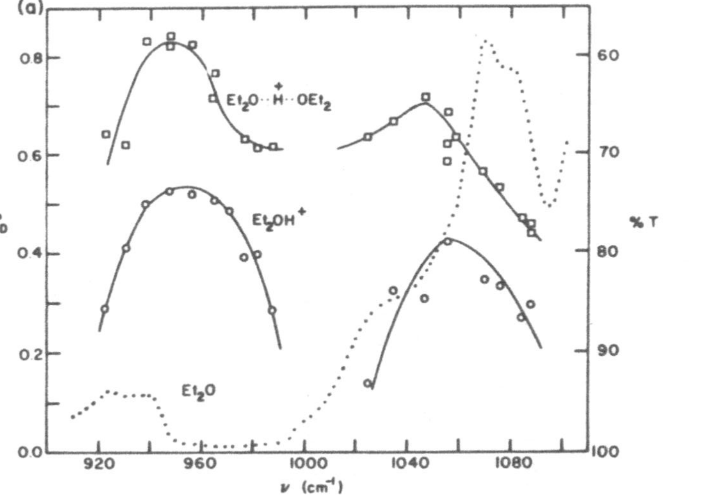

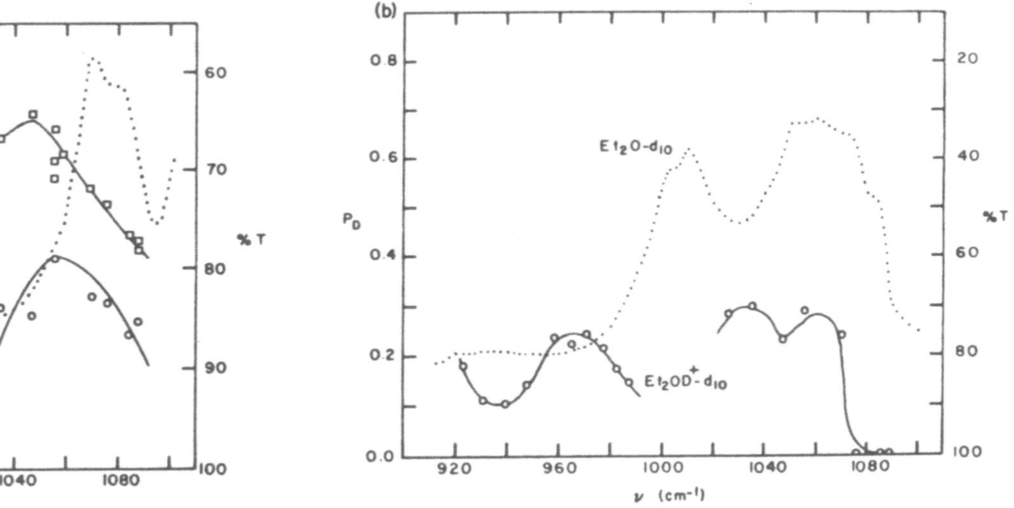

Fig. 9. (a) Photodissociation spectra of $(C_2H_5)_2OH^+$ (o) and $[(C_2H_5)_2O]_2H^+$ (□) over the CO_2 laser spectral range. For $(C_2H_5)_2OH^+$, P_D is the fraction of ions dissociated after 1.9 seconds of irradiation at 48 W/cm^2; $(C_2H_5)_2O$ pressure is 8.8×10^{-8} Torr. P_D for $[(C_2H_5)_2O]_2H^+$ is defined as the fraction of ions dissociated after 2.0 seconds of irradiation at 10 W/cm^2; $(C_2H_5)_2O$ pressure is 4.7×10^{-7} Torr. Ionization energy for both experiments is 14 eV. Dotted line is the infrared absorption spectrum of diethyl ether at 20 Torr in a 10 cm length cell. (b) Photodissociation spectrum of $(C_2D_5)_2OD^+$ over the CO_2 laser spectral range. Experimental conditions are the same as for photodissociation of $(C_2H_5)_2OH^+$ in (a). Dotted line is the infrared absorption spectrum of $(C_2D_5)_2O$ at 16 Torr in a 20 cm length cell.

the corresponding neutrals over the range of CO_2 laser wave lengths (note the change of scale in the axes for percent transmission for the neutrals). For both $(C_2H_5)_2OH^+$ and $(C_2D_5)_2OD^+$ experimental conditions are nearly identical; therefore differences in P_D values for the monomer species are a direct measure of differences in cross sections for multiphoton dissociation at each wavelength. No such direct comparison regarding dissociation cross sections can be made between the protonated ether and the proton bound dimer owing to differences in laser power and ether pressure for the two experiments.

Protonation of the ether molecules introduces three new degrees of freedom whose vibrational frequencies are all expected to lie outside the 925-1090 cm^{-1} region. The remaining vibrational bands should differ from those of the parent ether primarily as a result of perturbations introduced by a positive charge at oxygen and, secondly, from the effect of increased mass. Observed bands in the infrared spectrum of diethyl ether from 900-1100 cm^{-1} have been assigned to combinations of C-C stretches, C-O stretches and methylene wags [29]. Upon protonation, rehybridization at oxygen adds p-character to the newly formed O-H bond and results in increased s-character in the C-O bonds. The resulting blue shift in C-O stretching frequencies appears to be reflected in the photodissociation spectrum of $(C_2H_5)_2OH^+$. Specifically, the P_D maxima at 955 cm^{-1} and 1048 cm^{-1} correlate with the peak at 930 cm^{-1} and the shoulder at 1040 cm^{-1}, respectively. Addition of a second ether molecule to form the proton bound dimer distributes the charge between both ether moieties with the result that vibrational frequencies of the dimer are more like those of the neutral ether. This is verified in Figure 9a. The absorption spectrum of $(C_2D_5)_2O$ (Figure 9b) shows maxima at 1010 cm^{-1} (assigned to C-O stretch) and 1060 cm^{-1} (assigned to methylene bends). In the $(C_2D_5)_2OD^+$ photodissociation spectrum the former is blue shifted to 1035 cm^{-1} and no frequency change is observed for the methylene bending mode. This is in accord with arguments presented for $(C_2H_5)_2O$. Below 980 cm^{-1} the $(C_2D_5)_2O$ absorption spectrum exhibits a broad absorption which extends to 840 cm^{-1}. The photodissociation spectrum in the same region cannot be correlated with specific $(C_2D_5)_2O$ vibrational modes due to the lack of distinguishable features in the absorption spectrum.

Dimethylchloronium. Figure 10 shows the variation with laser wavelength of $(CD_3)_2{}^{35}Cl^+$ photodissociation yield, P_D. In this case, P_D is defined as the fractional change in ion intensity resulting from 2 s continuous laser irradiation at 51 W cm^{-2}. The band maximum is at $\approx$980 cm^{-1} and its width is $\approx$70 cm^{-1}. No chlorine isotope effect is observed; the $(CD_3)_2{}^{37}Cl^+$ photodissociation spectrum is identical to that shown in Figure 10.

To examine the effects of structural variation in addition to the effects of isotopic substitution an attempt was made to study infrared photochemistry of $C_2H_5ClH^+$, an isomer of dimethylchloronium. Multiphoton dissociation of $C_2H_5ClH^+$ (prepared by protonation of ethyl chloride) was observed but details of the laser induced chemistry could not be elucidated due to the competing ion-molecule reactions in this system [30].

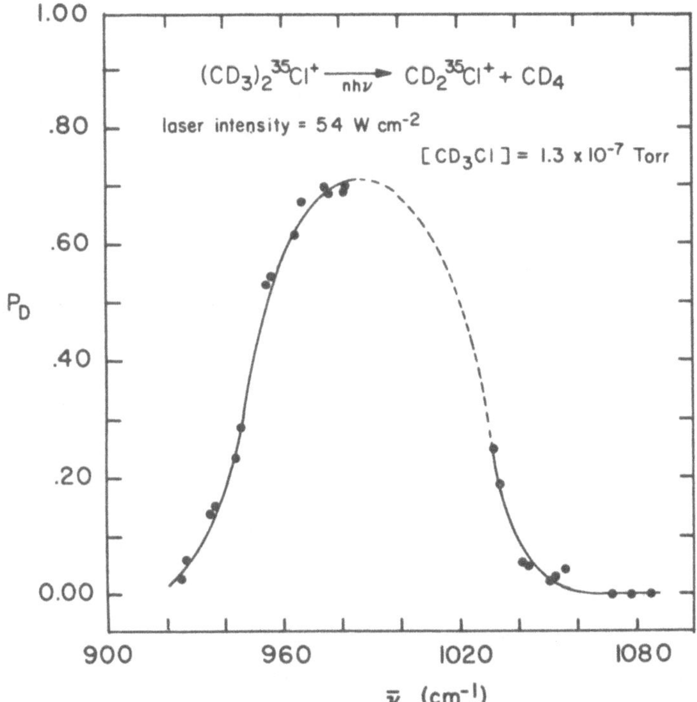

Fig. 10. Infrared photodissociation spectrum of $(CD_3)_2{}^{35}Cl^+$. CD_3Cl pressure is 1.3×10^{-7} Torr. Ionization as in Figure 1. Laser power is 54 W cm^{-2}. P_D denotes the fraction of ions dissociated during 2 s continuous laser irradiation.

Trifluoroiodomethane. The photodissociation spectrum of CF_3I^+ is shown in Figure 11. The band (solid circles) peaks at 960 cm^{-1}, within the frequency region expected for the ν_1 absorption of $CF_3I^+(\tilde{X}^2E_{1/2})$ and is about three times wider than the ν_1 absorption band of neutral CF_3I. This band is also 1.2 times wider and at 13 cm^{-1} higher energy than the same band previously reported for CF_3I^+ photodissociation in a fast ion beam [31]. There is no detectable photodissociation in the 1030-1085 cm^{-1} region where the ν_1 absorption $CF_3I^+(\tilde{X}^2E_{3/2})$ is expected to occur. The limits of detectability for CF_3I^+ loss in these experiments place an upper limit of 2 % on the extent of dissociation in this frequency region. This is in agreement with earlier results [31] for CF_3I^+ which showed photodissociation is 40 times more efficient near 940 cm^{-1} than near 1080 cm^{-1}. In the present case, a delay of 200 msec between ion formation and irradiation greatly reduces is unchanged as shown by the open circles in Figure 11.

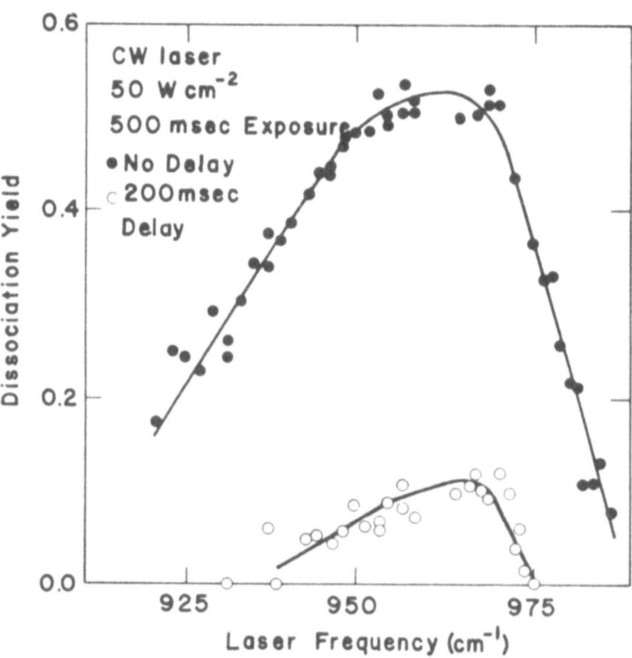

Fig. 11. Dependence of CF_3I^+ photodissociation on laser frequency using cw radiation. The extent of dissociation is monitored by CF_3I^+ signal intensity. Experimental conditions are: 5 x 10^{-6} Torr CF_3I pressure, 70 eV electron impact energy, 50 W cm^{-2} irradiance, 500 msec exposure time. Data are shown for no delay (●) and 200 msec (○) delay between ion formation and irradiation.

Cyanobenzene. Photodissociation of $C_6H_5CN^+$ by visible radiation is enhanced by cw infrared radiation. Dependence of the enhancement on laser frequency has a pronounced peak at 970 cm^{-1} as shown in Figure 12. Here, the enhancement is given by (I-I')/I where I is the ion population with visible irradiation only and I' is the population with irradiation by both lasers. Since only excited state molecules are sensitive to the infrared laser, the enhancement spectrum in Figure 12 is a photodissociation spectrum of a molecule with 2 eV of internal energy. The density of vibrational states at this level of excitation is estimated to be 2 x 10^{11} states per cm^{-1}, so the excitation is truly in a continuum of overlapping states. The distinct band in the photodissociation spectrum, however, indicates that although transitions are energetically possible at every wavelength, transition oscillator strengths remain spectrally localized and appear to be similar in shape to what might be expected for a ground state absorption profile.

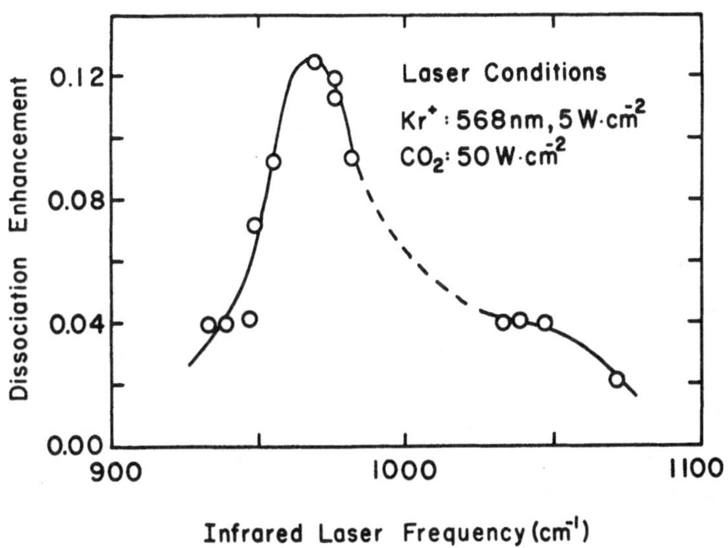

Fig. 12. Plot of dissociation enhancement versus infrared laser frequency. No laser lines capable of 50 W cm^{-2} were available in the region of the dotted curve.

3.1.3 Effects of Collisions on Photodissociation Yield

The role of collisions in multiphoton dissociation is not clearly understood. Both enhancement and reduction in photodissociation yield occur with increasing neutral gas pressure. However, for most ions studied the yield decreases with increasing pressure. In this case, the dominant mechanism involves transfer of internal energy from the excited ion to its collision partner thus reducing the energy ultimately available to the ion for dissociation. These collisional deactivation processes can be particularly efficient when the ion, M^+, collides with its corresponding neutral, M. For these collisions, symmetric charge transfer can occur and the close match between the vibrational frequencies of the two collision partners can enhance intermolecular vibrational energy transfer. In the case of increasing yield with increasing pressure, collisions may serve to repopulate the specific rotational levels which are depopulated by laser radiation, thus making absorption of subsequent photons more efficient.

Proton Bound Dimer of Diethyl Ether. Collisional effects both prior to and during irradiation have been studied in decompositions of $[(C_2H_5)_2O]_2H^+$. Since proton-bound dimer formation, eq 4, is bimolecular and exothermic a question arose as to whether or not $[(C_2H_5)_2O]_2H^+$ is in a highly vibrationally excited state prior to irradiation. An experiment similar to the one depicted in Figure 7 was carried out, modified such that the laser was delayed for up to 900 msec following onset of $(C_2H_5)_2OH^+$ ejection. At various diethyl ether pressures and buffer gas (SF_6 and $i\text{-}C_4H_{10}$) pressures no change in $[(C_2H_5)_2O]_2H^+$ photodissociation rate is observed with increasing laser delay. Invariance of dissociation rates with laser delay (at constant pressure) implies that $[(C_2H_5)_2O]_2H^+$ is vibrationally relaxed prior to irradiation.

The data in Figure 13 indicate that $[(C_2H_5)_2O]_2H^+$ multiphoton dissociation rates decrease with increasing pressure at constant laser intensity. In addition to diethyl ether, results obtained by adding either of two buffer gases, SF_6 and $i\text{-}C_4H_{10}$, to a small acount of diethyl ether are included. To allow for a direct comparison of deactivation efficiencies of the three gases, dissociation rates are plotted as a function of ion-molecule collision [32]. Both SF_6 and $i\text{-}C_4H_{10}$ appear to be more effective than diethyl ether at quenching dissociation.

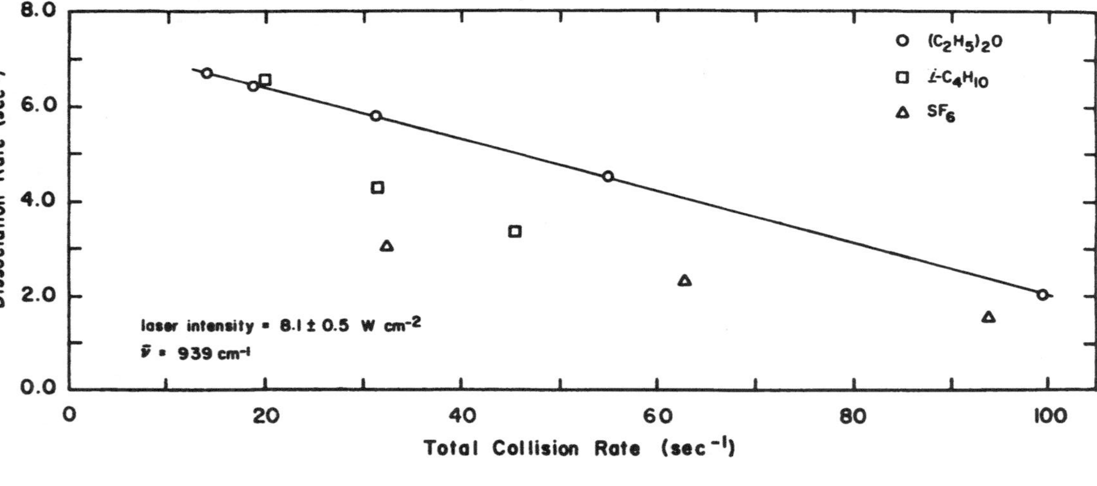

Fig. 13. $[(C_2H_5)_2O]_2H^+$ multiphoton dissociation rate as a function of added buffer gases: SF_6(Δ) and $i\text{-}C_4H_{10}$(□). Dissociation rate is plotted as a function of total collision rate $(C_2H_5)_2O$ plus buffer gas) to allow direct comparison of collision efficiencies. SF_6 or $i\text{-}C_4H_{10}$ are added to 3.7×10^{-7} Torr of diethyl ether. Ionization energy is 14 eV.

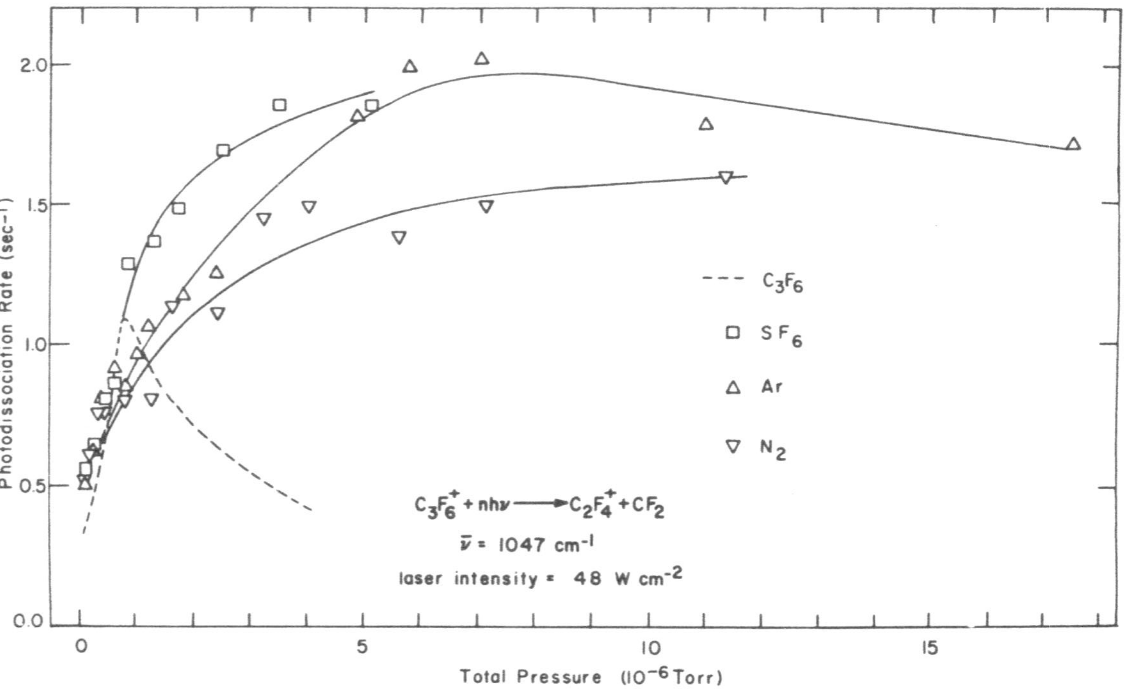

Fig. 14. $C_3F_6^+$ multiphoton dissociation rate as a function of added buffer gases: Ar (Δ), N_2 (∇) and SF_6 (□)[4]. The dashed line indicates $C_3F_6^+$ multiphoton dissociation rate with C_3F_6 as the added buffer gas. Ar, N_2, or SF_6 are added to 1.2×10^{-7} Torr C_3F_6. The abscissa indicates total pressure.

Perfluoropropylene. In sharp contrast to the collisional effects demonstrated in Figure 13 for $[(C_2H_5)_2O]_2H^+$, $C_3F_6^+$ multiphoton dissociation rates increase with increasing pressure up to 5×10^{-6} Torr as shown in Figure 14. These results are obtained for C_3F_6, Ar, N_2 and SF_6 as buffer gases. In order to minimize possible hot ion effects (as seen in Figure 8) and to avoid ionization of buffer gases other than C_3F_6, an electron impact ionization energy of 14 eV was used. Only C_3F_6 (dashed line in Figure 14) shows a marked tendency to quench dissociation, beginning at pressures above 1.5×10^{-6} Torr. This is attributed to deactivation by symmetric charge transfer

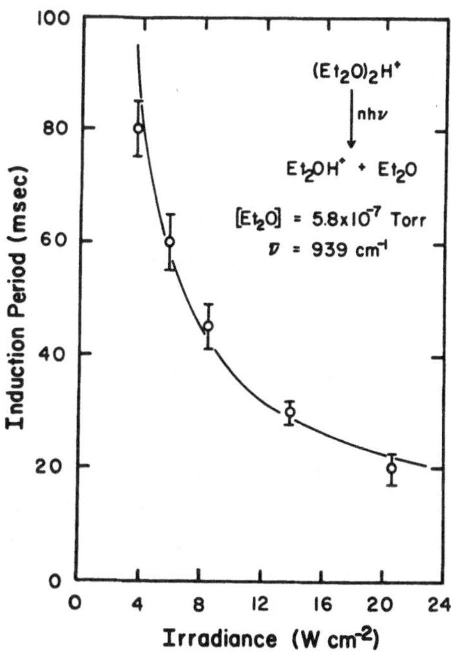

Fig. 15. Induction period for multiphoton dissociation of $[(C_2H_5)_2O]_2H^+$ as function of laser intensity. Induction period is defined as the time between shutter opening and initial measurable photodissociation. Diethyl ether pressure is 5.8×10^{-7} Torr, ionization energy is 14 eV and laser wavelength is 939 cm^{-1}. A close fit to the data is obtained by the equation given above (solid line) which takes into account the 5 msec shutter opening time. Induction periods (t) in msec and laser intensities (I_{las}) in J cm^{-2} s^{-1} give an energy fluence threshold of 300 mJ/cm^2.

processes. In Figure 14 photodissociation rates are plotted against buffer gas pressure rather than collision frequency because for all four gases used rate constants for collision with $C_3F_6^+$ are nearly identical [32].

3.1.4 Variation of Photodissociation Yield with Laser Irradiance

Variation of multiphoton dissociation yields with laser irradiance provides a useful probe of the excitation mechanism and offers a means of comparison with megawatt pulsed laser results. In studies of SF_6 at high laser irradiance the dissociation probability per pulse is proportional to $(I_{las})^n$ where I_{las} is the laser irradiance. Experimentally derived values [26] of n vary from 1.5 to 14. The use of focused laser beams leads to large uncertainties in measurements of power density and irradiated volumes, possibly accounting for the lack of agreement in reported values of n. The effects of laser power and sample pressure are intimately related since, at constant pressure, increased laser power reduces the time scale for excitation. Thus fewer collisions are experienced during excitation. Therefore the ideal experiment for determing the variation of photodissociation yield with I_{las} would utilize an unfocused laser beam and should allow extrapolation to a low pressure limit. Such conditions are readily obtained in the ICR studies of slow multiphoton dissociation.

Proton Bound of Diethyl Ether. Careful examination of multiphoton dissociation of $[(C_2H_5)_2O]_2H^+$ reveals an induction period prior to the onset of decomposition. Induction period is defined as the time delay between opening the shutter and the first observable photodissociation. The variation of induction period with laser irradiation is illustrated in Figure 15 for $[(C_2H_5)_2O]_2H^+$ at constant ether pressure. When corrected for a shutter rise time of 5 msec, the data in Figure 15 closely fit eq 20 (solid line in Figure 15),

$$(\text{Induction Period}) \times (I_{las}) = 0.3\ \text{J/cm}^2 \tag{20}$$

indicating an energy fluence threshold of 0.3 J/cm^2. Observation of an energy fluence threshold is in agreement with pulsed laser multiphoton dissociation experiments on SF_6 [33].

Assuming decomposition is rapid compared to excitation in the ICR experiments (section 3.1.5), the induction period represents the time required to pump ions to energies sufficient for dissociation. This arguments predicts that photodissociation rates are pressure independent at pressures such that

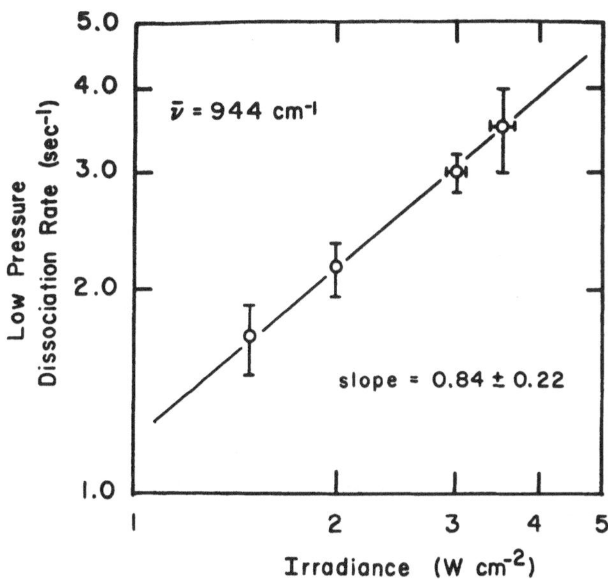

Fig. 16. Log-log plot indicating the dependence of photodissociation rate $[(C_2H_5)_2O]_2H^+$ on the first power of laser intensity. Extrapolation to obtain low pressure rates uses induction period data from Figure 15, as discussed in the following section.

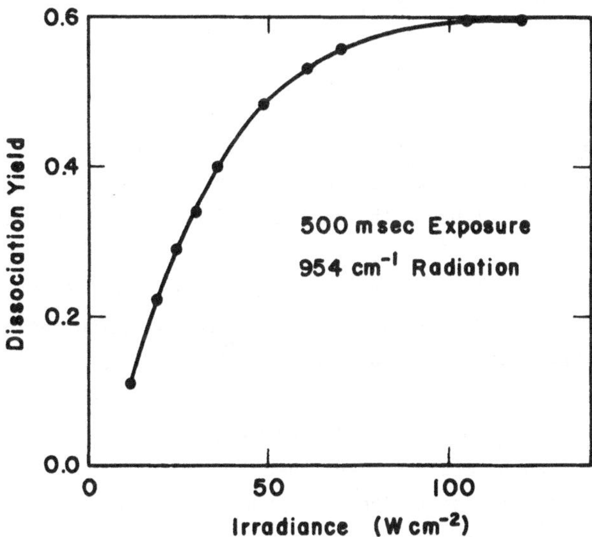

Fig. 17. Variation of the dissociation yield on laser irradiance as monitored by the CF_3I^+ signal. Experimental conditions are: 70 eV electron energy, 6 x 10^{-8} Torr CF_3I pressure, 500 msec exposure time and 954.6 cm^{-1} laser frequency. Longer exposure time does not increase the yield.

the time between collisions is long compared to the observed induction period. This criterion defines the low pressure limit. Extrapolation of data such as in Figure 14 to the low pressure limit yields the collision free photodissociation rate. A plot of the logarithm of collision free dissociation rate as function of ln (I_{las}) (Figure 16) is linear with slope 0.84 ± 0.22. Therefore, within experimental error, multiphoton dissociation of $[(C_2H_5)_2O]_2H^+$ is first order in laser intensity, that is n = 1. Over the range of laser irradiance (4-20 W/cm^2) which permits accurate measurements of dissociation rates, no saturation effects are observed. The ratio of the low power dissociation rate to the photon flux gives 2.0 ± 5 x 10^{-20} cm^2 for the multiphoton dissociation cross section of $[(C_2H_5)_2O]_2H^+$.

Perfluoropropylene. Since collisions in $C_3F_6^+$ are seen (section 3.1.3) to enhance multiphoton infrared dissociation, extrapolation of these data to a low pressure limit is not straightforward. $C_3F_6^+$ photodissociation has been studied as a function of laser power at several pressures. For C_3F_6 pressures $\gtrsim 1 \times 10^{-6}$ Torr and C_3F_6 plus buffer gas (N_2) mixtures at total pressures $\gtrsim 1 \times 10^{-6}$ Torr photodissociation rates increase linearly with increasing irradiance up to the highest attainable laser intensity, 100 W/cm^2. At C_3F_6 pressures $<1 \times 10^{-6}$ Torr the $C_3F_6^+$ signal does not exhibit a simple exponential decay with time at higher laser powers. The observed dissociation process can be described by the decay of two separate ion populations with different rate constants. While a complete quantitative analysis is not currently available, the mechanistic implications of this result are discussed in section 3.1.5.

Trifluoroiodomethane. Unlike the ions previously mentioned, the variation of CF_3I^+ photodissociation yield with laser irradiance shows saturation effects. As indicated in Figure 4 the photodissociation yield rises monotonically to a limiting value of 60 %. This is a result of electron impact on CF_3I producing CF_3I^+ in two, roughly equal populations CF_3I^+ ($\tilde{X}^2E_{1/2}$) and CF_3I^+ ($\tilde{X}^2E_{3/2}$) as mentioned previously. The $\tilde{X}^2E_{1/2}$ population rapidly dissociates while the $\tilde{X}^2E_{3/2}$ population dissociates slowly or not at all.

3.1.5 Infrared Multiphoton Dissociation Mechanism

While it is tempting to apply current theories [34-38] for megawatt pulsed laser multiphoton dissociation to the low intensity IR photolysis of gas phase ions, there are inherent differences between the two types of experiments which necessitate modification of the existing developed theories. In parti-

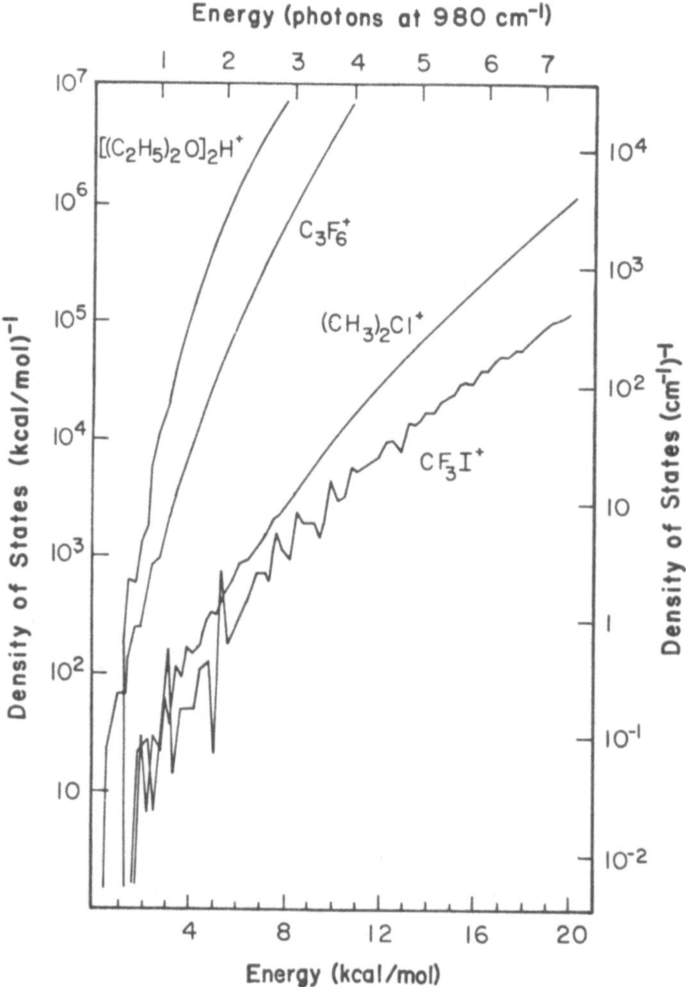

Fig. 18. Calculated vibrational state density as a function of vibrational energy content for $[C(C_2H_5)O_2]H^+$, $C_3F_6^+$, CH_3Cl^+ and CF_3I^+.

cular, the time scale of the ICR experiments requires consideration of spontaneous emission as a viable deactivation mechanism. Equally important, the low laser intensities obviate power broadening [38] as a mechanism for overcoming anharmonicities in hot band absorptions. In addition, any mechanistic model must allow for collisional enhancement of dissociation rates (as observed in $C_3F_6^+$).

Excitation Processes. Several authors have pointed out that a high levels of vibrational excitation, the spectrum of transitions is nearly continuous [39,40]. This realm of molecular exictation is generally known as the quasi-continuum of vibrational states. Onset of the quasi-continuum begins when the molecular vibrational state density is roughly equal to the laser bandwidth, typically 1000 states per cm^{-1} for cw CO_2 laser radiation. Figure 18 shows the vibrational state density as a function of vibrational energy for several of the ions mentioned above. A direct count of states was used for internal energies below 4.5 kcal $mole^{-1}$ except for $(CH_3)_2Cl^+$ and CF_3I^+ where the direct count was used up to 8 kcal $mole^{-1}$ and 20 kcal $mole^{-1}$ respectively. The Whitten-Rabinovitch approximation [41] was used for higher energies. Because vibrational frequencies are not known for $(CF_3)_2Cl^+$ the frequencies of $(CH_3)_2S$ [42] were used to approximate the state density of $(CH_3)_2Cl^+$. Similarly, C_3F_6 [28] frequencies were used for $C_3F_6^+$, CF_3I [43] frequencies for CF_3I^+ and $(C_2H_5)_2O$ [29] frequencies for $[(C_2H_5)_2O]_2H^+$. In the latter case, all of the $(C_2H_5)_2O$ frequencies were used twice and the nine additional frequencies were estimated from the vibrational modes which result from addition of H^+ and the relative motions of the two $(C_2H_5)_2O$ groups. Theories of exitation through the quasi-continuum treat the process as a sequence of incoherent single photon events. Photoacoustic experiments indicated that, for SF_6, absorption cross sections in the quasi-continuum decrease exponentially with increasing internal excitation [44]. Model calculations [45], using cross sections obtained from the photoacoustic results, agree well with pulsed laser multiphoton dissociation yields. Applying eq 20 to megawatt powers and short pulse durations predicts induction periods commensurate with the model calculations mentioned above. In a complementary fashion, scaling the model calculations [45] to low laser intensities yields results very similar to experimentally observed multiphoton dissociation reactions with cw radiation. Thus absorption through the quasi-continuum is expected to be similar for both high power pulse laser excitation and low power cw experiments. The only difference may arise from deactivation due to spontaneous emission in the low irradiance photolysis. Estimates of spontaneous emission rates are 1-100 sec^{-1} [46].

The primary difference between megawatt multiphoton activation and the ICR experiments involves the mechanism of initial excitation to the quasi-continuum. The ease of populating the quasi-continuum will depend on the number of photons required to reach this region. Molecules possessing many degrees of freedom will have a significant amount of internal energy at ambient temperature. The combination of many vibrational modes and appreciable thermal energy content serves to locate such molecules very nearly in the quasi-continuum prior to laser excitation. It is thus natural to consider large molecule and small molecule limits. For $[(C_2H_5)_2O]_2H^+$, the densities of states at energies corresponding to absorption of one and two IR photons (1000 cm^{-1}) are 120 states/cm^{-1} and 9 x 10^4 states/cm^{-1}, respectively. Thus, following resonant absorption of one IR photon, a near continuum of states (where the level separations and the laser bandwidth are comparable) is available for successive absorptions. Furthermore, at ambient temperatures the thermal energy content (3-4 kcal/mole) in excess of the zero point energy for $[(C_2H_5)_2O]_2H^+$ is comparable to excitation by a single photon. This puts the proton bound dimer of diethyl ether in the large molecule category, where the density of vibrational states is large at low energies and multiphoton excitation is expected to be facile.

Density of state calculations for $C_3F_6^+$ are 5 states/cm^{-1} and 213 states/cm^{-1} at energies corresponding to absorption of one and two IR photons, respectively. The quasi-continuum of vibrational states in $C_3F_6^+$ is not reached until an ion has absorbed approximately 4 photons; therefore $C_3F_6^+$ is classified as a "small" molecule. The mechanism by which these four photons are absorbed is significant in determining the $C_3F_6^+$ multiphoton dissociation rate. Even with the highest laser irradiance available sequential excitations within one vibrational mode is not possible without energy transfer out of the excited mode. This is because the power broadening, typically 5 x 10^{-3} cm^{-1} or less, is insufficient to overcome the vibrational anharmonicities. Among available theoretical treatments an excitation mechanism based on the formalism of a single resonant absorption appears best suited to these experiments [36, 37]. A two-level system assumed to be in resonance with the laser field, is pumped and then depopulated by intramolecular vibrational energy transfer. When the resonant mode returns to the ground state it is free to absorb another photon. The process is repeated until the quasi-continuum is reached.

The requirement of a single resonant vibrational mode is consistent with the sharp frequency dependence observed in the $C_3F_6^+$, photodissociation spectrum. The slow step in the excitation process is intramolecular V-V transfer at low excitation levels, since at all laser powers used resonant transitions from the ground state are saturated [47]. Therefore, this simplified model predicts photodissociation rates to be independent of photon flux because intramolecular V-V transfer is a nonradiative process. Even though photolysis of $C_3F_6^+$ shows a change in behavior at low pressures and high laser power, it does not exhibit the predicted saturation effect, indicating that modifications to the theory are needed to describe low intensity infrared multiphoton dissociation processes.

Effects of Collisions on Multiphoton Excitation. For species at the large molecule limit, excitation occurs exclusively via sequential photon absorptions through the quasi-continuum and collisions act only to depopulate excited vibrational states. This is observed in the decrease in $[(C_2H_5)_2O]_2H^+$ dissociation rates with increasing pressure, Figure 13.

Collisions have three possible effects on the excitation of small molecules. First, and most obvious, is simple deactivation as described in the above paragraph. Second, collisions can enhance the rate of intramolecular V-V transfer, thus increasing the rate at which the molecule is pumped to the vibrational quasi-continuum. The third effect results from the narrowness of the laser linewidth. At a given laser frequency only a few rotational states lead to a resonant transition from the ground vibrational level. These states are depleted by photodissociation and collisional repopulation of crucial rotational levels may be involved in the rate-limiting process.

Both collision-induced intramolecular V-V transfer [48,49] and collisional repopulation of depleted rotational states would account for an increase in photodissociation rate with increasing pressure [50]. To date no experimental results conclusively distinguish between either mechanism.

The two-population decay observed in $C_3F_6^+$ decomposition at low pressure and high laser intensity (section 3.1.4) results from either the presence of vibrationally excited ions prior to irradiation, or varying rates for collisional redistribution of rotational states. In the first case, the fraction of $C_3F_6^+$ with appreciable vibrational excitation decomposes rapidly because it has already passed the intramolecular V-V transfer bottleneck. At the lowest pressure used (8×10^{-8} - 2×10^{-7} Torr) there are not sufficient numbers of collisions to deactivate vibrationally "hot" $C_3F_6^+$ formed by electron impact ionization, Figure 8. The alternative explanation for the two population decay is derived from observations by Polanyi and Woodall [51,52] who reported that probabilities for collisional induced rotational transitions vary inversely with the change in rotational energy. During the course of infrared laser photolysis the involvement of species originally in rotational states energetically far removed from the resonant state leads to a slowing of repopulation rates, and hence a change in the observed dissociation rate.

The obvious complexities in low intensity, infrared multiphoton dissociation preclude further detailing of the excitation process. It is reasonable to expect that modeling calculations similar to those performed for megawatt pulsed laser photolyses of SF_6 will provide new insight into the nature of the phenomena.

Unimolecular Dissociation of Activated Molecules. Observation of a large isotope effect in the photodissociation of $(C_2H_5)(C_2D_5)OH^+$ (section 3.1.1) establishes that dissociation occurs before the molecule can absorb an appreciable amount of energy above threshold. The relatively slow time scale for low intensity IR laser excitation implies that standard statistical treatments of unimolecular reactions (i.e., RRKM theory [41]) can be utilized to describe the decomposition step and that decomposition is generally much more rapid than sequential photon absorption.

Kinetic Model for Multiphoton Dissociation. One of the objectives in studying infrared laser-induced chemical reactions is to better understand the mechanism by which many photons are sequentially deposited in a molecule. Individual absorption and deactivation events are difficult to probe directly, however, and usually the only physical observable is the overall dissociation of the molecules. With this limitation in mind, it is often instructive to construct simple models which can provide some insight to the excitation mechanism by reproducing various aspects of the observed dissociation kinetics.

One approach which has met with some success is the development of an energy grained master equation following the basic ideas put forth in Figure 1. Each molecule is assigned energy levels which are spaced by the energy of one laser photon. If n photons are required to reach the dissociation threshold, there are n + 1 levels and $n^2 + n$ possible rate constants between the various levels. These are reduced in number by assuming that up-pumping can only occur between adjacent levels and that collisional deactivation reduces the internal energy of a molecule to the lowest level. Crude estimates for infrared radiative emission rates may be calculated from the known infrared spectra of model compounds (such as the neutral parents of molecular ions) following the basic method outlined by Dunbar [53].

Analytical solutions to this master equation are difficult to find for $n > 2$, so a numerical stochastic approach to the solutions was carried out using the method developed by Gillespie [54] and applied to multiphoton dissociation kinetics by Barker [55]. The relative populations of the energy levels are determined by a series of trajectories, or random walks, through the various energy levels. Each trajectory is stopped when a dissociation event occurs, or when a set time limit is exceeded. The transition probabilities are determined by the rate constants and the results can be made arbitrarily precise merely by increasing the number of trajectories.

This method has been successfully used to account for behavior such as the exponential decay of ion population with irradiation period, the linear dependence of the dissociation rate on laser intensity, the variation of the induction period with laser intensity and the effects of collisional deactivation on the dissociation yield.

It must be noted that the available data do not allow a unique determination of each of the rate constants in the master equation. This type of modelling, however, is useful for gaining a qualitative understanding for the importance of various rate constants in determining the overall dissociation kinetics. In the case of $[(C_2H_5)_2O]_2H^+$, for instance, it is believed that radiative emission plays an important role in the deactivation of molecules in the upper energy levels [56]. The dissociation rates and induction periods are successfully accounted for using a model in which the rate constants for up-pumping between levels is independent of the energy content of the molecule.

As the infrared multiphoton dissociation kinetics of additional molecules are characterized, it is hoped that a generalized model will be developed which will be of value in predicting the behavior in unknown systems. Further work in this area is currently in progress.

3.2 Applications of Multiphoton Dissociation

3.2.1 Multiphoton Excitation as a Probe of Bimolecular and Unimolecular Reaction Energetics [8]

Proton bound alcohol dimers provide an opportunity to investigate the reaction potential energy surface for ion-molecule bimolecular reactions where the dimer is considered to be the intermediate. In the general case where two different product ions are possible from an ion-molecule reaction [21,22], the reaction will follow one of the three potential energy surfaces given in Figure 19.

$$A^+ + B \rightarrow [AB^+]^* \rightarrow C^+ + D \qquad \Delta H_1 < 0 \qquad (21)$$

$$A^+ + B \rightarrow [AB^+]^* \rightarrow F^+ + G \qquad \Delta H_2 < 0 \qquad (22)$$

For most ion-molecule systems the values of activation energies E_{a1} and E_{a2} are not known. When reaction of A^+ with B yields both C^+ and F^+ all that can be inferred is that no point along paths followed on the potential energy surface reactants to products exceeds the energy E

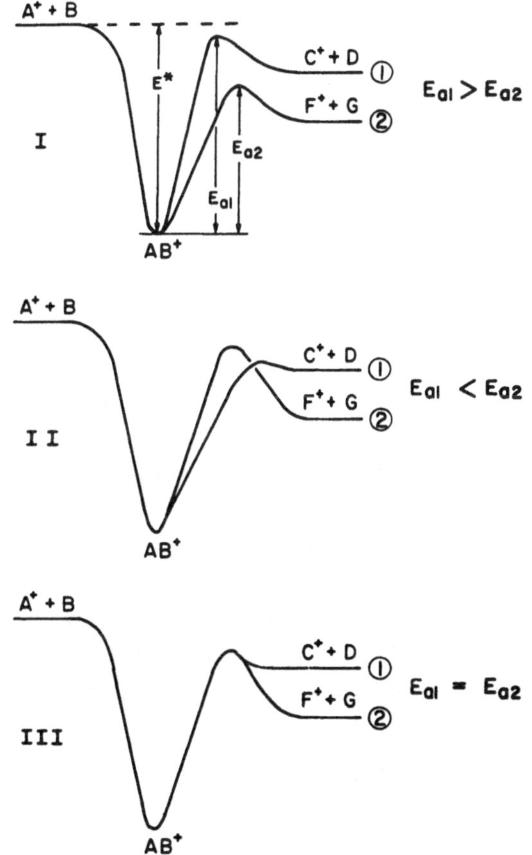

Fig. 19. Three possible potential energy surfaces for the exothermic ion-molecule reaction of $A^+ + B$ leading to products $C^+ + D$ and $F^+ + G$.

available to the reactive intermediate AB^+. No information is available to distinguish among cases I, II and II in Figure 19. However, gas phase ions irradiated by low power cw infrared radiation undergo slow, sequential multiphoton dissociation (see section 3.1.5). Thus the potential energy surface available to a decomposing ion is sampled thoroughly below (but not above) the lowest energy decomposition pathway. Consequently, in cases where AB^+ can be formed initially with little or no internal energy, as for the proton bound alcohol dimers, laser photodissociation of AB^+ identifies the lowest energy decomposition reaction.

Table 2 lists the alcohol proton bound dimers, enthalpy changes for reactions corresponding to processes 16-18, and the observed multiphoton dissociation pathways. Infrared activation of dimers *1*, *2*, *4*, *7* and *9* leads to formation of the lowest enthalpy products as would be expected for type I potential energy surfaces (Figure 1). $(t\text{-}C_5H_{11}OH)_1H^+$,*8* , is the one example which is described by a type II potential energy surface [18]. The remaining proton bound alcohol dimers listed in Table 1 (*3*,*5* and *6*) dissociate yielding two or more sets of products. Significantly, these are also the first examples of slow multiphoton dissociation following more than one reaction pathway.

Low power infrared laser activation does not open up any additional reaction channels available to $(ROH)_2H^+$. Products of multiphoton dissociation of $(ROH)_2H^+$, are the same as products generated by exothermic reaction of ROH_2^+ and ROH. The only difference is that laser activation is more selective. For example, multiphoton dissociation of $(C_2H_5OH)_2H^+$ yields exclusively $(C_2H_5)_2OH^+$ and H_2O (Table 1) which is also the only set of products formed at pressures 10^{-5} Torr by exothermic ion-molecule reaction of $C_2H_5OH_2^+$ and C_2H_5OH [16,18].

Proton bound alcohol dimers *3*, *5* and *6* each yield more than one set of products from slow multiphoton dissociation (Table 2). The measured branching ratios are invariant to changes in laser wavelength (942 - 1079 cm^{-1}) or laser power (27 to 63 W cm^{-2}). Dimers *3*, *5* and *6* contain n-propanol or 2-butanol. Of the alcohols used in this study, only these two can isomerize to more stable alcohols. We infer that more than one set of dissociation products is formed because an isomerization occurs in the transition state which releases sufficient energy to make accessible two or three reaction channels. This corresponds to case III in Figure 19. Results of a previous study indicate slow multiphoton dissociation is extremely sensitive to

Table 2: Unimolecular Decomposition Reactions of Proton Bound Alcohol Dimers[a]. Underlined Numbers Denote Observed IR Laser-Driven Reactions

	dimer ion $(R_1OH)H^+(R_2OH)$	ΔH reaction 16	ΔH reaction 17	ΔH reaction 18	ΔH'[b] reaction 18
1	$R_1=R_2\%CH_3$ [c]	31	$\underline{18}$	89	-
2	$R_1=R_2=C_2H_5$	32	$\underline{15}$	19	-
3	$R_1=R_2=n\text{-}C_3H_7$	$\underline{33}$(71%)	$\underline{17}$(17%)	$\underline{19}$(12%)	-
4	$R_1=R_2=i\text{-}C_3H_7$	31	$\underline{19}$	22	-
5	$R_1=R_2=s\text{-}C_4H_9$	32	$\underline{15}$(38%)	$\underline{20}$(62%)	-
6	$R_1=i\text{-}C_3H_7;R_2=s\text{-}C_4H_9$	33 [d]	$\underline{19}$(37%)	$\underline{18}$ [e](63%)	24 [f]
7	$R_1=R_2=t\text{-}C_4H_9$	31	$\underline{20}$	23	-
8	$R_1=R_2=t\text{-}C_5H_{11}$	31	19	$\underline{22}$	-
9	$R_1=i\text{-}C_3H_7;R_2=t\text{-}C_4H_9$	30 [d]	$\underline{13}$	22 [g]	23 [h]

a) Thermochemical data, in kcal/mole, taken from reference 15. Numbers in parentheses are product distributions for laser driven reactions where more than one product is observed.

b) ΔH' refers to enthalpy change for the alternate reaction 18 pathway available when R_1 and R_2 are not identical.

c) Reference 57.

d) Assumes more basic alcohol is protonated.

e) Products are $(i\text{-}C_3H_7OH)H^+(OH_2)$ + trans-2-butene

f) Products are $(s\text{-}C_4H_9OH)H^+(OH_2)$ + propene

g) Products are $(t\text{-}C_4H_9OH)H^+(OH_2)$ + propene

h) Products are $(i\text{-}C_3H_7OH)H^+(OH_2)$ + isobutylene

Table 3: Photoabsorption Cross Sections (in 10^{-19} cm^{-2}) for the Sequential Two-Photon Dissociation of the Cyanobenzene Cation

Wavelength (nm)	1	2
568.2	0.85	6.8
530.9	3.9	120

small differences in activation energies. More than one set of products can arise from different transition states only if the corresponding activation energies are almost identical [2]. It is highly unlikely that an accidental degeneracy exists in three different proton bound dimers.

A candidate for a transition state which allows for alkyl isomerization comprises an alkyl cation R^+, water, and alcohol ROH, in which R^+ isomerizes to R'^+ (*22* → *22'*).

$$\left[H_2O\cdot\cdot R^+\cdot\cdot O\begin{smallmatrix}H\\ R'\end{smallmatrix}\right] \longrightarrow \left[H_2O\cdot\cdot R'^+\cdot\cdot O\begin{smallmatrix}H\\ R\end{smallmatrix}\right] \qquad (22)$$

22 *22'*

Loss of H_2O results in formation of protonated ether $RR'OH^+$. The other products result from deprotonation of R'^+ to yield $(R'OH)H^+(OH_2)$ plus an olefin or recombination of R'^+ and H_2O to form $R'OH_2^+$ and ROH. In fact, all observed alcohol dimer decompositions, including those giving multiple sets of products, are consistent with the formation of a "solvated" alkyl cation as an electrostatically bound intermediate.

Proton bound dimers *3* and *4* have identical charge-to-mass ratios, and even identical infrared photodissociation spectra. However, the isomers yield different characteristic sets of products from multiphoton dissociation. Thus multiphoton dissociation reactions can be used to distinguish structural isomers of ions.

To date most detailed information about reaction coordinate profiles in ion-molecule reactions has been obtained from studies of metastable ion fragmentations [58,59,60]. In those experiments the decomposing species are ions initially formed with internal energy at, or slightly above, threshold for unimolecular decomposition. Infrared activation is an additional and complementary technique for the study of unimolecular decomposition of gas phase ions with internal energy slightly in excess of dissociation thresholds.

In light of current interest in mass spectrometric techniques [61] used to analyze complicated mixtures and to "fingerprint" ions by their secondary fragmentation patterns, infrared multiphoton excitation represents a novel method for ion decomposition. Although low power (< 100 W cm^{-2}) infrared radiation leads to decomposition exclusively by the lowest energy pathway [2,3,6] pulsed laser irradiation at high peak powers (≥ 1 MW cm^{-2}) is known to make multiple reaction channels energetically accessible [45,62,63]. The only requirement is that infrared absorption rates exceed unimolecular decomposition rates of the irradiated molecule.

3.2.2 Multiphoton Dissociation as a Probe of Molecular Relaxation Rates [7]

Multiphoton dissociation provides a sensitive probe of the internal energy content of CF_3I^+ since only those ions with sufficient energy before irradiation can be dissociated. With pulsed CO_2 laser radiation, photoproducts are produced on a submicrosecond time scale and can be quantitatively detected before they can react with neutral CF_3I. Increasing the time delay between ion formation and laser irradiation thus reduces the photodissociation yield due to relaxation of $CF_3I^+(\tilde{X}^2E_{1/2})$ to the ground state. The exponential decrease in photodissociation yield can be related to a relaxation rate as shown in Figure 20 for two different CF_3I pressures. Similar data obtained at various pressures indicate the relaxation of CF_3I^+ can be described by a pressure dependent, first-order rate constant, k_r, which is proportional to the CF_3I pressure. These data, shown in Figure 21, give $k_r = 3.1 \times 10^{-9}$ molecule^{-1} sec^{-1} with a zeropressure intercept of 11.3 ± 1.3 sec^{-1}. The value for k_r is three times ion-molecule collision rate constant indicating that collisions are highly effective in causing $CF_3I^+(\tilde{X}^2E_{1/2})$ relaxation. The zero-pressure intercept sets an upper limit to the spin-orbit relaxation rate which is comparable to the same rate in I (5.9 sec^{-1}) [64] and Xe^+ (18 sec^{-1}) [65].

3.2.3 Multiphoton Dissociation as a Probe of the Vibrational Quasi-Continuum [10].

As mentioned in section 3.1.1, cyanobenzene ions are dissociated by absorbing two visible photons. The decay of the cyanobenzene ion population with time has the functional form

$$I/I_0 = (r_+ - r_-)^{-1}\left[r_+\exp(r_-t) - r_-\exp(r_+t)\right],$$

$$r_\pm = -\frac{1}{2}\Big\{\Phi\sigma_1 + \Phi\sigma_2 + k_3 + \left[(\Phi\sigma_1 + \Phi\sigma_2 + k_3)^2 - 4\Phi^2\sigma_1\sigma_2\right]^{1/2}\Big\}, \qquad (23)$$

where Φ is the photon flux, σ_1 and σ_2 are the photoabsorption cross sections for the first and second photons and k_3 is the charge-transfer (deactivation) rate at the measured pressure of cyanobenzene neutrals. The photoabsorption

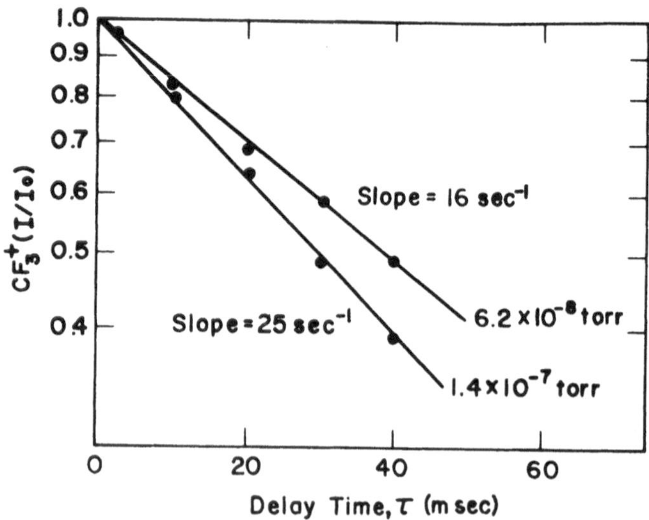

Fig. 20. Relaxation rate of CF_3I^+ at two different pressures as measured by CF_3^+ signal intensity. The delay time is the time between ion formation and irradiation. The slopes of the lines give the relaxation rates. Experimental conditions: 70 eV electron impact energy, CF_3I pressure as indicated, 1.6 J cm^{-2}, 40 nsec laser pulse at 952.9 cm^{-1}.

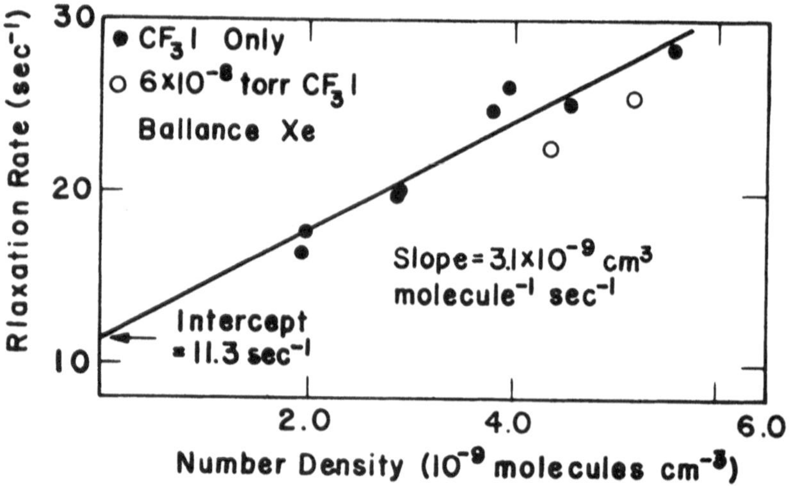

Fig. 21. Dependence of the CF_3I^+ relaxation rate on total pressure for CF_3I (●) and 6 x 10^{-8} Torr CF_3I with the balance Xe (○). The slope of the line gives a collisional relaxation rate constant of 3.1 x 10^{-9} cm^3 molecule^{-1} sec^{-1} and the intercept gives the zero-pressure relaxation rate constant of 11.3 sec^{-1}. Same experimental conditions as for Figure 20.

cross sections at several of the argon ion laser wavelengths were reported in a previous studiy [66]. Table 3 gives the analogous cross sections determined in the present experiments for two of the krypton ion laser wavelengths. The visible photodissociation is enhanced by CO_2 laser radiation showing peak enhancement at 970 cm^{-1} (see Figure 12). There are two possible mechanisms which explain the dissociation enhancement. These are illustrated in Figure 22.

Ground-state molecules may gain enough vibrational energy from the infrared laser to significantly increase the Franck-Condon overlaps between electronic states and thereby increase the photoabsorption cross sections of the visible laser. Alternatively, molecules having absorbed 2.2 eV in the form of a visible photon may then absorb nine infrared photons to reach the dissociation limit.

The observation that no dissociation takes place with irradiation by the infrared laser alone is evidence that only excited-state molecules may absorb an appreciable number of infrared photons. This restriction may be due to a combination of factors. The intramolecular vibrational energy transfer rate in ground-state (cold) molecules is not fast enough to relocate the absorbed energy out of the pump modes so that up-pumping cannot compete effectively with radiative and collisional relaxation. Secondly, the absorption profile in cold molecules is probably inhomogeneously broadened so that only a fraction of the molecules may absorb at a given infrared laser frewuency. Both these problems may be overcome in molecules which contain 2.2 eV of energy since the density of vibrational states is extremely high. Using the Whitten-Rabinovitch approximation [41] and the vibrational frequencies of neutral cyanobenzene [67], a density of states of 2×10^{11} per cm^{-1} is calculated. Extremely fast intra-molecular energy transfer rates at this level cause spectroscopic transitions to be dominated by homogeneous broadening as evidenced by the Lorentzian profiles of overtone spectra in benzene [68]. Thus mechanism (C) of Figure 22 most closely describes the observed results.

It is now possible to extend the kinetic model represented by eq 23 to include dissociation of the intermediate by the infrared laser. The term $\Phi\sigma_2$ is simply replaced by the sum of dissociation rate constants $\Phi\sigma_2 + \Phi_{IR}\sigma_{IR}$ where σ_{IR} is a phenomenologically defined dissociation cross section for the infrared laser radiation, which is dependent on wavelength. The new equation is used to fit the intensity data. The result is a cross section of $\sigma_{IR} = 6 \times 10^{-21}$ cm^2 at 969 cm^{-1}, which is comparable to cross sections for

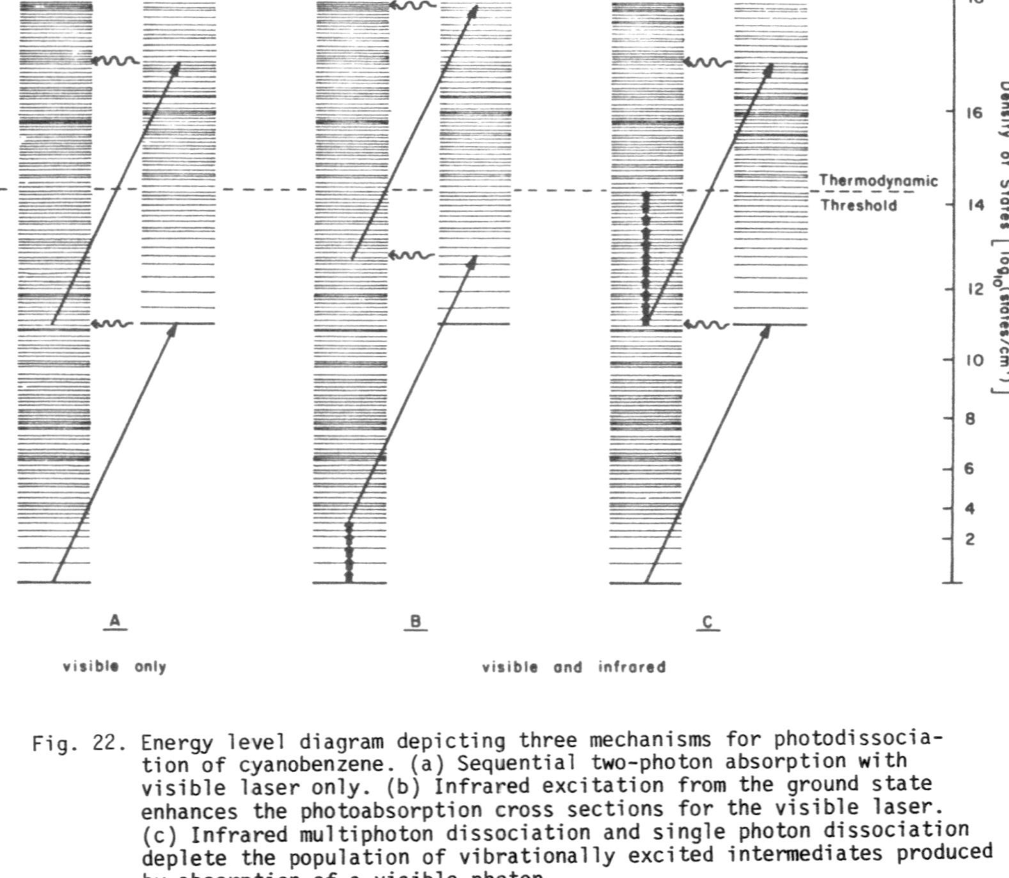

Fig. 22. Energy level diagram depicting three mechanisms for photodissociation of cyanobenzene. (a) Sequential two-photon absorption with visible laser only. (b) Infrared excitation from the ground state enhances the photoabsorption cross sections for the visible laser. (c) Infrared multiphoton dissociation and single photon dissociation deplete the population of vibrationally excited intermediates produced by absorption of a visible photon.

low-power infrared multiphoton dissociation of other molecules. Thus the infrared-enhanced visible dissociation of cyanobenzene provides a measure of the infrared absorption cross section in the quasi-continuum. Frequency dependence of the enhancement (Figure 12) gives a measure of the uniformity of the density of states in the quasi-continuum. Indeed the observed peak may be related to an overtone of ring skeletal mode at 488 cm^{-1} [68].

3.2.4 Isotopic Selectivity in Multiphoton Dissociation [9]

One key result obtained from low power multiphoton dissociation is that the dissociation spectra exhibit linewidths comparable to those observed in infrared spectra of corresponding neutrals. Thus low power multiphoton dissociation may be selective with respect to isotopic substitutions if the spectral shifts are sufficient to separate the dissociation bands of the isotopic components.

In a 1:1 mixture of CH_3Cl and CD_3Cl, all six isotopic variants of dimethyl chloronium ion are present, as shown in Figure 6. Only $(CD_3)_2{}^{35}Cl^+$ and $(CD_3)_2{}^{37}Cl^+$ photodissociate using the CO_2 laser. In this case, deuterium containing molecules may be selectively dissociated, however the isotope shift induced by the central chlorine atom is not enough to selectively dissociate either ^{35}Cl or ^{37}Cl containing ions. There are two plausible explanations for the observed selectivity. One possibility is that $(CH_3)_2Cl^+$ and $(CH_3)Cl^+(CD_3)$ are both transparent in the 925-1080 cm^{-1} region. Alternatively, the distribution and the density of vibrational states in $(CH_3)_2Cl^+$ might make it uniquely suited for multiphoton excitation. Unfortunately, there are no reported infrared absorption spectra of gas-phase dimethylchloronium to allow a test of the first hypothesis.

Assuming deuterium substitution alters photodissociation band positions but not lineshapes, a shift of $\geq$160 cm^{-1} to higher energy is necessary to move the peak shown in Figure 10 out of the CO_2 laser wavelength region. Dimethyl sulfide is isoelectronic with dimethylchloronium. Examination of reported vibrational frequencies of $(CH_3)_2S$, $(CH_3)S(CD_3)$ and $(CD_3)_2S$ shows five fundamental modes with isotope shifts in excess of 160 cm^{-1} [42]. Furthermore almost all $(CD_3)_2S$ vibrational frequencies of $\approx$1000 cm^{-1} exhibit large shifts to higher energy in $(CD_3)S(CH_3)$ or $(CH_3)_2S$. These changes suggest one possibility for the observed isotopic selectivity, namely that $(CD_3)_2Cl^+$ is the only species which absorbs strongly in the CO_2 laser wavelength region. However, it appears likely that all the ions have at least one infrared active vibrational mode within the tuning range of the laser. A second factor which may influence multiphoton dissociation probabilities are intramolecular energy transfer rates. The uniqueness of $(CD_3)_2Cl^+$ may reflect its relatively large density and, with more speculation, a favorable distribution of states which lead to large statistical factors facile energy transfer. Magnetic field induced level splittings may also enhance multiphoton excitation of $(CD_3)_2Cl^+$ [69].

In contrast to the special isotopic selectivity described above, a special case of kinetic isotope selectivity is observed in the decomposition of protonated diethyl ether. As mentioned above, photodissociation occurs by way of a four center intermediate eq 6 and yet photodissociation of $(C_2D_5)(C_2H_5)OH^+$ yields exclusively $C_2D_5OHD^+$ and not $C_2H_5OHD^+$. Since both products have the same precursor, isotopic selectivity must occur in the decomposition step as a result of differing activation energies for β-D or β-H transfer. This observation further demonstrates that decomposition competes favorably with further absorption of photons above the dissociation threshold as such a process would significantly dilute the kinetic isotope selectivity.

3.2.5 Isomeric Selectivity in Multiphoton Dissociation

In light of current interest in mass spectrometric techniques [27] used to analyze complicated mixtures and to "fingerprint" ions by their secondary fragmentation patterns, infrared multiphoton excitation represents a novel method for ion decomposition and identification.

Isomeric ions may be differentiated if they have different photodissociation spectra or if their photoproducts differ. For example, $C_3F_6^+$ molecular ions may be formed from either perfluoropropene or perfluorocyclopropane, affording the possibility of observing cyclic or acyclic ions of the same mass to charge ratio. Ions formed by both methods have identical photodissociation spectra, however, indicating that ring opening accompanies ionization of perfluorocyclopropane.

The proton bound alcohol dimers, $(n\text{-}C_3H_7OH)_2H^+$ and $(i\text{-}C_3H_7OH)_2H^+$ have identical charge to mass ratios and even identical photodissociation spectra. However, the isomers yield different characteristic sets of photoproducts, see Table 2 .

Perhaps the best example of isomeric selectivity to date is found in multiphoton electron detachment spectra of $C_7H_7^-$ isomers [74](see section 3.3). Proton abstraction from cycloheptatriene and toluene produces two isomeric $C_7H_7^-$ anions which have different multiphoton electron detachment spectra. A third $C_7H_7^-$ isomer may be formed by proton abstraction from norbornadiene by NH_2^-. This isomer does not undergo detachment upon CO_2 laser irradiation. Thus the three isomers may be conveniently distinguished on the basis of their spectra (or lack thereof), and easy identification is permitted where the structure of a $C_7H_7^-$ isomer may be uncertain.

3.3 Multiphoton Electron Detachment from Negative Ions [75]

Molecular anions present an interesting case for the study of infrared multiphoton excitation because for many, the lowest energy dissociation pathway is detachment of the extra electron forming the corresponding neutral molecule. Because vibrationally "hot" anions become relatively "cool" neutrals, multiphoton electron detachment (MED) may be thought of as a type of inverse electronic relaxation. It is thus conceptually distinct from multiphoton dissociation and single photon electron detachment since vibrational energy must be converted to electronic energy.

Infrared multiphoton electron detachment was first reported by Brauman and coworkers using a high power (10 MW cm^{-2}) pulsed CO_2 laser. Photodestruction of benzyl anion trapped in an ICR spectrometer was accompanied by an increase in the number of free electrons in the ion trap.

We have recently shown that electron detachment from $C_7H_7^-$ ions may also be induced by a low power (< 10 W cm^{-2}) cw CO_2 laser. Because the overall detachment kinetics are governed by laser-induced vibrational excitation, many parallels may be drawn with the kinetics of multiphoton dissociation of positive ions. Exponential decay of the ion population is preceded by a slight induction period which is evident at laser powers less than about 20 W cm^{-2}.

A multiphoton electron detachment spectrum is obtained by monitoring the extent of detachment as a function of infrared laser wavelength. Spectra of cycloheptatrienyl and benzyl anions are shown in Figure 23. The spectra provide a convenient means of identifying $C_7H_7^-$ ions whose structures may be uncertain.

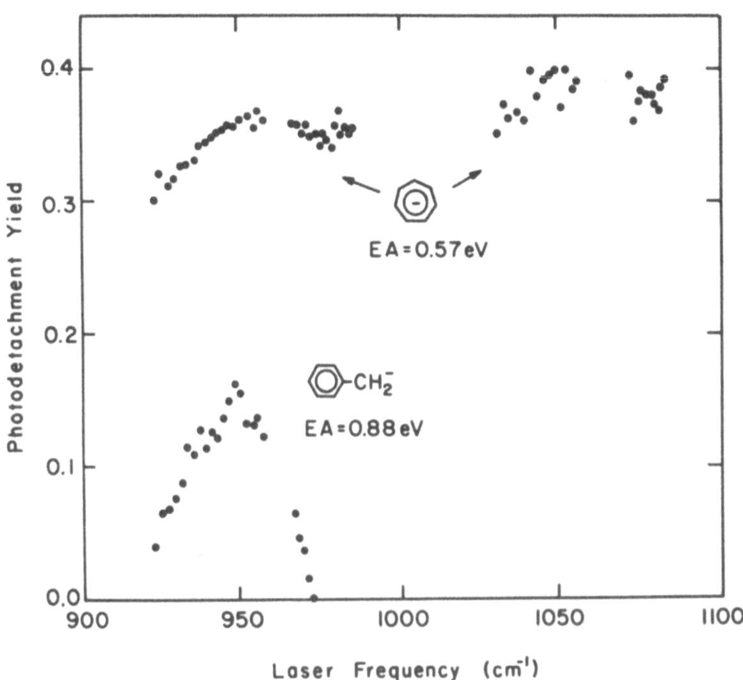

Fig. 23. Multiphoton electron detachment spectra of cycloheptatrienyl and benzyl anions showing yield vs. laser frequency for a 200 ms irradiation at 9 W cm^{-2}. The yield for benzyl anion above 1000 cm^{-1} is zero.

With the development of new lasers in other regions of the infrared spectrum, more complete spectra of both positive and negative ions may be obtained, yielding details of structure and bonding in molecular ions which have been previously inaccessible.

3.4 Selective Enhancement of Bimolecular Reaction Rates Using Low Intensity cw Laser Radiation

The equilibrium

$$(CH_3OH)H^+(OH_2) + CH_3OH \underset{k_r}{\overset{k_f}{\rightleftharpoons}} (CH_3OH)_2H^+ + H_2O \tag{24}$$

is characterized by forward rate constant $k_f = 5.0 \times 10^{-10}$ cm^3 mol^{-1} s^{-1}

and reverse rate constant $k_r = 8.2 \times 10^{-15}$ cm^3 mol^{-1} s^{-1}. k_f has been measured using ICR techniques, whereas k_r is calculated from k_f and the equilibrium constant K. The value $\Delta G = -6.5 \pm 1.0$ kcal/mol [70,71] for reaction 1 gives $K = 6.1 \times 10^4$ in favor of proton-bound methanol dimer, $(CH_3OH)_2H^+$, at room temperature.

During cw laser irradiation ($\tilde{\nu} = 947$ cm^{-1}) at 34 W/cm^2, the reverse reaction rate is enhanced by more than three orders of magnitude to $k_r^{IR} = 2.6 \times 10^{-11}$ cm^3 mol^{-1} s^{-1}. The forward reaction is unaffected by laser irradiation. Both the forward and backward reactions in equilibrium 24 proceed through a common intermediate, $(CH_3OH)_2H^+(OH_2)$ [73]. The competitive dissociation of this species can be evaluated using RRKM theory [41], where the internal energy is taken as absorbed infrared energy added to a 300 K Boltzmann distribution of vibrational energy. At an added energy of 10.5 kcal/mol the calculated ratio of H_2O to CH_3OH loss is equal to the observed value of $k_f/k_r^{IR} = 19.8$. This implies $(CH_3OH)_2H^+$ absorbs an average of 3.9 infrared photons ($\tilde{\nu} = 947$ cm^{-1}) prior to bimolecular reaction with H_2O.

Selective excitation of reactants not only represents an interesting tool for experimental chemical dynamics, it offers the possibility of using measured changes in reaction rates as a spectroscopic probe. The use of infrared excitation to alter bimolecular reaction rates should provide a general technique for obtaining spectra of ions and transient molecules.

4. Prognosis

Although additional experiments are in progress to further elucidate the photolysis of infrared multiphoton excitation, a reasonably clear picture of the mechanism of this process, involving both pulsed and cw lasers, has emerged over the past several years. Of the studies currently in progress, we are particularly excited about the use of multicolor excitation with continuously tunable sources to provide new insights into this phenomenon. All of the experiments described above have utilized only discrete line sources.

The development of infrared laser technology is continuing at a rapid pace, and it should be feasible to eventually extend the experiments described above over most of the infrared spectrum. This will facilitate the identification of molecular structures, making it possible to obtain infrared spectra with sensitivity which previously allowed only a mass spectrum to be obtained. By appropriately ionizing a mixture of compounds (e.g. by Li^+ attachment) it should be possible to obtain an infrared spectrum of each component without any need for preseparation. This is distinct from conventional infrared absorption spectroscopy, where the spectra of components are summed together. The slow accumulation of energy in increments of ~2 kcal/mole^{-1} allows a molecule to thoroughly sample the potential energy surface for the system at or below the activation energy and find the lowest energy process. This is distinct from collisional activation where energy is deposited in a comparatively short time period, normally with a distribution which allows for several competitive dissociation processes. It is not difficult to imagine many future applications which derive advantage from this aspect of infrared multiphoton activation. For example, a complex species can be sequentially fragmented, giving only a single product ion at each stage of dissociation. In this manner it might be possible to sequence a suitably ionized polypeptide. Combined with other newer developments in ion cyclotron resonance spectroscopy (utilizing Fourier transform detection and high field superconducting magnets) such experiments are entirely feasible.

5. Acknowledgement

The instrumental facilities were provided in part by a grant from the National Science Foundation. Support from the United States Department of Energy and the President's Fund of the California Institute of Technology is gratefully acknowledged.

6. References

1. T.A. Lehman, M.M. Bursey, in "Ion Cyclotron Resonance Spectroscopy", Wiley-Interscience: New York, 1976; J.L. Beauchamp, Ann. Rev. Phys. Chem. 22(1971)527.
2. D.S. Bomse, R.L. Woodin, J.L. Beauchamp, J. Am. Chem. Soc. 101(1979)5503.
3. R.L. Woodin, D.S. Bomse, J.L. Beauchamp, J. Am. Chem. Soc. 100(1978)3248.
4. D.S. Bomse, R.L. Woodin, J.L. Beauchamp, in "Advances in Laser Chemistry" A.H. Zewail, Ed., Springer: Berlin, Heidelberg, New York, 1978, pp 362-373.
5. R.L. Woodin, D.S. Bomse, J.L. Beauchamp, in "Chemical and Biochemical Applications of Lasers", C.B. Moore, Ed., Academic Press: New York, 1979, Vol. IV, p 355.
6. R.L. Woodin, D.S. Bomse, J.L. Beauchamp, Chem. Phys. Lett. 63(1979)630.
7. L.R. Thorne, J.L. Beauchamp, J. Chem. Phys. 74(1981)5100.
8. D.S. Bomse, J.L. Beauchamp, J. Am. Chem. Soc. 103(1981)3292.
9. D.S. Bomse, J.L. Beauchamp, Chem. Phys. Lett. 77(1981)25.
10. C.A. Wight, J.L. Beauchamp, Chem. Phys. Lett. 77(1981)30.
11. C.W. Tsang, A.G. Harrison, Org. Mass Spectrom. 3(1970)647.
12. D.W. Berman, D.S. Bomse, J.L. Beauchamp, unpublished photoionization results.
13. J.L. Beauchamp, D. Holtz, S.D. Woodgate, S.L. Pratt, J. Am. Chem. Soc. 94(1972)2798.
14. D.W. Berman, J.L. Beauchamp, L.R. Thorne, Int. J. Mass Spectrom. Ion Phys. 39(1981)47.
15. D.W. Berman, J.L. Beauchamp, J. Phys. Chem. 84(1980)2233.
16. K.R. Ryan, L.W. Sieck, J.H. Futrell, J. Chem. Pyhs. 41(1964)111; L.W. Sieck, F.P. Abramson, J.H. Futrell, J. Chem. Phys. 45(1966)2859.
17. E.P. Grimsurd, P.J. Kebarle, J. Am. Chem. Soc. 95(1973)7939.
18. D.R. Ridge, J.L. Beauchamp, J. Am. Chem. Soc. 93(1971)5925.
19. J.L. Beauchamp, R.C. Dunbar, J. Am. Chem. Soc. 92(1970)1477.

20. J.L. Beauchamp, M.C. Caserio, J. Am. Chem. Soc. 94(1972)2638.

21. J.L. Beauchamp, J. Am. Chem. Soc. 91(1969)5925; J.L. Beauchamp, M.C. Caserio, T.B. McMahon, J. Am. Chem. Soc. 96(1974)6243.

22. Much of the thermochamical data are obtained from appropriate proton affinity data, neutral heats of formation, and tabulated ion thermochemistry contained in J.R. Wolf, R.H. Staley, I. Koppel, M. Taagepara, R. T. McIver, Jr., J.L. Beauchamp, R.W. Taft, J. Am. Chem. Soc. 99(1977) 5417. H.M. Rosenstock, K. Draxl, B.W. Steiner, J.T. Herron, J. Phys. Chem. Ref. Data Suppl. 1(1977)6. J.D. Cox, G. Pilcher in "Thermochemistry of Organic and Organometallic Compounds", Academic Press, New York, 1970. For some alcohols proton affinities are not experimentally attainable. Methods for estimating these numbers as well as experimentally determined ΔH_f values for proton bound alcohol dimers are to be reported in another publication (D.S. Bomse, J.L. Beauchamp, J. Phys. Chem. 85(1981)488).

23. T.E. Orlowski, B.S. Freiser, J.L. Beauchamp, Chem. Phys. 16(1976)439.

24. B.J.M. Neijzen, C.A. deLange, J. Elektron. Spectrom. 14(1978)187.

25. J.W. Rabalais, R.J. Colton, J. Electron. Spectrom. 1(1972/1973)83.

26. R.W. Ambartzumian, V.S. Tetokhov in "Chemical and Biochemical Applications of Lasers". C.B. Moore, Ed., Academic Press: New York, 1977, Vol. III.

27. A. Hartford Jr., Chem. Phys. Lett. 53(1978)503.

28. J.R. Nielsen, H.H. Claassen, D.C. Smith, J. Chem. Phys. 20(1952)1916.

29. H. Wieser, W.G. Laidlaw, P.J. Krueger, H. Fuhrer, Spectrochim. Acta, 24A(1968)1055.

30. J.L. Beauchamp, D. Holtz, S.D. Woodgate, S.L. Pratt, J. Am. Chem. Soc. 94(1972)2798.

31. M.J. Coggiola, P.C. Crosby, J.R. Peterson, J. Chem. Phys. 72(1980)6507.

32. L. Bass, T. Su, W.J. Chesnavich, M.T. Bowers, Chem. Phys. Lett. 34(1975) 119.

33. P. Kolodner, C.W. Winterfeld, E. Yablonovitch, Opt. Commun. 20(1977)119.

34. D.M. Golden, M.J. Rossi, A.C. Baldwin, J.R. Barker, Acc. Chem. Res. 14 (1981)56.

35. P.A. Schulz, Aa.S. Sudbø, D.J. Krajnovich, H.S. Kurok, Y.R. Shen, Y.T. Lee, Ann. Rev. Phys. Chem. 30(1979)379.

36. M.J. Quack, Chem. Phys. 69(1978)1294.

37. C.D. Cantrell, S.M. Freund, J.L. Lyman in "Laser Handbook", North Holland Publishing: Amsterdam, Vol. III, to appear.

38. S. Mukamel, J.J. Jortner, J. Chem. Phys. 65(1976)5204.

39. N.R. Isenor, V. Merchant, R.S. Hallsworth, M.S. Richardson, Ca. J. Phys. 51(1973)1281.

40. V.M. Akulin, S.S. Alimprev, N.V. Kalov, L.A. Shelepin, Zh. Eksp. Theor. Fiz. 69(1975)836 (Sov. Phys. JETP 42(1975)427).

41. P.J. Robinson, K.A. Hollbrook in "Unimolecular Reactions", Wiley: New York, 1972.

42. J.W. Ypenburg, J. Gerding, Rec. Trav. Chim. 90(1971)885; J.W. Ypenburg, Rec. Trav. Chim. 91(1972)671.

43. W.F. Edgell, C.E. May, J. Chem. Phys. 22(1954(1808).

44. J.G. Black, E. Yablonovitch, N. Bloembergen, S. Mukamel, Phys. Rev. Lett. 38(1977)1131.

45. E.R. Grant, P.A. Schulz, Aa.S. Sudbø, Y.R. Shen, Y.T. Lee, Phys. Rev. Lett. 40(1978)115.

46. R.C. Dunbar, Spectrochim. Acta 31A(1975)797.

47. L. Allen, J.H. Eberly in "Optical Resonance and Two Level Atoms", Wiley-Interscience: New York, 1975.

48. D.S. Frankel Jr., T. Manuccia, J. Chem. Phys. Lett. 54(1978)451.

49. J.T. Knudtson, G.W. Flynn, J. Chem. Phys. 58(1973)1467.

50. G.P. Quigley in "Advances in Laser Chemistry", A.H. Zewail, Ed., Springer Series in Chemical Physics, Springer: Berlin, Heidelberg, New York, 1978, and references contained therein.

51. J.C. Polanyi, K.B. Woodall, J. Chem. Phys. 56(1972)1563.

52. A.M.G. Ding, J.C. Polanyi, Chem. Phys. 10(1975)39.

53. R.C. Dunbar, Spectrochim. Acta 31A(1975)797.

54. D.T. Gillespie, J. Comput. Phys. 22(1976)403; J. Phys. Chem. 81(1977) 2340.

55. J.R. Barker, J. Chem. Phys. 72(1980)3686.

56. R.L. Woodin, unpublished results.

57. Laser irradiation leads to an accelerated bimolecular reaction in addition to unimolecular decomposition reaction. See ref. 8.

58. D.H. Williams, Acc. Chem. Res. 10(1977)280.

59. R.D. Bowen, D.H. Williams, J. Am. Chem. Soc. 102(1980)2752.

60. D.H. Williams, I. Howe in "Principles of Organic Mass Spectrometry", McGraw Hill: New York, 1972. K. Levsen, J. Schwarz, Angew. Chem. 88 (1976)589, Angew. Chem., Int. Ed. Engl. 15(1976)509. R.G. Cooks, J.H. Benyon, R.M. Caprioli, G.R. Lester in "Metastable Ions", Elsevier: Amsterdam, 1973.

61. F.W. McLafferty, Acc. Chem. Res. 13(1980)33; G.A. McClusky, R.W. Kondrat, R.G. Cooks, J. Am. Chem. Soc. 100(1978)6045.

62. A.J. Collussi, S.W. Benson, R.S. Huang, Chem. Phys. Lett. 52(1977)349; Aa.S. Sudbø, P.A. Schulz, E.R. Grant, Y.R. Shen, Y.T. Lee, J. Chem. Phys. 68(1978)1306. Aa.S. Sudbø, P.A. Schulz, E.R. Grant, Y.R. Shen, Y.T. Lee, J. Chem. Phys. 70(1979)912 and references contained therein.

63. Reported results of high power pulsed infrared laser photolysis of trapped ions all indicate only one set of products formed. However, only a few systems have been studied to date: R.N. Rosenfeld, J.M. Jasinski, J.I. Brauman, J. Am. Chem. Soc. 101(1979)3999; R.N. Rosenfeld, J.M. Jasinski, J.I. Brauman, J. Chem. Phys. 71(1979)1030.

64. F.J. Comes, S. Pionteck, Chem. Phys. Lett. 42(1976)558.

65. P.B. Armentrout, D.W. Berman, J.L. Beauchamp, Chem. Phys. Lett. 53(1978) 255.

66. T.E. Orlowski, B.S. Freiser, J.L. Beauchamp, Chem. Phys. 16(1976)439.

67. G. Varsanyi in "Assignments for Vibrational Spectra of Seven Hundred Benzene Derivatives", Wiley: New York, 1974, p 73.

68. R.G. Bray, M.J. Berry, J. Chem. Phys. 71(1979)4909.

69. R. Duperrex, J. van den Bergh, J. Chem. Phys. 73(1980)585.

70. Reaction thermochemistry was compiled from: J.F. Wolf, R.H. Staley, I. Koppel, M. Taagepera, R.T. McIver, J.L. Beauchamp, R.W. Taft, J. Am. Chem. Soc. 99(1977)5417; H.M. Rosenstock, K. Draxl, B.W. Steiner, J.T. Herron, J. Phys. Chem. Ref. Data 1(1977)6. J.D. Cox, G. Pilcher in "Thermochemistry of Organic and Organometallic Compounds", Academic Press: New York, 1970; W.R. Davidson, J. Sunner, P. Kebarle, J. Am. Chem. Soc. 101(1979)1675.

71. Cluster formation $(CH_3OH)_nH^+$ (n = 1-8), is observed using high pressure mass spectrometry with total neutral (methanol plus unreactive buffer) pressures of ≤ 5 Torr: E.P. Grimsrud, P. Kebarle, J. Am. Chem. Soc. 95 (1973)7939.

72. Excitation of neutral CH_3OH does not contribute to the observed laser-induced chemistry because of the brief residence time of neutrals in the ion storage and irradiation region. This is discussed in detail elsewhere. Since both CH_3OH and $(CH_3OH)_2H^+$ absorb in the 10-μm region, it might be expected that $(CH_3OH)H^+(OH_2)$ also absorbs. However, no evidence is obtained for laser excitation enhancing the reverse of reaction 3 or for multiphoton dissociation of $(CH_3OH)H^+(OH_2)$ to $CH_3OH_2^+$ and H_2O (25 kcal/mol). Both results suggest that $(CH_3OH)H^+(OH_2)$, in comparison with $(CH_3OH)_2H^+$, is not heated significantly by the infrared laser. Based on the proposed kinetic scheme, heating might result in a slight decrease in k_1 for reaction 1, which would be difficult to detect.

73. At the laser powers used, the typical rate for absorbing a single photon (10^3 s^{-1} for a transition with absorption cross section of 10^{-17} cm^2) is very much slower than the decomposition rate ($>10^7 s^{-1}$) of the intermediate. Thus the enhanced reaction rate is due to excitation of reactants, not to the intermediate.

74. C.A. Wight, J.L. Beauchamp, J. Am. Chem. Soc., submitted for publication.

75. C.A. Wight, J.L. Beauchamp, manuscript in preparation.

STUDY OF ATOMIC METAL IONS
GENERATED BY LASER IONIZATION

R.C. Burnier, G.D. Byrd, T.J. Carlin
M.B. Wiese, R.B. Codym and B.S. Freiser
Department of Chemistry, Purdue University
West Lafayette, Indiana 47907

1. Metal Ion Chemistry

The development of a combined pulsed laser source-ion cyclotron resonance spectrometer in our laboratory has proven to be a convenient and powerful method of generating metal ions and for studying their subsequent chemistry in the gas phase. Based on this technique we have embarked on a wide range of studies on metal ions [1-7]. In particular our main emphasis this past year has been on the applications of metal ions as selective chemical ionization reagents and our progress in this area will dominate this discussion. In addition we report the first results combining laser ionization with a Nicolet FTMS prototype spectrometer recently put into operation and discuss some of the advantages of this approach.

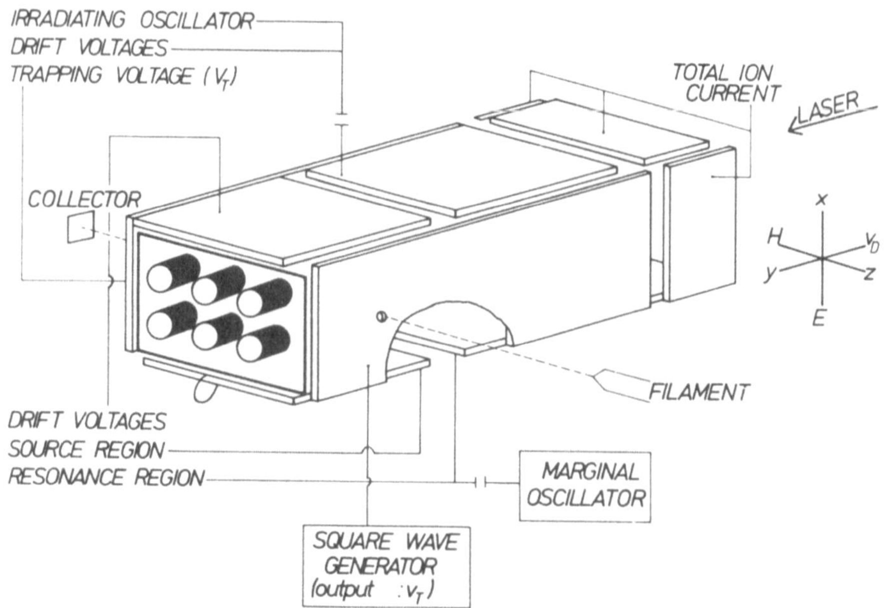

Fig. 1. Modified trapped ion cell for studying metal ion chemistry.

To facilitate these studies we have modified our ICR cell by placing several different metal rods in the end plate as shown in Figure 1. By moving the laser beam, a particular metal ion may be readily selected without having to break vacuum. The rods have been found to give fairly stable signals and are, of course, more resistant to laser damage than metal plates. Using these techniques we have begun systematic studies on the chemistry of simple atomic transition metal ions with various classes of organic compounds. The goal of this research is to identify trends in reactivity i.e. reaction mechanisms useful in interpreting the chemical ionization spectra of unknown compounds and to test for the functional group selectivity of the various metal ions. We have demonstrated the feasibility of these goals, thus far, in extensive studies on Cu^+ with ester and ketones [3], on Fe^+ with ethers [5], ketones [5] and hydrocarbons [6], and on Ti^+ with hydrocarbons [7]. In addition, preliminary results on sulfur containing compounds [4] and on a variety of other metallic ions have been obtained and will be discussed here.

Our first detailed study was on Cu^+ which was observed to display definite patterns of reactivity for different classes of oxygenated compounds [3]. Dissociative attachment reactions of Cu^+ with ester, for example, occur in which the ester is cleaved either to alcohols and ketones generalized by equation 1 or to carboxylic acids and alkenes

$$Cu^+ + RCO_2R' \longrightarrow \begin{cases} Cu(R'OH)^+ + \text{ketene} \\ Cu(\text{ketene}) + R'OH \end{cases} \tag{1}$$

$$Cu^+ + RCO_2R' \longrightarrow \begin{cases} Cu(RCO_2H)^+ + \text{alkene} \\ Cu(\text{alkene})^+ + RCO_2H \end{cases} \tag{2}$$

generalized by equation 2. The reaction pathways was shown to be strongly influenced by the thermodynamics of the dissociation channels of the free ester. The products of these reactions react further with the substrate ester by ligand displacement. The reactions of Cu^+ with methyl acetate and ethyl formate are typical of the generalized reactions 1 and 2, respectively. The reaction sequence for methyl acetate is given in equations 3-6 and for ethyl formate in equations 7-10. These sequences

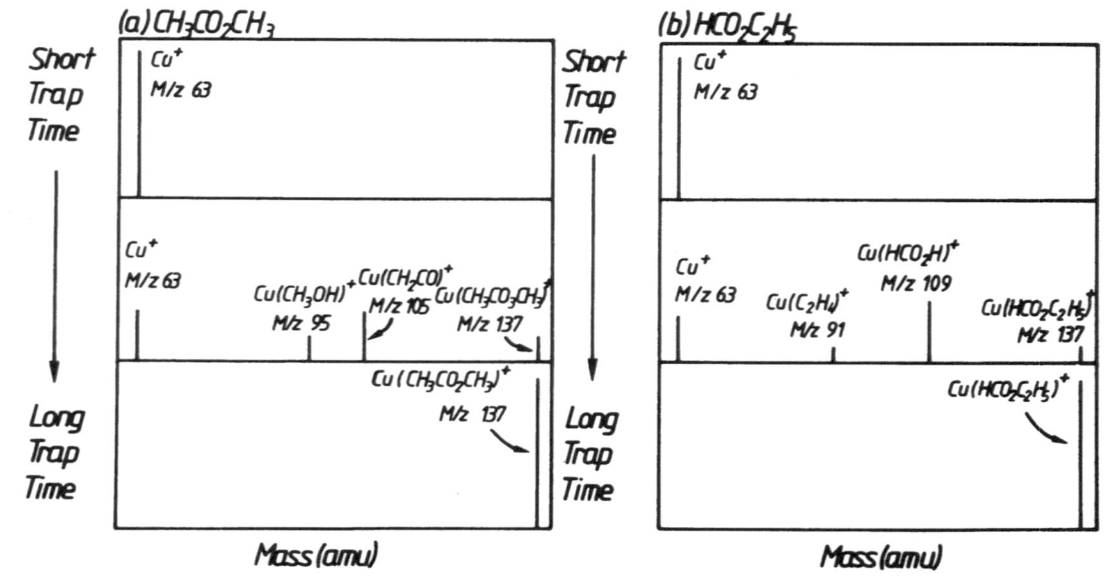

Fig. 2. Ion intensity at three arbitrary trapping times for Cu^+ in the presence of (a) methyl acetate and (b) ethyl formate.

$$Cu^+ + CH_3CO_2CH_3 \longrightarrow Cu(CH_3OH)^+ + CH_2CO \quad (3)$$

$$Cu^+ + CH_3CO_2CH_3 \longrightarrow Cu(CH_2CO)^+ + CH_3OH \quad (4)$$

$$Cu(CH_3OH)^+ + CH_3CO_2CH_3 \longrightarrow Cu(CH_3CO_2CH_3)^+ + CH_3OH \quad (5)$$

$$Cu(CH_2CO)^+ + CH_3CO_2CH_3 \longrightarrow Cu(CH_3CO_2CH_3)^+ + CH_2CO \quad (6)$$

$$Cu^+ + HCO_2C_2H_5 \longrightarrow Cu(C_2H_4)^+ + HCO_2H \quad (7)$$

$$Cu^+ + HCO_2C_2H_5 \longrightarrow Cu(HCO_2H)^+ + C_2H_4 \quad (8)$$

$$Cu(C_2H_4)^+ + HCO_2C_2H_5 \longrightarrow Cu(HCO_2C_2H_5)^+ + C_2H_4 \quad (9)$$

$$Cu(HCO_2H)^+ + HCO_2C_2H_5 \longrightarrow Cu(HCO_2C_2H_5)^+ + HCO_2H \quad (10)$$

of reactions are illustrated in Figures 2a and 2b which show schematically ion intensity at three arbitrary trapping times for Cu^+ in the presence of methyl acetate and ethyl formate, respectively. The variation in the spectra demonstrates clearly the "tunability" afforded by varying the trapping times in the ICR, as well as the type of information available by Cu^+ chemical ionization. The spectra are simple, with the pseudo molecular ions for the isomers (both at m/z 137) dominating at long times. These isomers are readily distinguished and identified by their characteristic dissociative attachment products at intermediate times.

The widely varying chemistry of different metal ions becomes more apparent with each new experiment. Reactions of Cu^+ and Fe^+ with ketones examplifies this point. Ketones, like ester, are observed to be cleaved by Cu^+ followed by a competitive loss of a neutral fragment [3]. With few exceptions ketones may be cleaved to form H_2O and a diene generalized in equation 11. H_2O is not effective in competing with the diene, however, and $Cu(H_2O)^+$ is not observed. Alternatively, ketones may be cleaved to

$$Cu^+ + RC(=O)R' \longrightarrow \begin{cases} \not\longrightarrow Cu(H_2O)^+ + \text{diene} \\ \longrightarrow Cu(\text{diene})^+ + H_2O \end{cases} \quad (11)$$

an alkene and an enol species generalized in equation 12. Although the enol probably rearranges to a ketone, a few transition metal complexes

$$Cu^+ + RC(=O)R' \longrightarrow \begin{cases} Cu(enol)^+ + alkene \\ Cu(alkene)^+ + enol \end{cases} \qquad (12)$$

of Fe and Pt containing the vinyl alcohol moiety π-bonded to the metal have been reported [8]. Further studies will include probing these types of ion structures. The specific reactions of Cu^+ with 2-butanone was observed as written in equation 13 and illustrates the generalized reaction

$$Cu^+ + CH_3C(=O)CH_2CH_3 \longrightarrow Cu(C_4H_6)^+ + H_2O \qquad (13)$$

type 11. In contrast to this, the reaction of 2-butanone with Fe^+ is found to be considerably more complex as indicated in the reaction sequence 14-18. To further visualize the differences of reactivity of Cu^+ and Fe^+

$$Fe^+ + CD_3C(=O)CD_2CH_3 \longrightarrow \begin{cases} Fe(CO)^+ + C_3D_5H_3 & (14) \\ Fe(CD_2CH_2)^+ + CD_3CHO & (15) \\ Fe(CD_2CHO)^+ + CD_2CH_2 & (16) \\ Fe(C_3D_2H_2O)^+ + CD_3H & (17) \\ Fe(C_4D_4H_2O)^+ + HD & (18) \end{cases}$$

with 2-butanone, Schemes I and II show the respective reaction mechanisms. These schemes both follow the rules of a more general mechanism for each metal ion, obtained by studying their chemistry with a series of ketones [3,5]. The salient features are that Cu^+ is not observed to insert into the carbon framework, but rather only activates the molecule, while Fe^+ undergoes oxidative addition followed by reductive elimination. This chemistry is in accord with the d^{10} configuration of Cu^+ being resistant to oxidation, while Fe^+ may readily be oxidized to what is formally Fe^{3+}.

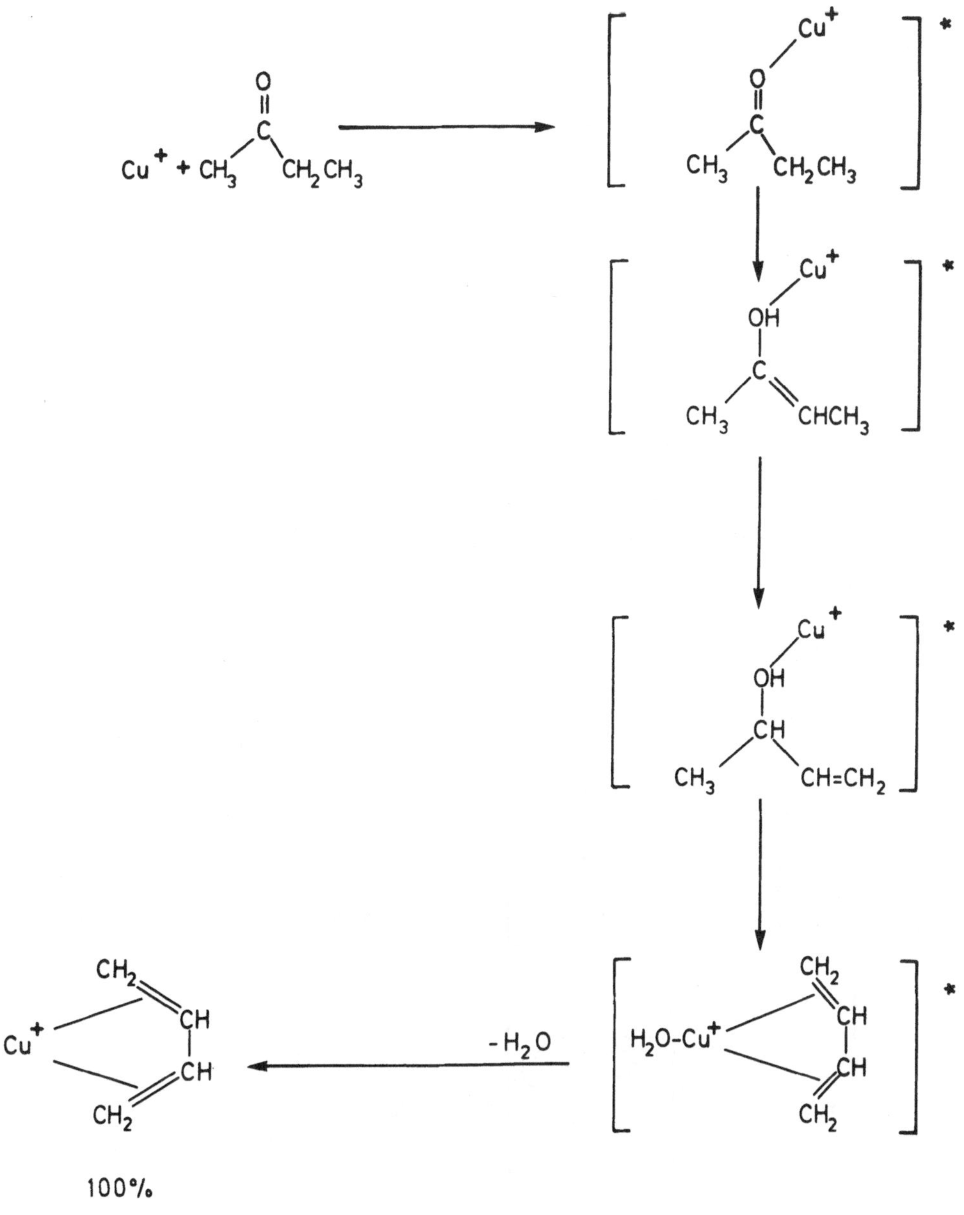

$$Cu^+ + CH_3COCH_2CH_3 \longrightarrow [CH_3C(=O\text{-}Cu^+)CH_2CH_3]^* \longrightarrow [CH_3C(OH\text{-}Cu^+)=CHCH_3]^* \longrightarrow [CH_3CH(OH\text{-}Cu^+)CH=CH_2]^* \longrightarrow [H_2O\text{-}Cu^+(CH_2=CH\text{-}CH=CH_2)]^* \xrightarrow{-H_2O} Cu^+(CH_2=CH\text{-}CH=CH_2)$$

100%

SCHEME I

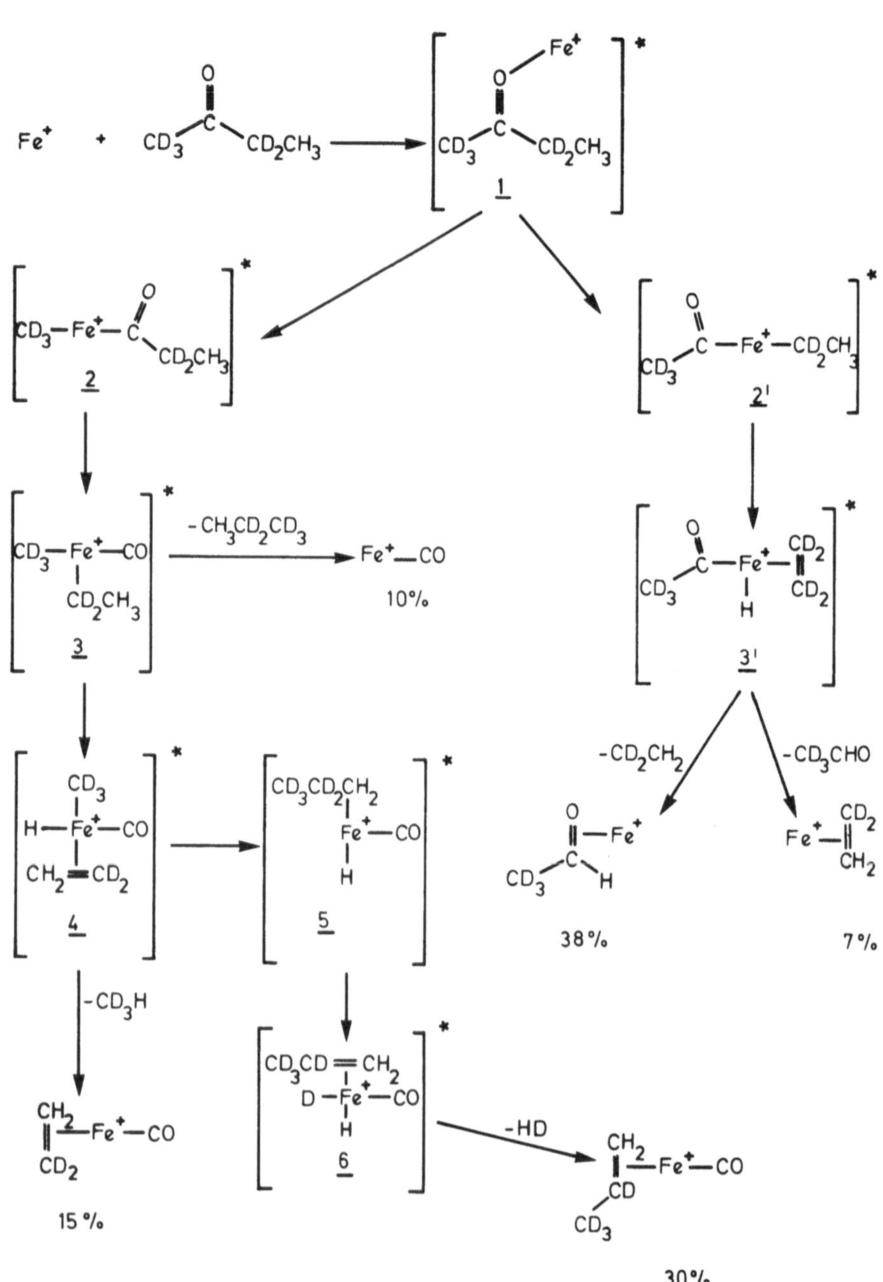
Fe⁺ + CD3 C O CD2CH3
1
2
3
4
5
6
2'
3'
- CH3CD2CD3
Fe⁺—CO
10%
-CD3H
15%
-HD
30%
-CD2CH2
-CD3CHO
38%
7%

SCHEME II

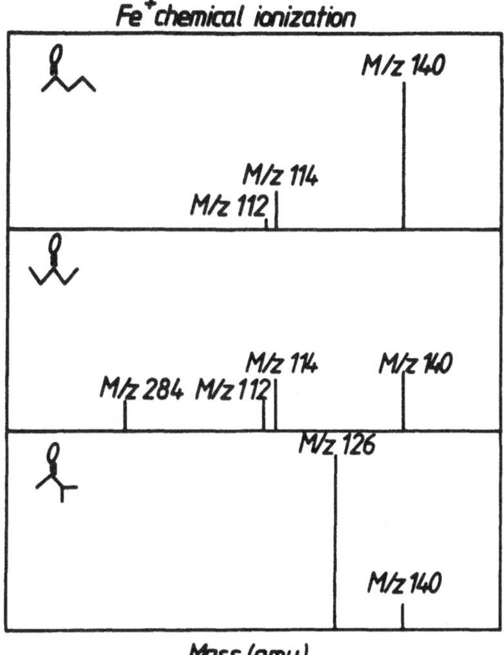

Fig. 3. Fe^+ chemical ionization spectra of 2-pentanone, 3-pentanone, and 3-methyl-2-butanone. Not shown is the ion at M/z 142 corresponding to the Fe^+ bound to the parent compound.

Table 1. Periodic Reactivity Plot for 2-Butanone for the process:

$$M^+ + \text{(2-butanone)} \longrightarrow [M(\text{2-butanone})^+]^* \xrightarrow[\text{Loss}]{\text{Mass}} \text{Product ions}$$

M^+ =	Al^+	Ti^+	TiO^+	V^+	VO^+	Co^+	Fe^+	Cu^+
Mass Loss								
2					x		x	
16						x	x	
18					x			x
28			x		x	x	x	
44					x	x	x	
54			x					
56		x		x				

The complexity of Fe^+ reactions is again quite applicable to the problem of distinguishing isomeric molecules as shown in Figure 3 for 2-pentanone, 3-pentanone, and 3-methyl-2-butanone. These three isomers may clearly be differentiated by monitoring m/z 84, which is characteristic of 3-pentanone, and m/z 126, which is characteristic of 3-methyl-2-butanone. If a particular metal ion is not sufficient for sample identification, however, one can readily take advantage of a series of different metal ion chemical ionization spectra. Table 1 illustrates this point by showing what we have termed a periodic reactivity plot for 2-butanone. The periodic reactivity plot is a "fingerprint" or "computer punch card" for a particular species which, if matched, identifies its presence in the spectrometer. Again, the widely differing reactivities shown in Table 1 are striking. Al^+ does not undergo any dissociative reactions with 2-butanone. The strength of the TiO bond is evident in that TiO^+ is the only reaction product observed when Ti^+ reacts with the butanone. Interestingly, the reactivity of TiO^+ is observed to be completely different than that of Ti^+ and suggests that these reagent ions may be further tailored for specific needs. The V^+ reacts similarly to Ti^+, forming VO^+ exclusively and again VO^+ is observed to react differently than V^+ and so forth.

We have seen in the previous chemistry that functional groups containing heteroatoms such as oxygen create a source of polarity in an organic molecule and serve to direct some of the reactions that occur with metal ions. The hydrocarbon portions, however, may also play some auxiliary role in the overall mechanism. This became particularly evident to us in our studies on Fe^+ with ketones and we, therefore, have also actively started studying metal ion reactions with hydrocarbons [6,7].

We have made considerable progress on the reactions of Fe^+ and Ti^+, and the story which emerges provides an interesting comparison. Table 2 summarizes our results for the reactions of Fe^+ with various acyclic hydrocarbons. From the observed product distribution ratios, the alkanes appear to be randomly cleaved along the chain [6]. The general reaction for Fe^+ with acyclic alkanes is given by reaction 19,

$$Fe^+ + C_nH_{2n+2} \longrightarrow \begin{cases} Fe[C_{n-m}H_{2(n-m)}]^+ + C_mH_{2m+2} \\ Fe[C_nH_{2n}]^+ + H_2 \end{cases} \qquad (19)$$

Table 2. Distribution of product intensities from Fe^{+}/acyclic alkane reactions given in percentages of total products[a]

Alkane	Product Mass =	84	98	112	126	140	154
	Obs	70	30				
	Pre	80	20				
	Obs	59 (75)	29 (10)	12 (6)[b]			
	Pre	52	24	24			
	Obs	57	8	24	11		
	Pre	44	11	19	25		
	Obs	35	12	12	17	24	
	Pre	39	9	9	15	27	
	Obs	27	19	8	14	6	26
	Pre	32	8	8	10	13	29
	Obs	0 (0)	62 (84)	38 (16)			
	Pre	0	72	28			
	Obs	-	29	9	22	40	
	Pre	-	23	27	27	23	
	Obs	5	13	66	12	4	
	Pre	12	-	28	55	5	

Obs = the experimentally observed intensity

Pre = the predicted value based on the random insertion mechanism

[a]Values in parentheses are from R.B. Freas and D.P. Ridge, J. Am. Chem. Soc., 102 (1980) 7129

[b]Reference in "a" also reports 9% of m/z 110

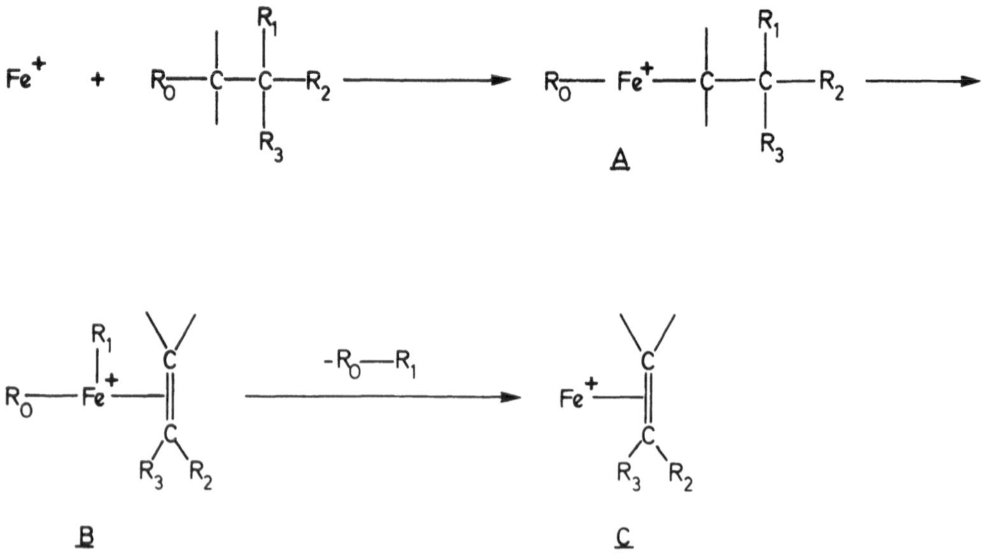

$$Fe^+ + R_0\text{—}C\text{—}C(R_1)(R_3)\text{—}R_2 \longrightarrow R_0\text{—}Fe^+\text{—}C\text{—}C(R_1)(R_3)\text{—}R_2 \;\; (\underline{A}) \longrightarrow$$

$$R_0\text{—}Fe^+(R_1)\text{—}[C{=}C(R_3)(R_2)] \;\; (\underline{B}) \xrightarrow{-R_0\text{—}R_1} Fe^+\text{—}[C{=}C(R_3)(R_2)] \;\; (\underline{C})$$

SCHEME III

where the iron cleaves the alkane into a smaller alkane or H_2. The mechanism may be viewed as shown in Scheme III where the initial insertion of Fe^+ into the alkane to form intermediate A is followed by a β-hydrogen or β-alkyl shift onto the iron to form intermediate B. The two R groups attached to the iron then combine to produce a new, smaller alkane or H_2 which departs leaving a charged iron-alkene complex. No reactions were observed with either methane or ethane and, while methane has no β-hydrogens, the lack of reaction with ethane is peculiar and suggests that H_2 loss from a terminal C-H insertion may not be favorable for Fe^+. Taking this fact into account and assuming random insertion, the overall mechanism can be used to predict the product ratios. As shown in Table 2 these predicted values match the observed ratios quite well with the exception of 2,2-dimethylbutane which forms $FeC_3H_6^+$ (13% of total reaction) not predicted to occur at all. We believe that this arises due to scrambling of the carbon skeleton while attached to the iron in analogy to the scrambling observed in conventional electron impact mass spectra of alkanes [9]. Finally in answer to our earlier question, these results indicate that although the presence of polar functional groups can greatly influence the site of insertion,

Table 3. Titanium (+)/alkane reactions

Reaction		% Product
$Ti^+ + CH_4$	→ N.R.	
$Ti^+ + C_2H_6$	→ $TiC_2H_4^+ + H_2$	100
$Ti^+ + C_3H_8$	→ $TiC_3H_6^+ + H_2$	100
$Ti^+ + n\text{-}C_4H_{10}$	→ $TiC_4H_6^+ + 2H_2$	100
$Ti^+ + i\text{-}C_4H_{10}$	→ $TiC_4H_8^+ + H_2$	84
	→ $TiC_4H_6^+ + 2H_2$	16
$Ti^+ + n\text{-}C_5H_{12}$	→ $TiC_5H_8^+ + 2H_2$	83
	→ $TiC_4H_6^+ + CH_4 + H_2$	12
	→ $TiC_3H_6^+ + C_2H_6$	4
$Ti^+ + n\text{-}C_6H_{14}$	→ $TiC_6H_{10} + 2H_2$	33
	→ $TiC_6H_8^+ + 3H_2$	28
	→ $TiC_5H_8^+ + H_2 + CH_4$	5
	→ $TiC_4H_8^+ + C_2H_6$	10
	→ $TiC_4H_6^+ + C_2H_6 + H_2$	24
$Ti^+ + (CH_3)_2CHCH(CH_3)_2$	→ $TiC_6H_{10}^+ + 2H_2$	38
	→ $TiC_5H_8^+ + CH_4 + H_2$	44
	→ $TiC_4H_8^+ + C_2H_6$	6
	→ $TiC_4H_6^+ + C_2H_6 + H_2$	12
$Ti^+ + (CH_3)_3CCH_2CH_3$	→ $TiC_6H_{12}^+ + H_2$	7
	→ $TiC_6H_{10}^+ + 2H_2$	8
	→ $TiC_5H_8^+ + CH_4 + H_2$	54
	→ $TiC_4H_8^+ + C_2H_6$	8
	→ $TiC_4H_6^+ + C_2H_6 + H_2$	22

the alkyl portion may also compete as a separate functional group. Results from our laboratory, for example, have shown that in reactions of Fe^+ with ketones and ethers the loss of H_2 and small alkanes from the iron-molecule complex becomes more important as the alkyl portions of these molecules are increased [5]. That this conclusion cannot be generalized to other metal ions, however, is illustrated by our studies on Ti^+. In carbonyl compounds having fewer than five carbons, Ti^+ extracts an O atom to form TiO^+ and it is unreactive toward larger carbonyl species [10]. Contrasting this behavior, Ti^+ is observed to be quite reactive toward hydrocarbons, although in a much more selective manner than Fe^+ as shown in Table 3 [7]. The major reaction products are loss of one or more H_2 molecules with only small amounts of alkane loss products observed. Ti^+ is not observed to react with methane but, unlike Fe^+, does react with ethane to give the ethylene complex, $TiC_2H_4^+$. From a further comparison of Fe^+ and Ti^+ with alkanes we may conclude that Ti^+ appears to be the better Lewis acid favoring the creation of many sites of unsaturation, i.e. double bonds, in the hydrocarbon chain. One final interesting example is the dehydrogenation reaction observed for both Fe^+ and Ti^+ with cyclohexane, reactions 20 and 21, respectively, leading to the formation of benzene complexes. Similar reactions are observed for other cyclic alkanes.

$$Fe^+ + \text{cyclo-}C_6H_{12} \longrightarrow \begin{cases} \xrightarrow{55\%} FeC_6H_{10}^+ + 1\ H_2 \\ \xrightarrow{24\%} FeC_6H_8^+ + 2\ H_2 \\ \xrightarrow{21\%} FeC_6H_6^+ + 3\ H_2 \end{cases} \quad (20)$$

$$Ti^+ + \text{cyclo-}C_6H_{12} \longrightarrow TiC_6H_6^+ + 3\ H_2 \quad (21)$$

We are also interested in sulfur containing compounds as analogs of oxygenated species for investigation of "hard-soft" effects on metal reactivity. In addition, sulfur remains a problem in the petroleum industry and several works have indicated that metal complexes may be effective desulfurization agents. In particular we have just begun examining the reactions of metal ions with mercaptans and sulfides [7]. A typical sequence of reactions is shown for the case of Co^+ with methyl sulfide. One interesting result here is that CH_2S is normally

Table 4. Periodic Reactivity Plot for C_2H_5SH for the process:

$$M^+ + C_2H_5SH \qquad [M(_2H_5SH)^+]^* \xrightarrow[\text{Loss}]{\text{Mass}} \text{Product ions}$$

Mass Loss	$M^+ = Al^+$	Ti^+	TiS^+	V^+	VS^+	Co^+	Fe^+
2		x		x			
16						x	
28						x	
30		x	x	x	x		
34						x	x

$$Co^+ + CH_3SCH_3 \longrightarrow Co(CH_2S)^+ + CH_4 \qquad (22)$$

$$Co(CH_2S)^+ + CH_3SCH_3 \longrightarrow Co(CH_3SCH_3)^+ + CH_2S \qquad (23)$$

$$\longrightarrow Co(CH_2S)(H_2S)^+ + C_2H_4 \qquad (24)$$

$$\longrightarrow Co(CH_2S)\ (CH_3SCH_3)^+ \qquad (25)$$

$$Co(CH_3SCH_3)^+ + CH_3SCH_3 \longrightarrow Co(CH_2S)(CH_3SCH_3)^+ + CH_4 \qquad (26)$$

$$\longrightarrow Co(CH_3SCH_3)_2{}^+ \qquad (27)$$

$$Co(CH_2S)(CH_3SCH_3)^+ + CH_3SCH_3 \longrightarrow Co(CH_3SCH_3)_2{}^+ + CH_2S \qquad (28)$$

unstable to polymerization, but my be stabilized by complexation. Another interesting result is that complexed-Co^+ reacts differentially than Co^+ itself. In reaction 22 and 24, for example, CH_4 loss occurs with Co^+ and C_2H_4 loss occurs with $Co(CH_2S)^+$. Further studies are underway and, for example, Table 4 shows a periodic reactivity plot for ethyl mercaptan, CH_3CH_2SH. Another particularly interesting example is the series of reactions observed for Ti^+ with SF_6. As written in equations 29-34 and shown in Figure 4, this system is unusual in that the major product ions, $SF_n{}^+$, do not involve the metal.

$$Ti^+ + SF_6 \longrightarrow \begin{cases} \xrightarrow{13\%} SF_5^+ + TiF & (29) \\ \xrightarrow{11\%} SF_4^+ + TiF_2 & (30) \\ \xrightarrow{40\%} SF_3^+ + TiF_3 & (31) \\ \xrightarrow{23\%} SF_2^+ + TiF_4 & (32) \\ \xrightarrow{9\%} TiF_2^+ + SF_4 & (33) \\ \xrightarrow{4\%} TiF_3^+ + SF_3 & (34) \end{cases}$$

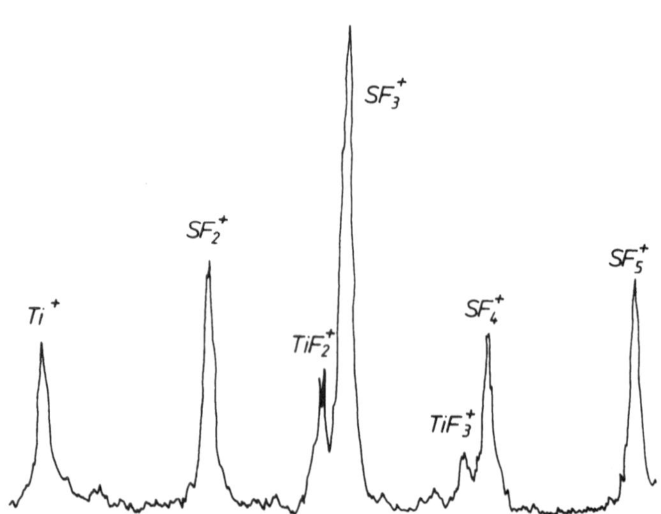

Fig. 4. Ti^+ chemical ionization mass spectrum of SF_6.

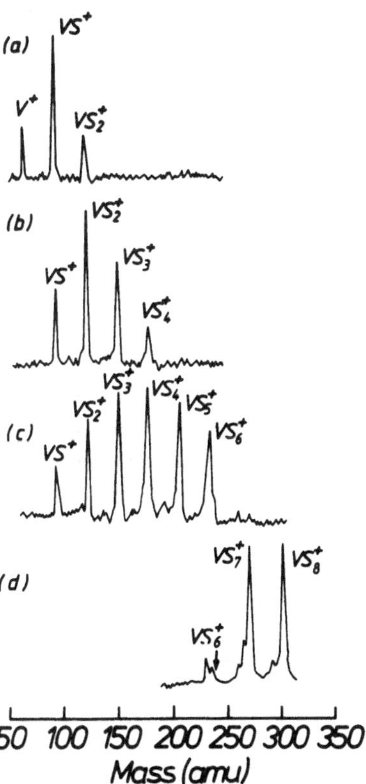

Fig. 5. V^+ chemical ionization mass spectra of ethylene suflide at neutral gas pressure of approximately (a) 1×10^{-7} torr (b) 2×10^{-7} torr (c) 5×10^{-7} torr and (d) 5×10^{-6} torr. Other conditions were a trapping of 250 ms and an observing frequency of 65.2 kHz.

Furthermore, the SF_n^+ product ion intensities observed for reactions 29-32 differ dramatically from the electron impact spectrum of SF_6. While SF_3^+ is the major product at 40% relative intensity in the Ti^+ spectrum, for example, it is only 15% in the 70 eV spectrum and continues to decrease relative to SF_5^+, the predominant ion, at lower electron energies. In addition to the thermodynamic implications of these results, it is evident that such reactions may permit altering a mass spectrum for a particular purpose in a way inaccessible by electron impact.

Finally, the most unexpected result so far during these studies involving sulfur compounds has been from reactions with ethylene sulfide [4]. Successive attachment of sulfur to Fe^+, Co^+, V^+, and Ti^+ was observed to be the dominant process in each case. Al^+ and Cu^+ were found only to attach directly at high pressures.

As a specific example, V^+ was found to attach at least eight sulfur atoms by reaction 35 (VS_9^+ was beyond our mass range) Figure 5 shows mass spectra of the V^+, ethylene sulfide system

$$VS_n^+ + CH_2\text{-}CH_2\,(\text{-S-}) \longrightarrow VS^+_{n+1} + C_2H_4 \quad (n = 0 \text{ to } 7) \tag{35}$$

at several neutral gas pressures. A variety of organometallic complexes with polysulfide ligands are known, many of them having ring structures. A ring structure may also apply for these gas phase ions. We plan to pursue these polysulfide reactions further now that we have extended mass capabilities as discussed below. In addition, we are pursuing possible catalytic cycles suggested by these results as speculatively written in equations 36 and 37, where CS is a sulfur containing molecule and C is a sulfur-free molecule. Ion catalyzed reactions have been

$$M^+ + CS \longrightarrow MS^+ + C \tag{36}$$

$$MS^+ + H_2 \longrightarrow M^+ + H_2S \tag{37}$$

previously demonstrated in a few limited examples, and could have potential for sulfur chemistry [11].

2. Kinetic Energy of Laser Generated Ions. A Comparison of Techniques

It is well known that laser generated metal ions may be produced with considerable amounts of kinetic energy [14]. In order to minimize this problem, our experiments are performed with low cell potentials permitting kinetically hot ions to escape the ion trap. One measure of the effectiveness of this approach is to compare our results to those obtained from metal ions generated by electron impact on volatile transition metal complexes. While the results are generally qualitatively similar, quite often they differ quantitatively. Results from an elegant beam experiment have recently been reported in which the reactivity of metal ions are studied as a function of an accurately selected kinetic energy [15]. In a limited comparison of laser and beam results for Fe^+ on hydrocarbons, for example, laser generated Fe^+ was found to have less than about 0.5 eV kinetic energy. Further comparisons of these related techniques are clearly needed.

Multiply charged ions may be formed by laser ionization and, therefore, electronically excited species as well as kinetically excited ions may also be produced and should be considered. The only evidence for electronically excited ions produced by laser ionization to date in our laboratory has been for $[Cu^+]^*$

which is observed to charge exchange with species having IP's up to about 2.5 eV above that of ground state Cu. Electron impact can also produce ions in electronically excited states and an example has recently been documented for Cr^+ [14]. Interestingly no excited state Cr^+ is observed by laser ionization. The differences in reactivity of excited and ground state metal ions is important from both a fundamental and practical standpoint and further work in this area will undoubtedly be actively pursued.

3. Fourier Transform Mass Spectrometry

All of the results described above were obtained on a conventional ion cyclotron resonance spectrometer. We have just recently put into operation a Nicolet FTMS-1000 prototype spectometer which greatly facilitates these studies. Fourier transform ion cyclotron resonance (FT-ICR) spectroscopy, also called Fourier transform mass spectrometry or FTMS, is a new method for conducting ICR experiments which was first demonstrated by Comisarow in 1974 [15]. The method involves the simultaneous excitation of the entire frequency domain ICR spectrum, sampling the resultant transient response of the system, and numerical Fourier transformtion to obtain the ICR frequency or mass spectrum. This holds several key advantages for the laser ionization experiment. A complete mass spectrum may be obtained rapidly from one laser pulse as demonstrated by Figure 6. Several thousand laser shots would normally be required to obtain a similar spectrum using a conventional ICR. Thus, a smaller sample may be studied or, alternatively less laser damage is sustained by the sample or target. In addition, because the entire mass spectrum is obtained from each laser shot, the FT experiment is far less susceptible to problems arising from pulse to pulse variations of the laser signal and signal averaging may readily be accomplished. Two other major advantages of the FT instrument are the increased mass range and resolution. To date using a magnetic field of 9kG, for example, we have been able to identify ion-molecule product peaks as high as m/z 1100 and have achieved a resolution of about 200,000 at m/z 100.

As a final demonstration of the laser ionization-FTMS technique, some results are presented from a strudy we have initiated on the cluster chemsitry of the transitional metal carbonyls. Beauchamp et al. have shown that the primary ions from $FeCO_5$ undergo sequential clustering reactions to produce multi-metal species with various numbers of carbonyls attached [16]. In particular Fe^+ is observed to react with $FeCO_5$ to produce $Fe_2CO_4{}^+$ which subsequently produces $Fe_3CO_7{}^+$, $Fe_3CO_8{}^+$, and eventually $Fe_4CO_{10}{}^+$. What effect would, for example, Cu^+ have on this cluster chemistry? The data in Figure 7 answer the question. Figure 7a shows the trapped ion spectrum arising from electron impact on $FeCO_5$. Fe^+ and the cluster ions which arise from its subsequent reactions are labelled. Figure 7b shows the Cu^+ chemical ionization spectrum of $FeCO_5$ under otherwise similar conditions. With the exception of $CuFeCO_3{}^+$, whose analog $Fe_2CO_3{}^+$ was not reported by the Beauchamp group, the cluster chemistry of the mixed Cu-Fe species looks remarkably similar to the pure Fe analogs. Further work is underway in this area varying both the metal ion reagent as well as the metal carbonyl. Of particular interest will be the effect of mixed metal clusters on ligand exchange reactions.

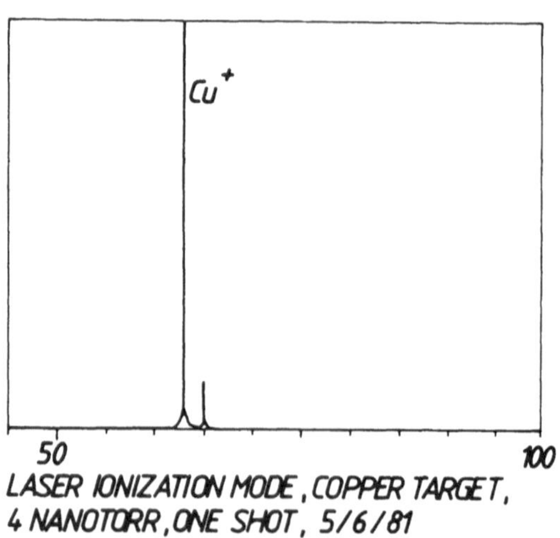

Fig. 6. FTMS spectrum of Cu^+ arising from the transform of one laser shot.

4. Concluding Remarks

Chemical ionization mass spectra are in general greatly dependent on a source conditions. The ratio of sample to reagent gas, for example, is often not precisely known. limiting reproducibility. An unusual teature of laser ionization CI is that no reagent gas is required. The sample pressure in the source may, therefore, be more precisely known. In addition, the absence of a reagent gas is especially attractive for chemical io-ization studies using ion cyclotron resonance spectroscopy because the mass resolution varies inversely with the pressure [17,18].

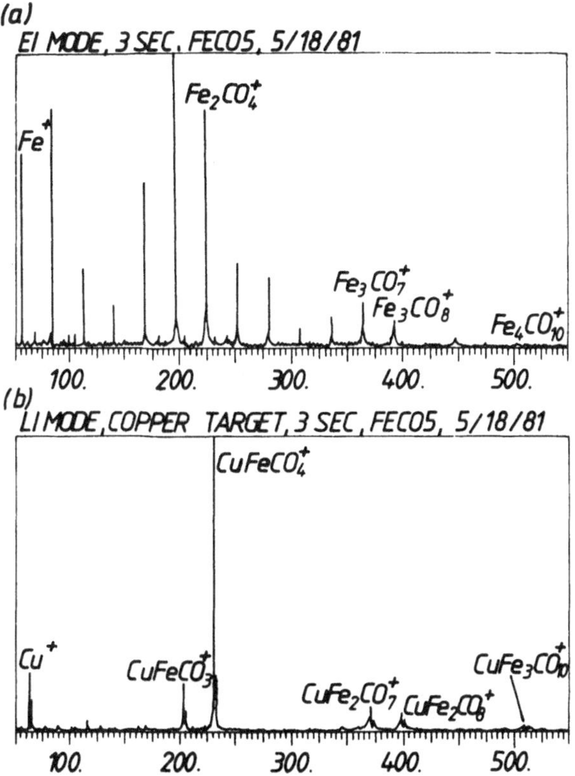

Fig. 7. (a) 70 eV electron impact spectrum of $FeCO_5$ (b) Cu^+ chemical ionization spectrum of $FeCO_5$ under otherwise similar conditions in a. Spectra obtained on the FTMS.

In summary, it is evident that the wide variety of metal ion reactivities, which may be further tailored by addition of substituents on the metal center, can provide not only fundamental information on organometallic chemistry, but also information and flexibility beyond that available from conventional proton transfer CI spectra. We are therefore, optimistic that the method may be extended to more complex samples.

5. Acknowledgements

This work was supported principally by the Department of Energy (DE-AC02-80ER10689). Acknowledgement is also made to the donors of the Petroleum Research Fund, administered by the American Chemical Society and to the National Science Foundation for providing funds for the purchase of the FTMS spectrometer.

Finally the authors wish to thank Professor Dallinger, chemistry department, Purdue University and Nicolet Instrument Corporation for their technical support.

6. References

1. R.B. Cody, R.C. Burnier, W.D. Reents, Jr., T.J. Carlin, D.A. McCrery, R.K. Lengel, and B.S. Freiser, Int. J. Mass Spec. Ion Phys. 33 (1980) 37.

2. R.C. Burnier, T.J. Carlin, W.D. Reents, Jr., R.B. Cody, R.K. Lengel, and B.S. Freiser, J. Am. Chem. Soc., 101 (1979) 7127.

3. R.C. Burnier, G.D. Byrd, and B.S. Freiser, Anal. Chem., 52 (1980) 1641.

4. T.J. Carlin, M.B. Wise, and B.S. Freiser, Inorg. Chem., in press.

5. R.C. Burnier, G.D. Byrd, and B.S. Freiser, J. Am. Chem. Soc., in press.

6. G.D. Byrd, R.C. Burnier, and B.S. Freiser, J. Am. Chem. Soc., submitted for publication.

7. R.C. Burnier, G.D. Byrd, T.J. Carlin, and M.B. Wise, unpublished results.

8. J. Frances, M. Ishaq, and M. Tsutsui in "Organotransition-Metal Chemistry", Y. Ishii and M. Tsutsui, Eds., Plenum, New York, 1975, pp. 57-64.

9. K. Levsen, "Fundamental Aspects of Organic Mass Spectrometry", Verlag Chemie, Weinham, West Germany, 1978.

10. J. Allison and D.P. Ridge, J. Am. Chem. Soc., 100 (1978) 163.

11. M.M. Kappes and R.H. Staley, J. Am. Chem. Soc., 103 (1981) 1286.

12. J.F. Ready, "Effects of High-Power Laser Radiation", Academic Press, New York, 1971.

13. P.B. Armentrout and J.L. Beauchamp, J. Am. Chem. Soc., 103 (1981) 784.

14. L.F. Halle, P.B. Armentrout, and J.L. Beauchamp, J. Am. Chem. Soc., 103 (1981) 962.

15. M.B. Comisarow in "Transform Techniques in Chemistry"; P.R. Griffiths, Eds.; Plenum Press, New York, 1978, p. 282.

16. M.S. Foster and J.L. Beauchamp, J. Am. Chem. Soc., 97 (1975) 4808.

17. R.L. Hunter and R.T. McIver, Jr., Anal. Chem. 51 (1979) 699.

GAS-PHASE ATOMIC METAL CATIONS. LIGAND BINDING ENERGIES, OXIDATION CHEMISTRY AND CATALYSIS

Manfred M. Kappes[a] and Ralph H. Staley[b]
Massachusetts Institute of Technology
Cambridge, MA o2139

1. Introduction

Ion cyclotron resonance (ICR) spectroscopy with a pulsed laser volatilization/ionization source of atomic metal cations has recently been applied to studies of the reactions of gas-phase metal ions with neutral molecules [1,2]. Initial results from this laboratory have included a variety of mechanistic studies [1-4] as well as measurements of gas-phase ligand binding energies for a number of different metal cations [5-9]. These mechanistic and thermodynamic studies yield data which is useful in developing models for understanding molecular interactions.

2. Ligand Dissociation Enthalpies

When a metal cation is generated in the presence of a neutral molecule elimination-displacement and direct condensation pathways may lead to the formation of metal-ligand cation complexes with one or two ligand molecules. Equilibrium constants for the ligand-exchange reaction of these complexes when two types of ligand are present can be determined. Combining the corresponding free energies for an interlocking series yields a scale of relative ligand binding energies. Such scales which are essentially basicity scales with the metal cation as the reference acid, have been obtained for Al^+ [5] and Mn^+ [6] with one ligand and Co^+ [7], Cu^+ [8] and Ni^+ [9] with two ligands. Figure 1 shows an example of such a relative ligand dissociation enthalpy scale for two-ligand complexes of Cu^+. Comparisons of these scales among themselves and with previously determined gas-phase basicity scales for the other reference acids H^+ [10,11] Li^+ [12], N^+ [13] and $CpNi^+$ [14] can be quite informative and often reveal interesting points about metal-ligand interactions.

Figure 2 shows a plot of the relative ligand binding energies to Al^+, $\delta D(Al^+\text{-}L)$, versus the results for Mn^+, $\delta D(Mn^+\text{-}L)$. The 15 oxygen bases show a good linear correlation. A least squares fit to this data gives: $\delta D(Al^+\text{-}L) = 1.22\delta D(Mn^+\text{-}L) - 6.1$ kcal/mol (correlation coefficient $r = 0.97$). The oxygen bases include alkyl

---

Present addresses: (a) Institute for Inorganic and Physical Chemistry, University of Bern, CH-3012 Bern, Switzerland, (b) Central Research Department, Experimental Station, DuPont Company, Wilmington, DE 19898.

alcohols, ethers, aldehydes, ketones and esters. These fall approximately on the line. The slope of this line, 1.22, shows that as basicity increases the strength of the bond to Al^+ increases slightly faster than that to Mn^+. This suggests that the ligand-metal bond distance for $Al(ligand)^+$ complexes is slightly shorter than for $Mn(ligand)^+$ complexes. With a shorter bond distance the effect of a larger alkyl substituent may be expected to be greater since it is closer to the charge center which remains on the metal atom.

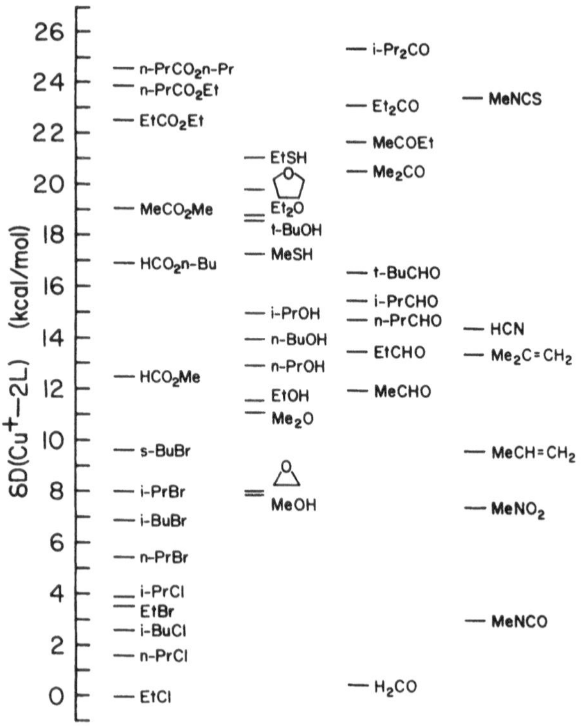

Fig. 1. Relative two-ligand dissociation enthalpies, $\delta D(Cu^+\text{-}2L)$, for the 43 molecules studied with Cu^+, arranged by functional group, relative to $D(Cu^+\text{-}2EtCl) = 0$.

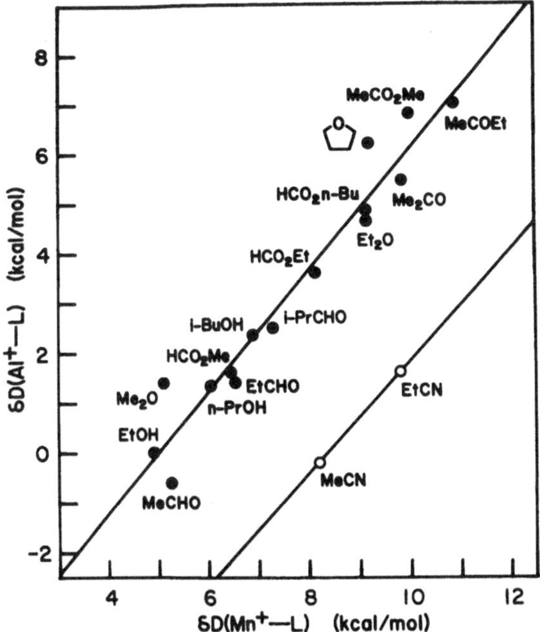

Fig. 2. Comparison of relative binding energies of molecules to Al^+ and Mn^+. The upper solid line is a least squares fit to the data for the oxygen bases: $\delta D(Al^+\text{-}L) = 1.22\delta D(Mn^+\text{-}L) - 6.1$ kcal/mol (correlation coefficient r = 0.97). The line through the two data points for nitriles is given by $\delta D(Al^+\text{-}L) = 1.13\delta D(Mn^+\text{-}L) - 9.4$ kcal/mol. This corresponds to an offset on the $\delta D(Mn^+\text{-}L)$ axis of 3.4 kcal/mol.

Two data points for nitriles in Figure 2 fall on a line given by $\delta D(Al^+\text{-}L) = 1.13\delta D(Mn^+\text{-}L) - 9.4$ kcal/mole. This has the same slope as the line for oxygen bases but is offset by 3.4 kcal/mole on the $\delta D(Mn^+\text{-}L)$ axis. This offset can be attributed to π-backbonding by 3d electrons on Mn^+ into empty π* CN orbitals on the nitriles. Offsets of 6 kcal/mole for nitriles compared to oxygen bases have been observed in plots of binding energies to $CpNi^+$ versus binding energies to H^+ [10] and to Al^+[5]. The smaller effect by about half which is observed for Mn^+ compared to $CpNi^+$ is reasonable for π-backbonding since the 3d orbitals on Mn^+ have only one electron each whereas for $CpNi^+$ they nearly full.

Comparison of the results for Mn^+ with data for Li^+ [12], Figure 3, also shows a good linear correlation for oxygen bases. Remarkably, the two nitriles in this data set are on the line with the oxygen bases. A least squares fit on the seven oxygen bases and two nitriles gives $D(Li^+\text{-}L) = 0.93\delta D(Mn^+\text{-}L) + 35.3$ kcal/mole (r = 0.98). Here the slope shows slightly stronger binding to Mn^+ compared to Li^+ with increasing basicity suggesting that the metal-ligand bond distance for Mn^+ is slightly shorter than for Li^+.

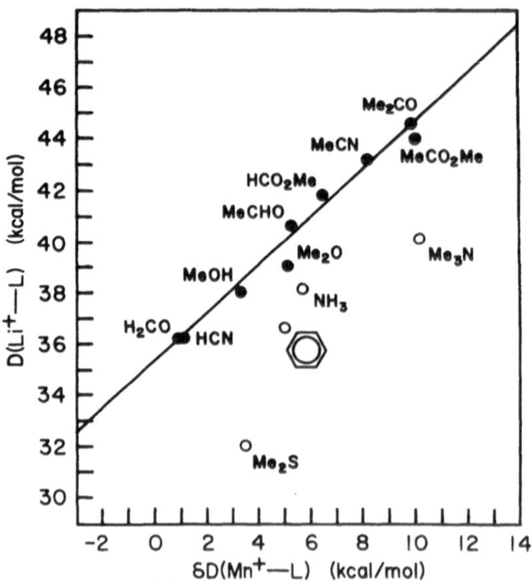

Fig. 3. Comparison of relative binding energies of molecules to Li^+ and Mn^+. The solid line is a least squares fit to the data for the seven oxygen bases and two nitriles (solid circles): $D(Li^+\text{-}L) = 0.93\ D(Mn^+\text{-}L) + 35.3$ kcal/mol ($r = 0.98$).

Several bases fall off the linear correlation in Figure 3: Me_2S (7.0), benzene (3.6), NH_3 (2.7), and Me_3N (5.0) where the offset on the $\delta D(Mn^+\text{-}L)$ axis from the linear correlation is given for each molecule in parentheses in kcal/mol. The offsets toward stronger bonding to Mn^+ for Me_2S and benzene could be explained in terms of π-backbonding which certainly does not occur for Li^+ but could be of some importance for Mn^+ as suggested by the preceeding discussion of the Al^+ versus Mn^+ comparison in Figure 2. However, the lack of offset for the nitriles in Figure 3 must then be regarded as bizarre since an offset is seen for the nitriles in the Al^+ versus Mn^+ comparison of Figure 2.

It is therefore reasonable to conclude that other factors in addition to π-backbonding and of comparable magnitude contribute to determining the offsets for functional group correlation lines. Such factors might include the strength of the ionic bond, bond distance, and covalent σ-bonding.

In a given correlation plot of binding energies of molecules to one metal versus those to another, it is thus to be expected that the molecules for each functional group may in general fall on different lines. Some of these lines may happen to be approximately the same such as alcohols and esters in Figure 2 and nitriles and the oxygen bases in Figure 3. Some evidence for this view has been previously noted. In the comparison of $D(B-H^+)$ to $\delta D(Al^+-L)$ it was seen that esters fall on a line somewhat apart from the general correlation line for all bases [5]. Correlations for $CpNi^+$ [14] versus Mn^+ data and for H^+ versus Mn^+ data lend further support to this idea [6].

The variation of slopes of the linear correlations in the comparison studied to date follow a systematic pattern. The slopes of the lines for oxygen bases for H^+, Al^+, Li^+, and $CpNi^+$ versus Mn^+ are 2.08, 1.22, 0.93 and 0.87 resp. For nitriles these slopes are 1.63, 1.30, 0.93, and 0.79 in the same order. Slopes for some of the linear correlations obtained to date are given in Table 1. The results all follow the same order $H^+ < Al^+ < Mn^+ < Li^+ < CpNi^+$.

Table 1. Slopes for Linear Correlations in Plots of Ligand Binding Energies to One Metal versus Another for Oxygen Bases.

Metal Cation	Versus				
	H^+	Al^+	Mn^+	Li^+	$CpNi^+$
H^+	1.00	1.53	2.08	2.71	3.38
Al^+	0.65	1.00	1.22	--	1.68
Mn^+	0.48	0.82	1.00	1.08	1.15
Li^+	0.37	--	0.93	1.00	--
$CpNi^+$	0.30	0.60	0.87	--	1.00

As discussed above this order may reflect the ligand-metal bonding distance for the atomic metal cations: H^+ having the shortest effective bonding distance and Li^+ the longest. Where there is significant delocalization of charge away from the metal, as with H^+, the distance to the charge center rather than the bond distance to the center of the metal atom may be the controlling parameter. For $CpNi^+$ it seems likely that the slope of the correlation plot is due to another factor, the presence of a second ligand. Polarization of $CpNi^+$ group produces an internal dipole moment which opposes the dipole moment induced in the $CpNi(ligand)^+$ complex. This tends to cancel the favorable interaction of $CpNi^+$ charge with the induced ligand dipole. No such cancellation occurs for the atomic metal cation complexes with ligands since the atomic metal cations do not have internal dipole moments.

Whereas Mn^+ forms complexes with one ligand molecule, Co^+, Cu^+ and Ni^+ form two-ligand complexes under the same conditions. It is still possible to measure ligand dissociation enthalpies, however, the situation is a little more complicated. A simple ideal model of the bonding in two-ligand complexes would assume that the bonding at each site is independent. It would follow that the dissociation energies for the first and second ligands would be equal: $D(LM^+-L) = D(M^+-L)$. For the real complexes, however, the dissociation energy for the first ligand should be significantly less than that for the second: $D(LM^+-L) < D(M^+-L)$. This arises because interaction of the partial charges on the ligands increases the total energy of the $L-M^+-L$ complex compared to twice the energy of the M^+-L complex. This interaction also favors a linear geometry for the $L-M^+$ species. Covalent effects will further complicate the bonding in the two-ligand complexes. The relative two-ligand dissociation energies, $\delta D(M^+-2L)$, measured for Co^+, Cu^+ and Ni^+ are therefore not simply related to the one-ligand dissociation energies, $D(M^+-L)$.

Relative two-ligand dissociation enthalpies for the 43 molecules studied with dissociation enthalpies for the 43 molecules studied with Cu^+, arranged by functional group, are plotted in Figure 1 [7]. Within each functional group series, substitution of a larger alkyl group for a smaller one is seen to lead to systematic increase in $\delta D(Cu^+ -2L)$. The results are consistent with and illustrate an increase in basicity in the sequence H < Me < Et < n-Pr < n-Bu < i-Bu < t-Bu. Similar systematic alkyl substituent effects have been observed for proton affinities [10,11], $PA(B) \equiv D(B-H^+)$, and for one-ligand binding energies to Li^+ [12], Al^+ [5], and Mn^+ [6]. Evidently, differences in bonding in the two-ligand complexes compared to the one-ligand complexes do not affect the qualitative behavior of substituent effects.

Atomic copper cation, Cu^+, has a ^1S ground state corresponding to an $[Ar]\,(3d)^{10}$ configuration [15]. The next higher states, $[Ar]\,(3d)^9\,(4s)\ ^3D$ at 2.7 eV and $[Ar]\,(3d)^9(4s)\ ^1D$ at e.e eV, should not be involved in ligand bonding. Some evidence was observed in studies of the inital reactions of Cu^+ with neutrals that a fraction of electronically excited species might be present [1]. The ^3D state might likely be the state involved if the observed phenomena indeed involve internal excitation. It seems unlikely that this excitation, if present, whatever its form, would affect the ligand exchange equilibria examined in the present work. Several dozen reactive and nonreactive collisions with various neutrals intervene between the production of Cu^+ and the point measured.

The bonding interaction of Cu^+ with ligand molecules may be expected to reflect both ionic and covalent factors. In bonding to Cu^+, ligands are attracted by

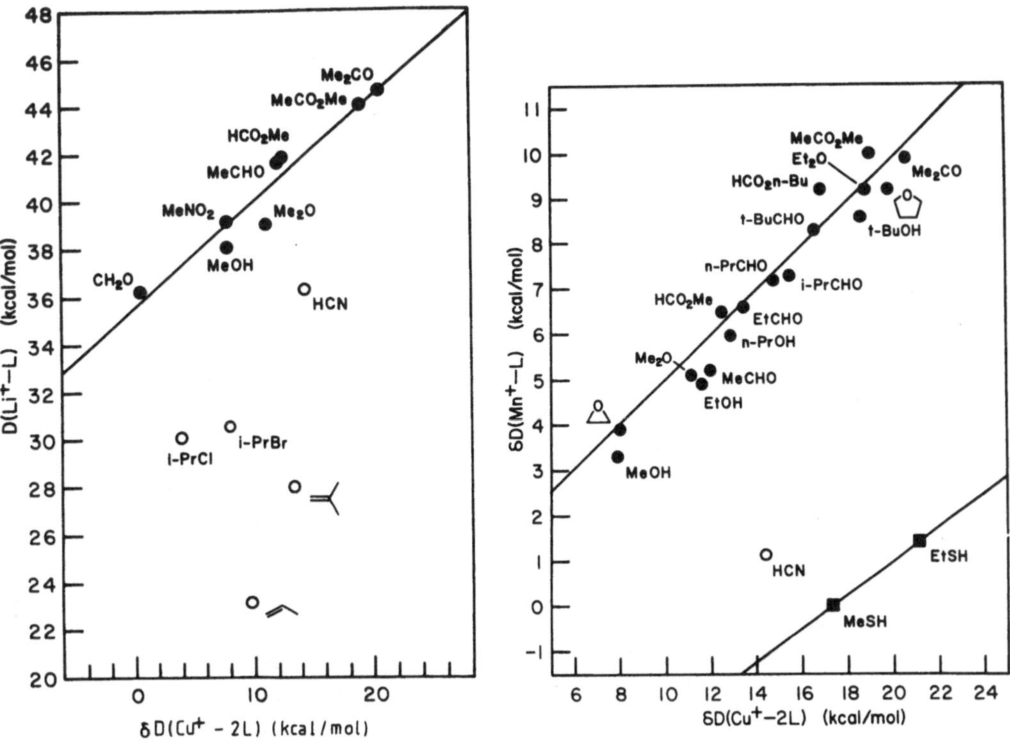

Fig. 4.

Fig. 5.

Fig. 4. Comparison of ligand dissociation enthalpies for Li^+ to relative two-ligand dissociation enthalpies for Cu^+. The solid line is a least squares fit to the data for the eight oxygen bases (solid circles): $D(Li^+\text{-}L) = 0.443\delta D(Cu^+\text{-}2L) + 35.5$ kcal/mol (correlation coefficient, r = 0.96).

Fig. 5. Comparison of relative ligand dissociation enthalpies for Mn^+ to relative two-ligand dissociation enthalpies for Cu^+. The upper solid line is a least squares fit to the data for the 17 oxygen bases: $\delta D(Mn^+\text{-}L) = 0.524\delta D(Cu^+\text{-}2L) - 0.631$ kcal/mol (r = 0.98). The line through the two mercaptans is given by $\delta D(Mn^+\text{-}L) = 0.368\delta D(Cu^+\text{-}2L) = 6.374$ kcal/mol.

interaction of intrinsic and induced ligand dipoles with the charge on Cu^+. Covalent bonding can occur by delocalization of electrons from occupied ligand orbitals into the empty 4s and 4p orbitals on Cu^+. This delocalization will be limited by repulsion by the electrons in the filled 3d and core orbitals. Covalent bonding can also occur by delocalization of 3d electrons on Cu^+ into unoccupied π^* or d orbitals, π-backbonding. This can occur only for ligand molecules with suitable orbitals available.

Comparison of the relative two-ligand dissociation enthalpies for Cu^+ with available results for other reference acids reveals some interesting points about the nature of the metal-ligand bonding interaction [7]. Figure 4 shows a plot of ligand binding energies to Li^+ [12], $D(Li^+\text{-}L)$, versus the results for Cu^+, $\delta D(Cu^+\text{-}2L)$. The eight oxygen bases show a good linear correlation. A least squares fit to this data gives: $D(Li^+\text{-}L) = 0.443\delta D(Cu^+\text{-}2L) + 35.5$ kcal/mol (correlation coefficient, $r = 0.96$). A similar plot of Al^+ [5] versus Cu^+ results for 18 oxygen bases gives $\delta D(Al^+\text{-}L) = 0.770\delta D(Cu^+\text{-}2L) - 8.72$ kcal/mol ($r = 0.97$). The zero of the $\delta D(Al^+\text{-}L)$ scale is arbitrarily chosen to be $D(Al^+\text{-}EtOH) = 0$. The oxygen bases include alkyl alcohols, ethers, aldehydes, ketones and esters. In the Li^+ vs. Cu^+ comparison, Figure 4, i-PrCl, i-PrBr, HCN, and two alkenes, all soft bases, are seen to fall off the line for oxygen bases, bonding relatively more strongly to Cu^+. This is consistent with the expectation that the soft acid Cu^+ should bond relatively more strongly to soft bases compared to the hard acid Li^+. Comparison of results for Mn^+ [7] with those for Cu^+, Figure 5, shows similar effects. Seventeen oxygen bases fall around a line given by $\delta D(Mn^+\text{-}L) = 0.524\delta D(Cu^+\text{-}2L) - 0.631$ kcal/mol. Cu^+ is clearly also a softer acid than Mn^+ as HCN and the two mercaptans show a strong preference toward binding to Cu^+, Figure 5. The offset of HCN from the line for the oxygen bases is 11.0 kcal/mol on the $\delta D(Cu^+2L)$ axis. The offset of the line for the two mercaptans is 16.4 kcal/mol. Note that there is actually a EtSH compared to EtOH on the Cu^+ scale compared to the Mn^+ scale, Figure 5. There is only a slight preference of HCN and MeSH for Cu^+ when compared to $CpNi^+$ [14], Figure 6. The line for the 14 oxygen bases is given by $D(CpNi^+\text{-}L) = 0.454\delta D(Cu^+\text{-}2L) + 42.1$ kcal/mol. HCN is offset by 2.2 and MeSH by 6.1 kcal/mol on the $\delta D(Cu^+\text{-}2L)$ axis.

Comparison of the Cu^+ results to proton affinities [10.11], $PA(B) \equiv D(B\text{-}H^+)$, reveals two separate lines for oxygen bases. Sixteen oxygen bases fall on a line given by $PA(L) = 1.074\delta D(Cu^+\text{-}2L) + 176.9$ kcal/mol ($r = 0.99$). These bases include alkyl alcohols, aldehydes, ketones, and esters. Four ethers fall along a separate line given by $PA(L) = 0.890\delta D(Cu^+\text{-}2L) + 182.8$ kcal/mol ($r = 0.99$).

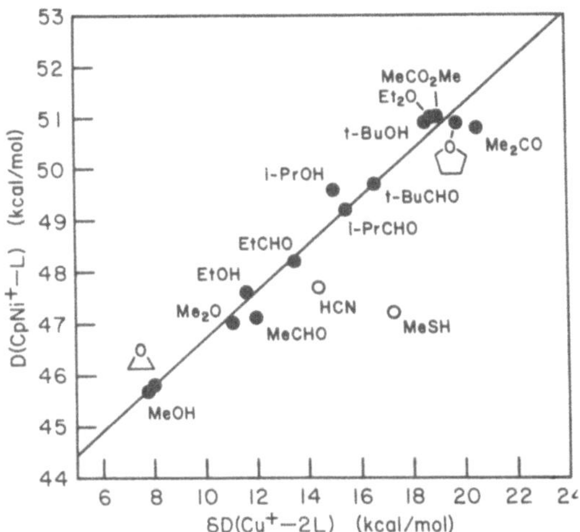

Fig. 6. Comparison of relative ligand dissociation enthalpies for $CpNi^+$ to relative two-ligand dissociation enthalpies for Cu^+. The solid line is a least squares fit to the data for the 14 oxygen bases: $D(CpNi^+\text{-}L) = 0.454\delta D(Cu^+\text{-}2L) + 42.1$ kcal/mol (r = 0.99).

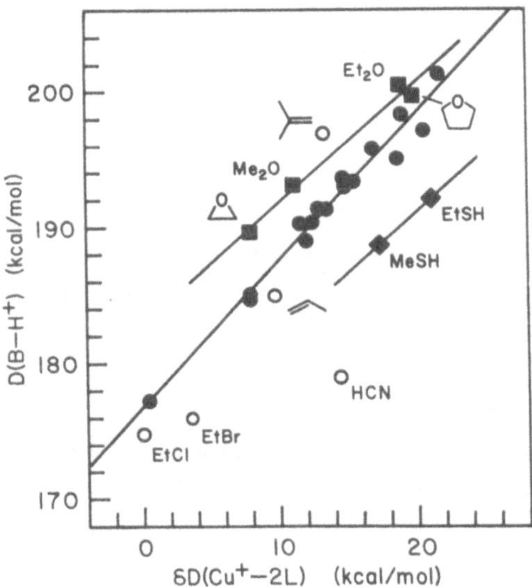

Fig. 7. Comparison of proton affinities $PA(B) \equiv D(B\text{-}H^+)$ to relative two-ligand dissociation enthalpies for Cu^+. The upper solid line is a least squares fit for the four ethers: $PA(L) = 0.890\delta D(Cu^+\text{-}2L) + 182.8$ kcal/mol. The middle solid line is a least squares fit for 16 other oxygen bases: $PA(L) = 1.074\delta D(Cu^+\text{-}2L) + 176.9$ kcal/mol. The lower solid line through the two mercaptans is given by $PA(L) = 0.895\delta D(Cu^+\text{-}2L) - 173.121$ kcal/mol.

A number of soft bases fall off the oxygen lines toward stronger binding to Cu^+; these include EtCl, EtBr, MeSH, EtSH and HCN, Figure 7, indicating as expected that Cu^+ is a softer acid than H^+. The two alkenes behave erratically, one falling above and the other below the lines for the oxygen bases. Bonding of the proton to alkenes is unlike complex formation of alkenes with metals, however. Protonation of isobutylene yields tert-butyl cation, for example. A simple correlation of proton affinities of alkenes with alkene-metal cation binding energies is therefore not expected.

The slopes of the correlation plots of the other reference acids versus Cu^+ for oxygen bases vary in the order H^+ (1.074), Al^+ (o.770), Mn^+ (o.524), and Li^+ (0.443). This same order was noted above for correlation plots against Mn^+ [6]. It likely reflects the relative metal-ligand bond distance, H^+ having the shortest bond distance and Li^+ the longest. With a shorter bond distance the effect of a larger alkyl substituent is greater since it is closer to the charge center. The slope of the line thus favors the metal with the shorter bond distance as basicity increases with increasing alkyl substitution. Dividing the $\delta D(Cu^+ -2L)$ scale by two to give the relative dissociation enthalpy per ligand for Cu^+, the slopes become H^+ (2.15), Al^+ (1.54), Mn^+ (1.05), and Li^+ (0.89). The ligand bonding distance to Cu^+ is thus seen to be shorter than that to Li^+ since the slope for Cu^+ versus itself is 1.00. Actually, the one-ligand dissociation enthalpies are not known from the average dissociation energy available from the present work and may as well be higher as discussed above. This would give a higher effective slope for comparison of Cu^+ to H^+, Mn^+, and Al^+. It is thus not clear whether bonding distance to Cu^+ may be shorter than that to these references acids or not. The significance of the slope in the correlation of $CpNi^+$ to Cu^+ is also not clear since two-ligand effects are present in both data sets in this comparison in different ways.

3. Cooperative Bonding Effects

Metal complexes having two or more ligand molecules can exhibit synergistic effects in which the properties and behavior of one ligand show a dependence on the identity of the other ligand or ligands [15]. Such phenomena were first observed in the kinetic behavior of square planar and octahedral transition metal complexes and have been generally referred to as the trans-effect (or in some cases the cis-effect) [16-18]. Ground state properties such as bond distance, vibrational frequencies, and NMR parameters also exhibit such phenomena. For the ground state properties these phenomena are referred to as trans-influence [16-18]. These phenomena should also be manifest in thermodynamic properties but little data has been available to establish this point 16-18 .

The ideal measure of trans-influence in thermodynamic properties would be provided by determination of gas-phase metal-ligand bond strengths, the enthalpy for reaction 1. Reaction 1 has been written for the case of a unipositive two-

$$AMB^+ \longrightarrow AM^+ + B, \qquad \Delta H = D(AM^+\text{-}B) \qquad (1)$$

ligand, linear complex, one of the simplest systems which would be expected to exhibit trans-influence. Relative bond strengths for two-ligand complexes are given by enthalpy differences for the ligand exchange reactions 2 and 3. The enthalpy expressions for these reactions follow from the enthalpy expression

$$AMA^+ + B \overset{K_2}{\rightleftharpoons} AMB^+ + A \qquad \Delta H_2 = D(AM^+\text{-}A) - D(AM^+\text{-}B) \qquad (2)$$

$$AMB^+ + B \overset{K_3}{\rightleftharpoons} BMB^+ + A \qquad \Delta H_3 = D(BM^+\text{-}A) - D(BM^+\text{-}B) \qquad (3)$$

for reaction 1. If there were no interaction between the two ligand sites on the metal, then the enthalpies for reactions 2 and 3 would be equal, $\Delta H_2 = \Delta H_3$, Scheme 1. If there is synergistic enhancement of the mixed species AMB^+ compared to the pure species AMA^+ and BMB^+, then this can be measured as a deviation Q of the relative enthalpy for the mixed species, where $Q = (\Delta H_3 - \Delta H_2)/2 = [D(AM^+\text{-}B) + D(BM^+\text{-}A) - D(AM^+\text{-}A) - D(BM^+\text{-}B)]/2$, Scheme 1.

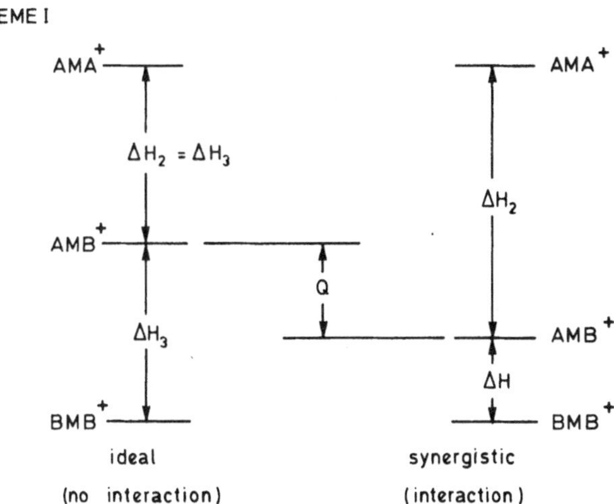

In equilibrium measurements of ligand dissociation enthalpies for two-ligand complexes of Co^+, Cu^+ and Ni^+, synergistic enhancement of the population of the mixed AMB^+ species was seen for certain combinations of the two ligands A and B. The ligands studied fall into four groups: (1) σ-bases (alkyl halides, alcohols, ethers, aldehydes, ketones, esters, isocyanates and nitro compounds), (2) S-bases (alkyl mercaptans and sulfides), (3) N-bases (alkyl amines and cyanides), and (4) π-bases (olefins and aromatics). Complexes with both ligands form the same group show no special stability for the mixed AMB^+ species, Table 2. The mean and standard deviation for the distribution of Q for 85 intragroup ligand-pair complexes is $Q = 0.01 \pm 0.09$ kcal/mol. The σ-base/π-base pairs show the largest stabilization, $Q \simeq 0.8$, 1.2, and 0.9 kcal/mol for Co^+, Ni^+, and Cu^+ complexes respectively, Table 3. Smaller stabilizations are seen for σ-base/S-base, σ-base/N-base, S-base/π-base, and N-base/π-base complexes, Table 3.

The observation of synergistic stabilization of the mixed ligand complex for σ-base/π-base and other intergroup pairs constitutes a direct measurement of thermodynamic trans-influence in the nearly ideal case of gas-phase two ligand metal ion complexes. Emperical trans-influence orders predict [16-18] π-bases > S-bases $\simeq$ N-bases > σ-bases, consistent with the observed order in the present work. Theories of trans-influence are based on the idea that with unlike trans-ligands having differing degrees of π-bonding a synergistic enhancement is obtained from asymmetric distortion of the metal orbitals involved in bonding. For example, with a pure σ-donor/π-acceptor ligand pair, the empty metal σ-orbital distorts toward the σ-donor ligand and an unoccupied metal d-orbital distorts toward the empty ligand π-acceptor orbital, Stucture 1 [16-18].

(1)

With like ligands on both sites, a symmetric structure results, giving no enhancement. Attempts at quantitative theoretical treatment of the effect have met with only very limited success [16-18]. The quantitative thermochemical data provided by the present work for relatively simple systems should facilitate development of more exact theoretical models.

Table 2. Stabilization Energies for Intergroup Two-Ligand Gas-Phase Transition-Metal Complexes Showing Synergistic Stabilization[a]

M	A[b]	B[b]	Q[c]	$D(M^+-2B)$ $-D(M^+-2A)$
(A) σ-Base/π-Base Complexes, AMB^+				
Co	EtOH	$EtCH{=}CH_2$	0.75	0.51
Co	Me_2O	$EtCH{=}CH_2$	0.86	0.59
Co	n-PrOH	$Me_2C{=}CH_2$	0.75	0.51
Co	$Me_2C{=}CH_2$	EtCHO	0.72	0.54
Ni	$MeCH{=}CH_2$	Me_2O	1.27	1.65
Ni	$Me_2C{=}CH_2$	n-PrOH	1.18	0.44
Ni	Me_2O	$EtCH{=}CH_2$	1.18	0.06
Ni	$CH_2{=}C{=}CH_2$	MeOH	1.11	0.81
Cu	$Me_2C{=}CH_2$	n-PrCHO	0.72	0.52
Cu	$MeCH{=}CH_2$	Me_2O	1.16	1.46
(B) σ-Base/S-Base Complexes, AMB^+				
Co	Me_2O	MeSH	0.27	0.12
Co	Me_2S	Me_2CO	0.49	0.07
Co	Et_2O	n-BuSH	0.23	0.01
Co	EtSH	n-PrCHO	0.34	0.09
Ni	Me_2S	Et_2CO	0.53	0.86
Ni	EtCOMe	Me_2S	0.36	0.83
Cu	Me_2CO	EtSH	0.75	0.53
Cu	t-BuCHO	MeSH	0.97	0.73
Cu	Et_2CO	MeNCS	0.30	0.30

(C) σ-Base/N-Base Complex, AMB^+				
Co	HCN	n-PrCHO	0.20	0.56
Ni	n-PrCHO	HCN	0.63	0.11
Ni	NH_3	Et_2CO	0.51	0.54
Cu	HCN	n-PrCHO	0.73	0.33
(D) S-Base/π-Base Complex, AMB^+				
Ni	C_6H_6Cl	EtSH	1.18	1.23
(E) N-Base/π-Base Complex, AMB^+				
Ni	$EtNH_2$	C_6H_5CN	0.35	1.23

[a]All data in kcal/mol. [b]B is always the stronger ligand and A the weaker ligand. [c]$Q = [D(AM^+\text{-}B) + D(BM^+\text{-}A) - D(AM^+\text{-}A) - D(BM^+\text{-}B)]/2$, see Scheme 1.

Table 3. Examples of Stabilization Energies, Q, For Intragroup Two-Ligand Gas-Phase Transition-Metal Complexes Showing Ideal (Statistical) Behavior[a]

M	A[b]	B[b]	Q[c]	$D(M^+\text{-}2B)$ $-D(M^+\text{-}2A)$
(A) σ-Base/σ-Base Complexes, AMB^+				
Co	Me_2O	n-PrOH	0.01	1.66
Co	Me_2CO	MeCOEt	0.04	1.33
Co	n-PrCHO	i-PrOH	0.02	0.15
Ni	$EtCO_2Et$	Et_2CO	-0.10	2.30
Ni	MeCHO	EtOH	0.09	0.46
Ni	t-BuCHO	Me_2CO	-0.11	1.39
Cu	EtBr	i-PrCl	0.03	0.34
Cu	EtCl	CH_2O	0.04	0.53

Cu	EtCHO	n-BuOH	-0.05	0.49
Cu	Et_2CO	$n\text{-}PrCO_2Et$	0.02	0.67
Cu	MeNCO	EtBr	0.08	0.43
Cu	$MeNO_2$	MeOH	0.10	0.01
(B) π-Base/π-Base Complexes				
Ni	C_6H_5Cl	C_6H_6	0.14	1.33
Ni	C_2H_2	$CH_2{=}C{=}CH_2$	0.14	0.19
(C) N-Base/N-Base Complexes, AMB^+				
Ni	MeCN	$MeNH_2$	0.07	0.14
Ni	EtCN	Me_3N	0.12	1.20

[a] All data in kcal/mol. [b] B is always the stronger ligand and A the weaker ligand. [c] $Q = [D(AM^+\text{-}B) + D(BM^+\text{-}A) - D(AM^+\text{-}A) - D(BM^+\text{-}B)]/2$, see Scheme 1. The mean and standard deviation of the distribution of Q for 85 intragroup equilibria is $\bar{Q} = 0.01 \pm 0.09$ kcal/mol.

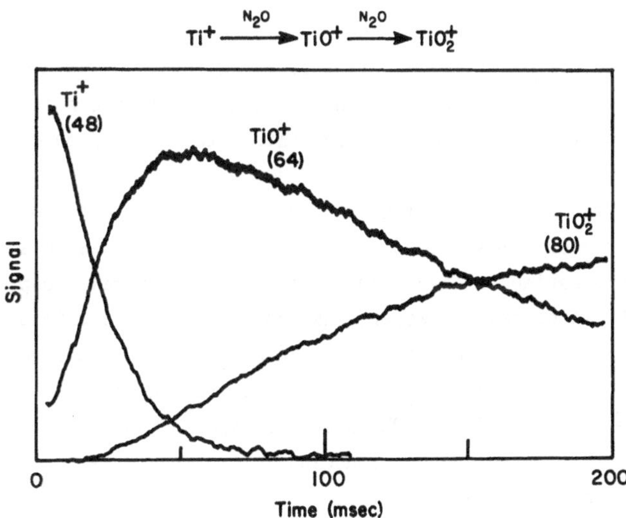

Fig. 8. Variation of ion abundance with time for the species seen when Ti^+ is generated in the presence of 7.5×10^{-6} torr of N_2O. Rapid reaction ($k = 4 \times 10^{-10}$ $cm^3 molec^{-1} sec^{-1}$) occurs to produce TiO^+. This is followed by a slower reaction which results in the formation of TiO_2^+ ($k = 4 \times 10^{-11}$ $cm^3 molec^{-1} sec^{-1}$). The m/z value for the isotopic peak which was followed for each ion species is given in parentheses.

4. Metal Oxide Cations

Observation of oxygen atom transfer from various oxidants to transition metal cations can be used to establish upper and lower limits for metal-oxide-cation bond dissociation energies, $D(M^+-O)$ [4]. Figure 8 gives a typical experimental result. Shown is a plot of ion abundance with time for the species observed when Ti^+ is generated in the presence of N_2O. Ti^+ rapidly reacts to form TiO^+, which can react further to TiO_2^+. Reaction of Ti^+ with N_2O implies a lower limit for $D(Ti^+-O)$. For MgO^+, AlO^+, MnO^+, CoO^+, NiO^+, CuO^+ and ZnO^+, similar results bracket $D(M^+-O)$ between $D(O_2-O) = 25.5$ kcal/mol and $D(N_2-O) = 40.0$ kcal/mol [19]. For V^+ and Fe^+ the results also bracket the bond energy in relatively narrow ranges: $D(V^+-O) = 135 \pm 16$ kcal/mol and $D(Fe^+-O) = 101 \pm 18$ kcal/mol. The VO^+ produced by reaction of V^+ with N_2O reacts with H_2 or CH_4 to give VOH^+. However, when VO^+ is produced by reaction of V^+ with O_2, $D(O-O) = 119.2$ kcal/mol [19] reaction to give VOH^+ does not occur. These results imply that VO^+ is produced in an excited state when it is generated by reaction of V^+ with N_2O. The lower oxygen bond dissociation energy of N_2O compared to O_2 provides an additional 80 kcal/mol to the products in the N_2O system compared to the O_2 case [19]. Since the excited VO^+ species reacts to give VOH^+ with a halftime of up to about 150 ms in typical experiments, the lifetime of the VO^+ excited states involved must be at least this long. This state is also insensitive to quenching by up to about 70 collisions by N_2 or CO molecules. [20]. The excited state of VO^+ in these processes could involve either vibrational or electronic excitation. The available evidence does not provide a basis to choose between these possibilities. Reaction of Ti^+ with N_2O or O_2 also appears to give an excited state species of TiO^+. With Fe, Zr and Nb there is no evidence for production of a long-lived excited state metal oxide product in the present results.

The VO^+ system affords an interesting opportunity to explore the photochemistry and spectroscopy of a gas phase ionic species. Photodissociation and photodetachment experiments have been carried out by several groups using ICR techniques [21]. In no case, however, has photoexcitation led to observable changes in bimolecular reaction chemistry. Irradiation of VO^+ produced by reaction of O_2 with V^+ in the presence of CH_4 could provide an example of this type of process. Reaction of VO^+ to VOH^+ would be expected in this system only if VO^+ absorbs the photon energy producing an excited state species similar to the one involved in the present results.

The ion chemistry of FeO^+ with seventeen different neutrals, encompassing a wide range of functional groups has recently been extensively studied in this laboratory. Work is presently underway to further extend this survey. A comparison of the results presently availble for FeO^+ to the reactions of Fe^+ with the same neutrals, reveals that FeO^+ is a much more reactive species than Fe^+. Mechanism involving insertion of the neutral atom into C-O, C-C or C-H bonds are proposed to explain the observed reactions of both Fe^+ and FeO^+. Of the many interesting reactions of FeO^+, one class of reaction is particularly striking and is presented in detail in the next section.

5. Catalysis

Certain reaction systems containing Fe^+, N_2O and an oxygen acceptor manifest catalytic reaction cycles [3]. For example, oxidation of CO to CO_2 by N_2O, is catalyzed by iron cations, Scheme 2. This occurs in a two-step process.

Scheme 2

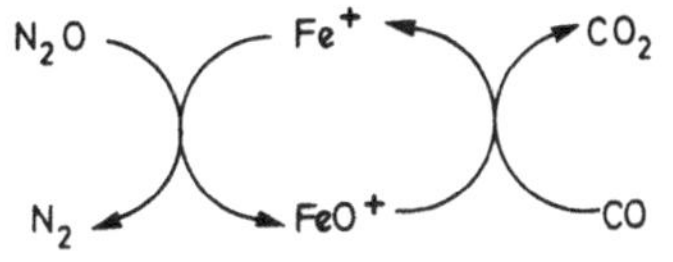

N_2O Fe^+ CO_2

N_2 FeO^+ CO

Fe^+ accepts an oxygen atom from N_2O to give FeO^+, reaction 4, and FeO^+ transfers the oxygen atom to CO to produce CO_2, regenerating Fe^+,

$$Fe^+ + N_2O \longrightarrow FeO^+ + N_2,\ k = 0.7 \times 10^{-10}\ cm^3 molecule^{-1} sec^{-1} \qquad (4)$$

reaction 5. The net result is transfer of an oxygen atom from

$$FeO^+ + CO \longrightarrow Fe^+ + CO_2,\ k = 9 \times 10^{-10}\ cm^3 molecule^{-1} sec^{-1} \qquad (5)$$

N_2O to CO, reaction 6. This overall process is exothermic by

$$N_2O + CO \longrightarrow CO_2 + N_2,\ \Delta H^o = -107\ kcal/mol^{-1} \qquad (6)$$

107 kcal mol^{-1} but does not occur directly at room temperature to any measurable extent [22].

When Fe^+ is generated in the presence of N_2O, first order decay of the Fe^+ signal is observed with a corresponding increase of the FeO^+ signal. The observed halftime indicates a rate constant of $k = 0.7 \times 10^{-10}$ $cm^3 molecule^{-1} sec^{-1}$ for reaction 4. Thus, at 4×10^{-6} torr of N_2O 80% of the Fe^+ is converted to FeO^+ after 180 ms. Addition of CO to this system increases the Fe^+ signal and decreases the FeO^+ signal with respect to their previous values with only N_2O present. At 180 ms, for example, the Fe^+ signal is 50% larger than with only N_2O present when 0.7×10^{-6} torr of CO is added. Double resonance spectra, Figure 9, show that the relative increase in Fe^+ is due to a reaction by FeO^+ that occurs only in the presence of CO.

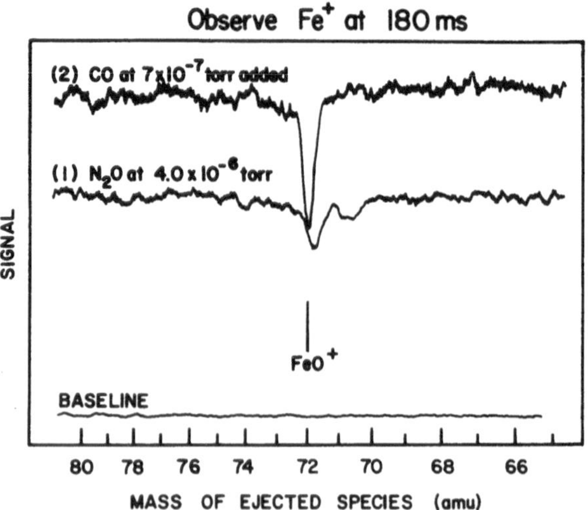

Fig. 9. Shown is the ICR signal obtained for Fe^+ while scanning the double resonance oscillator to eject ions of a given mass from the cell. Trace 1 is obtained with only N_2O present ($P_{N_2O} = 4.0 \times 10^{-6}$ torr). Trace 2 results when CO is added to the system ($P_{N_2O+CO} = 4.7 \times 10^{-6}$ torr). The increase in Fe^+ signal after adding CO, is due to the regeneration of Fe^+ by reaction 5. This is evidenced by the double resonance at m/z 72 (FeO^+) which indicates that FeO^+ is reacting to Fe^+.

Reactions 4 and 5 involve simple transfer of an oxygen atom from N_2O to Fe^+ and from FeO^+ to CO. The homolytic bond dissociation energy for FeO^+, $D(Fe^+-O)$, therefore must lie between $D(N_2-O)$ = 40 kcal mol^{-1} and D(CO-O) = 127 kcal mol^{-1}. These limits are consistent with our observations that Fe^+ will also react with O_3 to give FeO^+, $D(O_2-O)$ = 26 kcal mol^{-1} but that FeO^+ is not produced when Fe^+ is generated in the presence of O_2, D(O-O) = 119 kcal mol^{-1} [19].

FeO^+ will transfer oxygen atoms to other acceptors in addition to CO. Thus, evidence for catalytic cycles analogous to Scheme 2 is also seen in mixtures of N_2O with ethylene, propylene, allene, ethane and propane. In all of these systems other minor ionic products in addition to Fe^+ are formed when FeO^+ reacts.

With acetylene an interesting three-step catalytic cycle, Scheme 3, is observed to occur. FeO^+ reacts in part directly to give Fe^+, reaction 7, as in

Scheme 3

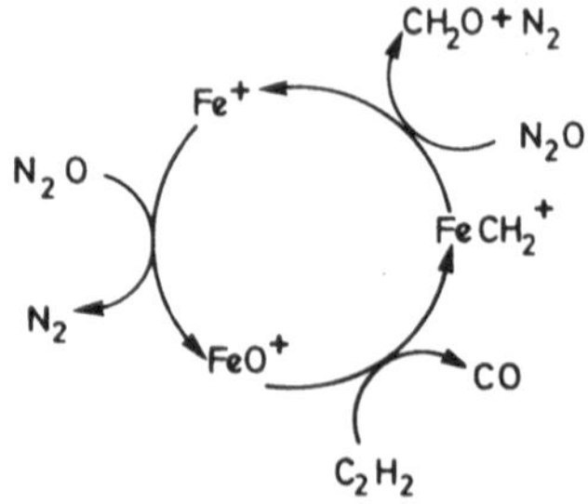

$CH_2O + N_2$
Fe^+
N_2O
N_2O
$FeCH_2^+$
N_2
FeO^+
CO
C_2H_2

Scheme 2, but also produces $FeCH_2^+$ eliminating CO, reaction 8. $FeCH_2^+$ reacts with N_2O to regenerate Fe^+, reaction 9. The net result of this catalytic cycle, reaction 4,8 and 9, is oxidation of acetylene by two N_2O molecules to give CO

$$FeO^+ + C_2H_2 \longrightarrow Fe^+ + C_2H_2O \qquad (7)$$

$$FeO^+ + C_2H_2 \longrightarrow FeCH_2 + CO \qquad (8)$$

$$FeCH_2^+ + N_2O \longrightarrow Fe^+ + N_2 + CH_2O \qquad (9)$$

and formaldehyde, reaction 10. This overall reaction is exothermic by 148 kcal mol^{-1} [22]. The branching ratio for reactions 7 and 8 cannot be determined ac-

$$C_2H_2 + 2N_2O \longrightarrow 2N_2 + CO + CH_2O \qquad (10)$$

curatly but appears to be approximately 1:1.

Transition metal oxide cations can be produced by reaction of N_2O with Ti^+, Zr^+, V^+, Nb^+, and Cr^+ in addition to Fe^+. Catalytic cycles involving M^+ and MO^+ are not observed for these five metals when CO or other simple oxygen acceptors are added. This is because $D(M^+\text{-}O)$ is too large in these cases. In fact, for Ti, Zr and Nb, Co_2 reacts with M^+ to generate MO^+, reaction 11. In other cases neither the forward nor reverse of reaction 11 is observed, indicating that even in the

$$M^+ + CO_2 \longrightarrow MO^+ + CO \qquad (11)$$

exothermic direction the reaction is slow. However, further reaction of MO^+ with N_2O occurs in these five systems to give metal dioxide cations, reaction 12. When simple oxygen acceptors, A, are added, evidence for a catalytic cycle, Scheme 4, involving MO^+ and MO_2^+ is observed in each case.

$$MO^+ + N_2O \longrightarrow MO_2^+ + N_2 \qquad (12)$$

Scheme 4

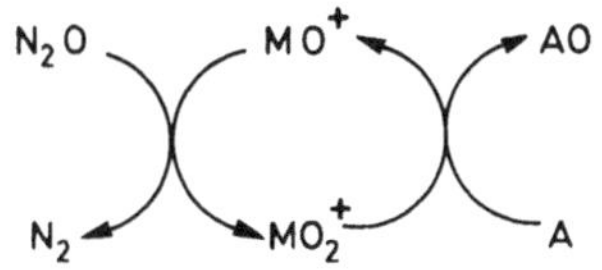

N_2O MO^+ AO

N_2 MO_2^+ A

6. References

1. R.W. Jones, R.H. Staley, J. Am. Chem. Soc., 102 (1980) 3794.

2. J.S. Uppal, R.H. Staley, J. Am. Chem. Soc. 102 (1980) 4144.

3. M.M. Kappes, R.H. Staley, J. Am. Chem. Soc. 103 (1981) 1286.

4. M.M. Kappes, R.H. Staley, J. Phys. Chem. 85 (1981) 942.

5. J.S. Uppal, R.H. Staley, J. Am. Chem. Soc., in press.

6. J.S. Uppal, R.H. Staley, J. Am. Chem. Soc., in press.

7. R.W. Jones, R.H. Staley, J. Am. Chem. Soc., in press.

8. R.W. Jones, R.H. Staley, J. Am. Chem. Soc., in press.

9. M.M. Kappes, R.H. Staley, J. Am. Chem. Soc., submitted for publication.

10. J.F. Wolf, R.H. Staley, I. Koppel, M. Taagepera; R.T. McIver, J.L. Beauchamp; R.W. Taft, J. Am. Chem. Soc. 99 (1977) 5417.

11. D.H. Aue, M.T. Bowers in "Gas Phase Ion Chemistry", Volume 2, M.T. Bowers, Ed., Academic Press: New York, 1979, Chapter 9.

12. R.H. Staley, J.L. Beauchamp, J. Am. Chem. Soc. 97 (1975) 5920.

13. W.D. Reents, Jr., B.S. Freiser, J. Am. Chem. Soc. 103 (1981) 2791.

14. R.R. Corderman, J.L. Beauchamp, J. Am. Chem. Soc. 98 (1976) 3998.

15. M.M. Kappes, R.W. Jones, R.H. Staley, J. Am. Chem. Soc., in press.

16. J.E. Huheey, "Inorganic Chemistry", Second Edition, Harper and Row: New York, 1978, pp. 489-498.

17. F.R. Hartley, Chem. Soc. Rev. 2, (1973) 163.

18. T.G. Appleton, H.C. Clark, L.E. Manzer, Coord. Chem. Rev. 10 (1973) 335.

19. Bond energies from Benson, S.W. "Thermochemical Kinetics", 2nd Edition, Wiley: New York, N.Y., 1976.

20. Assuming a Langevin rate of 5×10^{-10} $cm^3 molec^{-1} sec^{-1}$, quench gas pressures of 2×10^{-5} torr correspond to about 70 collisions within a typical 200 ms experiment.

21. For examples, see (a) D.S. Bomse, R.L. Woodin, J.L. Beauchamp, in "Advances in Laser Chemistry", A.H. Zewail: ed. Springer Series in Chemical Physics, Springer: Berlin, 1978. (b) R.C. Dunbar, H.H. Teng, E.W. Fu, J. Am. Chem. Soc. 101 (1979) 6506. (c) B.K. Janousek, K.J. Reed, J.I. Brauman, J. Am. Chem. Soc. 102 (1980) 3125.

22. The standard enthalpy change for this reaction was calculated from heats of formation given in ref. 20.

TRANSITION METAL IONS IN THE GAS PHASE

Douglas P. Ridge
Department of Chemistry
University of Delaware, Newark, Delaware 19711

1. Introduction

For several years we have been studying the chemistry of atomic transition metal ions with simple organic molecules. This research was stimulated by our interest in examining the consequences of oxidation and reduction of transition metals in their gas phase ion molecule chemistry. This is an area of chemistry, redox chemistry, which has received little attention from those of us interested in gas phase chemical dynamics. An initial discovery in our investigations was that atomic transition metal ions are quite subject to oxidative addition; a concerted process in which XY adds to M so that the metal is inserted into the XY bond to form XMY. As Pearson notes [1], it has only been in the last 15 years that oxidative addition has been recognized as an important elementary reaction. In fact our results appear to be the first direct observation of oxidative addition to transition metals in the gas phase. We have been particularly interested in oxidative addition processes that involve formation of metal carbon bonds. We have obtained evidence that such processes occur in alkyl halides and alcohols [2-4], in halobenzenes [5] and in alkanes [6,7].

The present report will describe recent results turned up in three lines of inquiry related to our investigations of oxidative addition processes. First a reaction of an excited state of an atomic transition metal ion will be described. Second, reactions of two diatomic transition metal ions will be discussed. Third, reactions of atomic transition metal ions with cyclic ketones will be described. The ketone reactions include an example of what is apparently a metal induced electrocyclic process.

2. Electronically Excited Cr^+

For purpose of these studies atomic transition metal ions have been formed by 70 eV electron impact on volatile metal compounds, mostly metal carbonyls. Given the time scale of the ion cyclotron resonance experiment the most reasonable initial assumption seemed to be that the atomic ions were in their ground state by the time they had an opportunity to react. The several milliseconds typically available between ion formation and reaction collision exceeds radiative lifetimes of all but the most long-lived excited states. In several systems we looked for chemical evidence of excited states. Changing the energy of ionizing electron beam produced no detectable change in apparent rate constants in the several systems where it was tried. Forming the same ion from two different compounds was also tried in several systems and the observed chemistry and apparent kinetics were again not discernibly altered. The report of Foster and Beauchamp on ion molecule reactions in $Fe(CO)_5$ includes a plot showing that the decay of Fe^+ with reaction time under trapped ICR conditions is clearly expo-netial [8]. This is important because if Fe^+ is formed in several long lived states it is probable that each state would react with $Fe(CO)_5$ at a different rate. The result would be a decidedly non-exponential temporal decay of Fe^+ in $Fe(CO)_5$. It is also true that the bond strength $D(Co^+\text{-}CH_3)$ deduced from reactions of Co^+ formed by electron impact on $Co(CO)_3NO$ and $Co_2(CO)_8$ (62±6 kcal/mol [2]) agrees well with that deduced from the reactions of Co^+ generated thermionically (61±4 kcal/mol [9]). All of this evidence suggests that our initial presumption is essentially correct, at least in the case of Fe^+ and Co^+. It would be surprising if more than 5 to 10% of the Fe^+ and Co^+ formed by 70 eV electron impact on the metal carbonyls were found to be in an excited state a millisecond after formation: Our examination of the chemistry of Ni^+ with $Ni(CO)_4$ has been less extensive, but there has been no indication there either of excited state involvement.

Of particular interest for a variety of reasons is the chemistry of the atomic ions with alkanes. Since there had been no report of oxidative addition of an alkane carbon-carbon bond to a transition-metal prior to ours, [6,7] it is important to determine whether excited states of the atomic metal ions are involved. As indicated above, they do not appear to be in the case of Ni^+, Fe^+ and Co^+. The reactions of those ions with butanes to from $M(C_2H_4)^+$, $M(C_3H_6)^+$ and $M(C_4H_8)^+$ appear to involve ground state ions (see ref. 7 and 9 for more detailed descriptions of the reactions). Mn^+ is not observed to react with alkanes so the question doesn't arise in that case [7]. Cr^+ does react with the butanes, however, forming products similar to those formed in the group 8 metal ion reactions. The involvement of an excited state of Cr^+ is suggested by several kinds of evidence.

First, the collision induced decompostion spectra of $CrC_4H_{10}^+$ ions suggest that the alkane retains its integrity in the complex [7]. The complex is formed by the reaction

$$CrCO^+ + C_4H_{10} \longrightarrow CrC_4H_{10}^+ + CO \; . \qquad (1)$$

$CrCO^+$ is formed by electron impact on $Cr(CO)_6$. As indicated in Table 1 collision induced decomposition of $CrC_4H_{10}^+$ gives predominantly Cr^+. This contrasts sharply with the CID spectra of $FeC_4H_{10}^+$ where the major fragments, $FeC_4H_8^+$, $FeC_3H_6^+$ and $FeC_2H_4^+$, correspond to the products of thermal energy reaction of Fe^+ with the butanes. Failure of alkane to fall apart in the Cr^+ complex suggests that the $CrC_2H_4^+$ and $CrC_4H_8^+$ products of thermal reaction between Cr^+ and butane are formed by reaction of excited Cr^+.

Table 1. Relative Abundances of Fragments in Collision Induced Decomposition Spectra of $MC_4H_{10}^+$ ions[a]

	$Cr(iC_4H_{10})^+$	$Cr(nC_4H_{10})^+$	$Fe(iC_4H_{10})^+$	$Fe(nC_4H_{10})^+$
M^+	0.84	0.68	0.17	0.13
MH^+				
MCH_3^+	0.04		0.11	0.05
$MC_2H_3^+$		0.07		
$MC_2H_4^+$		0.07		0.22
$MC_2H_5^+$				0.03
$MC_3H_5^+$	0.04	0.05		
$MC_3H_6^+$	0.01		0.28	0.26
$MC_4H_6^+$				0.04
$MC_4H_7^+$	0.04	0.09		
$MC_4H_8^+$	0.03	0.04	0.44	0.22

[a]From spectra in reference 7. Spectra were obtained with a VG Micromass ZAB-2F mass spectrometer through the courtesy of Drs. E. E. McEven and M. Rudat of the duPont Co.

Another indication that excited states are involved is that the reactivity of Cr^+ does not follow the expected periodic trend. The special stability of a $(3d)^5$ electronic configuration might be expected to render the Cr^+ less reactive than Mn^+ which has a $(3d)^5(4s)^1$ configuration. Instead Cr^+ reacts and Mn^+ does not. If an excited state of Cr^+ is responsible for the reactivity, then the behavior of both ions may be understood. The evidence indicates that the alkane reactions proceed by a mechanism involving oxidative addition of a carbon-carbon hydrogen bond to the metal [6,7,9]. For this to be energetically possible the metal must be able to form two reasonably strong σ bonds to carbon or hydrogen as it inserts into the C-C or C-H bond. The ground state of Cr^+ has the special stability of the $(3d)^5$ configuration. Bond formation would break into that stable configuration and may, therefore, be energetically unfavorable. An excited state with a $(3d)^4(4s)^1$ configuration, for example, could more readily form the requisite bonds and react. The Mn^+ may also be rendered unreactive because of the stability of the $(3d)^5$configuration. Mn^+ has a $(3d)^5(4s)$ configuration. The 4s electron is available to form the first bond, but formation of a second bond would involve one of the 3d electrons. If it is supposed that the stability of the $(3d)^5$ configuration renders that unfavorable, the inertness of Mn^+ towards alkanes is explained. Hence the involvement of excited Cr^+ in the observed alkane reactions is suggested by simple periodic arguments.

Participation of an excited state of Cr^+ is the only explanation for the reaction of Cr^+ with methane [7]:

$$Cr^+ + CH_4 \longrightarrow CrCH_2^+ + H_2 \qquad (2)$$

This reaction is surely endothermic by several eV [10]. If the ground state of Cr^+ attacks CH_4 in the manner indicated in reaction 2, then $CrCH_2^+$ or $CrCH_3^+$ might be expected to appear as fragments in the CID spectrum of $CrCH_4^+$. The collision induced decompostion spectrum of $CrCD_4^+$ has essentially only one peak, Cr^+ (see Figure 1), suggesting that an excited state of Cr^+ is responsible for reaction 2.

Formation of an excited state of Cr^+ by electron impact on $Cr(CO)_6$ is also indicated by the kinetics of [11]:

$$Cr^+ + Cr(CO)_6 \longrightarrow \begin{cases} Cr(CO)_5^+ + CrCO \\ Cr_2(CO)_n^+ + (6-n)CO \end{cases} \qquad (3)$$

$$n = 1,2,3,4$$

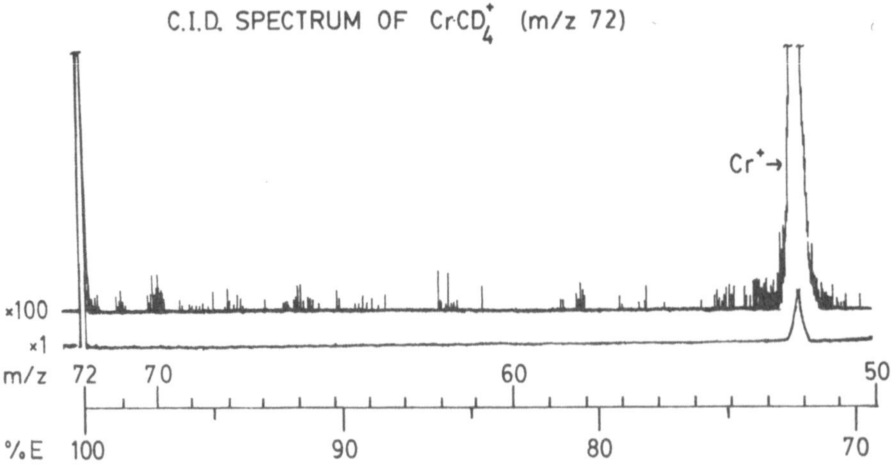

Fig. 1. The collision induced decomposition spectrum of $CrCD_4^+$ as described in the text. The accelerating voltage was 6kV and the collision gas was He. The spectrum was obtained on a VG Micromass ZAB-2F reversed geometry mass spectrometer with the cooperation of Drs. E. E. McEwen and M. Rudat.

This reaction results in the decay of Cr^+ with time illustrated in Figure 2. The decay is non-exponential suggesting that there are two or more states of Cr^+ reacting with different rates. In fact, the data can be fit as the sum of two exponentials. The fit indicates the rapidly reacting state to be 28% of the total Cr^+ population. The ratio of the rate constants is ∿4.8 and the slower rate is $\sim 2 \times 10^{-10}$ cm^3 $molecule^{-1}$ s^{-1}. No other primary ion in this mixture shows a non-exponential decay, which suggests that the effect is not instrumental.

Several kinds of evidence thus suggest that part of the Cr^+ formed by electron impact on $Cr(CO)_6$ is in a long-lived excited state. The excited state probably accounts for the observed reactions of Cr^+ with butanes to form $CrC_4H_8^+$ and $CrC_2H_4^+$. It certainly accounts for the reaction of methane to form $CrCH_2^+$. The $CrCH_2^+$ ought to be an excellent CH_2 donor. Preliminary evidence suggests that $CrCH_2^+$ reacts with CO to form Cr^+ and CH_2CO:

$$CrCH_2^+ + CO \longrightarrow Cr^+ + CH_2CO \qquad (4)$$

Further studies of $CrCH_2^+$ chemistry are underway in our laboratory.

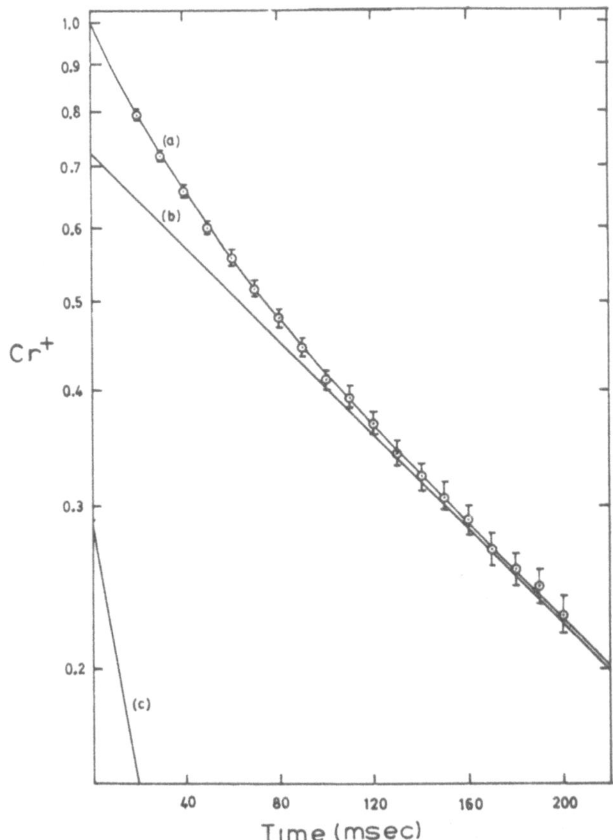

Fig. 2. Variation with time of the abundance of the Cr^+ ion plotted on a logarithmic scale. The Cr^+ is formed by a 1 msec pulse of a 70 eV electron beam on 9.4 x 10^{-7} Torr of $Cr(CO)_6$. The ion decays as a result of reaction 3. The line through the data (a) is the sum of the two exponential decays represented by lines (b) and (c).

3. Reactions of Diatomic Metal Ions

The reactions of Mn_2^+ and Co_2^+ formed by 70 eV electron impact on $Mn_2(CO)_{10}$ and $Co_2(CO)_8$ are summarized in Table 2. Several points are worth mentioning.

First, reactions similar to those of the atomic ions are observed. In particular, the dehydration of alcohols and dehydrohalogenation of alkyl halides both occur.

Second, Mn_2^+ reacts with RX to form $MnRX^+$ and loses an Mn atom. This suggests a weak bond in Mn_2^+. This is to be expected on the basis of simple M.O. theory, since Mn has a very stabel $(3d)^5(4s)^2$ configuration. Knudson cell measurements lead to the conclusion that $D(Mn_2^+) < 1eV$ [12]. Another thing suggested by the $MnRX^+$ product is that RX interacts initially with only one end of the metal dimer ion.

Table 2. Branchung Ratios[a] for Reactions of RX with Ions Formed by Electron Impact[b] on $Mn_2(CO)_{10}$ and $Co_2(CO)_8$

Reactants		Product Ions					
Ion	Neutral	MRX^+	M_2X^+	M_2HX^+	$M_2C_3H_6{}^+$	MX^+	$MC_3H_6{}^+$
Mn^+	iC_3H_7Br					60	40
$Mn_2{}^+$	C_2H_5Br	20	57	23			
$Mn_2{}^+$	nC_3H_7Br	10	63	27			
$Mn_2{}^+$	iC_3H_7Br	10	58	32			
$Mn_2{}^+$	tC_4H_9Br	4	31	65			
$Mn_2{}^+$	ROH[c]	100					
Co^+	iC_3H_7Br						100
$Co_2{}^+$	C_2H_5Br			100			
$Co_2{}^+$	nC_3H_7Br			60	40		
$Co_2{}^+$	iC_3H_7Br			40	60		
$Co_2{}^+$	iC_3H_7OH			50	50		

[a]Given as percent total product ions from indicated reactnat ion. Precision of branching ratios is 5 to 10% of total product from given reactant.

[b] 70 eV electrons.

[c] $R = CH_3$, C_2H_5 and iC_3H_7.

Third, the product of the Mn_2^+ induced dehydrobromination of RBr is always Mn_2HBr^+ and never $Mn_2(olefin)^+$. This is in contrast to the reaction of the atomic ions and transition metal ions (including Mn^+) which tend to form $M(olefin)^+$ in preference to $MHBr^+$. This may be because the Mn_2HBr^+ forms some especially stable structure such as

$$\left[\text{H} \diagdown \text{Mn-Mn} \diagup \text{Br} \right]^+$$

Fourth, the chemistry of Co_2^+ differs from that of Mn_2^+ in several respects. The Co_2^+ bond is never broken. This suggests that the strength in Co_2^+ exceeds that in Mn_2^+. This substantiates Knudsen cell measurements which lead to the conclusion that $D(Co_2^+) > 2.5$ kcal/mol [13] and $D(Mn_2^+) < 1$ eV [12]. It could be that the two dimers have different amounts of internal excitations, but the difference in chemistry reflects the expected difference in bond strength between ground state ions. Co_2^+ also differs from Mn_2^+ in that Co_2^+ reacts with RBr to form both Co_2HBr^+ and $Co_2(\text{olefin})^+$, rather than just Co_2HBr^+. This suggests that Co_2^+ has a higher affinity for the olefin than does Mn_2^+.

A final point of interest with regard to the dimer chemistry is the failure of either dimer to react at all with hydrocarbons. This contrasts with condensed phase results where metal dimers [14] and clusters [15] in a low temperature matrix are observed to react with hydrocarbons. It is worth noting, however, that we do observe the reaction:

$$Co_2CO^+ + i\,C_4D_{10} \rightarrow Co_2C_4D_8^+ + CO + D_2.$$

The neutral product could be D_2CO, of course, and the possibility that the product ion is $Co_2C_3D_6O^+$ has not been eliminated. Further study of this interesting reaction is underway in our laboratory.

The chemistry of the metal dimers can be summarized in terms of a mechanism that begins with formation of

$$\left[M\text{-}M\langle{}^{R}_{X} \right]^+$$

which can lose M or R depending on which is energetically favorable. If R has a low ionization potential, R^+ might be formed, but evidently it does not escape the collision complex without transferring a proton to the M_2X fragment. The resulting olefin might either escape the collision complex so that M_2HX^+ is the product or it might displace HX from the M_2HX^+ fragment so that $M_2(\text{olefin})^+$ is the product. Other mechanisms are possible, but this one is consistent with all the features of the dimer chemistry discussed above.

4. Reactions of Cyclic Ketones

Contrary to what might be expected, the major product of reaction of Fe^+ with cyclohexanone is a metal butadiene complex:

$$Fe^+ + \text{cyclohexanone} \longrightarrow FeC_4H_6^+ + CH_2CO + H_2$$

The indicated neutral products are those that would result from the following mechanism:

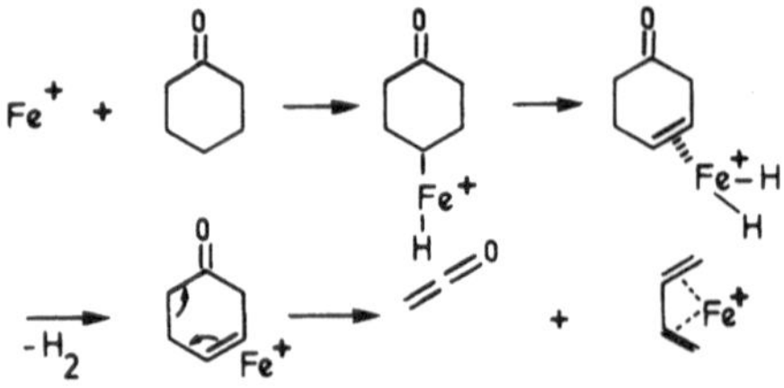

This mechanism is supported by several observations. First, α,α,α', α' d_4-cyclohexanone reacts with Fe^+ to give only $FeC_4H_4D_2^+$. The 2-methyl-cyclohexanone gives both $FeC_4H_6^+$ and $FeC_5H_8^+$. Co^+ undergoes a similar reaction but one of the observed products in that case is $CoCH_2CO^+$ in addition to $CoC_4H_6^+$.

These results thus suggest that the atomic transition metal atomic ions promote electrocyclic processes. This is an observation that could be of some importance for condensed phase chemistry. Ketene is not a particularly good dienophile. These results suggest that a metal ion reagent might be found which would promote the synthetically useful addition of ketene to dienes.

Another indication of metal ion promotion of an electrocyclic process is in the reaction of cyclopentanone with Fe^+:

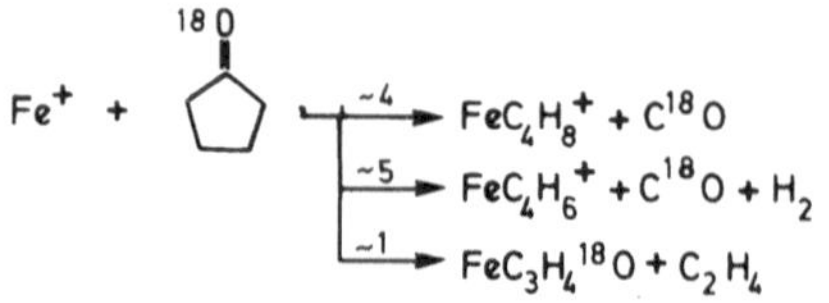

$$Fe^+ + \text{cyclopentanone-}^{18}O \xrightarrow{\sim 4} FeC_4H_8^+ + C^{18}O$$
$$\xrightarrow{\sim 5} FeC_4H_6^+ + C^{18}O + H_2$$
$$\xrightarrow{\sim 1} FeC_3H_4{}^{18}O + C_2H_4$$

The decarbonylation is evidently facile and probably energetically favorable, but it is the ethylene elimination that is of immediate interest. Noyori and coworkers have developed a very useful synthetic scheme to cyclopentanones involving an iron reagent [16]:

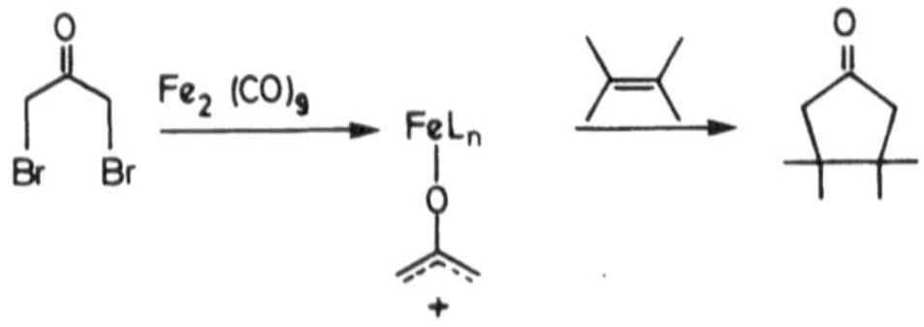

$Fe_2(CO)_9$ ... FeL_n

We believe that what we observe in the gas phase is the reverse of the ethylene addition in the Noyori reaction. The intermediate in the Noyori reaction has not been fully characterized so the gas phase observation is a useful confirmation of the proposed mechanism. The parallel between the Noyori reaction and the gas phase result supports our suggestion that a condensed phase process might be found in which a metal ion promotes addition of ketene to butadiene.

If ketene elimination is the dominant process in cyclohexanone and decarbonylation is the dominant process in cyclopentanone, then we expect to see both in norbornanone. The products observed are:

Fe^+ + [norbornanone] $\xrightarrow{\sim 1}$ $FeC_5H_6^+ + CH_2CO + H_2$

$\xrightarrow{\sim 2}$ $FeC_6H_{10}^+ + CO$

$FeC_6H_8^+ + CO + H_2$

$FeC_6H_6^+ + CO + 2H_2$

Other stoichiometries are possible for some of these products, but these seem the most probable.

The carbonyl functionality plays a crucial role in these reactions. Cyclohexane, for example, reacts with Fe^+ and Co^+ to lose one or more molecules of H_2 in an associative elimination. Even methylene cyclohexane reacts primarily to lose H_2. An interesting methane loss process is observed which probably proceeds by an initial series of allylic rearrangements:

Fe^+ + [methylenecyclohexane] → [FeH^+ allyl complex] → [Fe^+ complex] → [--Fe^+ complex] → [CH_3–Fe^+ complex] → [H–Fe^+–CH_3 complex] $\xrightarrow{-CH_4}$ $FeC_6H_8^+$ $\xrightarrow{-H_2}$ $FeC_6H_6^+$

These reactions of cyclic compounds lacking a carbonyl functionality or ring strain suggest that in such cases the metal attacks bonds external to the ring rather than carbon-carbon bonds in the ring. This is in contrast to the case of open alkanes where carbon-carbon bond attack is preferred. This in turn, suggests that there are geometric constraints on the metal insertion β-H atom shift mechanism proposed for the alkane reaction which cannot be accommodated by carbon-carbon bonds in a ring. The carbonyl functionality evidently facilitates attack of the ring by other mechanisms such as those discussed here.

One final interesting observation in the ketone chemistry is the direct addition of the metal ion to the ketone. There are several mechanisms by which metal ketone complexes might be found in these systems, but double addition occurs in a bimolecular process. A probable mechanism for stabilization of these complexes is radiation from charge transfer band of the metal-ketone collision complex.

5. Acknowledgements

I would like to acknowledge my coworkers, John Wronka, Kevin Kalmbach and Rob Freas, for doing the experiments, Drs. C.E. McEven and M. Rudat for assistance in obtaining the CID spectra and NSF and the the donors of the Petroleum Research Fund administered by the American Chemical Society for financial support.

6. References

1. R.G. Pearson, Symmetry Rules for Chemical Reactions, Wiley, New York 1976, p. 280.

2. J. Allison and D.P. Ridge, J. Organomet. Chem., 99 (1975) C11.

3. J. Allison and D.P. Ridge, J. Amer. Chem. Soc., 98 (1976) 7445.

4. J. Allison and D.P. Ridge, J. Amer. Chem. Soc., 101 (1979) 4998.

5. T.G. Dietz, D.S. Chatellier and D.P. Ridge, J. Amer. Chem. Soc., 100 (1978) 4905.

6. J. Allison, R.B. Freas and D.P. Ridge, J. Amer. Chem. Soc., 101 (1979) 1332.

7. R.B. Freas and D.P. Ridge, J. Amer. Chem. Soc., 102 (1980) 7129.

8. M.S. Foster and J.L. Beauchamp, J. Amer. Chem. Soc., 97 (1975) 4808.

9. P.B. Armentrout and J.L. Beauchamp, J. Amer. Chem. Soc., 103 (1981) 784.

10. The energy required for the reaction $CH_4 \rightarrow CH_2 + H_2$ is 112 kcal/mole (H.M. Rosenstock, K. Draxl, B.W. Steiner and J.T. Herron, J. Phys. Chem. Ref. Data, 6, Supplement No. 1 (1977)), far in excess of $D(Cr^+-CH_2)=$ 65 ± 6 kcal/mole (L.F. Halle, P.B. Armentrout and J.L. Beauchamp, private communication).

11. a) J. Allison, Ph.D. Thesis, University of Delaware, 1978;
b) R.B. Freas, J. Wronka and D.P. Ridge, to be submitted for publication.

12. A. Kant, S.-S. Lin and B. Strauss, J. Chem. Phys., 49 (1968) 1983.

13. A. Kant and B. Strauss, J. Chem. Phys., 41 (1964) 3806.

14. P.H. Barrett, M. Pasternak and R.G. Pearson, J. Amer. Chem. Soc., 101 (1979) 222.

15. Stephen C. Davis and K.J. Klabunde, J. Amer. Chem. Soc., 102 (1980) 1736.

16. R. Noyori, Accts. Chem. Res., 12 (1979) 61.

ELUCIDATION OF THE TRANSFER MECHANISM IN ION-MOLECULE REACTIONS BY ICR

Itzhak Dotan, Ze'ev Karpas* and Fritz S. Klein
Isotope Department. The Weizmann Institute of Science, 76100 Rehovot, Israel

1. Introduction

Ion cyclotron double-resonance (ICDR) techniques have been widely used to determine the pathways of ion-molecule reactions [1] . Methods using isotopically labelled reactants, mainly hydrogen isotopes, have also been previously applied to studies of reaction mechanisms [2] . These methods are not always straight-forward. In this work the mechanisms of some positive and negative ion-molecule reactions were studied employing ICDR techniques in combination with isotope labelling.

Consider a system containing two reactant ions, differing only in isotope label, at a relative abundance of X, and a neutral reactant which is similarly labelled. The observed intensity ratio of double resonance signals of a product ion when both primary ions are irradiated will depend only on the relative isotope abundance of the reactant ions in each case is determined directly by single-resonance measurements and thus can the reaction mechanism be elucidated from the ratio of the double resonance signals observed. This technique is des-described in detail in the experimental part.

The following reactions were studied and their mechanisms determined:

$$CO^+ + CO_2 \longrightarrow CO_2^+ + CO \qquad (1)$$

$$O^- + NO_2 \longrightarrow NO_2^- + O \qquad (2)$$

$$SO^- + SO_2 \longrightarrow SO_2^- + SO \qquad (3)$$

$$O^- + N_2O \longrightarrow NO^- + NO \qquad (4)$$

$$Cl_2^- + Cl_2CO \longrightarrow Cl_3CO^- + Cl \qquad (5)$$

$$H_2O^+ + H_2O \longrightarrow H_3O^+ + OH \qquad (6)$$

---

*Present address: Nuclear Research Center Negev, Beer Sheva, P.O. Box 9001, Israel

2. Experimental

a) Instrument: A commercial ion cyclotron resonance mass spectrometer, Varian V-5900, was used. It was equipped with a three section flat cell. Single resonance spectra were obtained by magnet field modulation. In double resonance experiments the irradiating oscillator was applied to the source region. The electron energy was set at 70 eV for the production of positive ions and at resonance energy (from 1 to 4 eV) for the production of negative reactant ions. Emission currents were less than 0.05 and 0.5 μA in the positive and negative ion spectra, respectively. Gas sample pressures were generally less than 3 x 10^{-5} torr.

Care was always taken to minimize possible ejection of primary ions, detuning effects and space-charge effects [3].

b) Materials: Nitrogen dioxide, nitrous oxide and phosgene, obtained from Matheson Co., and sulphur dioxide from Fluka, were further purified by repeated vacuum distillations. Heavy water, $H_2{}^{18}O$, obtained from the Institute's ^{18}O-enrichment plant, was the source of the oxygen labelled compounds, which were prepared either by isotope exchange or by standard procedures [4].

c) Method: An isotopically labelled reactant compound is introduced into the mass spectrometer. The isotope abundance ratios of the two reactant ion species, X, and the two neutral species (after ionization), X', are measured by single resonance, preferably by magnetic field modulation. Switching then to double resonance mode a single conveniently chosen [5] product ion species is focussed on and the double resonance oscillator scanned over the frequencies of the two selected reactant ions. The two double resonance signals are recorded, see Figure 1.

In order to evaluate the results consider the general reaction scheme:

$$A_nB^+ + A_mB \longrightarrow A_pB^+ + C \qquad (7)$$

and the equivalent isotopic reactions with the labelled neutral reactant $A^*A_{m-1}B$:

$$A_nB^+ + A^*A_{m-1}B \longrightarrow A_pB^+ + C \qquad (8)$$

and with the labelled reactant ion, $A^*A_{n-1}B^+$

$$A^*A_{n-1}B^+ + A_mB \longrightarrow A_pB^+ + C \qquad (9)$$

Here A and A* denote the normal and the isotopic species of atom A, respectively, n, m and p are the number of A in the reactant ion, reactant neutral and the product ion, respectively and C denotes all neutral products.

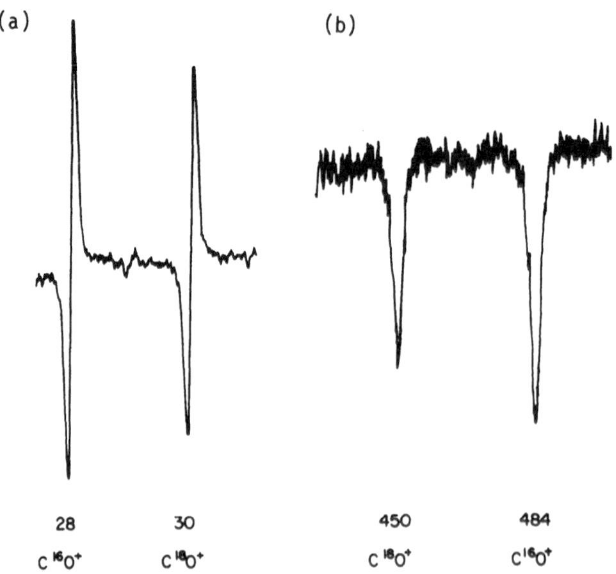

Fig. 1. Experimental results for reaction 1. (a) Single resonance spectrum of the two isotopically labelled primary ions; X = 0.75. (b) Double resonance spectrum obtained when monitoring $C^{16}O^{16}O^{+}$ ions and irradiating $C^{18}O^{+}$ and $C^{16}O^{+}$ ions; $DR(C^{18}O^{+})/DR(C^{16}O^{+}) = 0.77$.

We shall discuss now four main reaction transfer mechanisms, namely:

1. Charge transfer, where only electric charge is transfered from one (generally the charged) reactant to the other in the reactive collision.
2. Ion transfer - a charged atom or group of atoms is transferred from the reactant ion to the neutral reactant.
3. Atom transfer - an atom or group of atoms is transferred from the neutral to the charged reactant, and
4. Complex formation - during the reactive collision a complex is formed with a life time long enough to permit complete isotopic scrambling.

Assuming that the double resonance signal is proportional to the number of reactive collisions occurring in each case, we shall now examine the reaction probabilities of each mechanism for the three general reaction schemes, 7 to 9, listed above.

The reaction between reactant ion and a neutral, where none is labelled, reaction 7, will result in unlabelled A_pB^+ product ion formation in each case of a reactive collision, as shown in Table 1. Consequently, the reaction probabilities for all the mechanisms will be unity, see Table 1, line 1.

Table 1. The relative probability of the reactions leading to the formation of the unlabelled product ion when monitoring the unlabelled reactant ion and the ion labelled with one heavy atom. The mechanisms considered are Charge Transfer (CT), Atom Abstraction (AA), Ion Transfer (IT), And Complex Formation (CF).
(Number f = 1/[(m+n-p) (m+n)] of combinations giving required products divided by (old) number of combinations)

	Reactant ion + neutral	Reactant abundance ion	Reactant abundance neutral	Product ion	Mechanism CT	Mechanism IT	Mechanism AA	Mechanism CF
1.	$A_nB^+ + A_mB$	1	1	A_pB^+	1	1	1	1
2.	$A_nB^+ + A^*A_{m-1}B$	1	X'		0	0	$\frac{m-1}{m}X'$	fX'
3.	$A^*A_{n-1}B^+ + A_mB$	X	1		1	$\frac{n-1}{n}X$	0	fX
4.				$\frac{DR\ (labelled)}{DR\ (unlabelled)} =$	X	$\frac{n-1}{n}X$	0	$\frac{fX}{1+fX'}$

The probability of producing a nonlabelled reactant ion and an isotopically labelled neutral reactant, reaction 8, is summarized in line 2 of Table 1. The probability in case of:

1. Charge transfer is zero, because only labelled neutrals accept the transfered charge and no unlabelled product ions are formed.
2. Ion transfer, similar to charge transfer, is zero.
3. Atom transfer is (m-1)/m, because only m-1 nonlabelled atoms can be transfered to yield a non-labelled product ion.
4. Complex formation is a function f. The probability of f is given by the number of combinations the complex containing m + n - 1 unlabelled A atoms out of a total of m + n A atoms can split up to form unlabelled A_pB^+ product ions. The equation for f is of the form:

$$f = 1/[(m+n-p)\ (m+n)]. \qquad (10)$$

The probability of producing unlabelled product ions in reactive collisions of labelled reactant ions with unlabelled neutral reactants - reaction 9 is shown in line 3 of Table 1. This probability in the case of:

1. Charge transfer is unity, because all the neutral unlabelled reactants will become unlabelled product ions on receiving the transfered charge.
2. Ion transfer is (n-1)/n, since only in (n-1) reactive collisions out of a total of n collisions an unlabelled charged A atom will be transfered to an unlabelled neutral reactant to produce an unlabelled product ion.
3. Atom transfer is zero, since none of the reactant ions which are the precursors of the product ions is unlabelled.
4. Complex formation is a function f, because the complex formed is identical with that of reaction 2 and consequently the probability to give the required product ion is the same.

The general expressions for the ratios of double resonance signals expected for each type of mechanism can now be derived by dividing the respective reaction probability of reaction 9 (Table 1, line 3) multiplied by the isotope abundance X, of the ionic reactant by the combined probabilities of reaction 7 (Table 1, line 1) and that of reaction 8 (Table 1, line 2) multiplied by the isotope abundance of the neutral reagent. The results show that it is easy to distinguish precisely between the four mechanisms by the proposed method.

Taking into account the errors involved in measuring the relative abundance of the reactants and the relative intensity of the DR signals, the estimated accuracy of the results is better than 90%. Therefore, when the results indicate that a reaction proceeds by a given mechanism, minor contributions (up to 10%) from other mechanisms cannot be ruled out.

3. Results and Discussion

Six reaction systems have been investigated in the course of this work.

When carbon dioxide is ionized by electron impact in the ICR cell the major ion observed is CO_2^+. This ion is predominantly a primary ion, but it is also produced by the reaction

$$CO^+ + CO_2 \longrightarrow CO_2^+ + CO \qquad (1)$$

as can be easily verified by standard ICDR techniques. By using isotopically labelled reactants the mechanism of reaction 1 can be elucidated.

The three ion-molecule reactions in which $C^{16}O_2^+$ may be produced are listed in Table 2. The abundance ratio of the ionic reactants is $C^{18}O^+/C^{16}O^+ = X$ and that of the neutral reactants $C^{16}O^{16}O : C^{16}O^{18}O = X' = 2X$, as indicated in square brackets in Table 2. The predicted relative intensity of the DR signals arising from irradiation of $C^{16}O^+$ and $C^{18}O^+$, while monitoring the $C^{16}O_2^+$ signal, is shown at the bottom line of Table 2. Thus, if only a charge transfer mechanism is active in reaction 1 a ratio equal to X, the abundance ratio of the ionic reactants, is expected. A scrambling mechanism would give $X/(2X + 3)$, and an atom abstraction mechanism would give zero. An ion transfer mechanism is obviously not possible in this case since this could not give the required product ion.

In a typical experiment, the primary ion abundance ratio was $X = 0.75$ (Figure 1a) and the relative intensity of the DR signals was $DR(C^{18}O^+)/DR(C^{16}O^+) = 0.77$, as seen in Figure 1b. This is consistent with the charge transfer mechanism and in agreement with the beam experiment results of Klein et al. [6] .

The negative ion-molecule reaction:

$$O^- + NO_2 \longrightarrow NO_2^- + O \qquad (2)$$

is analogous to reaction 1. The primary ion ratio, $^{18}O^-/^{16}O^-$, was equal to the DR signal ratio for the same two ions. This indicates that reaction 2 proceeds by a charge transfer mechanism with some contribution from a scrambling process.

In the reaction system:

$$SO^- + SO_2 \longrightarrow SO_2^- + SO \qquad (3)$$

the ratio of the double resonance signals was found equal to the abundance ratio of the precursor ion $S^{18}O^{16}O^-$, indicating a charge transfer mechanism. To our knowledge no other work on mechanism of reaction 3 has been published.

Table 2. Relative probabilities of reactions leading to formation of $C^{16}O_2^+$ in reaction 1. (For explanation see text).

No.	Reaction	CT	AA	CF
1a.	$C^{16}O^+ + C^{16}O_2 \rightarrow C^{16}O_2^+ + C^{16}O$ [1] [1]	1	1	1
1b.	$C^{18}O^+ + C^{16}O_2 \rightarrow C^{16}O_2^+ + C^{18}O$ [X] [1]	X	-	X/3
1c.	$C^{16}O^+ + C^{16}O^{18}O \rightarrow C^{16}O_2^+ + C^{18}O$ [1] [2X]	-	1/X	2X/3
	$DR(C^{18}O^+)/DR(C^{16}O^+)$	X	0	$\frac{X}{2X+3}$

Table 3. Relative probabilities of reactions leading to formation of $N^{16}O^-$ in reaction 4. (For explanantion see text).

No.	Reaction	AA	IT	CF
4a.	$^{16}O^- + N_2{}^{16}O \rightarrow N^{16}O^- + N^{16}O$ [1] [1]	1	1	1
4b.	$^{18}O^- + N_2{}^{16}O \rightarrow N^{16}O^- + N^{18}O$ [X] [1]	-	X	X/2
4c.	$^{16}O^- + N_2{}^{18}O \rightarrow N^{16}O^- + N^{18}O$ [1] [X]	X	-	X/2
	$DR(^{18}O^-)/DR(^{16}O-)$	0	X	$\frac{X}{X+2}$

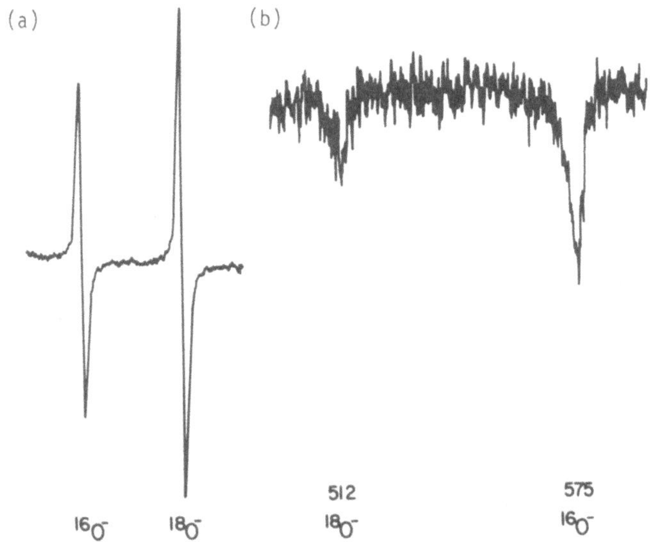

Fig. 2. Experimental results for reaction 4. (a) Single resonance spectrum of the two isotopically labelled primary ions; X = 1.3. (b) Double resonance spectrum obtained when monitoring $N^{16}O^-$ ions and irradiating $^{18}O^-$ and $^{16}O^-$ ions; $DR(^{18}O^-)/DR(^{16}O^-) = 0.44$.

The negative ion molecule reaction (cf. Figure 2)

$$O^- + N_2O \longrightarrow NO^- + NO \tag{4}$$

has been found to proceed by complex formation.

The possible reaction leading to the formation of $N^{16}O^-$ and their relative probabilities are shown in Table 3. Dissociative charge transfer can be ruled out since it is highly endothermic. The relative abundance of the reactant ions was $^{18}O^-/^{16}O^- = X = 1.3$ and the relative intensity of the corresponding DR signals, when $^{16}O^-$ is monitored, was $DR(^{18}O^-)/DR(^{16}O^-) = 0.44$. This is in good agreement with the prediction for a complex-formation mechanism (0.40 in this case). Futrell and Tiernan [8] and Paulson [9] have reported that reaction 4 proceeds by a complex formation mechanism.

In the reaction system:

$$Cl_2^- + Cl_2CO \longrightarrow Cl_3CO^- + Cl \tag{5}$$

both reactants are naturally labelled, the isotope abundance ratio for $^{35}Cl^{37}Cl/^{35}Cl_2$ is X = 0.67 and is the same for Cl_2CO.

The reactions in which the lightest product ion, $^{35}Cl_3CO^-$, is formed are shown in Table 4. Only two mechanisms are possible. Namely, chloride ion transfer and complex formation (scrambling of chlorine atoms).

In the experiment, a DR signal ratio of $DR(^{35}Cl^{37}Cl^-)/DR(^{35}Cl^{35}Cl^-) = 0.37$

Table 4. Relative probabilities for $^{35}Cl_3CO^-$ formation in phosgene by reaction 5. (For explanation see text)

No.	Reaction		IT	CF
5a	$^{35}Cl_2^- + {}^{35}Cl_2CO$ [1] [1]	$^{35}Cl_3CO^- + {}^{35}Cl$	1	1
5b.	$^{35}Cl_2^- + {}^{35}Cl^{37}ClCO$ [1] [X]	$^{35}Cl_3CO^- + {}^{37}Cl$	-	X/4
5c.	$^{35}Cl^{37}Cl^- + {}^{35}Cl_2CO$ [1] [1]	$^{35}Cl_3CO^- + {}^{37}Cl$	X/2	X/4
	$DR(^{37}Cl^{35}Cl^-)/DR(^{35}Cl^{35}Cl^-)$		X/2	X/X+4

Table 5. Relative probabilities of reactions leading to formation of $H_3{}^{18}O^+$ in reaction 6. (For explanation see text)

No.	Reaction		AA	IT	CF
6a.	$H_2{}^{18}O^+ + H_2{}^{16}O$ [X] [1]	$H_3{}^{18}O + {}^{16}OH$	X	-	X/2
6b.	$H_2{}^{18}O^+ + H_2{}^{18}O$ [X] [X]	$H_3{}^{18}O^+ + {}^{18}OH$	X^2	X^2	X^2
6c.	$H_2{}^{16}O^+ + H_2{}^{18}O$ [1] [X]	$H_3{}^{18}O^+ + {}^{16}OH$	-	X	X/2
	$DR(H_2{}^{18}O^+)/DR(H_2{}^{16}O^+)$		∞	X	1+2X

was obtained, indicating that a chloride ion is transferred from Cl_2^- to phosgene. (The expected value for anion transfer mechanism is X/2, which is here equal to 0.33). The data for this reaction have been reported earlier by Karpas and Klein [10].

The protonation reaction:

$$H_2O^+ + H_2O \longrightarrow H_3O^+ + OH \quad (6)$$

would generally be studied with hydrogen isotopes. But in this case we carried out the experiments with a mixture of $H_2{}^{16}O$ and $H_2{}^{18}O$ and observed the formation of $H_3{}^{18}O^+$. We monitored this heavy product ion in order to avoid interference between the $^{18}OH^+$ and the $H_3{}^{16}O^+$ ion signals, as both these ions have the same m/z 19. The reactions leading to the formation of $H_3{}^{18}O^+$ are summarized in Table 5.

The experimental values for the DR intensity ratio $DR(H_2{}^{18}O)/DR(H_2{}^{16}O^+)$ varied between 1.03X -1.3X as a function of the kinetic energy. These results indicate that reaction 6 proceeds via more than one mechanism - Ion (proton) transfer plus contributions from either atom abstraction or complex formation.

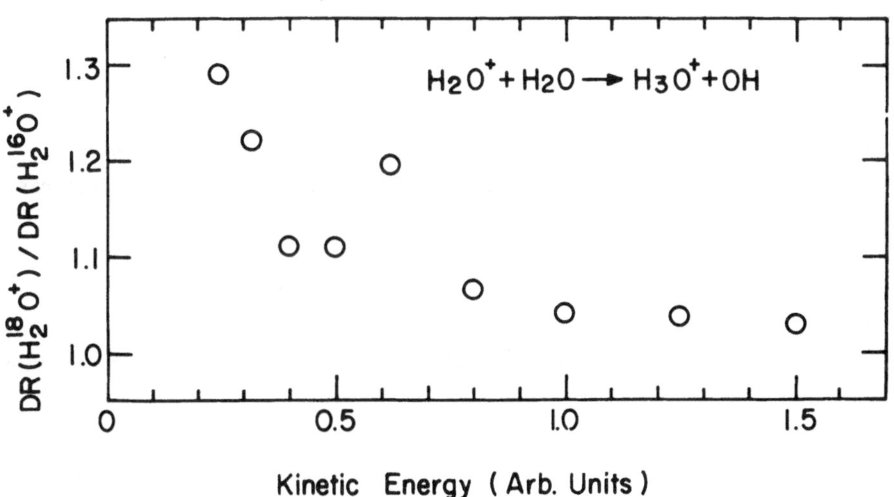

Fig. 3. Kinetic energy dependence of the ratio $DR(H_2{}^{16}O^+)/DR(H_2{}^{18}O^+)$ for reaction 6 . The results are normalized to X - the isotopic abundance ratio of the two ionic reactants.

The ratio $DR(H_2{}^{18}O^+)/DR(H_2{}^{16}O^+)$ as a function of kinetic energy is plotted in Figure 3. (The energy was varied by changing the amplitude of the double resonance oscillator). At high energies the values are very close to unity which means that the mechanism is almost entirely direct proton transfer. At lower energies this ratio increases to about 1.3X at the lowest energies measured, showing contributions of about 30% from one of the other possible mechanisms. A similar ratio for room temperature reaction was suggested by Huntress and Pinizatto [11].

In a separate experiment using a mixture of D_2O and $H_2{}^{18}O$ we found negligible contribution from complex formation mechanism and the second mechanism is probably atom abstraction. Further experiments are being carried out.

The change of mechanism with energy is a very typical one. For many reactions it has been found that at higher energies there is a shift towards a direct mechanism. (See for example [12] and references therein).

4. Conclusion

The mechanism of a number of positive and negative ion-molecule reactions have been determined by using ICDR techniques with isotopically labelled reactants. Using this technique one can determine whether reactions proceed mainly by charge transfer, ion transfer, atom abstraction or complex-formation mechanism. A reaction system which proceeds by more than one mechanism was also identified.

For the reactions reported here, heavy labelled atoms (oxygen and chlorine) were used, so, within the experimental accuracy, any isotope effect can be neglected. In cases where the atom labelled is hydrogen the isotope effect has to be taken into account.

The results obtained in this work are in agreement with the available published data. For two of the reactions (3 and 5) this is the first reported mechanism determination. The method is relatively simple; however, it is limited to systems in which all species (reactants and products) have at least one element in common.

5. Acknowledgement

We acknowledge the contribution of Mr. A. Pizem in carrying out the H_2O^+ experiments. F. Klein gratefully acknowledges the financial help from Hoechst A.G.

6. References

1. T.A. Lehman and M.M. Bursey, Ion Cyclotron Resonance Spectrometry, Wiley, New York, 1976.

2. See for example, J.M.S. Henis and M.K. Tripodi, J. Chem. Phys., 61 (1974) 4863.

3. G.C. Goode, A.J. Ferrer-Correia and K.R. Jennings, Int. J. Mass Spectrom. Ion Phys., 5 (1970) 229.

4. R.H. Herber, Inorganic Isotopic Synthesis, Benjamin, New York, 1962.

5. For increased sensitivity care should be taken to avoid overlap of mass/charge between the chosen product ion and various reactant ions and other fragment ions present.

6. F.S. Klein, G.D. Lempert, E. Murad and A. Persky, Adv. Mass Spectrom., 6 (1973) 749.

7. D. Vogt, Int. J. Mass Spectrom. Ion Phys., 3 (1969) 81.

8. J.H. Futrell and T.O. Tiernan, in J.L. Franklin (Ed.), Ion-Molecule Reactions Plenum Press, 1972.

9. J.F. Paulson, J. Chem. Phys., 52 (1970) 959.

10. Z. Karpas and F.S. Klein, Int. J. Mass Spectrom. Ion Phys., 22 (1976) 189.

11. W.T. Huntress and R.F. Pinizatto, J. Phys. Chem., 59 (1973) 4742.

12. I. Dotan, Chem. Phys. Lett., 75 (1980) 509.

EQUILIBRIUM STUDIES OF ELECTRON TRANSFER REACTIONS

Robert. T. McIver, Jr. and Elaine K. Fukuda
Department of Chemistry, University of California
Irvine, California 92717

1. Introduction

The Ion Cyclotron Resonance (ICR) technique has been utilized extensively in recent years to measure equilibrium constants for ion-molecule reactions. In our laboratory at the University of California at Irvine, most of our effort has been focused on bimolecular proton transfer reactions. Equilibrium constants for positive ion-molecule reactions such as

$$H_3O^+ + B = BH^+ + H_2O \tag{1}$$

provide a quantitative measure of the gas-phase basicity of the base B relative to water. Several hundred reactions of this type have been investigated for a wide variety of alcohols, amines, pyridines, esters and ethers [1,2]. Recently Michael Locke in our laboratory has measured the gas-phase basicity of several amino acids, purines, pyrimidines and nucleosides using a direct insertion probe to admit these low volatility compounds to the ICR analyzer cell [3,4].

Proton-transfer reactions involving negative ions have also been extensively studied. Equilibrium constants for reactions of the type

$$F^- + AH = A^- + HF \tag{2}$$

measure the gas-phase acidity (proton donor ability) of AH relative to hydrogen fluoride. The gas-phase acidity of over two hundred compounds have been measured in our laboratory [5-7]. The most recent work is that of Dr. Mizue Fujio on an extensive series of substituted phenols, carboxylic acids and phenylacetonitriles [8]. An important use of the gas-phase acidity data is to calculate heats of formation of negative ions. In a few cases, the data can also be used to calculate A-H bond dissociation energies and heats of formation of free radicals. We found that this approach to calculating thermochemical data was limited primarily by a paucity of reliable and precise electron affinity (EA) values. For this reason we began two years ago to investigate alternative routes for obtaining electron affinities for polyatomic radicals and molecules. This paper describes our progress in developing a scale of electron affinities.

2. Electron Affinities: Theoretical Considerations

One of the fundamental properties of negative ions is the lowest energy required to remove an electron [9]. This energy is called the electron affinity (EA) and is equal to ΔE for the process

$$M^- = M + e^- \tag{3}$$

where both the ion and neutral are gaseous species in their ground electronic, vibrational and rotational states, and the electron has zero kinetic and potential energy. The quantity EA is positive if a stable negative ion exists.

In principle electron affinities can be calculated by subtracting the total energy of the negative ion from that of the atom or molecule M.

$$EA\,(M) = E_{tot}(M) - E_{tot}(M^-) \tag{4}$$

Consider the simpliest case, a hydrogen atom consisting of one electron and one proton. In the ground state the electron is in a 1s orbital and the total binding energy is -13.598 eV. If another electron is added, hydride H^- is obtained. Since H^- is a three-particle system, the total energy cannot be calculated exactly, and one must resort to an approximation method such as the Hartree-Fock self-consistent-field theory. At first glance, one might expect H^- to have a $1s^2$ ground state electronic configuration similar to that of helium, but this view is incorrect. The calculated radial distribution func-

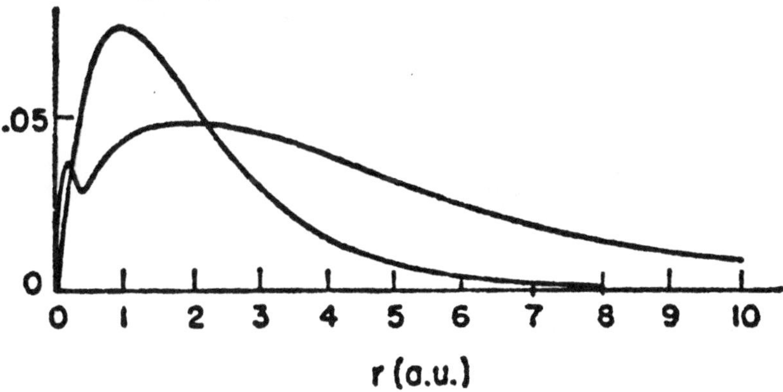

Fig. 1. Radial wave functions for the electrons of H^-. The two electrons do not occupy the same orbital as in helium. Instead, H^- is qualitatively like an electron in the field of a polarized hydrogen atom.

Table 1: Results of Theoretical Calculations of Electron Affinities (eV)[a]

M	$E_{HF}(M)$	$E_{HF}(M^-)$	EA_{HF}	$E_c(M) - E_c(M^-)$	EA_{tot}	EA_{exp}
H	-13.598	-13.270	0.328	1.083	0.754209	0.756 ±0.13
O	-2035.428	-2034.886	0.542	2.005	1.46	1.465 ±0.005
F	-2704.927	-2706.290	1.36	2.054	3.45	3.399 ±0.002
OH	--	--	0.02	1.78	1.76	1.825 ±0.002

[a] All data were taken from a review by H. Hotop and W.C. Lineberger, J. Phys. and Chem. Ref. Data, 4 (1975) 539.

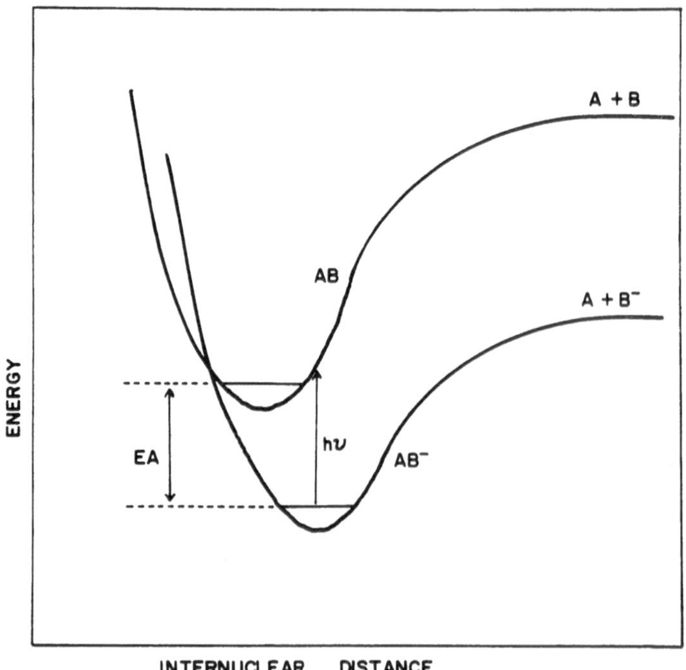

Fig. 2. Potential energy curves for a hypothetical diatomic molecule and its negative ion.

tions for hydride, shown in Figure 1, suggest that hydride can best be thought of as an electron interacting with the field of a polarized hydrogen atom. The best Hartree-Fock energy for H^- is -13.270 eV [10]. Combining this with the total energy of hydrogen atom gives -13.598 - (-13.270) = -0.328 eV for the electron affinity. This calculation predicts that H^- is not stable, which of course is incorrect.

This example illustrates a difficulty that has plagued theoretical calculations of electron affinities. Table 1 shows the results of more extensive theoretical calculations for several atoms and negative ions. The total energy of an atom can be written as

$$E_{tot} = E_{HF} + E_C + E_{SO} + \alpha \quad (5)$$

where E_{HF} corresponds to the (restricted) Hartree-Fock energy, E_C is the correlation energy describing the deviation of the many-electron system from the Hartree-Fock (HF) self consistent field model, E_{SO} is the spin-orbit energy for states with nonzero orbital angular momentum and spin, and α comprises several small correction terms due to relativistic effects [11].
The first two columns of Table 1 show that the Hartree-Fock energy accounts for most of the total energy. However, if just E_{HF} is considered, the calculated electron affinities for H, O and OH, shown in the third column, all come out negative. This is a fairly general observation; for atomic number $Z < 10$, only carbon and fluorine are predicted to form stable negative ions within the Hartree-Fock model. The fourth column of Table 1 shows that the difference in correlation energies plays a decisive role in the formation of the negative ion. When all these effects are carefully included, the best theoretical calculations (EA_{tot}) agree well with the experimental values listed in the last column. In conclusion, the task of calculating electron affinities is essentially that of obtaining an accurate number for the correlation energy difference of the negative ion and atom.

The situation is even more complicated for molecules because the structure of the negative ion may differ from that of the neutral molecule. Figure 2 shows the potential energy curves for a hypothetical diatomic molecule and its negative ion. An important feature of these potential energy curves is that electron repulsion causes the curve for the negative ion to be displaced to the right, the greater internuclear distance. Under these circumstances a Franck-Condon transition (vertical transition) from the ground state of the negative ion to the neutral molecule will give an energy threshold that is greater than the electron affinity. Thus the vertical detachment energy comes out larger than the adiabatic electron affinity.

3. Experimental Methods for Measuring Electron Affinities

Table 2 summarizes some of the most useful and accurate experimental methods for measuring electron affinities.

Table 2: Experimental Methods for Determining Electron Affinities

Method	Typical Reactions
Photodetachement and Photoelectron Spectroscopy	$OH^- + h\upsilon \longrightarrow OH + e^-$
Gas-Phase Acidities	$CH_3O^- + C_2H_2 = C_2H^- + CH_3OH$
Exothermic Charge Transfer	$SH^- + NO_2 \longrightarrow NO_2^- + SH$
Endothermic Charge Transfer	$(Cl^-)^* + NO_2 \longrightarrow NO_2^- + Cl$
Collisional Ionization	$Cs^* + M \longrightarrow M^- + Cs^+$
Equilibrium Electron Transfer	$NO_2^- + M = M^- + NO_2$

Photodetachment and photoelectron spectroscopy are optical methods which study the reaction

$$A^- + h\upsilon \longrightarrow A + e^- \quad (6)$$

In the photodetachment experiment ion abundance is monitored as a function of varying wavelength to determine the threshold for photodetachment [12,13]. In photoelectron spectroscopy a fixed frequency light source intersects a beam of negative ions and the resultant photoelectrons are energy analyzed [14]. A related method, which has proved most useful for halide ions, involves observation of the absorption and emission spectra of alkali halide salts heated in a shock tube [15]. All of these are optical methods which induce vertical transitions from the negative ion to the resulting neutral molecule. Since the intensities of the transitions are governed by the Franck-Condon principle, identification of the 0-0 transition in molecular systems (the one corresponding to the adiabatic electron affinity) may be difficult or even impossible if the structure of the neutral is substantially different from that of the negative ion.

There are several chemical methods shown in Table 2 which can be used to obtain electron affinities. Gas-phase acidities [7] can be used to calculate the enthalpy change ΔH^{o}_{acid} for heterolytic dissociation processes such as

$$AH \;=\; A^{-} \;+\; H^{-} \tag{7}$$

Using a simple thermodynamic cycle ΔH^{o}_{acid} can be written as

$$\Delta H^{o}_{acid} \;=\; DH^{o}(A\text{-}H) \;+\; IP(H) \;-\; EA(A) \tag{8}$$

where $DH^{o}(A\text{-}H)$ is the A-H bond strength, IP(H) is the ionization potential of hydrogen (313.6 kcal/mol), and EA(A) is the adiabatic electron affinity of the radical A. Electron affinities for a number of polyatomic radicals have been calculated in this way from bond strength and gas-phase acidity data.

Another useful chemical method involves observing the occurence of charge transfer reactions of the type

$$A^{-} + B \longrightarrow A + B^{-} \tag{9}$$

If ground state A^{-} ions react to give B^{-}, the reaction is exothermic and one can conclude that the order of electron affinities is EA(B) > EA(A). However, if charge transfer is not observed between A^{-} and B, no conclusions can be drawn since it could be either an endothermic reaction or a slow exothermic reaction with a high activation energy barrier. Exothermic charge transfer reactions have been studied with flowing afterglow [16] systems and ion cyclotron resonance spectrometers [17,18].

A third chemical method involves measuring the translational energy dependence for endothermic charge transfer reactions. One approach utilizes a tandem mass spectrometer to study reaction 9 in the endothermic direction [19,20]. The cross section for appearance of product negative ions is measured as a function of the translational energy of the reactant ions, and in favorable cases, the threshold for endothermic charge transfer can be taken as the difference in the electron affinities of A and B. Another approach utilizes beam techniques to study ion pair production in neutral collisions with alkali atoms [21,22].

$$Cs^* + M \longrightarrow Cs^+ + M^- \qquad (10)$$

In principle the electron affinity of M can be calculated from the ionization potential of Cs and the measured threshold for the reaction. There is a major uncertainty, however, with both of these methods. If the activation barrier happens to be greater than the endothermicity of the reaction, M^- will be formed with excess energy even at threshold and the calculated electron affinity value will be too low.

In this paper we report how equilibrium electron transfer reactions can be used to determine molecular affinities. Equilibrium constants for electron transfer reactions of the general type

$$A^- + B = B^- + A \qquad (11)$$

have been measured for a wide variety of benzoquinones, nitrobenzenes and aromatic ketones using a pulsed ion cyclotron resonance (ICR) spectrometer. From the equilibrium constants the standard free energy changes for electron transfer and the relative electron affinities can be calculated. A scale of relative electron affinities has been constructed using the method of multiple overlaps, and ion cyclotron double resonance experiments were carried out to test that the rates of the forward and reverse reactions are rapid relative to the ICR time scale of about 1 sec. Evidence is presented that these experiments provide accurate, adiabatic electron affinities even for large polyatomic molecules.

4. Pulsed ICR Techniques for Negative Ions

The electron affinity measurements reported in this paper were performed with a pulsed ion cyclotron resonance (IC) mass spectrometer which was constructed in our laboratory at the University of California at Irvine. The pulsed ICR technique utilizes the cyclotron resonance principle for mass analysis of gaseous ions stored in a one-region trapped ICR cell [5,23]. A pulsed mode of operation is used for ion formation, trapping, double resonance irradiation, and mass analysis. At a typical operating pressure of 5×10^{-6} torr, ions are stored for about 1 s by a uniform magnetic field of 13,000 Gauss and a weak electric field (0.5 V/cm). The ions suffer on the order of a hundred ion-neutral collisions while stored in the ICR cell, which is generally sufficient to thermalize them to the temperature of the neutral gas. Many of the experimental techniques utilized for this study are similar to those used for construction of the gas-phase acidity scale [5-7] and the proton affinity scale [1,2]. Only changes or additional procedures will be described here.

Scheme 1 describes the sequence of reactions which occur in a typical electron transfer experiment. An experiment is initiated by a 10 ms pulse of

Scheme 1:

$$A + e^- \rightleftharpoons [A^-]^* \longrightarrow A^- + h\nu \quad ; \quad [A^-]^* \xrightarrow{M} A^- + M^* \tag{12}$$

$$B + e^- \rightleftharpoons [B^-]^* \longrightarrow B^- + h\nu \quad ; \quad [B^-]^* \xrightarrow{M} B^- + M^* \tag{13}$$

$$A^- + B \rightleftharpoons B^- + A \tag{14}$$

$$A^- + M \longrightarrow \text{ion loss} \tag{15}$$

$$B^- + M \longrightarrow \text{ion loss} \tag{16}$$

a low energy electron beam (0.5 eV) through the ICR cell. Inelastically scattered electrons are trapped in the cell and captured by the neutral reactants to form excited negative ion radicals $[A^-]^*$ which can either autodetach an electron or be stabilized by spontaneous emission and collisions with a buffer gas M [18,24]. Negative ion radicals which are difficult to generate in a conventional mass spectrometer ion source can be readily produced in a trapped ICR cell because the autodetached electrons are trapped in the analyzer cell and recaptured to produce negative ions which are ultimately stabilized. After many cycles of the reversible electron transfer reaction 14 the ions become thermalized and an equilibrium constant K can be calculated where

$$K = \frac{[B^-][A]}{[A^-][B]} \tag{17}$$

The relative abundances of A^- and B^- are determined with a capacitance bridge detector [25,26] and the pressures of neutral species are measured with a Bayard-Alpert ionization gauge which has been calibrated with a Baratron type

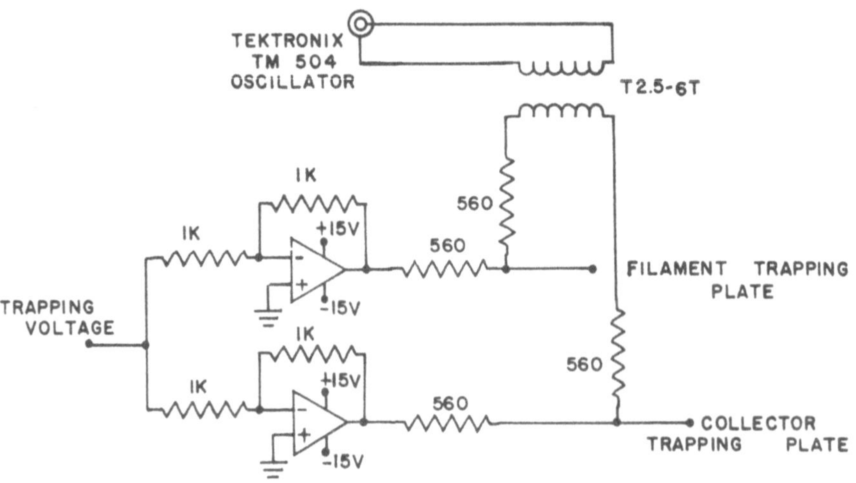

Fig. 3. Circuitry for ejecting electrons from the trapped ion analyzer cell. The high frequency (8 MHz) signal needed to eject the electrons is coupled to the trapping plates of the ICR cell by a transformer (Minicircuit Laboratories model T2.5-6T) and a resistaive divider network. This is necessary because the operational amplifiers (National Semiconductor type LF356N) used to provide DC voltage for the trapping plates will not pass such high frequency signals.

Fig. 4. Photograph of the sample inlet system used for low volatility compounds the variable leak valve and stainless steel sample cups.

145-AHS capacitance manometer. Electron transfer reactions involving benzophenones were noticeably slower than the others, and under these circumstances the equilibrium can be perturbed by ion loss processes, reactions 15 and 16 , which cause the ions to diffuse gradually out of the ICR cell. This problem was minimized by working at constant magnetic field to insure that the rates of ion loss were comparable. Another potential cause for concern is the occurence of alternate reaction channels which perturb the equilibrium. Fortunately, this did not cause difficulties with the systems reported here because electron transfer was the dominant reaction channel observed.

Ion cyclotron double resonance experiments were done on each reported equilibrium reaction to check that electron-transfer is fast compared to the time scale of the experiment, about 1 sec. Double resonance experiments were performed by scanning the magnetic field across the resonance for an ion, first with the double resonance rf turned off and then with it turned on. This procedure was used to avoid spurious double resonance results caused by shifts in the position of the resonance [27]. For an exothermic electron transfer reaction, the product ion signal height was observed to decrease when the reactant ion was ejected from the cell, thus confirming that the reactant ion produces the product. For the reverse experiment, observing the reactant ion signal and ejecting the product ion, low rf levels (0.1 V) sometimes caused a slight increase in the abundance of the reactant ion, while higher rf levels (0.5 V) sufficient for ejection caused a decrease in the height of the reactant ion. During the double resonance experiment electrons were ejected from the ICR cell using the circuitry shown in Figure 3. The frequency of the oscillator was normally set at 7.5 MHz, corresponding to the axial oscillation frequency for electrons in the electrostatic trapping well [28]. The oscillator was turned off during the electron beam pulse so that negative ion radicals could autodetach, reform and stabilized. After 20 ms it was gated on and kept on for the duration of the reaction period. This procedure minimizes space charge resonance shifts and distortions in the power absorption signals.

Most of the compounds investigated in this study are high boiling liquids or solids. It was necessary, therefore, to operate the inlet system and the ICR analyzer at elevated temperatures. Figure 4 is a photograph of the inlet system. It is made with 304 stainless steel tubing, all-metal bellows-sealed valves (Hoke model 4213Q6Y), and variable leak valves (Varian type 951-5106). Samples were placed in individual stainless steel cups which were connected to the inlet system using Cajon 6-VCR flanges. While the experiments were being run, the inlet system and the ICR analyzer were warmed to

80^0-85^0C with resistance heating tapes. Temperatures were measured with copper-constantan thermocouples attached to the ICR cell and to the wall of the analyzer. Normally there was a gradient of only a few degrees if the cell filament current was kept below 2A, and for the purposes of calculating free energy changes the average of the two temperatures was taken. The analyzer system was baked at 150^0C overnight and pumped by an 8 ℓ/s ion pump to achieve base pressures in the low-10^{-8} torr range.

Most compounds were obtained commercially and were checked for purity by their positive ion and negative ion mass spectra in the ICR instrument itself. Any needed purifications were effected by the appropriate distillation, recrystallization or sublimation procedures. Removing water from the benzophenone samples was a problem. Following recrystallization, molecular sieve beads (MCB type 3A) were added to the sample cup as it was put on the inlet system, and the ICR spectrum was monitored until the water signal was no longer detected. Each liquid or gas sample was subjected to several freeze-pump-thaw cycles on the ICR inlet system to remove air. The inlet ports containing solid samples were degassed to the diffusion-pumped foreline to assure the removal of air.

5. Results for Equilibrium Electron Transfer Equilibria

A typical pulsed ICR time plot of ion abundance vs. reaction time is shown in Figure 5 for an electron-transfer reaction between m-chloronitrobenzene and m-fluoronitrobenzene. The chloronitrobenzene sample was first added to the ICR analyzer system until its pressure had stabilized at 1.3×10^{-6} torr. Then m-fluoronitrobenzene was added to a partial pressure of 2.5×10^{-6} torr. The molecular anions M^- of each compound formed rapidly during the 10 ms pulse of the electron beam at an energy of about 0.5 eV. After the electron beam was turned off, m/z 141^- from m-fluoronitrobenzene decreases slightly and m/z 157^- and m/z 159^- from m-chloronitrobenzene increase until the ratio reaches a constant value. Ion cyclotron double resonance confirms that these ions are coupled by the following electron transfer reaction:

$$m\text{-}FC_6H_4NO_2^- + m\text{-}ClC_6H_4NO_2 \rightleftharpoons m\text{-}ClC_6H_4NO_2^- + m\text{-}FC_6H_4NO_2 \quad (18)$$

m/z 141 m/z 157, 159

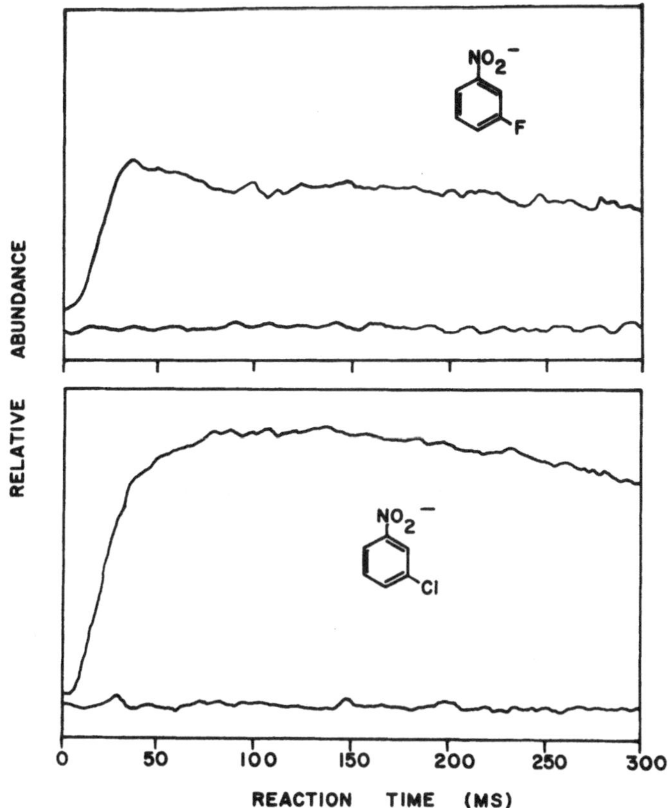

Fig. 5. Pulsed ICR time plots for m/z 141⁻ and m/z 157⁻ in a mixture of m-fluoronitrobenzene and m-chloronitrobenzene. The gaseous negative ion radicals were produced during the first 10 ms of the reaction period by attachment of slow electrons. Evaluation of the equilibrium constant at 300 ms gives ΔG^o = -1.1 kcal/mole for transfer of an electron between the two substituted nitrobenzenes.

Ejection of m/z 141⁻ from the ICR cell causes a decrease in the m/z 157⁻ signal and vice versa. Figure 6 shows a double resonance ejection time plot for m/z 157⁻. The upper trace is the normal time plot of ion abundance vs. reaction time. At a time delay of 260 ms a pulsed double resonance rf signal at the cyclotron frequency of m/z 141⁻ ejects it from the cell and indirectly causes the m/z 157⁻ ion to decay exponentially, just as it should if the two ions are coupled chemically. The rapid decay of the ion abundance curve shows that the reaction is rapid. By analyzing the slope of the decay curve an estimate of 2×10^{-10} cm^3/molecule sec is obtained for the bimolecular rate constant for the reverse reaction. The forward rate is just slightly smaller than the theoretical ADO rate constant of 1.5×10^{-9} cm^3/molecule sec [28].

Another important feature of Figure 5 is that the two signals level out after a reaction time of about 150 ms, which at the pressures used corresponds to about 20 ion-neutral collisions per ion. Evaluation of an equilibrium constant at 300 ms gives K = 4.9 for reaction 19. The free energy change for the reaction can be calculated using $\Delta G^o = - RT \ln K$ if the temperature of the ions and neutral molecules is known. Since the ions are rapidly thermalized by resonant charge transfer and inelastic ion-neutral molecule collisions, we have assumed that the temperature of the ions is 357 K, the same as the temperature of the ICR cell for this particular experiment. Therefore for reaction 18 our determination is ΔG^o = -1.1 kcal/mole.

Figures 7 and 8 show the results of additional experiments of this same type involving various substituted benzoquinones, nitrobenzenes and aromatic ketones. The compounds with the highest electron affinities are at the top of the scale. Each reported value is the average of at least 3 determinations

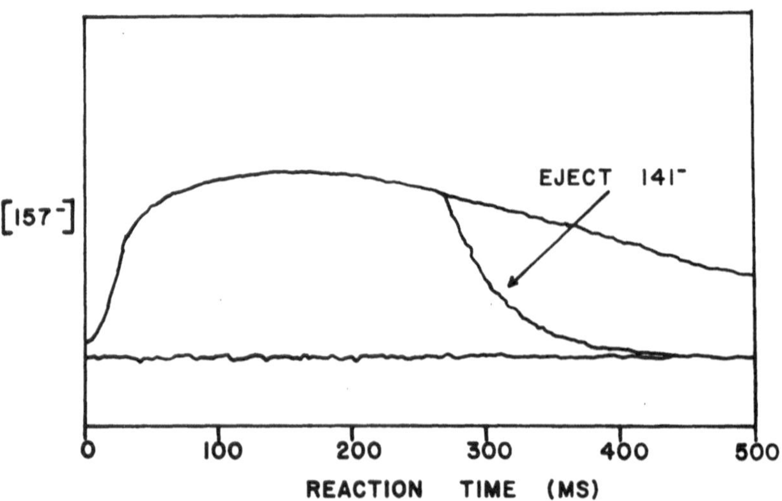

Fig. 6. Ion cyclotron double resonance ejection time plot for m/z 157^- with ejection of m/z 141^- from m-fluoronitrobenzene. The rapid decay of m/z 157^- indicates that electron transfer between the two nitrobenzenes is rapid.

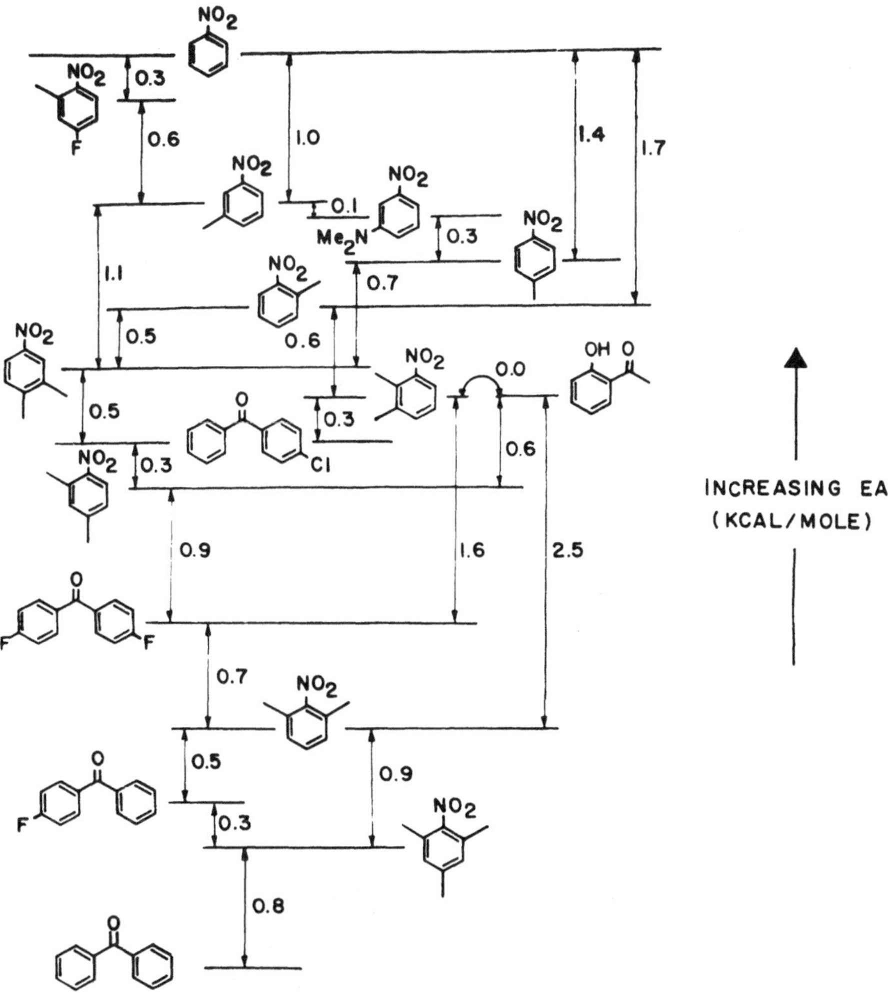

Fig. 7. Scale of ΔG^0 (kcal/mole) values for gas-phase electron transfer equilibria such as reaction 11 . Molecules with high electron affinities are at the top of the scale.

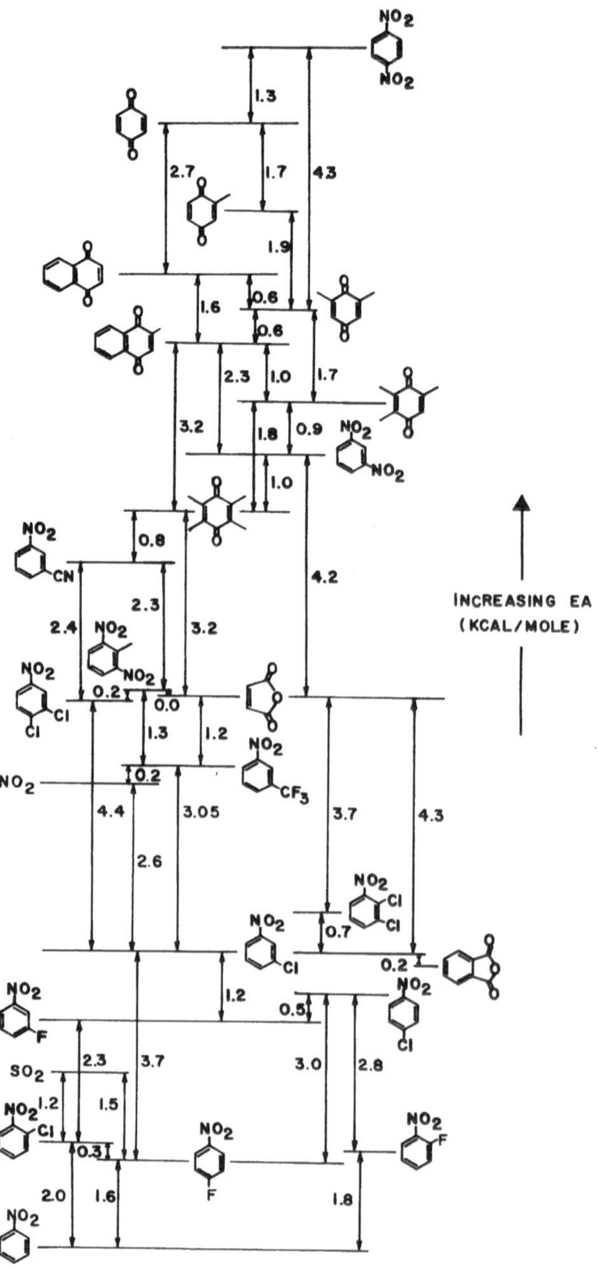

Fig. 8. Continuation of the scale of ΔG^{0} (kcal/mole) values for gas-phase electron transfer equilibria such as reaction 11 . All these compounds have higher electron affinities than those shown in Fig. 7.

of ΔG^o. Multiple overlaps were used to insure internal consistency of the data. For example, ΔG^o from nitrobenzene to p-fluoronitrobenzene was determined in a direct experiment to be -1.6 kcal/mol. This value agrees well with the sum of the separate determinations for nitrobenzene to o-chloronitrobenzene (-2.0 kcal/mole) and o-chloronitrobenzene to p-fluoronitrobenzene (+0.3 kcal/mole). The overall consistency of the scale is ± 0.2 kcal/mole.

6. Discussion

The systems shown in Figures 7 and 8 can be roughly classed into three categories of reactivity. First, rapid electron transfer is observed between compounds that form localized negative ions. Nitrobenzenes with electron-withdrawing substituents fall into this category. The rate constants for electron-transfer are generally greater than 4×10^{-10} cm^3/molecule sec and the cross sections for capture of electrons are also quite large. The second category includes reactions which are noticeably slower and molecules with electron attachment cross sections which are also much smaller. This category includes nitrobenzenes and benzophenones with electron-releasing substituents such as methyl. Reliable equilibrium constants can be measured for these reactions with the pulsed ICR technique, but more care must be taken to avoid artifacts caused by impurities and ion losses. We have not made quantitative measurements of cross sections for attachment or rates of electron transfer, but the effects of double resonance ejection and the ease of generation of the radical anions provide qualitative measured of these trends. The third category is reserved for SF_6, a most unusual molecule indeed. It has an enormous cross section for electron attachment [29], but SF_6^- is very slow to transfer its electron even to molecules which most certainly have a higher electron affinity. We had hoped to use SF_6^- as an efficient electron transfer reagent ion for generating other negative ion radicals, but all such attempts were unsuccessful.

One of the nice features of the equilibrium electron-transfer method is that one can be reasonably confident that it measures adiabatic electron affinities rather than vertical detachment energies. The main rationale for this presumption is that the measured rate constants agree with those expected for orbiting collisions and formation of an ion-molecule complex. Orbiting ion-molecule collisions are relatively slow processes, and there is ample time for the ion to adjust its structure to the lowest configuration. This is in contrast to the spectroscopic electron detachment experiments (Fig. 2) in which only Franck-Condon transitions are allowed from the potential energy surface of the negative ion to that of the neutral.

In another sense, however, this feature of the equilibrium method implies several limitations. First, detailed information on ion and neutral structures is not available since the experiment deals with a thermal distribution of ions. In addition, care must be taken in relating the experimental ΔG^o values to molecular electron affinities. Obviously the two quantities are related, but EA is defined strictly in terms of the energy separation of the ground state ion and neutral, not ΔG^o which is measured experimentally.

A useful starting point for relating our experimental results to molecular electron affinities is to examine the fundamental definition of an equilibrium constant in terms of molecular partition functions. For electron transfer between two molecules (reaction 11) the equilibrium constant is defined as

$$K = \frac{Q_{B^-} Q_A}{Q_B Q_{A^-}} e^{-\Delta E/RT} \tag{19}$$

where Q represents the molecular partition functions for each reactant and product. ΔE, the energy separation between the ground states of the products and the reactants, is the same as EA(A) - EA(B), the difference in their adiabatic electron affinities. Taking the logarithm of both sides of Eq. (19) and multiplying by -RT gives

$$\Delta G^o = -RT \ln K = \Delta E - RT \ln \frac{Q_{B^-} Q_A}{Q_B Q_{A^-}} \tag{20}$$

In the special case where the term involving the partition functions is zero, the experimental quantity -RT ln K is identical to the difference in electron affinities. This will be approximately true in cases where the molecule A and its negative ion A^- are similar in structure. The net contribution from the translational partition functions is neglibible because the mass of a negative ion is almost the same as the mass of its corresponding neutral. Rotational and vibrational contributions to the partition functions are very difficult to evaluate since so little is known about the structures of the negative ion radicals, but small effects are expected for reactants of similar structure such as two substituted halonitrobenzenes. Electronic spin degeneracy is important in reactions where there are odd-electron species on both sides of the equilibrium. For example, a correction of RT ln 4 must be applied to the experimental data for reactions involving odd-electron neutrals such as NO_2.

Additional work is in progress to establish reference points for the relative scale. When this is completed the absolute electron affinities for all these compounds will be available.

7. Acknowledgements

We would like to thank Professor H. W. Moore for providing many of the substituted benzoquinones used in the experiments. This work was supported by a grant from the National Science Foundation (CHE 8024269).

8. References

1. J. F. Wolf, R. H. Staley, I. Kopple, M. Taagepera, R. T. McIver, Jr., J. L. Beauchamp and R. W. Taft, J. Am. Chem. Soc. 99 (1977) 5417.

2. D. H. Aue and M. T. Bowers in "Gas Phase Ion Chemistry", Vol. 2, Chapter 9; M. T. Bowers, Ed., Academic Press, New York, 1979.

3. M. J. Locke, R. L. Hunter and R. T. McIver, Jr., J. Am. Chem. Soc. 101 (1979) 272.

4. M. J. Locke and R. T. McIver, Jr.,J. Am. Chem. Soc., in press

5. J. E. Bartmess, J. A. Scott and R. T. McIver, Jr., J. Am. Chem. Soc. 101 (1979) 6046.

6. J. E. Bartmess, J. A. Scott and R. T. McIver, Jr., J. Am. Chem. Soc. 101 (1979) 6056.

7. J. E. Bartmess and R. T. McIver, Jr., in "Gas Phase Ion Chemistry", Vol. 2, Chapter 11, M. T. Bowers, Ed., Academic Press, New York, 1979.

8. M. Fujio, R. T. McIver, Jr. and R. W. Taft, J. Am. Chem. Soc. 103 (1982) 4017.

9. For recent reviews of electron affinities see: (a) B. K. Janousek and J. I. Brauman in "Gas Phase Ion Chemistry" Vol. 2, Chapter 10, M. T. Bowers, Ed., Academic Press, New York, 1979, pp 53 - 86; (b) R. R. Corderman and W. C. Lineberger, Ann. Rev. Phys. Chem. 30 (1979) 347.

10. A. W. Weiss, Phys. Rev. 122 (1961) 1826.

11. For a review see: H. Hotop and W. C. Lineberger, J. Phys. Chem. Ref. Data, 4 (1975) 539.

12. L.M. Branscomb in "Atomic and Molecular Processes"; D.R. Bates, Ed.; Academic Press: New York, 1962; pp 100-140.

13. K.C. Smyth and J.I. Braunman, J. Chem. Phys., 56 (1972) 1132.

14. W.C. Lineberger in "Chemical and Biochemical Applications of Lasers", Vol.1 C.B. Moore, Ed.; Academic Press: New York, 1974; pp 71-101.

15. R.S. Berry and C.W. Reimann, J. Chem. Phys., 38 (1963) 1540.

16. D.B. Dunkin, F.C. Fehsenfeld and E.E. Ferguson, Chem. Phys. Lett., 15 (1972) 257.

17. J.L. Beauchamp, J. Chem. Phys., 64 (1976) 929.

18. L.J. Rains, H.W. Moore and R.T. McIver, Jr., J. Chem. Phys., 68 (1978) 3309.

19. C. Lifshitz, B.M. Hughes and T.O. Tiernan, Chem. Phys. Lett., 7 (1970) 469.

20. C. Lifshitz, T.O. Tiernan and B.M. Hughes, J. Chem. Phys., 59 (1973) 3182.

21. (a) S.J. Nalley and R.N. Compton, Chem. Phys. Lett., 9 (1971) 529; (b) S.J. Nalley, R.N. Compton, H.C. Schweinler and V.E. Anderson, J. Chem. Phys., 59 (1973) 4125.

22. (a) C.B. Leffert, W.M. Jackson, E.W. Rothe and R.W. Fenstermaker, Rev. Sci. Instrum., 43 (1972) 917; (b) C.B. Leffert, W.M. Jackson and E.W. Rothe, J. Chem. Phys., 58 (1973) 5801.

23. R.T. McIver, Jr., Rev. Sci. Instrum., 49 (1978) 111.

24. M.S. Foster and J.L. Beauchamp, Chem. Phys. Lett., 31 (1975) 482.

25. R.T. McIver, Jr., E.B.Ledford, Jr. and R.L. Hunter, J. Chem. Phys. 72 (1980) 2535.

26. R.T. McIver, Jr., R.L. Hunter, E.B. Ledford, Jr., M.J. Locke and T.J. Francl, Int. J. Mass Spectrom. Ion Phys., 39 (1981) 65.

27. D.J. DeFrees, W.J. Hehre, R.T. McIver, Jr. and D.H. McDaniel, J. Chem. Phys. 83 (1979) 232.

28. J.L. Beauchamp and J.T. Armstrong, Rev. Sci. Instrum., 40 (1969) 123.

29. F.C. Fehsenfeld, J. Chem. Phys., 53 (1970) 2000.

ICR STUDY OF NEGATIVE IONS PRODUCED BY ELECTRON IMPACT IN WATER VAPOR

Masao Inoue
Department of Engineering Physics,
The University of Electro-Communications
1-5-1, Chofugaoka, Chofu-shi, Tokyo, 182 Japan

1. Introduction

Negative Ions produced by electron impact in water vapor have been studied extensively. The primary ions are the H^- and O^- formed by the dissociative attachment processes:

$$H_2O + e \longrightarrow H^- + \text{products} \qquad (1)$$

$$H_2O + e \longrightarrow O^- + \text{products} \qquad (2)$$

The H^- ion is the most abundant primary ion whose cross section peaks at 6.4 and 8.6 eV. The maximum cross section at 6.4 eV is $6.4 \cdot 10^{-18}$ cm^2 [1]. The cross section for the O^- ion has three peaks at 6.4, 8.6, and 11.6 eV. The maximum cross section at 11.6 eV is $5.7 \cdot 10^{-19}$ cm^2 [1] which is smaller by a factor of ten than that of the H^- ion. The OH^- ion is hardly formed directly by dissociative attachment of an electron to a water molecule [2] or if even formed, the cross section of OH^- is negligibly small [1] compared with those of H^- and O^-. If the vapor pressure of water is high enough to allow ion-molecule reactions, both H^- and O^- ions react rapidly with water molecule to form OH^- ions [3]

$$H^- + H_2O \longrightarrow OH^- + H_2 \qquad (3)$$

$$O^- + H_2O \longrightarrow OH^- + OH \qquad (4)$$

Because in these reactions the H^- and O^- ions are produced by the dissociative attachment processes involving low energy electrons, most previous studies on the OH^- ion have been carried out using electron energies below 11 eV.

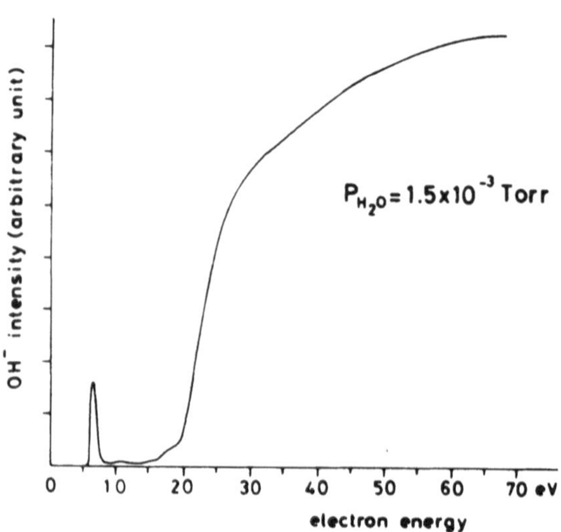

Fig. 1. OH^- ion intensity as a function of electron energy.

However, when pressure of water vapor in the ICR cell is greater than 10^{-4} Torr, OH^- ions are observed not only with low energy electrons but also with high energy electrons. Figure 1 shows the variation of OH^- ion intensity as a function of electron energy at $1.5 \cdot 10^{-3}$ Torr. The peak at electron energies from 5 to 10 eV is due to the OH^- ion produced by H^- and

O^- ions, but an appreciable amount of OH^- ions is also observed above 20 eV. According the previous studies, the formation of OH^- in the high energy region can be attributed to the inelastically scattered electrons which attach to water molecules to produce H^- and O^-. Since in the high energy region the positive ionization takes place and the cross section for H_2O^+ production is two orders larger than that for H^- production, we would like to propose another sequence for OH^- production starting with the ionization of water molecules:

$$H_2O + e \longrightarrow H_2O^+ + e + e' \tag{5}$$

$$H_2O^+ + H_2O \longrightarrow H_3O^+ + OH \tag{6}$$

$$OH + e \longrightarrow OH^{-*} \tag{7}$$

$$OH^{-*} + H_2O \longrightarrow OH^- + H_2O \tag{8}$$

Reaction 5 is the ionization of water molecule to produce H_2O^+ ion and secondary electrons. At high pressures H_2O^+ undergoes the fast proton transfer reaction 6 with a water molecule to an H_3O^+ ion and an OH radical. It is well known that secondary electrons produced by inelastic collisions are trapped and transmitted along the ICR cell [4-7], therefore we assume that these secondary electrons are captured by the OH radicals to form OH^- ions 7 . Because the OH radical has a positive electron affinity of 1.8 eV the OH^- ion produced by attachment of an electron to OH radical must have an excess energy and should be stabilized by collision with a third molecule (reaction 8).

Ion cyclotron resonance spectrometry has the unique capability of identifying the precursor ions of the corresponding product ions in complex ion-molecule reaction sequences by means of the double resonance technique [8]. In the proposed mechanism, OH^- is coupled to H_2O^+ by the ion-molecule reactions 6 and 7 . Since ion-molecule reaction rate constants are usually dependent on the relative ion-molecule velocities, acceleration of H_2O^+ by an irradiating rf field should give rise to a substantial change in the concentration of OH radicals and consequently in the number density of OH^-. However, this double resonance experiment is difficult because of the different charge polarities of precursor and product ions.

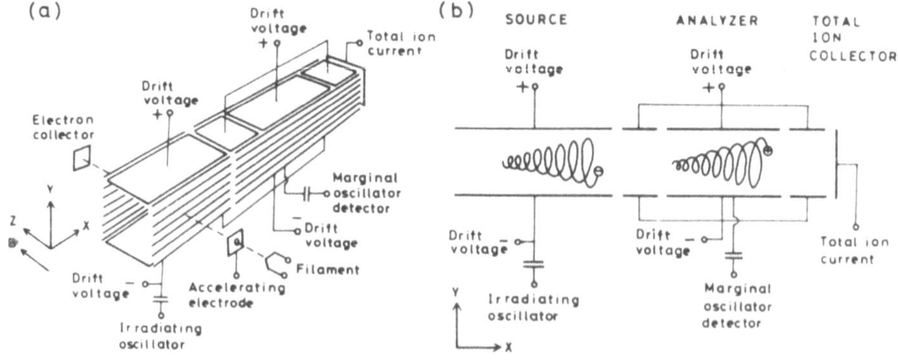

Fig. 2. (a) Schematic drawing of the improved cell. Cartesian co-ordinates are taken as shown in the figure. The electron beam is collinear with the magnetic field.
(b) Trajectories of the resonant positive and negative ions in the X Y plane.

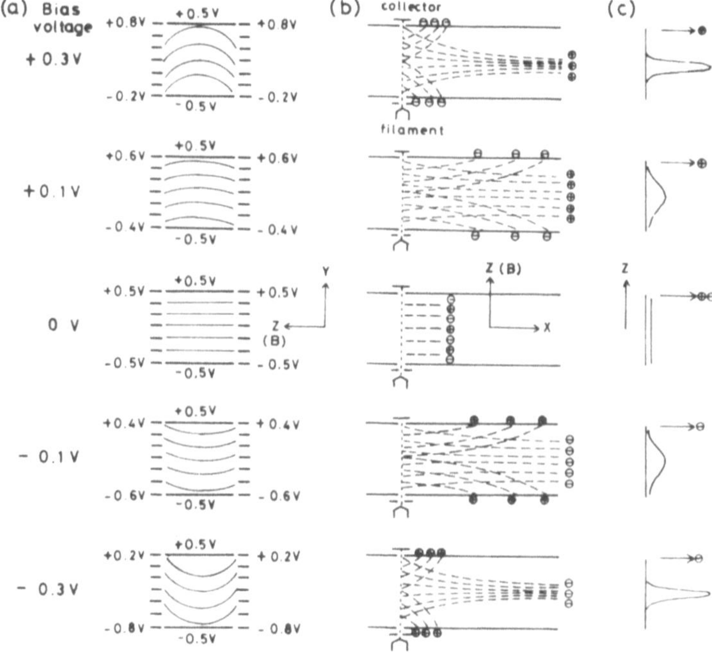

Fig. 3. (a) Equipotentials in the cross section of the cell (Y Z plane) with different bias voltages.
(b) Trajectories of the positive ions and the scattered electrons in the Z X plane.
(c) Distribution of the charged species in the Z-direction.

2. Experimental

In order to carry out the double resonance experiment we applied the trapping voltage modulation technique [9] developed for the conventional drift cell to our improved cell which is equipped with Venetian blind type trapping electrodes [10]. Figure 2(a) is a schematic drawing of the improved cell. It is essentially a three section drift cell composed of source, analyzer and total ion collector. The analyzer drift electrodes have two pairs of auxiliary electrodes on both sides to prevent penetration of electric fields from source and total ion collector and to make the electric field in the analyzer region as uniform as possible. X, Y, and Z axis are taken as shown in the figure. The Z axis is parallel to the direction of the magnetic field. An irradiating oscillator and a marginal oscillator are connected to the lower plates of the source and analyzer drift electrodes, respectively. Figure 2(b) shows the trajectories of resonant positive and negative ions in the X Y plane. The direction of rotation is opposite for positive and negative ions, the direction of drift motion of both ions is same for a given combination of magnetic and static electric fields.

The trapping electrodes of source and analyzer regions are separated. Each trapping electrode is divided into seven small plates instead of the one flat plate of the conventional cell. These small plates are connected in series by glass-sealed resistors so that the potential of each plate changes stepwise from top to bottom plate. The same potential difference between the top and bottom plates and the upper and lower drift electrodes is adjusted and in addition, a fixed bias voltage is added to all small plates so as to produce a potential well and to trap ions inside the cell. Because the sign of the bias voltage determines the polarity of ions trapped in the cell, the bias voltage was modulated to trap positive and negative ions alternatively.

Figure 3(a), (b) and (c) show the variation with bias voltage of the equipotential lines in the Y Z plane of the trajectories of positive and negative ions and electrons in the X Z plane, and their distribution in the Z direction. As seen in the top figure, where the bias voltage is positive and fairly high, positive ions are drifted in the X direction and at the same time moved toward the center of the cell under the influence of the trapping field. On the other hand, negative ions and scattered electrons are accelerated to the outside and quickly leave the cell. At high pressures the harmonic oscillatory motion of ions in the Z direction caused by the trapping potential is rapidly damped by collisions with neutral molecules. When the bias voltage is reduced, positive ions are spread in the Z direction and negative ions and electrons stay longer in the cell and leave more slowly. With no bias voltage all charged species move freely in the Z direction while drifting down the cell in the X direction. When a negative bias voltage is applied, positive ions start to leave the cell and negative ions and electrons are confined in the cell. So one would expect that at a suitable negative bias voltage positive ions stay for sufficiently long in the source region to absorb energy from the irradiating rf field, while negative ions are trapped and drifted from the source to the analyzer region where they are detected by the marginal oscillator detector. Therefore we modulated the trapping bias voltage and monitored OH^- ion in the analyzer region by means of the marginal oscillator detector and a phase sensitive detector; at the same time, we swept the frequency of the irradiating oscillator connected to the lower drift electrode in the source region in order to see a change in the OH^- signal intensity due to the acceleration of H_2O^+ ions.

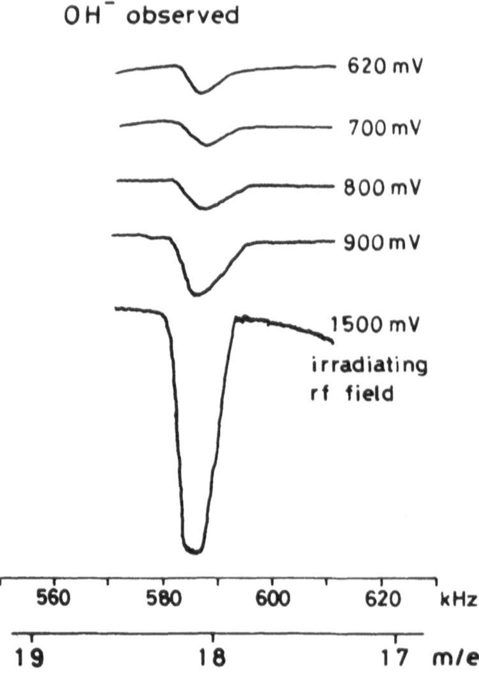

Fig.4.
Double resonance spectra of OH^-. Abscissas are the frequency of the irradiating oscillator and the equivalent mass number. Different peaks were recorded by changing the amplitude of the irradiating rf field.

3. Results

Figure 4 shows the double resonance spectrum of OH^-. The water vapor pressure was $3.8 \cdot 10^{-4}$ Torr. The ionizing electron energy and the electron current were 50 eV and 0.3 μA, respectively. The frequency of the marginal oscillator detector was 624 kHz. The abscissas indicate the frequency of irradiating oscillator and the equivalent mass number. Signals are seen at frequency corresponding to m/e 18. The different peaks were obtained by changing the amplitude of the irradiating rf field.

To prove the contribution of positive ionization to the OH^- production, we did the same experiment with a mixture of argon and water vapor. In this case following reaction sequence is assumed:

$$Ar + e \longrightarrow Ar^+ + e + e' \qquad (9)$$

$$Ar^+ + H_2O \longrightarrow H_2O^+ + Ar \qquad (10)$$

$$Ar^+ + H_2O \longrightarrow ArH^+ + OH \qquad (11)$$

$$H_2O + e \longrightarrow H_2O^+ + e + e' \qquad (5)$$

$$H_2O^+ + H_2O \longrightarrow H_3O^+ + OH \qquad (6)$$

$$OH + e \longrightarrow OH^{-*} \qquad (7)$$

$$OH^{-*} + H_2O \text{ or } Ar \longrightarrow OH^- + H_2O \text{ or } Ar \qquad (12)$$

When the mixture is bombarded with 50 eV electrons, both water and argon molecules are ionized. The Ar^+ ions transfer positive charge to water molecules to produce H_2O^+ or undergo the hydrogen transfer reaction 11 with water molecules to produce ArH^+ ions and OH radicals. Argon molecules also be take a part as a third body in the stabilizing collisions with the excited OH^- ions. The double resonance spectrum of OH^- obtained with the mixture is shown in Figure 5. The pressures of argon and water vapor were $1.2 \cdot 10^{-4}$ Torr per compound, the electron current was 2.5 μA. The frequency of the marginal oscillator detector was 624 kHz. The different signals were obtained again by changing the amplitude of the irradiating rf field.

At the highest pressure obtained without loosing the ICR signals a solvated OH^- ion was observed in the negative ion mass spectrum of water vapor. Two spectra obtained at electron energies of 70 and 6.4 eV are shown in Figure 6. The pressure of water vapor was $1.5 \cdot 10^{-3}$ Torr. Although the resolution of mass peaks is very poor due to collision-broadening, a peak in

OH$^-$ observed

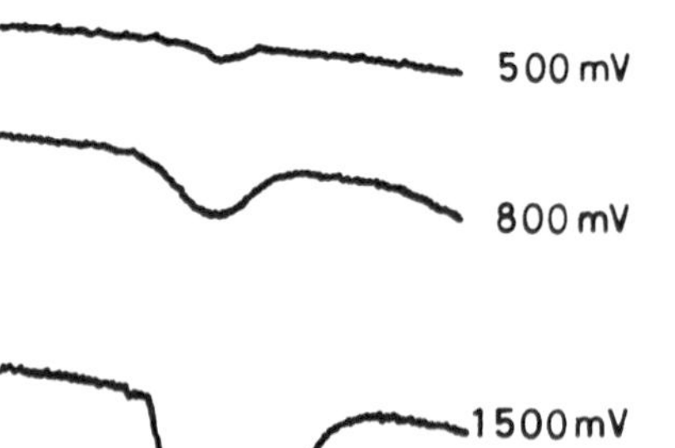

Fig. 5.
Double resonance spectra of OH$^-$ obtained with the mixture of argon and water vapor. Different peaks were recorded by changing the amplitude of the irradiating rf field.

electron energy 70 eV

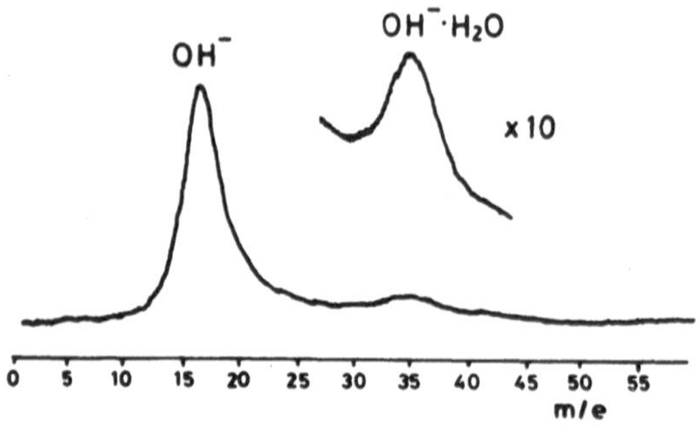

Fig. 6.
Negative ICR mass spectra of water vapor obtained with 6.4 and 70 eV electrons at a pressure of $1.5 \cdot 10^{-3}$ torr.

electron energy 6.4 eV

OH$^-$

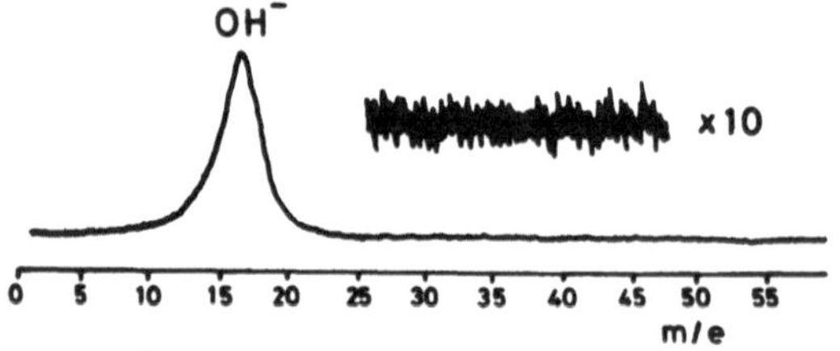

addition to the OH^- peak a peak at m/e 35 is detected which can be assigned to $OH^- \cdot H_2O$. In contrast to the spectrum at 70 eV, no ion is observed at m/e 35 at 6.4 eV, because at this energy the OH^- ions are formed by the ion-molecule reaction 3 involving H^- ions. It seems that the excess energy of OH^{-*} which results from the attachment of an electron to an OH radical promotes the solvation of OH^-. Solvated negative ions were also observed in deuterated water vapor bombarded with a 70 eV electron beam. Two peaks at m/e 38 and 58 corresponding to $OD^- \cdot D_2O$ and $OD^- \cdot (D_2O)_2$ ions were found in a negative ion spectrum of D_2O at $1.5 \cdot 10^{-3}$ Torr.

4. Discussion

Until now most studies of ion-molecule reactions in the gas-phase were concerned only with either positive or negative reaction sequences. But in real gases bombarded by electrons, different species such as positive and negative ions, secondary electrons, radicals and excited molecules are produced and interact with each other to give final products. The present study has demonstrated that the positive ionization is responsible for negative ion formation. Therefore it provides an example of an relationship between positive and negative ions.

The reaction sequence in which negative ions are formed by attachment of electrons to radicals produced in positive ion-molecule reaction has been also studied for the CN^- formation in HCN [11] and the sequence can be generalized as follows:

$$HR + e \longrightarrow HR^+ + e + e' \tag{13}$$

$$HR^+ + HR \longrightarrow H_2R^+ + R \tag{14}$$

$$R + e \longrightarrow R^{-*} \tag{15}$$

$$R^{-*} + HR \longrightarrow R^+ + HR \tag{16}$$

$$\phantom{R^{-*} + HR} \longrightarrow R^- \cdot HR$$

HR and R stand for a neutral molecule and a radical respectively. If R has a positive electron affinity, it can be expected that a negative ion R^- would be formed by attachment of an electron.

The so-called solvated or localized electrons produced by radiolysis in polar and nonpolar liquids and glasses have been of major interest to radiation chemists and a very large number of theoretical and experimental studies have been carried out on the subject since the early 1960s. Theoretical models for the solvation mechanism have been proposed, however, all the models have difficulties in interpreting, in a comprehensive manner, the different behaviors observed with the solvated electrons [12]. The present study on the negative ion formation seems to resolve the difficulties the previous models have encountered.

Gaathon, Czapski and Jortner [13] have studied the optical spectrum of solvated electrons produced by pulse radiolysis in water vapor in the density range 0.1 - 0.2 g cm^{-3}. According to their results, there is a striking similarity between the absorption spectra of the localized electron in the gas phase and in the liquid phase and, even in a wide range of densities, the spectrum is only moderately affected by density changes. On the other hand, the ICR experiment has shown that at pressures above 10^{-3} Torr the OH^- and solvated ions are formed by the attachment of electrons to OH radicals. These ions are the most abundant negatively charged ions in water vapor bombarded by electrons. If things are the same in water vapor in the density range of the experiments by Gaathon, Czapski and Jortner, and if we regard the attachment of secondary electrons to OH radicals as localization of the electrons, some physical properties of the solvated electrons seem to be conveniently explained.

In regard to the formation time of the solvated electrons in water at room temperature the absorption spectrum of the electrons is recorded instantaneously after the radiation pulse, and using a picosecond pulse radiolysis technique the solvation time is estimated as less than 0.2 psec [14]. In view of the relaxation time of water dipole (10 psec), this result is unexpected, if the electrons were trapped and solvated by the dipole orientation of water molecules. Therefore, most theoretical models presume that some molecular configurations or structural defects suitable for electron localization exist prior to the arrival of the electrons, however the detailed physical nature of these pre-existing traps is still unclear. The rate constant of the ion-molecule reaction 6 yielding the H_3O^+ ion and OH radical has been measured to be about $2 \cdot 10^9$ cm^3 $molecule^{-1}$ sec^{-1}, which is very large. In liquid water the reaction is considered to occur in 0.01 psec [15]. The rate constants of the subsequent reactions for the formation of OH^- have not been determined, but these reactions could also be rapid in water and the proposed mechanism may be consistent with the finding of the extremely fast formation of solvated electrons in water.

The localized electrons produced in irradiated liquids and glasses are characterized by their optical absorption spectra covering the spectral region from near-infrared to near-ultraviolet wavelengths. The major feature of the spectra is their broadness and asymmetry toward the high energy side. This shape of absorption spectra is not interpreted in terms of bound-bound transitions of the localized electrons. Many authors assumed that the relevant transitions are bound-continuum transitions of the photodetachment type [12]. If we regard the localized electrons as negative ions produced by the present mechanism, the characteristic shape of the absorption spectra could be interpreted in terms of photodetachment of electrons from the negative ions. As shown in Table 1 the agreement between photon energies at absorption maximum of the localized electrons in various liquids and photodetachment energies of the corresponding negative ions seems to support the proposed process for the localization of electrons.

Table 1 Comparison of the absorption maxima of the solvated electrons in various liquids and the photodetachment energies for the negative ion R^- (or the electron affinities of the radical R).

Molecule HR	H_2O	NH_3	CH_3OH	C_2H_5OH	C_3H_7OH
Absorption maximum of the solvated electron (eV)*	1.72	0.80	1.97	1.77	1.68
Photodetachment energy for the negative ion R^- (or electron affinity of the radical R) (eV)**	1.83 (OH^-)	0.78 (NH_2^-)	1.59 (CH_3O^-)	1.73 ($C_2H_5O^-$)	1.79 ($C_3H_7O^-$)

* data are from A. Ekstrom, Rad. Res. Rev. 2 (1970) 381.

** data are from B.K. Janousek and J.I. Brauman (Ch. 10) and J.E. Bartmess and R.T. McIver, Jr. (Ch. 11) in 'Gas Phase Ion Chemistry', Vol. 2, Ed. M. T. Bowers, Academic press, New York, 1979.

5. References

1. C.E. Melton, J. Chem. Phys. 57 (1972) 4218
2. C.E. Klots and R.N. Compton, J. Chem. Phys. 69 (1978) 1644
3. (a) E.E. Muschlitz, Jr. and T.L. Bailey, J. Chem. Phys. 60 (1956) 681

   (b) M. Cottin, J. Chim. Phys. 56 (1959) 1024
4. D.P. Ridge and J.L. Beauchamp, J. Chem. Phys. 51 (1969) 470
5. R.M. O'Malley and K.R. Jennings, Int. J. Mass Spectrom. Ion Phys. 2 (1969) Appendix 1-3
6. P. Kriemler and S.E. Butrill, J. Amer. Chem. Soc. 92 (1970) 1123
7. T. McAllister, J. Chem. Soc. Chem. Comm. 245 (1972); Chem. Phys. Lett. 13 (1972) 602; J. Chem. Phys. 57 (1972) 3353
8. J.D. Baldeschwieler, Science, 159 (1968) 263
9. T.B. McMahon and J.L. Beauchamp, Rev. Sci. Instrum. 42 (1971) 1632
10. J. Urakawa, H. Shibata and M. Inoue, 'Ion Cyclotron Resonance Spectrometry', Eds. H. Hartmann and K.-P. Wanczek, Springer Verlag, Heidelberg, 1978, p. 33.
11. M. Inoue, J. Chim. Phys. 63 (1966) 1061
12. A.M. Brodsky and A.V. Tsarevsky, Adv. Chem. Phys. 44 (1980) 483 and references therein.
13. A. Gaathon, Czapski and J. Jortner, J. Chem. Phys. 58 (1973) 2648
14. L. Gilles, J.E. Aldrich and J.W. Hunt, Nature Phys. Sci. 243 (1973) 70
15. J.L. Magee, Discussions Faraday Soc. 36 (1963) 232

REACTIONS WITH ALKOXIDE ANIONS

Geneviève Boand, Raymond Houriet and Tino Gäumann
Department of physical chemistry, Federal Institute of Technology
CH-1015 Lausanne

1. Introduction

The development of negative chemical ionization (NCI) as a technique in the field of analytical mass spectrometry has stimulated interest in detailed studies of processes involving negative ions such as $O^{-\cdot}$, OH^-, alkoxide and halide anions [1]. In a recent publication, we have demonstrated the potential use of the reactant ion $O^{-\cdot}$ to probe the structure of aliphatic alcohols [2] ; it was found that the product distribution in these systems allows for analysis of structural isomers. On the other hand, reaction of OH^- with alcohols resulted in formation of the alkoxide anion, thus allowing for the determination of molecular weights; only minor formation of the corresponding enolate anion was observed. It could be that the latter product ion might be formed via pyrolytic dehydrogenation of the alcohol molecule to the corresponding aldehyde or ketone that would in turn transfer a proton to a strong base such as OH^- to form the enolate anion. Preliminary results [2] indicated that under our conditions (very diluted alcohol/N_2O or alcohol/H_2O mixtures), pyrolysis plays a minor role and that the formation of enolate occurs in a collisionary induced loss of H_2 from the alkoxide anion, reaction 1 , in which M stands

$$C_nH_{2n+1}O^- \xrightarrow{M} C_nH_{2n-1}O^- + H_2 \qquad n > 1 \qquad (1)$$

for a bath gas molecule. It is an intriguing fact that reaction 1 which is endothermic by approximately 30kJ/mol should occur, in particular if we consider the low kinetic energy of the ions in the ICR cell. The purpose of the present work is to investigate the species involved in process 1 in more detail . On one hand, we have used ICR techniques for a kinetic study of this process with alkoxide anions formed from different precursors. On the other hand, we studied alkoxide and enolate anions formed in the high pressure source of a double sector mass spectometer of inverse geometry and we applied the technique of mass analyzed ion kinetic energy spectrometry to study collisionally induced decompostions (CID) of alkoxide and enolate anions.

2. Experimental

The ICR spectrometer used in this study is equipped with a four section cell and marginal oscillator detection. The time dependent experiments were carried out under the McMahon-Beauchamp trapped-ion conditions [3]. The hydroxyl anion was formed from water by dissociative attachment of electrons accelerated at about 6.5 V; this reaction produces in a first step H^- that rapidly reacts with water [4]:

$$H_2O + e^- \longrightarrow H^- + OH^{\cdot}$$

$$H^- + H_2O \longrightarrow OH^- + H_2$$

The oxygen radical anion was formed from pure N_2O by dissociative resonance capture of 1.6 eV electrons. It is known that under these conditions, O^- is formed with an average kinetic energy of 0.38 eV [5].

In all experiments, the total ion current (TIC) was minimized by ejection of the electrons from the reaction region of the ICR cell with a continuous radiofrequency of about 8 MHz. This value is within 20% of the calculated oscillation frequency along the y axis of the ICR cell (parallel to the magnetic field, see ref. 6.).

Mixtures of the product (alcohols, ethers) with either H_2O or N_2O were introduced via the dual inlet system at a total pressure of $3 \cdot 10^{-5}$ torr. The alkoxide anions can be generated very efficiently from alkyl nitrites by dissociative electron attachment at low electron energies, therefore they were studied without reactant gas. The third body (collision gas: He, Ne, Ar, C_2H_6) used in some ICR experiments was introduced via an extra inlet system connected to the introduction line. Pressure readings were corrected using total ionization cross section values [7].

A VG-Micromass ZAB-2F mass spectrometer coupled to a PDP-11 computer was used in this study, for CID measurements. The pressure of the target gas (He) was kept constant at a value which resulted in loss of 70% of the main beam intensity. The ionizing electrons were of 50 eV energy, the accelerating voltage 8 kV. The data represent computer averaged values of at least five scans. The pressure in the ion source was about 1 Torr with N_2O as reactant gas.

The following compounds were measured: EtOH, n-PropOH, i-PropOH, n-ButOH, s-ButOH, t-ButOH, $(i\text{-Prop})_2O$, n-ButONO, s-ButONO and butanone-2. The nitrites were sythesized by standard procedures [8]. The labelled compounds 1-$d_3$butanol-2

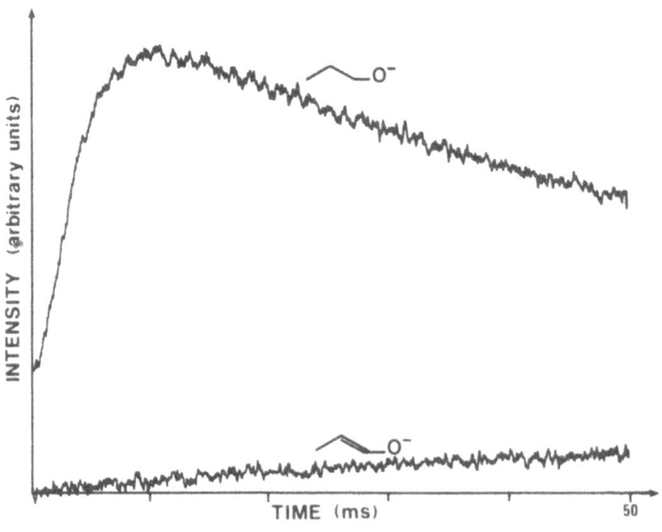

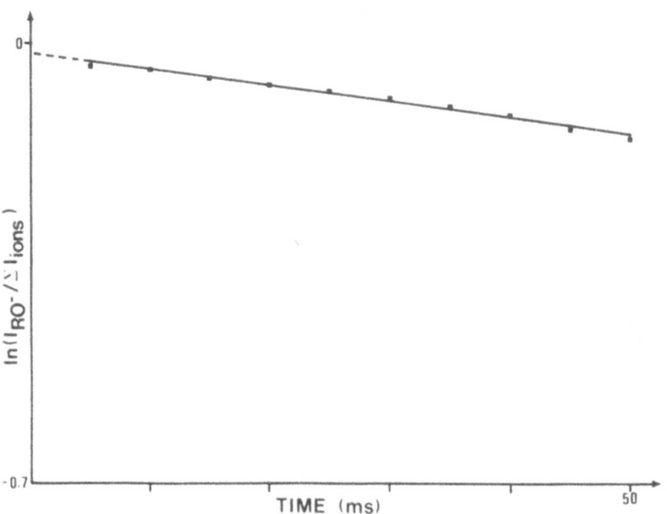

Fig. 1. Time dependence spectra for n-PropOH/H_2O system
$P_{ROH} = 2\cdot10^{-6}$ torr, $P_{H_2O} = 9\cdot10^{-6}$ torr (uncorrected ion intensities).

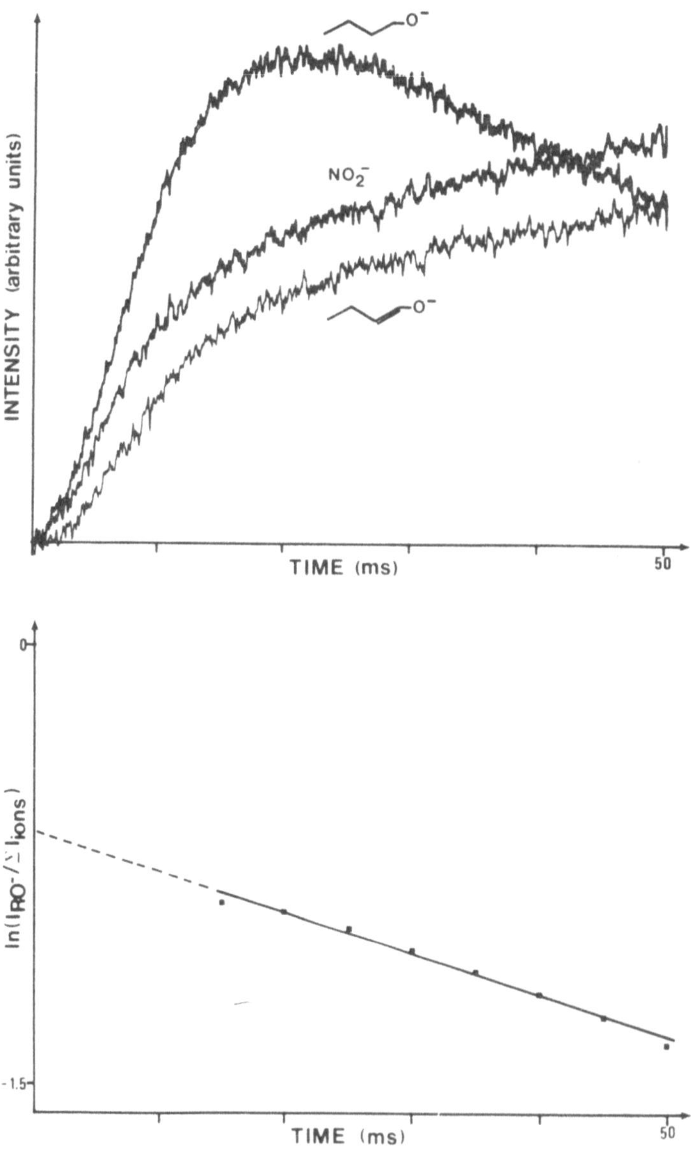

Fig. 2. Time dependence spectra for n-ButONO at $3 \cdot 10^{-7}$ torr (uncorrected ion intensities).

were synthesized in absolute ether with CD_3I + Mg + EtCHO and with CD_3CD_2Br +Mg + CH_3CHO, respectively.

3. Results

An example of the time dependent concentrations of alkoxide and enolate anions is reported in Figure 1 for the n-propano/H_2O system. Pseudo-first order kinetic treatment yields the bimolecular rate constant k for process 1' :

$$O^- \xrightarrow[k]{M} O^- + H_2 \qquad (1')$$

All alcohols studied (EtOH, i-PropOH, n-ButOH, s-ButOH) gave values of k within the limits given in Table 1. Alkoxide anions formed from tertiary alcohols do not exhibit loss of H_2 in accordance with previous observations [2] . Experiments were also carried out with various bath gases M, i.e. He, Ne, Ar and C_2H_6; the value of k did not change significantly in these systems.

The time dependent concentrations in n-ButONO are reported in Figure 2. In this system, the butoxide anion can react via two different pathways, i.e. collisionary induced loss of H_2, reaction 2a, and reaction with the nitrite molecule to form NO_2^-, reaction 2b;

$$O^- \xrightarrow{M} O^- + H_2 \qquad (2a)$$

$$O^- + ONO \longrightarrow NO_2^- + OH + \qquad (2b)$$

which has been previously reported [9] .

The time dependent concentrations in n-PropOH/n-ButONO/H_2O system are reported in Figure 3. The CID spectra of alkoxide and enolate anions formed under high pressure NCI conditions are reported in Figure 4 and 5. Unimolecular decompositions of metastable alkoxide anions (MIKES) were found to be of too low intensity in order to be recorded. Moreover we found that the CID spectra of the anions investigated were of very low intensity (approximately 10% of the value usually observed in spectra obtained from positive ions.

4. Discussion

The time dependence of the concentrations for alkoxide and enolate anions indicate that process 1 is induced by collisions of alkoxide anions with the bath gas molecules M. We have previously pointed out the specificity of the process 1 , viz. 1,2-elimination of H_2, as studied with deuterated reactants [2] .

Table 1. Mean Reaction Enthalpies and Rate Constants for Alkoxide Formation

	$\overline{\Delta H_r}$ a)	$\overline{k}$ b)	
$OH^- + ROH \xrightarrow[-H_2O]{} RO^-$	-71 ± 4	1 ± 0.5	(A)
$OH^- + ROR \xrightarrow[-(R-H)]{-H_2O} RO^-$	-8 ± 4	0.1	(B)
e^- c) + $RONO \xrightarrow[-NO]{} RO^-$	0 ± 4	50 ± 30	(C)

a) mean values for the enthalpy of reactions studied, from $R=C_2H_5$ up to $R=C_4H_9$, in kJ/mol; thermochemical values are given in Table 2

b) $\cdot 10^{-11}$ $cm^3/(molecules \cdot s)$; mean values (see text)

c) electron energy: 0.0-0.1 eV

These facts together with the present results allow us to minimize the role played by pyrolysis. An interesting feature of process 1 is that its rate constant differs by orders of magnitude according to the precursor of alkoxide anions, viz. ethers, alcohols, or nitrites (Table 1). These differences may be either due firstly to the internal energy content of alkoxide anions formed in processes A , B , or C , or secondly to consecutive reactions in which alkoxide anions formed via A, B, or C, react specifically with their neutral precursors, i.e. alcohols, ethers and nitrites, or thirdly to different structures for alkoxide and/or enolate anions formed via A, B, or C.

4.1. Internal Energy

We can observe that alcohols and ethers as precursors are cases in which the values for rate constant k follow the relative orders for ΔH_T. Moreover, the exothermicity of processes A and B will be partitioned between the reaction products; this contributes to lower the internal energy of alkoxide anions formed in reaction B relative to A. However reaction C does not fit into this kind of argument; nevertheless, in this cases loss of H_2 from alkoxide is a facile process as already observed. Noest and Nibbering [9] suggested that alkoxide anions formed from nitrites are formed in internally excited states. At this point it is worth considering a complicating effect that takes place in negative ions trapping experiments. Under these conditions it is difficult to evaluate to what extent electrons are being trapped together with anions. Furthermore, the efficiency of ejection by the radio-frequency applied along the y-axis (parallel to magnetic field) is difficult to estimate from measurements on the total ion current alone; thus there is a possibility that some contribution of dissociative electron attachment from non-ejected thermal electrons may take place in the case of nitrites. This could explain the slow increase in the concentration of butoxide ion shown in Figure 2. In cases in which the process of dissociative electron attachment requires higher kinetic energy electrons (N_2O, H_2O), this problem is considerably reduced; since we can trap only quasi-thermal electrons; only a minor contribution could occur during the acceleration of the electrons which are being ejected.

4.2. Enolate Formation

The problems arising with consecutive reactions of alkoxide anions are best illustrated in the case of n-butyl-nitrite, see Figure 2. It is seen that the formation of NO_2^-, reaction 2b takes place at a rate comparable to the formation of the $C_4H_7O^-$ enolate anion. Considering the thermochemistry of process 2a and

and 2b, ΔH_r=30 kJ/mol and -150 kJ/mol respectively, it seems improbable that reactions should occur with similar reaction rates.

Alternative pathways for the formation of enolate anions are provided by reactions of the alkoxide anions with their neutral precursors, reactions 3-5:

$$O^- + \text{ONO} \rightleftharpoons [\ ]^- \rightarrow O^- + \text{HONO} + \qquad (3)$$

$$O^- + \text{OH} \rightleftharpoons [\ ]^- \rightarrow O^- + H_2O + \qquad (4)$$

$$O^- + (\quad)_2O \rightleftharpoons [\ ]^- \rightarrow O^- + \text{OH} + \qquad (5)$$

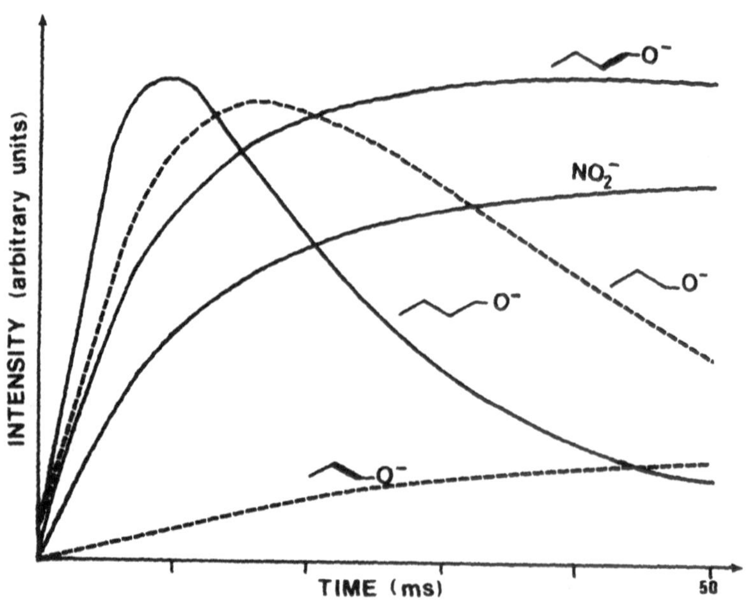

Fig. 3. Time dependence spectra for n-ButONO/n-PropOH/H_2O system P_{RONO} = $3\cdot10^{-7}$ torr, P_{ROH} = $2\cdot10^{-6}$ torr, P_{H_2O} = $9\cdot10^{-6}$ torr (uncorrected ion intensities).

All three processes are exothermic with values for the enthalpies of reaction, ΔH_r, of -23, -63, and -27 kJ/mol, respectively; our results suggest that reaction 3 proceeds more efficiently than reactions 4 and 5 (see Table 1). Further evidence is provided in Figure 3. The fast decrease in the concentration of propoxide anion formed from n-PropOH support the predominant role played by reaction 3':

$$\text{n-C}_3\text{H}_7\text{O}^- + \text{n-C}_4\text{H}_9\text{ONO} \rightarrow \text{CH}_3\text{CH}_2\text{CH}{=}\text{CHO}^- + \text{HONO} + \text{C}_3\text{H}_8 \qquad (3'a)$$

$$\rightarrow \text{NO}_2^- + \text{n-C}_3\text{H}_7\text{OH} + \text{CH}_3\text{CH}_2\text{CH}{=}\text{CH}_2 \qquad (3'b)$$

The fact that reaction 3 proceeds more readily than reactions 4 and 5 can be due to energy barriers that favor the backward dissociation of the intermediate complexes in the latter two cases.

4.3. CID experiments

In order to test the stuctures of alkoxide anions formed in processes A-C, alkoxide anions were formed in the high pressure source of a double sector mass spectrometer. The distributions of the major peaks in the NCI spectra of ethers, alcohols and nitrites were found to be similar to the low -pressure ICR spectra [2] . In addition, small intensities of low mass ions were observed. Moreover, an important formation of protonated alkoxide anion dimer, $ROHOR^-$, is observed in the NCI spectra of alcohols. The formation and decompostion of these species will be the subject of a forthcoming publication.

Figure 4a - 4c are typical examples for CID spectra of alkoxide anions. Structural features can be directly derived from these spectra; the major dissociation pathways are: loss of H_2 from primary alkoxide anions, Figure 4a, loss of H_2 and of alkane molecules from secondary alkoxide anions, 4b, and loss of alkane from tertiary alkoxide anions, 4c. This is an indication that structural analysis of alcohols can be performed by examining the CID spectra of their aloxide anions (see also Ref. 2 for analytical applications on alcohols).

We found that in each series of alkoxide anions (designated thereafter as primary, secondary and tertiary alkoxides), the CID spectra are identical (within experimental uncertainty) for all homologues investigated, whichever precursor (alcohol, ether or nitrite) was used. Therefore, we conclude that

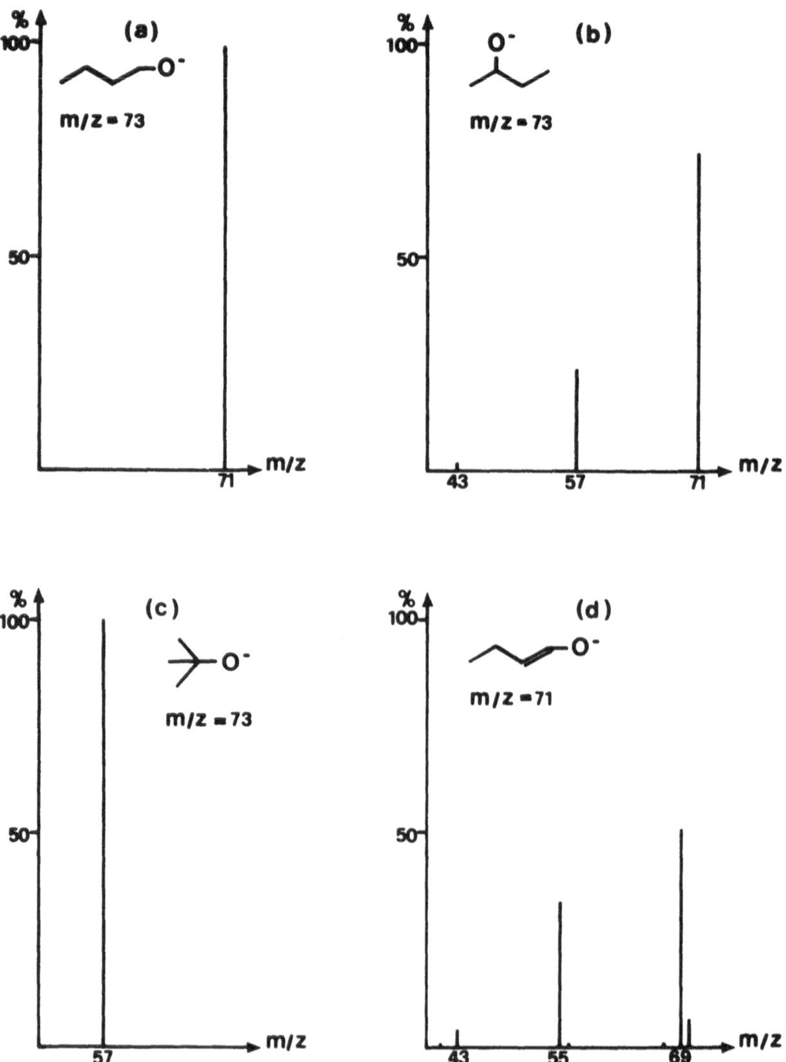

Fig. 4. CID spectra: (a) "Primary" alkoxide ion from n-ButOH, din-ButO and n-ButONO

(b) "Secondary" alkoxide ion from s-ButOH, dis-But and s-ButONO

(c) "Tertiary" alkoxide ion from t-ButOH

(d) "Primary" enolate ion from n-ButOH, din-ButO, n-ButONO and n-PropCHO

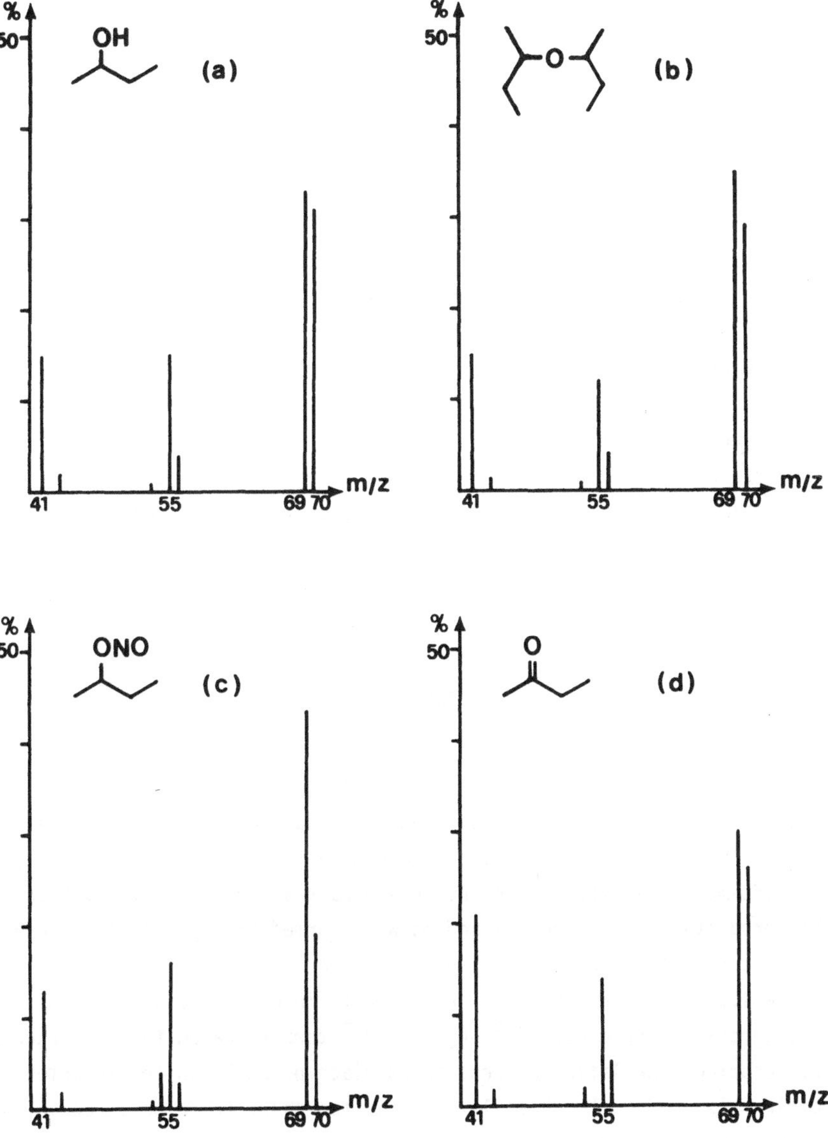

Fig. 5. CID spectra of "secondary" enolate ion: $C_4H_7O^-$ (m/z 71) from different precursors.

the structure of the alkoxide anions formed from different precursors are identical. We have also applied the CID method to the enolate anions in order to investigate the products of reaction 1 . "Primary" enolate anions formed from different neutrals (alcohols, ethers and nitrites) gave identical spectra, a typical example being shown in Figure 4d. In the case of enolates formed from neutrals with a non-linear structure, ions issued from alohols ethers and ketones showed to be similar, Figures 5a, 5b, and 5d, whereas enolate anions formed from nitrites gave a slightly different product distribution, Figure 5c. Low-lying excited states for enolate anions have recently been postulated [10] , but in the present stage it is not possible to conclude on the reasons for this difference.

4.4. Labelled Reactants

The loss of H_2 from the alkoxide anion formed from butanol-2 represents an interesting case since the reaction can occur to give two enolate isomers as shown in reaction 6:

$-H_2$ (6a)

(6b)

The details of this process were studied using samples of 1-d_3- and 3,4-d_5-butanol-2. The following observations were made:

a) in drift-mode ICR experiments, using N_2O as reactant gas, it was found that products of reactions 6a and 6b were formed in a ratio close to 3/2 which represents also the ratio of β-hydrogen atoms that can be involved in the 1,2 loss of H_2. However, experiments in which the pressure of alcohol changed from $0.5 \cdot 10^{-5}$ up to $5 \cdot 10^{-5}$ torr revealed that H/D exchange was occuring. The latter process takes place probably in the intermediate complex $[(s\text{-ButO})_2H]^-$ whose lifetime is too short to be observed in the low-pressure ICR conditions.

b) CID experiments performed on the two enolate isomers permitted to distinguish structure 6a from 6b: 6b loses a methane molecule whereas 6a loses both methane and ethane neutrals. The unexpected loss of methane from 6a requires the rearrangement of two hydrogen atoms; further experiments with labelled compounds are required to clear out this point.

Table 2. Enthalpies of Formation of Neutrals and Anions in kJ/mol.

Neutral (HA)	H^0_{f298} (HA)	H^0_{f298} (A^-)	Reference
EtOH	-235.1	-198.7	(a)
n-PropOH	-256.5	-225.1	(a)
i-PropOH	-272.8	-243.9	(a)
n-ButOH	-274.5	246.9	(c)-(a)
$(Et)_2O$	-252.2		(c)
$(n\text{-Prop})_2O$	-292.9		(c)
$(i\text{-Prop})_2O$	-318.8		(c)
$(n\text{-But})_2O$	-343.1		(d)
MeONO	-69.0		(b)
n-PropONO	-129.3		(b)
N-ButONO	-149.8		(b)
MeCHO	-158.6	-169.5	(a)
$MeCH_2CHO$	-190.4	-195.8	(a)
$EtCH_2CHO$	-205.0	-216.7	(c)-(d)
$(CH_3)_2CO$	-216.3	-209.6	(a)
NO	+90.2		(b)
H_2O	-241.8	-140.9	(b)
HNO_2	-76.7	-263	(e)-(c)

(a) "Gas Phase Ion Chemistry", ed. by M.T. Bowers, Vol. 2, 1979
(b) Nat. Stand. Ref. Data Ser., Nat. Bur. Stand. (US.) 1969
(c) J. Phys. Chem. Ref. Data, Vol. 6, Suppl. 1, 1977
(d) From group equivalents estimated by the authors
(e) JANAF Thermochemical Tables, 2nd Ed., 1971

5. Acknowledgements

The Swiss National Science Foundation is gratefully acknowlegded for financial support.

6. References

1. a) Environm. Health Perspect. 36 (1980),
   b) K.R. Jennings in "Mass Spectrometry", Vol. 4, ed. by R.A.W. Johnstone, The Chemical Society, London (1977),
   c) A.L.C. Smit and F.H. Field, J. Am. Chem. Soc. 99 (1977) 6471.

2. G. Boand, R. Houriet and T. Gäumann, in "Advances in Mass Spectrometry", Vol. 8, ed. by A. Quayle, Heyden, London (1980).

3. T.B. McMahon and J.L. Beauchamp, Rev. Sci. Instr. 43 (1972) 509.

4. J.A.D. Stockdale, R.N. Compton and P.W. Reinhardt, Phys. Rev., 184 (1969) 81.

5. R.J. Chantry, J. Chem. Phys. 51 (1969) 3369.

6. see for instance J.M.S. Henis, in "Ion Molecule Reactions", ed by J.L. Franklin, Vol. 2, Butterworths, London (1972).

7. Values for ionization cross sections taken from: "Mass Spectrometry", C.A. McDowell, ed. McGrawHill, New York (1963) or estimated by the method outlined by J.E. Bartmess, R. Georgiadis and G.W. Caldwell, 28th Ann. Conf. on Mass Spectrom., paper MAMP20, New York City (1980).

8. Organic Syntheses, Coll. Vol. II (1969) 108.

9. A.J. Noest and N.M.M. Nibbering, in ref. 2, p. 227

10. R.W. Westmore, H.F. Schaefer III, P.C.Hiberty and J.I. Braumann, J. Am. Chem. Soc. 102 (1980) 5470.

A FOURIER TRANSFORM ION CYCLOTRON RESONANCE STUDY OF NEGATIVE ION-MOLECULE REACTIONS OF PHENYL ACETATE, PHENYL TRIFLUOROACETATE AND ACETANILIDE

J.C. Kleingeld and N.M.M. Nibbering
Laboratory of Organic Chemistry, University of Amsterdam
Nieuwe Achtergracht 129, 1018 Amsterdam, The Netherlands

1. Introduction

Recently we have shown by ^{18}O-labelling [1] that in the gas-phase reaction of OH^- with methyl phenyl ether ($C_6H_5OCH_3$) 15% of the phenoxide anions ($C_6H_5O^-$) are formed via an ipso substitution reaction (1a), whereas the remaining 85% are formed via an S_N2 displacement [2] reaction 1b.

$$^{18}OH^- + C_6H_5OCH_3 \longrightarrow [H^{18}O(OCH_3)C_6H_5^-]^* \longrightarrow C_6H_5{}^{18}O^- + CH_3OH \quad (1a)$$

$$\Delta H_r^o = -147.7 \text{ kJ/mole}$$

$$^{18}OH^- + CH_3\text{-}O\text{-}C_6H_5 \longrightarrow C_6H_5O^- + CH_3{}^{18}OH \quad (1b)$$

These channels are of minor importance in the reaction of OH^- with ethyl phenyl ether [1] ($C_6H_5OC_2H_5$) where the chemistry is dominated by the occurrence of a facile E2 elimination reaction [3]:

$$OH^- + H\text{-}CH_2\text{-}CH_2\text{-}O\text{-}C_6H_5 \longrightarrow C_6H_5O^- + C_2H_4 + H_2O \quad (2)$$

$$\Delta H_r^o = -112.6 \text{ kJ/mole}$$

Another interesting example of a gas-phase anionic ipso substitution reaction has been published by Fukuda and McIver [4]. They have studied ion-molecule reactions of several bases with phenyl acetate ($C_6H_5OCOCH_3$). Surprisingly, they only found the formation of ions m/z 59 (CH_3COO^-) in the reactions of unclustered bases with phenylacetate. This has been explained by means of an ipso substitution reaction:

$$CH_3COOC_6H_5 + X^- \longrightarrow [CH_3COO(X)C_6H_5^-]^* \longrightarrow \underset{m/z\ 59}{CH_3COO^-} + C_6H_5X \quad (3)$$

$X^- = OH^-, CH_3O^-, CN^-, SH^-, CH_3S^-, C_6H_5O^-$

Ions m/z 93 ($C_6H_5O^-$), however, were also observed in the reactions of clustered bases with phenyl acetate [4]. It seemed interesting to study these reactions in more detail with our drift cell ICR instrument (the FT-ICR instrument was under construction at that time).

If the reaction of OH^- with phenyl acetate does proceed via an ipso attack of OH^- on the phenyl ring, ion a can be written as an intermediate structure (see also reaction 3).

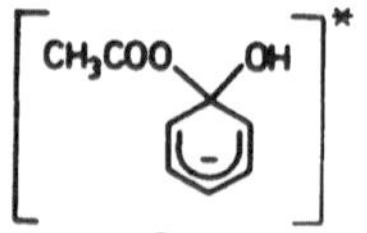

$$\left[\text{CH}_3\text{COO}\diagdown\!\!\diagup\text{OH} \;\; C_6H_5^- \right]^*$$

a

This excited species will dissociate into CH_3COO^- and neutral C_6H_5OH rather than into $C_6H_5O^-$ and neutral CH_3COOH, since acetic acid is more acidic than phenol [5]. If OH^- reacts with, for instance, para-fluorophenylacetate, the resulting intermediate would be ion b.

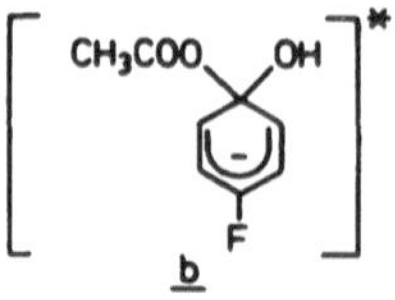

$$\left[\text{CH}_3\text{COO}\diagdown\!\!\diagup\text{OH} \;\; C_6H_4F^- \right]^*$$

b

This ion would be expected to dissociate into para-fluorophenoxide anions ($FC_6H_4O^-$) and neutral acetic acid, since para-fluorophenol is more acidic than acetic acid [5]. Indeed, the expected formation of ions m/z 111 ($FC_6H_4O^-$) has been confirmed experimentally. Furthermore, peaks due to ions m/z 59 (CH_3COO^-) are absent in the spectrum. However, no label incorporation into the para-fluorophenoxide anions has been observed with $^{18}OH^-$ as nucleophile. This is not in line with the proposed ipso substitution reaction (see also reaction 1a). Therefore, phenyl acetate itself was studied again. To our surprise the published results [4] could not be reproduced. An abundant peak at m/z 93 ($C_6H_5O^-$) was observed as well as some small peaks at m/z 135 ($(M-H)^-$) and m/z 111 ($C_6H_5O^- \ldots H_2O$). No ions m/z 59 (CH_3COO^-) could be detected. Similar observations have now been made in McIver's pulsed ICR instrument and also in DePuy's flowing afterglow instrument [6].

The question then arises how the phenoxide anions are formed. In theory three mechanisms can be suggested: a $B_{AC}2$ displacement reaction at the carbonyl carbon atom 4a, an S_N2 displacement reaction 4b and an E2 elimination reaction 4c.

$$X^- + CH_3-\overset{O}{\overset{\|}{C}}-OC_6H_5 \xrightarrow{B_{AC}2} \left[CH_3-\overset{O^-}{\overset{|}{\underset{X}{\underset{|}{C}}}}-OC_6H_5\right]^* \longrightarrow CH_3COX + C_6H_5O^- \quad (4a)$$

$$X^- + CH_3-\overset{O}{\overset{\|}{C}}-OC_6H_5 \xrightarrow{S_N2} CH_3X + CO + C_6H_5O^- \quad (4b)$$

$$X^- + H-CH_2-\overset{O}{\overset{\|}{C}}-OC_6H_5 \xrightarrow{E2} HX + CH_2CO + C_6H_5O^- \quad (4c)$$

To get further insight into these reactions the phenoxide formation has been studied with our new FT-ICR spectrometer [7,8]. Not only phenyl acetate, but also phenyl trifluoroacetate ($C_6H_5OCOCH_3$) and acetanilide ($C_6H_5NHCOCH_3$) have been included in the present study.

2. Results and Discussion

2.1. Phenyl trifluoroacetate ($C_6H_5OCOCF_3$)

This compound has been selected for the present study, because in this case an E2 elimination reaction (such as reaction 4c) will be impossible, thus simplifying the problem. Unfortunately, it does give abundant primary negative ions, which will be discussed first.

2.1.1. Primary Ions

The various primary negative ions of phenyl trifluoroacetate are listed in Table 1. The appearance energy curves of some of these ions have been measured by use of a rapid single ion monitoring computer program developed in our laboratory [7] and are given in Figure 1. Fragmentation mechanisms can be suggested to explain the formation of these ions, but are beyond the scope of this paper. Nevertheless, some of these primary ions can be expected to interfere with product ions resulting from ion-molecule reactions with phenyl trifluoroacetate. These ions are ejected therefore from the cell directly after the electron beam pulse. This is achieved by applying a radio frequency pulse of 20 ms duration (amplitude 7 V_{p-p}) and covering the mass range from m/z 60 to 200 to the excitation plates of the cell [7a]. Exciting and listening directly after this pulse shows that the ions have actually been removed from the cell. Moreover, exciting and listening later in time indicates that the ions do not reappear. This timing sequence is illustrated in Figure 2.

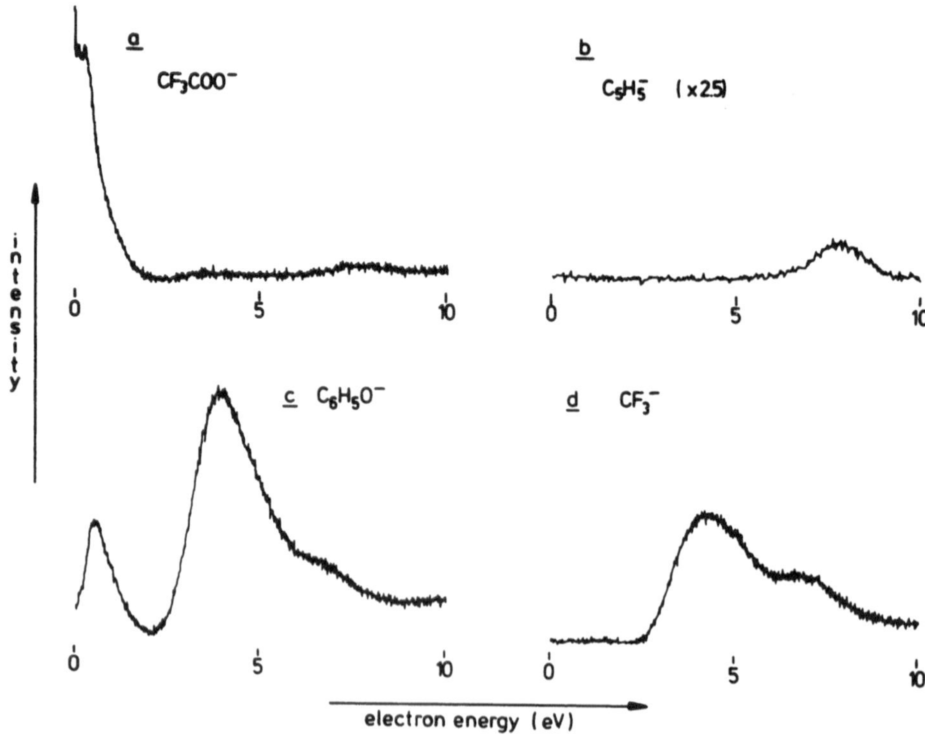

Fig. 1. Appearance energy curves of primary negative ions of phenyl trifluoroacetate. These curves were taken by slowly increasing the electron energy (1 eV/min). Each point of the curves a), c) and d) is an average of two consecutively measured values, while each point of curve b) is an average of eight values. The points of curve b) were multiplied by a factor of 2.5 with respect to the other curves.

Table 1. Primary negative ions of phenyl trifluoroacetate

m/z	assigned composition
170	$(M-HF)^{-\cdot}$
142	$(M-HF-CO)^{-\cdot}$
113	CF_3COO^-
93	$C_6H_5O^-$
69	CF_3^-
65	$C_5H_5^-$ = $(C_6H_5O - CO)^-$
19	F^-

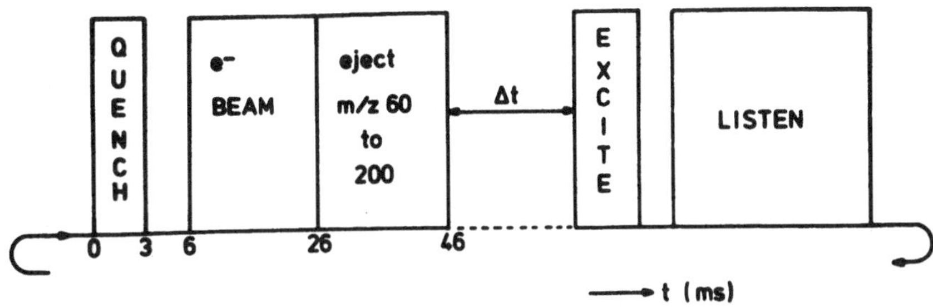

Fig. 2. Timing sequence to remove primary negative ions of phenyl trifluoroacetate from the cell. See for excitation parameters the experimental section.

Now the appropriate base can be admitted to study its ion-molecule reactions whith phenyl trifluoroacetate without any interfering primary ions.

2.1.2 $H_2O/C_6H_5OCOCF_3$

Four major peaks are present in the spectrum of water and phenyl trifluoroacetate (shown in Figure 3), namely m/z 189, 113, 93 and 69. A small peak is found at m/z 141. Double resonance ejection [9] shows that OH^- is the precursor of all these ions.

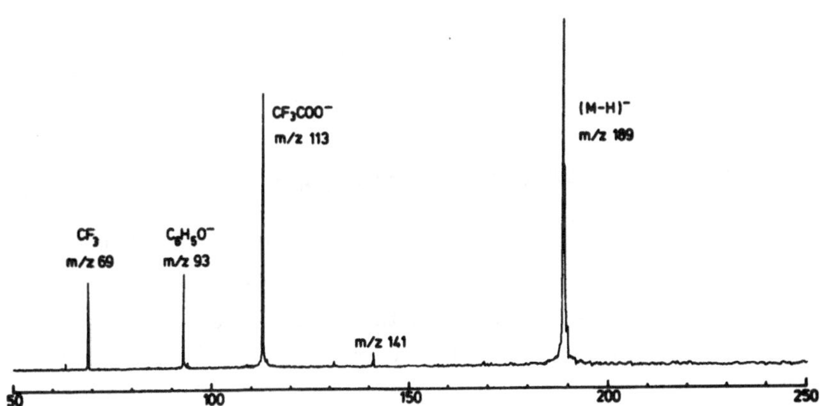

Fig. 3. Spectrum of a 2:1 mixture of water and phenyl trifluoroacetate. Total pressure 40 µPa. 256 Transients accumulated, 340 ms trapping time. Timing sequence as given in Figure 2 (Δt = 300 ms).

The ions m/z 189 are the $(M-H)^-$ ions, indicating that phenyl trifluoroacetate is more acidic than water.

The ions m/z 69 are probably formed as follows:

$$OH^- + C_6H_5{-}O{-}\overset{O}{\overset{\|}{C}}{-}CF_3 \longrightarrow \text{HO(H)}C_6H_4{=}O + CO + CF_3^- \qquad (5)$$

m/z 69

Of course, also ortho attack would result in CF_3^- formation.

More important, however, are the ions m/z 93 and 113. Use of $^{18}OH^-$ instead of OH^- shows that no ^{18}O-label is incorporated into the $C_6H_5O^-$ ions (m/z 93), whereas (90 ± 10)% of the ^{18}O-label is incorporated into the CF_3COO^- ions (m/z 113). This proves a $B_{AC}2$ mechanism in the case of CF_3COO^- formation:

$$OH^- + CF_3{-}\overset{O}{\overset{\|}{C}}{-}OC_6H_5 \xrightarrow{B_{AC}2} \left[CF_3{-}\underset{OH}{\overset{O^-}{C}}{-}OC_6H_5\right]^* \longrightarrow CF_3COO^- + C_6H_5OH \qquad (6)$$

c — m/z 113

$\Delta H_r^o = -263$ kJ/mole

An E2 elimination reaction is impossible in phenyl trifluoroacetate, so that for the formation of phenoxide anions two mechanisms do remain (see also reactions 4a and 4b):

$$OH^- + CF_3{-}\overset{O}{\overset{\|}{C}}{-}OC_6H_5 \xrightarrow{B_{AC}2} \left[CF_3{-}\underset{OH}{\overset{O^-}{C}}{-}OC_6H_5\right]^* \longrightarrow C_6H_5O^- + CF_3COOH \qquad (7a)$$

c — $\Delta H_r^o = -151$ kJ/mole

$$OH^- + CF_3{-}\overset{O}{\overset{\|}{C}}{-}OC_6H_5 \xrightarrow{S_N2} CF_3OH + CO + C_6H_5O^- \qquad (7b)$$

$\Delta H_r^o = -59$ kJ/mole

The same intermediate ion c is proposed both in reaction 7a and in reaction 6. It seems unreasonable, however, to assume that this species will dissociate into the phenoxide anion and neutral trifluoroacetic acid (reaction 7a) because the reaction pathway leading to neutral phenol and the trifluoroacetate anion via the same intermediate (reaction 6) is more exothermic by about 112 kJ/mole [5]. Thus, reaction 7b is considered to be correct and attempts have been made to support this.

Of course, a first check is the heat of reaction, since a highly endothermic ion-molecule reaction will be too slow to be observed in the gas phase. Estimating the heat of formation of phenyl trifluoroacetate as 904 kJ/mole [10] and of trifluoromethanol as 820 kJ/mole [10], the exothermicity of reaction 7b is calculated to be about 59 kJ/mole, so that in principle this reaction is possible.

Further support for this idea is given by the occurrence of a small peak at m/z 141 in the spectrum (see Figure 3). Double resonance ejection [9] shows that the corresponding ions are generated from the ions m/z 189 ($(M-H)^-$). A possible mechanism for their formation is given in reaction 8, in which the first step resembles reaction 7b.

$$\text{m/z 189} \xrightarrow{-CO} \left[C_6H_4(O^-)CF_3 \right] \xrightarrow{-HF} C_7H_3OF_2^- \;\; \text{m/z 141} \qquad (8)$$

The mechanism of the loss of HF in the second step is not known. However, it should be noted that hydrogen fluoride eliminations are not uncommon in negative ion gas-phase chemistry [11].

More evidence for the occurrence of reaction 7b is derived from the results of ejection of the collision complex from the cell. In all reactions 5, 6 and 7 a collision complex between OH^- and $C_6H_5OCOCF_3$ must be formed before dissociation into products. This complex will have a nominal mass of 207 daltons. Although no peak is observed at m/z 207, a radio frequency pulse has been applied to remove the corresponding ions. This does have a drastic effect on the abundance of all product ions as shown in Figure 4 and Table 2. Especially the abundance ratio of the trifluoroacetate and phenoxide anions is influenced and decreases from 3 to almost 1 under the applied experimental conditions (see Figure 4 and Table 2). This can be explained in terms of the lifetimes of the collision complexes involved in these reactions, but first other explanations should be considered.

One possible explanation is that the applied RF pulse is so large that it disturbs the ion motion completely. This explanation is ruled out by the application of RF pulses, having the same level as those to eject mass 207, but now to eject the ions m/z 189 and 170: the resulting spectra are hardly influenced and resemble closely the spectrum obtained without application of the RF pulse. This observation also rules out the possiblity of ion ejection by a tail of the RF pulse extending to the m/z 113 ions, since an application of pulses closer to these ions should increase such an ejection which is not observed.

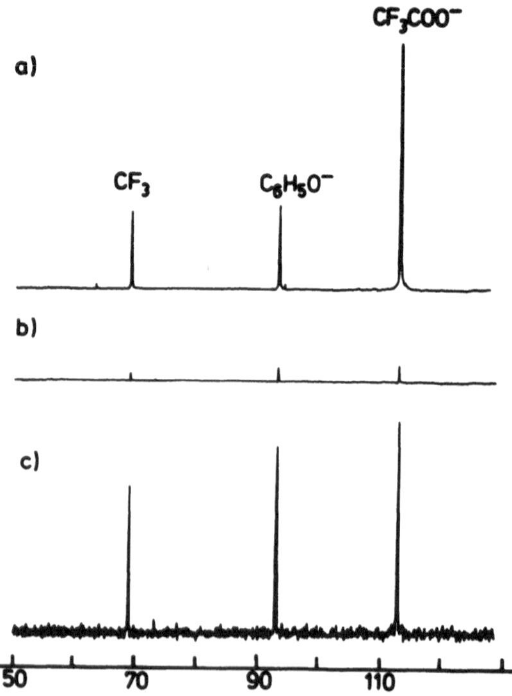

Fig. 4. (a) Spectrum of a 2:1 mixture of water and phenyl trifluoroacetate. Total pressure 40 µPa. 256 Transients accumulated. This spectrum is part of the spectrum shown in Figure 3. (b) As 4a, but an RF-pulse of 300 ms duration, 10 Vp-p and starting at 46 ms (see timing sequence given in Figure 2) is applied to the excitation plates of the cell. The frequency of this pulse corresponds with singly charged ions of mass 207. 256 Transients accumulated. The absolute gain factor is the same as in Figure 4a. (c) As 4b. 1800 Transients accumulated to increase the signal-to-noise ratio.

Table 2. Relative abundances of the product ions from the reaction of OH^- and $C_6H_5OCOCF_3$ upon ejection of the collision complex at m/z 207 at various amplitudes[a].

Amplitude[b]	$[CF_3COO^-]$[c] m/z 113	$[C_6H_5O^-]$[c] m/z 93	$[CF_3^-]$[c] m/z 69	$\frac{[113]}{[93]}$	$\frac{[93]}{[69]}$
0 [d]	5.1	1.7	1.6	3.0	1.1
5	2.3	0.95	0.75	2.4	1.3
7.1	0.95	0.55	0.40	1.7	1.4
10	0.33	0.28	0.21	1.2	1.3

[a] Conditions: see legend to figure 4.
[b] In Volts peak-to-peak (V_{p-p}).
[c] Arbitrary units.
[d] Pulse absent.

Another explanation is that the Rf-pulse to eject mass 207 will have a second harmonic peak at mass 103.5 (being twice the frequency corresponding with mass 207). Indeed, this peak can be observed upon listening to the frequency synthesizer output 7a. Its intensity is small (less than 1% of the main RF pulse), but it is large enough to eject ions, perhaps even the ions m/z 113, although normally the pulses are not that broad. When applying an RF pulse to "eject" mass 226, it is observed indeed that the second harmonic peak of this pulse at mass 113 does eject the CF_3COO^- ions. However, the original abundance of the m/z 113 ions is recovered when this pulse is applied to "eject" mass 220. So, the pulse to eject mass 207 certainly will not eject the m/z 113 ions.

Having ruled out instrumental artefacts, the explanation for the observed effect shown in Figure 4 and Table 2 may be as follows: the formation of CF_3COO^- anions proceeds via a $B_{AC}2$ mechanism as shown by ^{18}O-labelling experiments (vide supra) and must involve the tetrahedral type of ion c in reaction 6. Such an ion most probably corresponds with a minimum in the potential energy surface [12], since a real chemical bond is formed between the reactant ion and the neutral molecule [13]. In other words, the $B_{AC}2$ reaction should be described by a triple well potential model, where the outer minima will correspond with the clustered reactant and product ions and the central minimum with the tetrahedral intermediate ion c.

For an S_N2 reaction, however, a double well potential has been proposed by Brauman et al. [2]. The two minima in this model correspond again with the clustered reactant- and product ions, but now the species essential for the reaction will lie on a potential energy maximum. The transient ion corresponding with this maximum will resemble ion d below in the formation of $C_6H_5O^-$ anions via an S_N2 mechanism (reaction 7b).

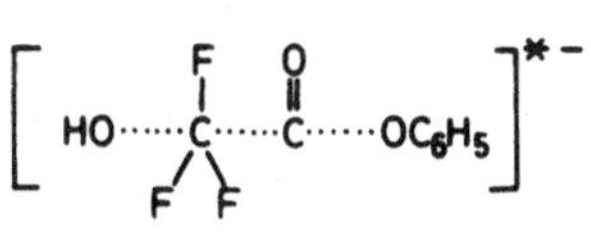

$$\left[HO\cdots\overset{F}{\underset{F\ \ F}{C}}\cdots\overset{O}{\overset{\|}{C}}\cdots OC_6H_5 \right]^{*-}$$

d

The difference between reactions 6 and 7b is therefore an additional minimum in the potential energy surface for the former reaction. On the basis of this additional minimum the tetrahedral intermediate ion c in the $B_{AC}2$ reaction 6 is expected to live longer than the transient complex ion d in the S_N2 reaction 7b. Therefore, during the time that the ions are accelarated by the RF-pulse to such large orbits that they will hit the cell plates [14], relatively more of the collision complexes leading to phenoxide anions will dissociate compared to the complexes leading to trifluoroacetate anions, as is observed.

2.2. Acetanilide ($C_6H_4NHCOCH_3$)

In the reaction of NH_2^- (generated from NH_3) with acetanilide two types of product ions are formed, namely $(M-H)^-$ ions (m/z 134) and $C_6H_5NH^-$ ions (m/z 92) [15]. In the reaction of OH^- (generated from H_2O) with acetanilide only the $(M-H)^-$ ions are formed. Peaks due to ions m/z 58 (CH_3CONH^-), m/z 59 (CH_3COO^-), m/z 109 ($C_6H_5NH^- \ldots NH_3$) or m/z 110 ($C_6H_5NH^- \ldots H_2O$) are not observed in any of the spectra.

The formation of the $(M-H)^-$ ions is not surprising since acetanilide is much more acidic than ammonia and water [5]. The formation of $C_6H_5NH^-$ ions can be explained in three ways (see also reactions 4a to c):

$$X^- + CH_3-\overset{O}{\overset{\|}{C}}-\underset{H}{N}-C_6H_5 \xrightarrow{B_{AC}2} CH_3-\overset{O}{\overset{\|}{C}}-X + C_6H_5NH^- \qquad (9a)$$

$$X^- + CH_3-\overset{O}{\overset{\|}{C}}-\underset{H}{N}-C_6H_5 \xrightarrow{S_N2} CH_3X + CO + C_6H_5NH^- \qquad (9b)$$

$$X^- + H-CH_2-\overset{O}{\overset{\|}{C}}-\underset{H}{N}-C_6H_5 \xrightarrow{E2} HX + CH_2CO + C_6H_5NH^- \qquad (9c)$$

The $B_{AC}2$ mechanism given in reaction 9a can in principle also result in other products shown in reactions 10 and 11 for $X^- = NH_2^-$ and OH^-, respectively. Thermodynamic data of these reactions are given in Table 3

$$NH_2^- + CH_3-\overset{O}{\overset{\|}{C}}-\underset{H}{N}-C_6H_5 \xrightarrow{B_{AC}2} \left[CH_3-\underset{NH_2}{\overset{O^-}{C}}-\underset{H}{N}-C_6H_5 \right]^* \longrightarrow \underset{m/z\ 58}{CH_3-\overset{O}{\overset{\|}{C}}-NH^-} + C_6H_5NH_2 \qquad (10)$$

$$OH^- + CH_3-\overset{O}{\overset{\|}{C}}-\underset{H}{N}-C_6H_5 \xrightarrow{B_{AC}2} \left[CH_3-\underset{OH}{\overset{O^-}{C}}-\underset{H}{N}-C_6H_5 \right]^* \longrightarrow \underset{m/z\ 59}{CH_3-\overset{O}{\overset{\|}{C}}-O^-} + C_6H_5NH_2 \qquad (11)$$

From these data it can be concluded that in the reaction of OH^- with acetanilide a barrier exists in the potential energy surface of the $B_{AC}2$ mechanism [12,17] (reactions 9a and 11), since no products formed via such a reaction are observed, although the overall reaction is calculated to be highly exothermic. Note that reaction 11 is even more exothermic than the observed proton transfer reaction.

In the reaction of NH_2^- with acetanilide, however, no distinction can be made between the three mechanisms 9a, b and c on the basis of the thermodynamic data in Table 3.

Table 3. Heats of reaction [16] (in kJ/mole) of some ion-molecule reactions of acetanilide.

Reaction[a] type	Product ion	Neutral product(s)	ΔH_r^o
Reactant anion NH_2^-			
$B_{AC}2$ (9a)	$PhNH^-$	CH_3CONH_2	-157.4
S_N2 (9b)	$PhNH^-$	CH_3NH_2, CO	-49.0
E2 (9c)	$PhNH^-$	NH_3, CH_2CO	-22.6
$BB_{AC}2$ (10)	CH_3CONH^- [b]	$PhNH_2$	c
p.t.[d]	PhN^-COCH_3	NH_3	-208.5
Reactant anion OH^-			
$B_{AC}2$ (9a)	$PhNH^-$ [b]	CH_3COOH	-100.5
S_N2 (9b)	$PhNH^-$ [b]	CH_3OH, CO	+22.6
E2 (9c)	$PhNH^-$ [b]	H_2O, CH_2CO	+31.0
$B_{AC}2$ (11)	CH_3COO^- [b]	$PhNH_2$	-178.4
p.t. [d]	PhN^-COCH_3	H_2O	-159.1

[a] The numbers between parentheses refer to the reactions in the text.

[b] This product ion has not been observed under the present experimental conditions.

[c] Heat of reaction can not be calculated, because the heat of formation of CH_3CONH^- is not known.

[d] p.t. = proton transfer reaction.

2.3. Phenyl acetate ($C_6H_5OCOCH_3$)

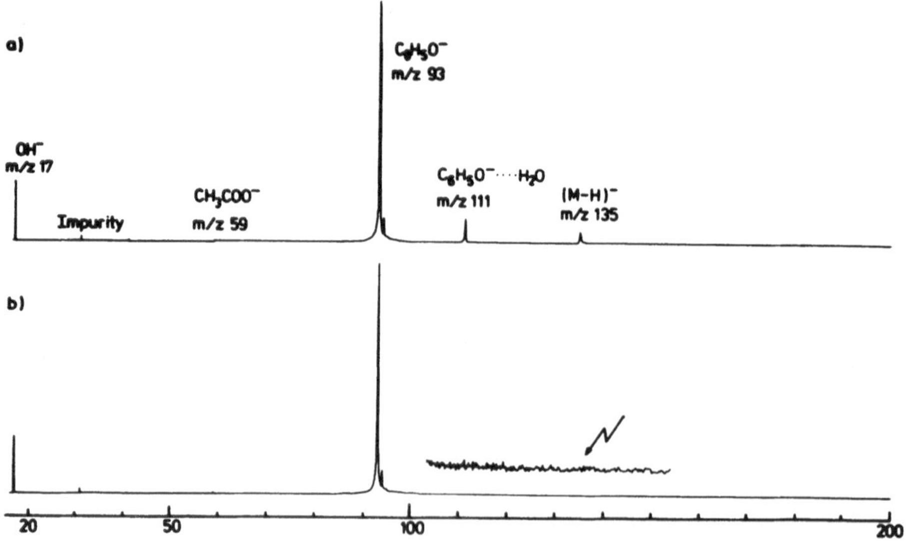

Fig. 5. (a) Spectrum of a 1:1 mixture of water and phenyl acetate. Total pressure 50 µPa. 128 Transients accumulated, 250 ms trapping time. (b) As 5a, but starting at 5 ms the ions m/z 135 are ejected from the cell by applying an RF-pulse of 250 ms duration, 0.9 Vp-p, to the excitation plates to see double resonance from the ions m/z 135. (The observed peak at m/z 31 is due to an impurity of methanol, which was used before taking these spectra to examine the solving switching reactions (see text)).

2.3.1. $H_2O/C_6H_5OCOCH_3$

As stated in the introduction, the major peaks in the spectrum of water and phenyl acetate taken with our drift cell ICR instrument originate from the ions m/z 135 ($(M-H)^-$), m/z 111 ($C_6H_5O^-...H_2O$) and m/z 93 ($C_6H_5O^-$). A similar spectrum has been obtained with our new FT-ICR instrument [7] (Figure 5a). A very small peak at m/z 59 (CH_3COO^-) is now present, but this is most probably due to hydrolysis of a small fraction of the molecules at the walls of the inlet system.

The cluster ions m/z can be formed in three ways:

$$OH^- + H{-}CH_2{-}\overset{O}{\overset{\|}{C}}{-}OC_6H_5 \longrightarrow C_6H_5O^- \cdots H_2O + CH_2CO \qquad (12a)$$

$$C_6H_5O^- + H_2O + M \longrightarrow C_6H_5O^- \cdots H_2O + M \qquad (12b)$$

$$C_6H_5OCOCH_2^- + H_2O \longrightarrow C_6H_5O^- \cdots H_2O + CH_2CO \qquad (12c)$$

Reaction 12a resembles the base induced carbon monoxide loss from alkyl formates reported by Riveros et al. [18]. Reaction 12b would correspond with a three-body collision process. This is not very likely because of the low total pressure used (about 50 μPa). Double resonance ejection [9] shows that the $(M-H)^-$ ions (m/z 135) are the precursors of the m/z 111 ions (see Figure 5b, ejection of m/z 135 completely removes m/z 111). Increasing the time during which the ions are trapped in the cell also shows that the $(M-H)^-$ ions react to give the ions m/z 111, this reaction being complete in approximately one second. Thus, reaction 12c is the correct one. This indicates that the $(M-H)^-$ ions behave as cluster ions of phenoxide ketene.

It can be understood why in acetanilide no cluster ions $C_6H_5NH^-...M$ ($M = NH_3$, H_2O) are observed (vide supra). In this compound the proton will not be abstracted from the methyl group, but from the nitrogen atom [5]. In phenyl acetate, however, the proton will be abstracted from the methyl group. This is in line with our and other's observations that the acidities of phenyl acetate and acetanilide differ by about 85 kJ/mole [4,5], whereas the acidities of acetanilide and phenylacetone ($C_6H_5CH_2COCH_3$) differ by only 1.3 kJ/mole [5].

The various solvent switching reactions can be seen very nicely in the system water/methanol/phenyl acetate. The following reaction scheme accounts for the observations made and has been confirmed by double resonance ejection:

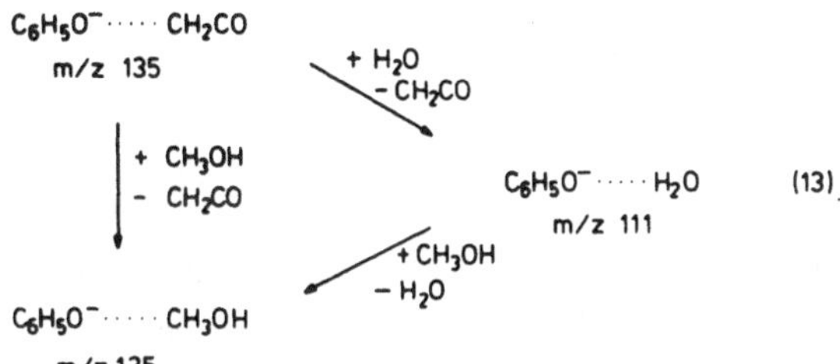

(13)

Returning to the subject of this paper: what is the mechanism for the phenoxide formation from phenyl acetate? The three proposed mechanisms 4a, b and c in the introduction are all exothermic for the reaction of OH^- with this compound, but the most exothermic one would be reaction 14a rather than 14b which is similar to reaction 4a (see Table 4).

The product ions of reaction 14a (CH_3COO^-) can hardly be observed. This raises the question whether the somewhat less exothermic reaction 14b will occur via the same intermediate. In other words, the intermediate ion e would be expected to dissociate into the acetate anion and neutral phenol rather than into the phenoxide anion and neutral acetic acid, since acetic acid is more acidic than phenol [5]. The very low abundance of the acetate anions, therefore, indi-

$$OH^- + CH_3-\overset{O}{\overset{\|}{C}}-OC_6H_5 \xrightarrow{B_{AC}2} \left[CH_3-\overset{O^-}{\overset{|}{\underset{OH}{\underset{|}{C}}}}-OC_6H_5 \right]^* \quad (14)$$

$\underline{e}$

a: $CH_3COO^- + C_6H_5OH$ — m/z 59 — $\Delta H_r^o = -181.7$ kJ/mole

b: $C_6H_5O^- + CH_3COOH$ — m/z 93 — $\Delta H_r^o = -176.7$ kJ/mole

Table 4: Heats of reaction [16] (in kJ/mole) of some ion-molecule reactions of phenyl acetate.

Reactant anion	Reaction[a] type	Product ion	Neutral product(s)	ΔH_r^o
OH^-	$B_{AC}2$ (4a,14b)	PhO^-	CH_3COOH	-176.7
OH^-	S_N2 (4b)	PhO^-	CH_3OH, CO	- 53.6
OH^-	E2 (4c)	PhO^-	H_2O, CH_2CO	- 45.2
OH^-	$B_{AC}2$ (14a)	CH_3COO^-	PhOH	-181.7
CH_3O^-	$B_{AC}2$ (4a)	PhO^-	$CH_3OCOCOCH_3$	-144.4
CH_3O^-	S_N2 (4b)	PhO^-	CH_3OCH_3, CO	- 29.3
CH_3O^-	E2 (4c)	PhO^-	CH_3OH, CH_2CO	+ 3.3
CH_3O^-	$B_{AC}2$	$CH_3OCOCH_2^-$ [b]	PhOH	- 55.7
$CH_3COCH_2^-$	$B_{AC}2$ (4a,16b)	PhO^-	$(CH_3CO)_2CH_2$	- 54.4
$CH_3COCH_2^-$	S_N2 (4b,17)	PhO^-	$CH_3COC_2H_5$, CO	- 24.7
$CH_3COCH_2^-$	E2 (4c)	PhO^-	CH_3COCH_3, CH_2CO	+ 45.6
$CH_3COCH_2^-$	$B_{AC}2$ (16a)	$(CH_3CO)_2CH^-$	PhOH	- 80.0

[a] The numbers between parentheses refer to the reactions in the text.

[b] This product ion has not been observed under the present experimental conditions.

cates that the phenoxide anions are probably not formed via a $B_{AC}2$ mechanism. At the present no distinction can be made between the S_N2 and E2 mechanisms (reactions 4b and c respectively) when OH^- is used as nucleophile. It may be that both are responsible for the phenoxide formation.

Attempts have been made to show that an S_N2 reaction is occuring under conditions where an E2 reaction can be excluded. Phenyl trifluoroacetate has been the first example (vide supra). Another example is described in the following section.

2.3.2 $H_2O/CH_3COCH_3/C_6H_5OCOCH_3$

Reaction of $CH_3COCH_2^-$ with phenyl acetate according to the E2 mechanism is too endothermic to be observed (see Table 4). The $B_{AC}2$ and S_N2 reactions, however, are still exothermic, so that the reactions of $CH_3COCH_2^-$ with phenyl acetate were chosen to be studied in more detail.

The major product ions of these reactions are phenoxide anions (m/z 93) and cluster ions m/z 151 ($C_6H_5O^-...(CH_3)_2CO$). The latter must have been formed by reaction 15, because no $(M-H)^-$ ions are present and double resonance ejection [9] shows that the m/z 151 ions are not formed from a reaction of the m/z 93 ions (compare reaction 12b). Reaction 15 resembles the base induced carbon monoxide loss from alkyl formates reported by Riveros et al. [18].

$$CH_3\overset{O}{\overset{\|}{C}}CH_2^- + H{-}CH_2{-}\overset{O}{\overset{\|}{C}}{-}OC_6H_5 \longrightarrow \underset{m/z\ 151}{C_6H_5O^-\cdots(CH_3)_2CO} + CH_2CO \qquad (15)$$

Reaction of $CH_3COCH_2^-$ with phenyl acetate also generates ions m/z 99 to a very small extent (∿1%). This points to a $B_{AC}2$ reaction 16a:

$$CH_3\overset{O}{\overset{\|}{C}}CH_2^- + CH_3{-}\overset{O}{\overset{\|}{C}}{-}OC_6H_5 \xrightarrow{B_{AC}2} \left[CH_3{-}\underset{CH_2COCH_3}{\overset{O^-}{C}}{-}OC_6H_5 \right]^* \qquad (16)$$

(a) $\longrightarrow$ $CH_3\overset{O}{\overset{\|}{C}}CH^-\overset{O}{\overset{\|}{C}}CH_3 + C_6H_5OH$, m/z 99, $\Delta H_r^o = -80.0$ kJ/mole

(b) $\longrightarrow$ $C_6H_5O^- + CH_3\overset{O}{\overset{\|}{C}}CH_2\overset{O}{\overset{\|}{C}}CH_3$, m/z 93, $\Delta H_r^o = -54.4$ kJ/mole

As discussed above for reaction 14, the low abundance of the product ions of reaction 16a indicates that reaction 16b will probably not occur, since acetylacetone ($CH_3COCH_2COCH_3$) is more acidic than phenol [5].

In this case an E2 elimination reaction is not possible (see Table 4), so that it is very likely that the phenoxide anions are now formed via an S_N2 displacement reaction:

$$CH_3\overset{O}{\overset{\|}{C}}CH_2^- + CH_3-\overset{O}{\overset{\|}{C}}-OC_6H_5 \longrightarrow CH_3\overset{O}{\overset{\|}{C}}CH_2CH_3 + CO + C_6H_5O^- \quad (17)$$

$$\Delta H_r^0 = -24.7\,kJ/mole$$

3. Conclusion

On the basis of the various observations and arguments reported in this paper it is plausible that some nucleophiles may react with phenyl acetate via an S_N2 mechanism to form phenoxide anions, especially when an E2 elimination reaction can not occur on energetic grounds. The $B_{AC}2$ mechanism is almost negligible for phenyl acetate. This is not true for phenoxide anion formation from phenyl trifluoroacetate, where the $B_{AC}2$ and S_N2 mechanisms are operative in the ratio of 3 to 1. From solution chemistry it is known that halogen substituents retard an S_N2 reaction [19],

$$X^- + CH_3Y \xrightarrow{k_1} CH_3X + Y^- \quad (18a)$$

$$X^- + CF_3Y \xrightarrow{k_2} CF_3X + Y^- \quad (18b)$$

$$k_1 >> k_2$$

whereas they activate displacements at carbonyl groups [20]:

$$X^- + CH_3COY \xrightarrow{k_3} CH_3COX + Y^- \quad (19a)$$

$$X^- + CF_3COY \xrightarrow{k_4} CF_3COX + Y^- \quad (19b)$$

$$k_3 << k_4$$

Both effects can account for the observations that the $B_{AC}2$ mechanism dominates in phenyl trifluoroacetate, whereas the S_N2 mechanism does in phenyl acetate.

4. Experimental Section

The drift cell ICR instrument has been described elsewhere [11b]. The home-made FT-ICR instrument will also be described elsewhere [7].

The amide and hydroxide anions were generated by dissociative electron attachment from ammonia (∿5.5 eV) and water (∿6.5 eV), respectively. False interpretations of the spectra because of interfering products of reactions of OH^- itself with phenyl acetate were avoided by ejection of the hydroxide ions from the cell after 80 ms of reaction time by a radiofrequency pulse of 10 ms duration and an amplitude of 1.25 V_{p-p}. This reaction time was long enough to form an appropriate amount of methoxide or enolate anions. Product ions generated in this time by reaction of OH^- with phenyl acetate were ejected by an RF-pulse of 15 ms, 10 V_{p-p}, covering the mass range from m/z 90 to 200 and applied directly after the ejection of OH^-. Because the signal from the phenoxide anions was very large, these ions were ejected by an extra RF-pulse of 10 ms, 1.25 V_{p-p}, to be sure that none of these ions were left in the cell. Exciting and listening directly after these pulses showed that the only ions present in the cell were the methoxide or enolate anions, respectively, so that the system was ready to study reactions of these anions with phenyl acetate.

Double resonance ejection was normally done by applying an RF-pulse of 0.5 to 1 V_{p-p} at a frequency corresponding with the mass-to-charge ratio of the ion to be expelled. This pulse started during the electron beam pulse and finished just before the excitation pulse.

The conditions of the excitation pulse were varied depending upon the mass range to be excited (i.e. the mass range to be observed). Typical parameters were: For the excitation of a large frequency range, for instance, from 1.85 MHz down to 75 kHz corresponding with a mass range of m/z 10 to 250 at 1.2 T, a frequency scan rate of about 1 MHz/ms and an amplitude of 7.1 V_{p-p} was used.

For the excitation of a smaller frequency range, for example from 460 to 75 kHz corresponding with a mass range of m/z 40 to 250 at 1.2 T (see Figure 3), a scan rate of 250 kHz/ms and an amplitude of 3.5 V_{p-p} was used. The scan rate was calculated by the computer from the range to be excited and the length of the pulse. This length is normally chosen to be 1.5 ms.

The trapping time is defined as the time between the start of the electron beam pulse and the beginning of the excitation pulse.

For the generation of negative ions the electron beam was switched on for 20 ms, the emission current being about 600 nA. During and shortly after the electron beam pulse the electrons were ejected from the cell to avoid spurious signals. Ions were stored in the cell by a differential trapping voltage of about 1.2 V.

The partial pressures of the precursors of the reactant ions and of the substrates were about 15 to 25 µPa.

Phenyl acetate, acetanilide and parafluorophenyl acetate were synthesized according to standard methods [21]. Phenyl trifluoroacetate was synthesized by adding 10 ml of trifluoroacetic acid anhydride and a few drops of sulphuric acid to 3 g phenol. The reaction mixture was stirred for about 45 minutes and extracted with tetrachloromethane. The compound was purified by preparative GC as were phenyl acetate and para-fluorophenyl acetate. Acetanilide was purified by recrystallization from toluene.

5. Acknowledgements

Both authors are very grateful to Dr. J.H.J. Dawson and Drs. A.J. Noest for designing and constructing the FT-ICR spectrometer. They further wish to express their sincere thanks to Prof. Dr. ir. B.M. Wepster of the Technical University of Delft, the Netherlands, for the generous gift of para-fluorophenol from which para-fluorophenyl acetate was synthesized. They also wish to thank the Netherlands organization for Pure Research (SON/ZWO) for the purchase of the basic ion cyclotron resonance mass spectrometer.

6. References

1. J.C. Kleingeld and N.M.M. Nibbering, Tetrahedron Letters, 21 (1980) 1687.

2. W.N. Olmstead and J.I. Brauman, J. Am. Chem. Soc., 99 (1977) 4219 (Only reaction 1b has been considered by these authors in the rate constant measurement for the reaction of OH^- with $C_6H_5OCH_3$).

3. See for other examples of anionic E2 eliminations in the gas phase, a) D.P. Ridge and J.L. Beauchamp, J. Am. Chem. Soc., 96 (1974) 637; b) D.P. Ridge and J.L. Beauchamp, J. Am. Chem. Soc., 96 (1974) 3595; c) S.A. Sullivan and J.L. Beauchamp, J. Am. Chem. Soc., 98 (1976) 1160.

4. E.K. Fukuda and R.T. McIver, Jr., J. Am. Chem. Soc., 101 (1979) 2498.

5. J.E. Bartmess and R.T. McIver, Jr. in "Gas Phase Ion Chemistry", Vol. 2, (M.T. Bowers, Ed.) Academic Press, New York, 1979, Chapter 11.

6. J.C. Kleingeld, N.M.M. Nibbering, J. Grabowski, C.H. DePuy, E.K. Fukuda and R.T. McIver, Jr., to be published.

7. a) See for a description of the essential hardware the chapter by J.H.J. Dawson in this volume of "Lecture Notes in Chemistry". b) The essential software will be published by A.J. Noest and C.W.F. Kort from our laboratory.

8. See for general references to the FT-ICR technique and its applications:
a) M.B. Comisarow and A.G. Marshall, Chem. Phys. Letters, 25 (1974) 282,
b) M.B. Comisarow and A.G. Marshall, Chem. Phys. Letters, 26 (1974) 489,
c) M.B. Comisarow and A.G. Marshall, Can. J. Chem., 52 (1974) 1997.
d) M.B. Comisarow in "Transform Techniques in Chemistry" (P.R. Griffiths, Ed.) Plenum Press, New York, 1978, Chapter 10; e) M.B. Comisarow, Adv. Mass Spectrom., 8B (1980) 1698, f) E.B. Ledford, Jr., S. Ghaderi, C.L. Wilkins and M.L. Gross, Adv. Mass Spectrom., 8B, (1980) 1707; g) E.B. Ledford, Jr., S. Ghaderi, R.L. White, R.B. Spencer, P.S. Kulkarni, C.L. Wilkins and M.L. Gross, Anal. Chem., 52 (1980) 463, h) E.B. Ledford, Jr., R.L. White, S. Ghaderi, M.L. Gross and C.L. Wilkins, Anal. Chem., 52 (1980) 1090;
i) M. Alleman, Hp. Kellerhals and K.-P. Wanczek, Chem. Phys. Letters, 75 (1980) 328.

9. M.B. Comisarow, V. Grassi and G. Parisod, Chem. Phys. Letters, 57 (1978) 413

10. a) S.W. Benson, F.R. Cruickshank, D.M. Golden, G.R. Haugen, H.E. O'Neal, A.S. Rodgers, R. Shaw and R. Walsh, Chem. Rev., 69 (1969) 279;
b) H.K. Eigenmann, D.M. Golden and S.W. Benson, J. Phys. Chem., 77 (1973) 1687.

11. a) J.M. Riveros and K. Takashima, Can. J. Chem., 54 (1976) 1839; b) J.H.J. Dawson, A.J. Noest and N.M.M. Nibbering, Int. J. Mass Spectrom. Ion Phys., 29 (1979) 205

12. a) M.J. Pellerite and J.I. Brauman, J. Am. Chem. Soc., 103 (1981) 676;
b) N.M.M. Nibbering, Recl. Trav. Chim.Pays-Bas, in press.

13. N.M.M. Nibbering in "Kinetics of Ion-Molecule Reactions", NATO ASI, Vol. B40 (P. Ausloos, Ed.) Plenum Press, New York, 1979, p. 165.

14. Using the equation derived by M.B. Comisarow (J. Chem. Phys., 69 (1978) 4097 and A.G. Marshall and D.C. Roe (J. Chem. Phys., 73 (1980) 1581) this time can be calculated to be 77 μs in our cell, where the distance between the plates is 1 inch (2.54 cm), when the amplitude of the pulse is 10 V_{p-p} and the magnetic field is 1.2 T.

15. The $(M-H)^-$ ions are about 25 times more abundant than the $C_6H_5NH^-$ ions.

16. Heats of formation of the anions have been taken from ref. 5. The heat of formation of phenyl acetate has been taken from ref. 4. Heats of formation of CH_3CONH_2, CH_3COOH and $C_6H_5NHCOCH_3$ have been taken from ref. 10a. Heats of formation of all the other compounds have been taken from H.M. Rosenstock, K. Draxl, B.W. Steiner and J.Z. Herron, J. Phys. Chem. Ref. Data, Vol. 6, Suppl, 1, 1977.

17. O.I. Asubiojo and J.I. Brauman, J. Am. Chem. Soc., 101 (1979) 3715.

18. a) L.K. Blair, P.C. Isolani and J.M. Riveros, J. Am. Chem. Soc., 95 (1973) 1057; b) J.F.G. Faigle, P.C. Isolani and J.M. Riveros, J. Am. Chem. Soc., 98 (1976) 2049.

19. See for example: J. Hine, C.H. Thomas and S.J. Ehrenson, J. Am. Chem. Soc., 77 (1955) 3886.

20. See for example: I. Ugi and F. Beck, Chem. Ber., 94 (1961) 1839.

21. A.I. Vogel, "A Text-Book of Practical Organic Chemistry", third edition, Longmans, London, 1956.

SITE OF PROTONATION IN GASEOUS FIVE-MEMBERED RING SYSTEMS $C_4H_4X(X=NH,O,S,CH_2)$

Raymond Houriet* and Helmut Schwarz**
Institut de Chimie Physique EPF-Lausanne, Switzerland, and
Institut für Organische Chemie der Technischen Universität
Berlin, Strasse des 17. Juni 135, Berlin, West Germany

1. Introduction

The study of acid-base properties of isolated systems have resulted in a fast growing output of data obtained in high pressure mass spectrometers and in low pressure ion cyclotron resonance spectrometers (ICR), (see Ref. 1 for recent reviews). These studies provide the basis for evaluating intrinsic substituent effects and Taft et al. [2] have recently stressed the predominance of polarization effects over polar (inductive) effects in stabilizing charged species. Subsequent comparison of gas phase and condensed phase data have the potentiality of disclosing solvent effects operating in the condensed phase [3]. While the understanding of the basic properties of monofunctional compounds is a relatively unambiguous matter, the situation becomes evidently more complex in multifunctional systems. Recent examples of the latter are provided by the ketene molecule in which it was found that the carbonyl group is less basic than the methylene group by 18 ± 8 kcal/mol [4], and also by enamine systems where the N-site was found to be less basic than the β-C-site by approximately 9 kcal/mol [5].

In this study we want to determine the site of protonation in a series of unsaturated five-membered ring compounds consisting of pyrrole (1) furan (2), thiophene (3) and cyclopentadiene (4). ICR techniques are especially well suited for these purposes since the method permits the determination of properties of molecules (and ions) under equilibrium conditions [6]. Moreover, double resonance experiments permit to investigate properties under non-equilibrium conditions. In this publication we will describe essentially the experimental approach that we have considered in trying to determine the reactivity of unsaturated five-membered ring compounds towards a proton. A full paper together with results from semi-empirical molecular orbital calculations on these systems will be published elsewhere [7].

---

* Lausanne

** Berlin

2. Experimental

The ICR instrument built at the EPF-Lausanne utilises a Varian V 7300 magnet and a four-section flat cell operated in the trapped mode [8]. Equilibrium proton transfer reactions were carried out under the same conditions as previously described [9]. Determination of the basicities for the β-position in furan and thiophene was carried out in ejection experiments (double resonance) using a 1:1 mixture of β-deuterated furan (or thiophene) with a reference base (total pressure ca. $2 \cdot 10^{-6}$ Torr) to which CH_4 was added in excess up to ca. $5 \cdot 10^{-5}$ Torr. For pyrrole and cyclopentadiene, an excess of CD_4 was used (see text next section). The following deuterated samples were used: α-D and β-D-furan, α-D and β-D-thiophene (for synthesis of these compounds, see Ref. 7)

Table 1. Proton Affinities for Five-membered Ring Compounds

	neutral	α-PA [a]	x = β-PA	y = ΔPA(α-β)
(1)	 N	208.9 [b]	$206<x<208.9$	$0\leq y<2.9$
(2)	 O	196.0	$191.4<x<193.1$	$2.9\leq y<4.6$
(3)	 S	195.9	$191.4<x<193.1$	$2.8\leq y<4.6$
(4)		200.0 [b]	$191.4<x<193.1$	$6.9\leq y<8.6$

[a] kcal/mol, reference value PA (NH_3) = 205.0

[b] value from Ref. 1

3. Results and Discussion

3.1. Equilibrium Proton Affinities (PA)

The equilibrium PA values for compounds (1-4) are reported in the first column of Table 1. PA values for pyrrole and cyclopentadiene were taken from a recent compilation [1] based on the reference value PA(NH_3) = 205.0 kcal/mol. PA for furan and thiophene were determined in equilibrium proton transfer reactions using acetone (PA = 197.2 [1]) as a partner, reaction 1:

$$C_4H_5X^+ \quad + \quad CH_3COCH_3 \rightleftharpoons C_4H_4X \quad + \quad CH_3COHCH_3^+ \ ; \ X=O,S \qquad (1)$$

Similar experiments conducted with samples of furan and thiophene monodeuterated in the α- and in the β-position show that D^+ is transferred from $C_4H_4DX^+$ (X=O,S) to acetone only from the α-deuterated samples. This establishes the α-position to be the exclusive site of protonation at equilibrium. As already noted in the case of furan [9] this behaviour parallels the reactivity of furan towards electrophiles in the condensed phase and agrees with the results from molecular electrostatic calculations [10].

3.2. Basicity at the β-carbon atom

The monolabelled samples of furan and thiophene in the β-position have been used to determine the basicity of the β-position in the following fashion (see Scheme I): we used the strong Brønsted acids CH_5^+ and $C_2H_5^+$ derived from CH_4 to protonate the β-D-samples in a series of reference bases of lower basicity was observed by double resonance techniques. As illustrated in Scheme I, we can consider three different structures for the protonated furan, respectively thiophene, i.e.:

1. structure a which is the β-protonated furan (thiophene) is unable to transfer H^+ (hence D^+) to bases B_i because of the relative PA values.
2. structure b which is protonated on the heteroatom might possibly transfer a H^+ to bases B but not D^+ (see later for a discussion on this point)
3. structure c which is the β-protonated furan (thiophene) is able to transfer D^+ to bases B as long as the latter has a higher basicity than the β-position in furan, respectively in thiophene. The results for this bracketing-type experiments are given in Table 2. For the pyrrole and cyclopentadiene cases in which no deuterated samples were available, non-specific deuteration was performed by reacting the compounds with CD_5^+ and $C_2D_5^+$ from CD_4 and the subsequent transfer of D^+ from $C_4H_4XD^+$ (X=N,CH_2) to bases B_i was followed in an analogous manner to the aformentioned double resonance experiments.

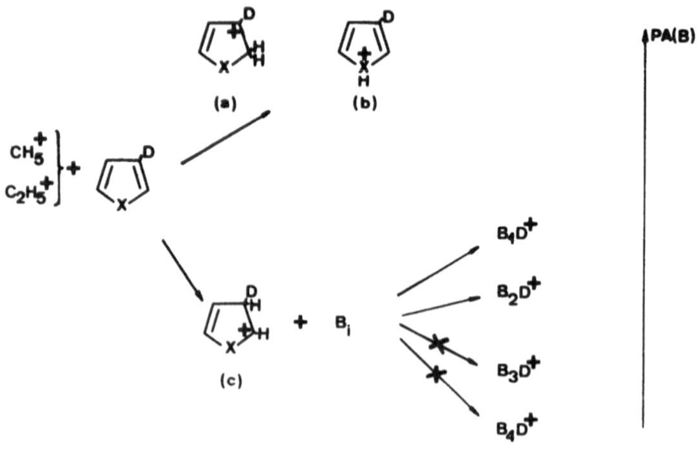

Scheme I

Table 2. Occurence, (+), and non-occurence (-) of double resonance signals indicating D^+ transfer from $C_4H_4DX^+$ to the reference bases B (see text).

Reference base B	PA(B)[1] kcal/mol	X = α-PA(X) =	NH 208.9	O 196.0	S 195.9	CH_2 200.0
$(i\text{-prop})_2O$	206		(-)			
NH_3	205		(-)			
t-butOH	195			(+)	(+)	(+)
$(CH_3)_2O$	193.1			(+)	(+)	(+)
n-propOH	191.4			(-)	(-)	(-)
CH_3CH_2CHO	191.4			(-)	(-)	(-)
CH_3CH_2OH	190.3			(-)	(-)	(-)
$(CH_2)_2O$	189.6			(-)	(-)	(-)

These results are reported in Table 2. It has to be mentioned that while double resonance results for cyclopentadiene are clearly related to the basicity of the β-carbon center, in the case of pyrrole it is not possible at this point to distinguish unambiguously between the basicity of the N-atom and that of the α- or β-carbon atom. However, we can notice that the PA (pyrrole) = 208.9 kcal/mol is reduced significantly with respect to PA (dimethylamine) = 220.5 kcal/mol [1] and PA (pyrrolidine) = 224.3 [1]. In the latter two systems, protonation undoubtly occurs on the N-atom, therefore we deduce that the protonation of pyrrole involves the carbon centers (for further discussion, see Ref. 7).

It has to be noted that the results of the ejection experiments may be subject to the possible errors inherent in such bracketing techniques [11]. In our case the most severe limitation is the inablity to detect processes occuring with low efficiences when the basicities of a base pair lie close together [12]. Another important point concerns the reactions in which the β-protonated form (such as structure c in Scheme I) is formed. Given the PA for CH_4 and C_2H_4 [1], respectively 128.2 and 163.5 kcal/mol, therefore the values for ΔGB (α-β) must be considered as representing minimum difference values. We can also note that in order to measure ΔGB(α-β) experimentally, the possible isomerization processes via 1,2-hydride shift must have barriers higher than the ΔGB(α-β). We want now to discuss our experimentally determined ΔGB(α-β) in terms of stabilities of the products formed in each process, therefore we shall assume ΔGB(α-β) = ΔPA(α-β), and justify this approximation later in the discussion by considering possible structures for the protonated systems.

The following conclusion can be drawn from the data in Table 1: The proton affinities at C(α) for (2),(3),(4) are higher than at C(β). Energy differences ranging from 2.8-4.6 kcal/mol for systems (2) and (3) and from 6.9-8.6 kcal/mol for system 4 favor protonation at C(α) over C(β) (see Scheme II). This implies that the gas phase protonation of furan and thiophene, which can be considered as electrophilic aromatic substitution par excellence, occurs preferentially on the α-position. This observation is in agreement with the reactivity of unsaturated five-membered cycles in solution where the α/β ratio for substitution decreases in the order furan ≫ thiophene ≫ pyrrole [10]. This agrees qualitatively with the lower ΔPA(α-β), 0-2.9 kcal/mol, determined in the case of pyrrole. It is interesting to note the surprisingly low ΔPA value for 4, (6.9-8.6 kcal/mol), and to compare this value with that for the allyl resonance stabilization energies of the system cyclopentenyl cation (4α), the "homoconjugated" [13] cation (4β) and the cyclopentyl cation (5). The stabilization energy (SE) for (4α):(5) can be derived from equation 2 by the isodesmic substitution procedures using the experimentally derived values for the heats of formation

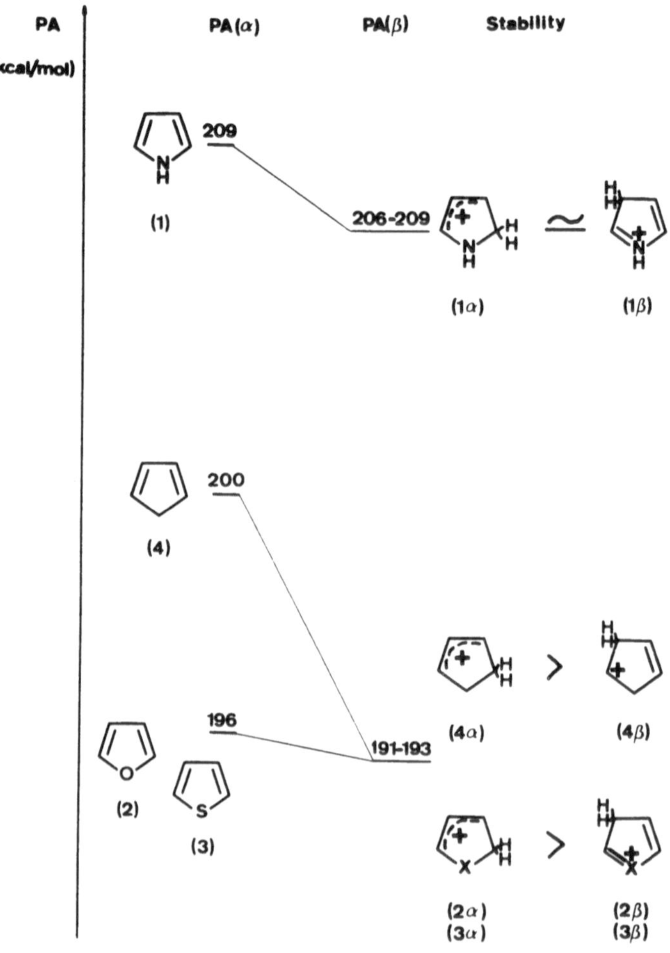

Scheme II

(in kcal/mol) of (4α)= 200.7 [14]. (5) = 198.2 [14], (6) = -18.4 [15] and (7) = 7.7 [15], respectively. SE is found to be 23.6 kcal/mol. From our PA measurements it has to be concluded that (4α)is only 6.9-8.6 kcal/mol more stable than the isomeric, homoconjugated cyclopentenyl cation (4β), thus leaving a stabilization

$$\mathbf{6} + \mathbf{4\alpha} \rightleftharpoons \mathbf{7} + \mathbf{5} \qquad (2)$$

energy of ca. 16 kcal/mol for (4β) compared with (5). There is now strong evidence that the product (4β) is a non-classical bridged ion having the bishomocyclopropenyl structure (4β')(see Ref. 7 for details of this structure determined by MINDO/3 calculations).

(4β')

4. Acknowledgements

Support of this work by the Fonds National Suisse de la Recherche Scientifique, the Deutsche Forschungsgemeinschaft and the Fonds der Chemischen Industrie is gratefully acknowledged.

5. References

1. "Gas-Phase Ion Chemistry", M.T. Bowers, Ed., Chapters 9-11, Academic Press, New York (1979).

2. R.W. Taft, M. Taagepera, J.L.M. Abboud, J.F. Wolf, D.J. DeFrees, W.J. Hehre, J.E. Bartmess and R.T. McIver, Jr., J. Am. Chem. Soc. 100 (1978) 7767.

3. E.M. Arnett, Acc. Chem. Res. 6 (1973) 404; b) P. Kebarle, W.R. Davidson, J. Sunner and S. Meza-Höjer, Pure Appl. Chem. 51 (1979) 63.

4. J. Vogt, A.D. Williamson and J.L. Beauchamp, J. Am. Chem. Soc., 97 (1975) 6682.

5. R. Houriet, J. Vogt and E. Haselbach, Chimia 34 (1980) 277.

6. J.F. Wolf, R.H. Staley, I. Koppel, M. Taagepera, R.T. McIver, Jr., J.L. Beauchamp and R.W. Taft, J. Am. Chem. Soc. 99 (1977) 5417.

7. R. Houriet, H. Schwarz, W. Zummack, J. G. Andrade and P.v.R. Schleyer, Nouv. J. Chim., in press.

8. T.B. McMahon and J.L. Beauchamp, Rev. Sci. Instr., 43 (1972) 509.

9. R. Houriet, H. Schwarz and W. Zummack, Angew. Chem. 92 (1980) 934; Angew. Chem. Int. Ed. Engl. 19 (1980) 905.

10. P. Politzer and H. Weinstein, Tetrahedron, 31 (1975) 915.

11. See for example, T.A. Lehman and M.M. Bursey, "Ion Cyclotron Resonance Spectrometry", Wiley-Interscience, New York (1976).

12. See for example D.K. Bohme, G.I. Mackay and H.I. Schiff, J. Chem. Phys. 73 (1980) 4976.

13. For leading articles on "homoconjugation" see P.R. Story and B.C. Clark, Jr., in "Carbonium Ions", p. 1007, G.A. Olah and P.V.R. Schleyer, eds., Wiley-Interscience, New York (1972) and references cited therein.

14. F.P. Lossing and J.C. Traeger, J. Am. Chem. Soc. 97 (1975) 1579.

15. J.D. Cox and G. Pilcher, "Thermochemistry of Organic and Organometallic Compounds", Academic Press, New York (1970).

GAS-PHASE RADICAL-ION CYCLOADDITIONS : EXPERIMENT AND THEORY

J. O. Lay, Jr. and M. L. Gross
Department of Chemistry, University of Nebraska-Lincoln
Lincoln, NE 68588, U.S.A.

1. Introduction

The importance of cycloaddition pathways in ion-molecule reactions has been the subject of recent investigations [1]. This is in some measure because of the importance of cycloaddition reactions in the synthesis of organic neutrals [2]. Indeed, the selection rules which govern the reactivity as well as regio- and stereo-selectivity of organic neutrals in solution are well established [3]. There are now enough data available to test applicability of some of these selection rules to the cycloaddition reactions of radical cations and neutrals. Some of these concepts have been incorporated into a simple theory, a Frontier Molecular Orbital approach modified to take into account the unique features of radical cations [4].

In the following sections we will

1. describe some of the unique features of radical ion chemistry,
2. introduce the methods used to establish the existence of cycloaddition pathways for radical cation systems,
3. set forth the basic tenets of the theory governing ionic cycloadditions,
4. show correlations of the data presently available, and
5. discuss some of the problems with the theory.

2. Radical-Ion Versus Neutral-Neutral Chemistry

Ion-neutral interactions in the gas-phase have some notable features compared to similar solution-phase reactions, and even to gas-phase neutral-neutral reactions. First, the associative interactions (either ion-dipole or ion-induced dipole) are strong, unmitigated by solvent and operate over a long distance. Consequently, the collision rates are far greater than for neutral-neutral and/or solution-phase reactions [5]. Second, many reactions occur without any significant activation barrier, often resulting in a very high efficiency of reactive collisions [6]. For example, many reactions of radical cations and neutral molecules occur with at least one percent collision efficiency, and, for some of the cycloaddition reactions discussed here, almost every collision results in a reaction. Third, selection rules governing radical cation-neutral cycloadditions, although similar to the principles governing neutral-neutral reactions, should take into account the fact that the reactant is an "open shell" system (i.e., a system containing an unpaired electron).

The reaction profiles for dilute gas-phase ion-molecule reactions are surprising in shape to chemists who are more accustomed to neutral-neutral condensed phase reaction [7]. The intermediate complex often lies in a deep potential well separated from the reactants by little or no activation barrier. Although this ion-neutral adduct is generally a stable species on the reaction profile, it will dissociate to products in the absence of stabilizing collisions. Thus, the intermediate is termed a "chemically activated" species with internal energy equal to the exothermicity for production of the intermediate from the starting materials, and generally will not be observed.
In contrast, neutral-neutral solution phase reactions often proceed over an activation barrier to products which may be stabilized by collisions with surrounding solvent molecules. Finally, endothermic reactions are often not observed in the dilute gas-phase, because of the lack of an energy source.

3. Cycloaddition Pathways in Ion-Neutral Reactions

Solution-phase cycloaddition reactions generally result in measurable quantities of cycloaddition products which can be separated from the reaction mixture; indeed, synthesis of the cycloadduct is often the principal object. Dilute, gasphase ion-neutral cycloaddition typically produce no intact adduct, but rather fragmentation products characteristic of the "activated complex". For this reason, proof of the existence of an ion-neutral cycloaddition product is generally much more difficult. However, a number of schemes have been developed to approach this problem. They include

1. analysis of product distributions to provide information about the nature of the cativated complex,
2. analysis of reaction products by collision induced dissociation (CID) 8 to ascertain the existence of cyclic fragments of the activated cycloaddition complex,
3. interpretation of isotope label distributions of product ions in a manner similar to simple product distributions, and
4. comparison of CID spectra from mass selected, collision stabilized ion-molecule adducts with those of ionized model or reference compounds suspected to have the same structure. An example of each method follows.

3.1 Product Distributions

Product distributions are particularly informative when the reactants are unsymmetrically substituted. K.N. Houk, for example, has shown a good correlation exists between predicted and observed regio- and stereo-selectivity as well as reaction rate in condensed phase cycloaddition reactions [9]. Using Frontier Molecular Orbital (FMO) theory, he has shown that a preference for a single regio- or stereo-isomer may be predicted, and experimentally confirmed, based largely on experimental values such as the ionization potential (IP) and electron affinity (EA) of the reactants.

Similar experiments with dilute gas-phase ion-neutral systems, although complicated by the absence of a stable adduct, may be studied indirectly from the distribution of decomposition products of the intermediate complex. Gross, Lin, and Franklin [10] studied the reactions of ionized 1,3-butadiene and various isomeric 1- and 2-pentenes. This dilute gas-phase reaction, if preceeding via and activated Diels-Alder cycloaddition complex, should fragment to give product ions characteristic of the position of the double bond in the neutral. Specifically, they observed significant differences in product distributions when mixtures of 1,3-butadiene and C_5H_{10} isomers were allowed to react in an ion cyclotron resonance (ICR) spectrometer. For example, the three terminal olefins, 1-pentene, 3-methyl-1-butene, and 2-methyl-1-butene produce a collision complex which predominately loses C_3H_6 to form $C_6H_{10}^{+}$. For cis- and trans-2-pentene, loss of C_2H_5 to form $C_7H_{11}^{+}$ was found. While 3-methyl-1-butene shows more ethane loss via a rearrangement process, direct loss of ethyl radical is about four times larger in 2-methyl-1-butene, consistent with an exposed ethyl group to an unrearranged cycloadduct.

The differences observed in the reactions of butadiene ions and pentene neutrals, are suggestive of a reaction in which each pentene neutral forms a uniqué intermediate complex with 1,3-butadiene. The complex seems to be formed at least initially, as a cycloaddition product, which fragments directly and may in some instances rearrange before fragmentation. However, evidence from product distributions is usually not entirely convincing, and often acyclic intermediate complexes cannot be completely ruled out.

3.2 Detection of Cyclic "Activated Adduct"-Fragment Ions

It is expected that a cyclic intermediate would decompose to give cyclic product ions. Therefore, the detection of product ions having a cyclic structure strongly implicates a cycloaddition mechanism. However, it is also important to establish that the reacting ion and product ions do not undergo any type of ring closing isomerization.

Van Doorn, Nibbering, Ferrer-Correia and Jennings [11] have used Collision Induced Dissociation (CID) spectrometry to establish the existence of a cyclic fragment from the intermediate complex formed by 1,3-butadiene radical cation reacting with methyl vinyl ether. The 11 eV ICR spectrum from the reaction mixture is quite similar to the 15 eV mass spectrum of 4-methoxycyclohexene, the expected Diels-Alder cycloadduct. The intermediate complex in the ion-neutral reaction loses methanol, as does ionized 4-methyloxycyclohexene, presumably from a 4-methyloxycyclohexene cycloadduct ion (Eq. 1) to give an ion at m/z 80.

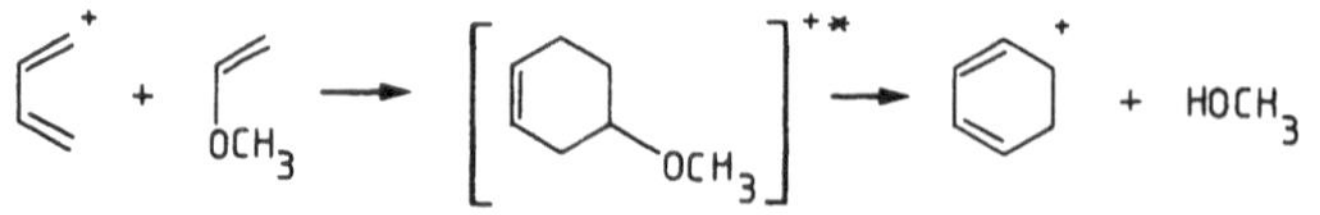

$$^{+} \; + \; OCH_3 \; \longrightarrow \; \left[\quad OCH_3 \right]^{+*} \longrightarrow \; ^{+} \; + \; HOCH_3 \qquad (1)$$

To prove this, the ion-molecule reaction was run in a high pressure source. The structure of the ion at m/z 80 was determined by comparing its CID spectrum using linked [12] scan on a magnetic sector instrument with that of m/z 80 from model compounds. The CID spectrum of the ion-neutral reaction product was most similar to the molecular ion of 1,4-cyclohexadiene, but was also similar to m/z 80 from ionized 1,3-cyclohexadiene and from 4-methoxycyclohexene. The m/z 80 from an acyclic isomer, ionized 1,3,5-hexatriene, gave a distinctly different CID spectrum. The small differences observed in the spectra of the intermediate and the reference compounds were deemed not to be significant in this context because the CID spectra of all of the cyclic m/z 80 ions were sufficiently similar to that of the m/z 80 fragment from the "activated complex" to indicate a ring closed product ion and, hence, an intermediate complex formed via a cycloaddition mechanism.

3.3 Isotope Label Distributions in Product Ions

Label distributions in ion-neutral reaction products provide information regarding the intermediate complex in a manner similar to that obtained using product distributions. For this experiment, the appropriate isotopically labeled compounds, rather than suitably substituted reactants, are required. In principle, this is an easier and more reliable method, because substitution of an isotopic atom into a reactant should have less effect on the rate or mechanism of a reaction than addition of substituents to the reactant(s). For example, some [4 + 2] cycloaddition reactions only proceed with (or without) a particular substituent on the diene/or dienophile (see below). These considerations are not intended to rule out the use of product distributions, but rather to suggest the advantages of isotopic labeling.

Russell and Gross [13] have used extensive ^{13}C and ^{2}H labeling to elucidate the nature of the ion-neutral adduct when fulvene radical cation reacts with 1,3-butadiene in an ICR spectrometer. The intermediate complex has been found to be an ionic analog of a neutral [6 + 4] cycloaddition. Although this is not a common mode of cycloaddition with neutral fulvenes, it can become the dominant mechanism under certain conditions as suggested by Houk and coworkers [14]. The [6 + 4] cycloaddition of dimethylamino-1,3-butadiene with substituted fulvenes [15] is an example.

Ionized fulvene reacts with neutral 1,3-butadiene according to equations 2 - 4, as verified by the correct pressure dependence of relative ion intensities and by double resonance.

$$C_6H_6^{+\cdot} + C_4H_6 \longrightarrow \begin{cases} C_9H_{11}^{+} + H^{\cdot} & (2) \\ C_9H_9^{+} + CH_3^{\cdot} & (3) \\ C_8H_8^{+\cdot} + C_2H_4 & (4) \end{cases}$$

With 1,3-butadiene-1-^{13}C, loss of labeled methyl radical indicated that the two terminal carbons originating in butadiene are involved in nearly half of the methyl loss. Using 1,3-butadiene-2,3-$^{13}C_2$, it was established that one third of the methyl loss originated from the interior butadiene carbons while 80 percent came from some unspecified butadiene carbon. Label distributions for ethylene loss in the above reactions also indicated that 80 % of the loss included a carbon originating from the butadiene terminal positions. With the $^{13}C_2$-labeled material, 11 % $^{13}C_2$-ethylene was observed along with $^{12}C^{13}CH_4$ and $^{12}C_2H_4$, indicating that a fulvene carbon is also lost as part of the departing neutral ethylene.

These results indicate formation of a cyclic intermediate or transition state for at least some of the adduct ions. Otherwise the interior carbons in butadiene would not participate in methyl and ethylene loss.

Numerous cycloaddition products can be postulated for the reaction of ionized fulvene and 1,3-butadiene. Since each reactant can provide more than a single double bond, a number of combinations are possible, including [2 + 2], [2 + 4], [4 + 4] and [6 + 4] cycloaddition pathways (using the terminology for neutrals to simplyfy the considerations). The first three adducts listed above were ruled out for a variety of reasons including isotope label distribution of C-13 atom loss from labeled adducts and difficult steric requirements for some adducts.

The experimental results were best described by a single intermediate structure, 1, which can arise by a stepwise or sychronous process analogous to a [6 + 4] cycloaddition of neutrals (eqs. 5,6).

$$\rightarrow C_9H_9^+ \quad CH_3 \qquad (5)$$

$$\rightarrow C_8H_8^+ \quad C_2H_4 \qquad (6)$$

1

The measured loss of C-13 is consistent with structure 1. For example, methyl loss from five available carbon atoms of the seven membered ring should involve each position equally because of the mobility of the double bond. Hence, the measured values for loss from each position are equal to the expected value of 20 % with an error of ±4 % absolute. Loss of $^{13}C_2H_4$ from the adduct of fulvene ion and 1,3-butadiene-2,3-$^{13}C_2$ is impossible to explain using other likely adducts but is compatible from this intermediate structure after a double bond migration.

3.4 Collision Induced Dissociation of Collision Stabilized Intermediates

Clearly the most powerful method for elucidating the nature of an ion - neutral adduct should involve examination of the intact adduct species. This species, formed as an activated complex in an ICR, can be examined after sufficient stabilization via non-reactive ion-neutral collisions. More useful are the unimolecular and collision induced fragmentations observed using the technique of mass spectrometry-mass spectrometry (ms-ms) [16]. The collision stabilized ion-neutral adduct may be mass selected free of additional ions or ion-neutral products using the initial separation capability (the first analyzer in an ms-ms instrument). The unimolecular or bimolecular (CID) spectra of the adduct are then obtained and compared with the spectra of reference compounds using MS 2. Although this experiment provides the most direct evidence concerning the intermediate, isomerization reactions may occur prior to or during stabilization or during acceleration and transit to the collision cell. Thus, the structure of the intermediate examined in this experiment may not be the same as the initially formed intermediate.

Such an experiment has been conducted by Chess, Lin and Gross [17] using a triple sector mass spectrometer. The collision complex of o-quinodimethane radical cation 2 and neutral styrene was prepared in a high pressure chemical ionization source. The "activated" collision complex was then stabilized by collisions with inert background gas, and the stable adduct was separated by mass analysis from the reactant ions, their fragements and other ion-neural products. High energy (kilovolt) collisions with neutral Helium after double focussing mass selection activated the stable adduct to fragment more extensively than the chemically activated intermediate possessing only the exothermicity of the ion-neutral reaction. The collision induced fragements characteristic of the adduct ion structure were then obtained by scanning an electrostatic analyzer (MS-2). Spectra were also obtained using low energy electron impact ionized neutrals to give $C_{16}H_{16}^{+}$ ions whose characteristic CID spectra could be compared with the pattern obtained for the adduct ion. Additional confirmation was obtained using deuterium labeled styrene, and suitably labeled model compounds.

The CID spectra of the ion-neutral adduct corresponded to the spectra of ionized 2-phenyltetralin, the analogous neutral [4 + 2] cycloaddition product (eq. 7), rather than to 1-phenyltetralin or any of the other seven likely adduct structures considered.

$$\text{2}^{+\bullet} + \text{C}_6\text{H}_5\text{CH=CH}_2 \longrightarrow [\text{2-}C_6H_5\text{-tetralin}]^{+\bullet} \qquad (7)$$

Because measurements of the adduct are made directly using this technique, the results regarding this elusive species are generally more informative. This direct comparison method should be compared with many of the previous method which rely on more indirect or inferential information.

4. Frontier Molecular Orbital Theory of Radical-Ion-Neutral Cycloaddition

There are two compelling reasons for developing a theory. First, it is well known that many ion-neutral reactions that are potentially cycloadditions occur, but it is not yet possible to predict a priori the course of these reactions. Secondly, these dilute gas-phase reactions are a measure of the "intrinsic" properties of molecules and ions because the reactions occur more or less free of any solvation. In that sense, they are an appropriate test of molecular orbital (MO) theory, which seeks to quantify "intrinsic" properties.

Several features of these reactions must be considered in developing a theory for predicting gas-phase ionic cycloadditions. First, the ion-neutral reaction of interest must occur for about 1 % or more of the ion-neutral collisions to be observed using mass spectrometric methods. In addition, there must be little or no activation barrier. If these criteria are not satisfied for a cycloaddition, for example, the observed reaction pathway will involve some other mechanism which more nearly satisfies these requirements.

These requirements are consistent with strong interaction of the bonding orbitals in the intermediate. It is well known that such strong stabilizing interactions due to overlapping orbitals tend to increase reaction rates by lowering activation energy barriers [18].

The most important orbitals, and the ones we will consider, are the highest occupied MO's (HOMO) and the lowest unoccupied MO's (LUMO) in each reactant (ion and neutral). As the ion approaches the neutral, the species are drawn together to produce an intermediate adduct. If initial orbital overlap produces highly stabilized orbitals of a cycloadduct, a cycloaddition reaction will occur. Otherwise some other, or perhaps, no reaction will occur. These ideas are drawn from Frontier Molecular Orbital (FMO) Theory, a perturbation theory originally proposed by Fukui [19] and extended to neutral cycloadditions by Houk [20]. The basic tenant of this theory is that the HOMO and LUMO of each reacting species play the dominant role in determining the course of the reaction. As noted by Houk [14] the "frontier orbital" approximation can be used with remarkable success to rationalize reactivity and regio-selectivity phenomena, in spite of the fact that interactions of extra frontier orbitals, closed shell repulsions, and coulombic terms may also contribute to energy changes. Omission of these other interactions, particularly extrafrontier orbitals, is based on the observation that the lower level orbitals interact in each species to produce bonding and antibonding pairs, and hence, they do not strongly affect the incipient adduct. Furthermore, it is the nature of many organic polyenes that if the outermost orbitals (FMO's) interact in a symmetry allowed manner to produce energy stabilized adduct orbitals, the remaining lower level extrafrontier orbitals will follow accordingly. The stabilization energy, ΔE, resulting from orbital overlap is given by the second order perturbation expression:

$$\Delta E = \frac{H_{ij}^{2}}{E_i - E_j}$$

where H_{ij} is a measure of the effective overlap of the HOMO and LUMO and $E_i - E_j$ is the energy level difference between the two orbitals. Thus, as ΔE increases for a particular mechanism, the likelihood that this will be the dominant mechanism also increases. Therefore, it is proposed that a necessary conditon for an ion-neutral cycloaddition is sufficient stabilization (large ΔE) of the new, incipient molecular orbitals in the adduct ion, otherwise another reaction pathway will be observed, or the ion neutral system may not react at all.

4.1 Requirements for a Cycloaddition Pathway

Based on the requirement of large stabilization of the incipient molecular orbitals, we suggest three criteria for a cycloaddition mechanism:

1. Orbital symmetry must be conserved. This provides good overlap, large H_{ij} and, hence, large ΔE.
2. The relevant HOMO-LUMO orbital interaction must produce large stabilization through the interaction of orbitals of similar energy. Thus, for a large ΔE, $E_i - E_j$ must be small to produce "sufficient" stabilization. An additional

complication arises due to the fact that the reactant is a radical ion with a single electron in its HOMO. Therefore, we propose to ignore the one electron ion-HOMO/neutral-LUMO interaction because it leads to a bond with bond order one half. Thus, the relevant orbital interaction is considered to be the two electron ion-LUMO/neutral-HOMO interaction (Figure 1) with a possible exception when E_i approaches E_j.

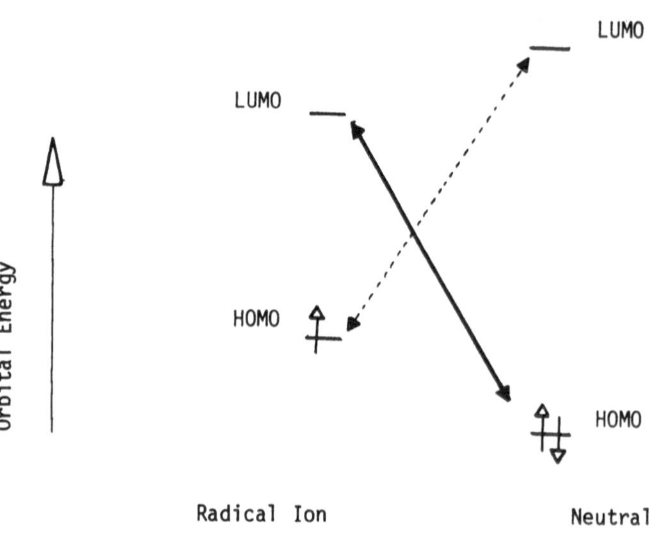

Fig. 1. The two electron HOMO-neutral / LUMO-ion and the weaker one electron HOMO-ion / LUMO-neutral interactions.

3. The sites of charge localization in the reactant must be consistent with cycloaddition. Because of the presence of the charge, ion/neutral reactions are generally quite efficient. Hence, it is proposed that at least one and preferably two of the charge sites participate in new bond formation in order that a cycloaddition may procede.

The extent of stabilization can be estimated using MO theory, or by the method of Fukui [21] or Epiotis [22]. Orbital energy levels may be calculated using computational methods, but ionization potential and electron affinity data on neutral molecules may also be used in accord with Koopman's Theorem [23]. The sites of charge localization are easily determined using simple Huckel calculations.

4.2 Application of the Theory to the Reaction of Fulvene Radical Cation and 1,3-Butadiene

As discussed previously, fulvene radical cation reacts with 1,3-butadiene via a mechanism analogous to a neutral [6 + 4] cycloaddition. We will now consider this reaction in terms of the three criteria proposed above.

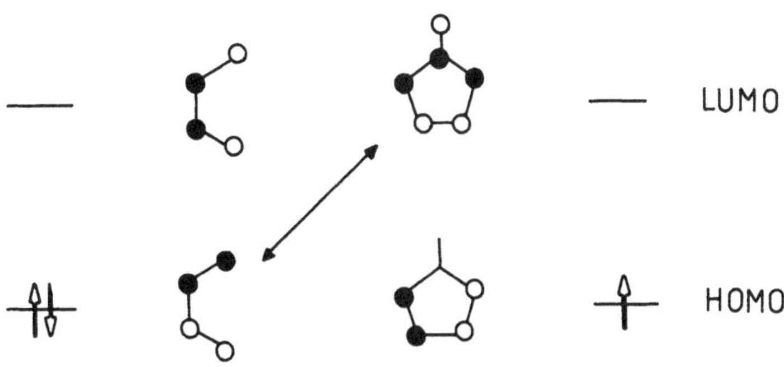

Fig. 2. The symmetry of the fulvene/butadiene HOMO's and LUMO's showing the two electron neutral-HOMO / ion-LUMO interaction.

First, the relevant two electron HOMO-LUMO interaction leads to overlap of orbitals with the same phase (see Figure 2). The other, one electron HOMO-LUMO interaction is neither "allowed" or "disallowed" with regard to the phase of the orbitals because of the node on the exposed carbon.

Second, the sites of major charge localization are situated at the points of new bond formation in the incipient adduct (see Table 2). The values are calculated from the sum of the products of the MO occupancies and the square of the AO coefficients evaluated by the Huckel Molecular Orbital (HMO) approach.

The final factor, the extent of stabilization due to the energy match of the relevant overlapping orbitals (here the ion LUMO and neutral HOMO) can be approximated by considering the difference in the IP of the neutral and the EA of the neutral corresponding to the ion. Since the relationship is inverse, relavtively small orbital energy level differences (IP-EA) provide sufficient stabilization for cycloaddition to occur. For this example, the value of about 10.5 eV is relatively small, as will be shown later, and the reaction does indeed occur.

In summary, charge considerations are in accord with cycloadduct formation involving two of the three atoms labeled 1, 3 or 6 (see Figure 3). Furthermore we may suggest that the terminal butadiene carbons would react with carbon atoms 1 and 3 or the equivalent combination 1 and 6 based on orbital symmetry. These two considerations dictate how the ion and neutral orient themselves with respect to one another in the adduct. The remaining consideration, whether or not

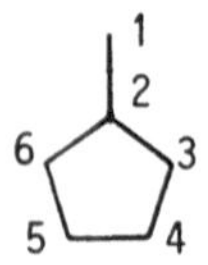

Fig. 3. The fulvene carbon skeleton.

Table 1. Correlation of energy level differences (IP - EA) for the two electron HOMO-LUMO interaction and reaction mechanism. Data taken from references 14, 24, 25.

System (Ion/Neutral)	Orbital Stabilization[1] (IP - EA)	Mechanism	Reference
1-methoxybutadiene/ ethylene	12.9 eV	no reaction	29
butadiene/cyanoethylene	11.5 eV	no reaction	29
butadiene/ethylene	11.1 eV	reaction,no cycloaddition	26
furan/butadiene	11.0 eV	cycloaddition	26
butadiene/1-pentane	10.0 eV	cycloaddition	10
butadiene/2-pentane	9.7 eV	cycloaddition	10
butadiene/methyl vinyl ether	9.5 eV	cycloaddition	27
fulvene/butadiene	∿9.5 - 10.7 eV[2]	cycloaddition	13
o-quinodimethane/butadiene	∿9 eV[2]	cycloaddition	17
styrene/styrene	8.8 eV	cycloaddition	28

notes

(1) Orbital stabilization increases as IP - EA decreases.

(2) Approximate values, the appropriate EA values are estimated.

the stabilization gained because of the similar energy level of the orbitals, (E_i-E_j) or (IP-EA), is sufficient to make cycloaddition competitive with other potential mechanisms must be answered empirically at this time based on observations of numerous similar reaction and non-reacting systems. This will be discussed in the following section.

4.3 Correlation of Experiment and Theory

Using the available experimental data, it is possible to correlate the known radical cation-neutral cycloadditions with predictions based on the theoretical model presented. Orbital stabilization is considered separtely from orbital symmetry and charge localization. We believe orbital stabilization arising from the two electron HOMO-LUMO interaction is most relevant to whether or not a cycloaddition mechanism can occur and/or be competitive with alternate mechanisms. On the other hand, orbital symmetry for the two electron FMO pair, and charge distribution on the ion seem more likely to direct the orientation of cycloaddition reactions.

The correlation between orbital energy level differences (IP-EA) and reactivity (see Table 1) indicates that when the ion LUMO and neutral HOMO are separated in energy by about 11 eV, or less, the cycloaddition reaction occurs. As for neutral-neutral systems, this value is approximate and best used as a general guide. The electron affinities (and occasionally ionization potentials) are sometimes not well established; thus, the correlation between (IP-EA) and orbital energy level differences is only approximate. More importantly, any correlation between orbital energy levels and reactivity is most appropriate when the reacting systems being compared are identical except for substituents around the reacting multiple bonds. Because of the lack of additional data regarding ion-neutral cycloadditions, the ion-neutral reactions in Table 1 are compared as if the reaction systems were indeed identical. This generalization illustrates that the important orbital stabilization is due to the two electron FMO's as postulated.

The major sites of charge localization and the symmetry of the two electron FMO interaction are shown in Table 2. If one requires one of the two incipient sigma bonds to form at the locale of greatest positive charge, and the other at one other major charge site, and then further requires that the location of these new sigma bonds be such that the symmetry of orbital overlap for the two electron FMO interaction is conserved, the predicted adducts shown in the

Table II. Predicted and observed orientation of cycloaddition.

System	Major Charge Locations	Sym. Ion LUMO	Sym. Neut. HOMO	Predicted	Observed
	.36 .14				
	.14 .36				
	.14 .36				
	.14 .36				
	.16 .35 .11 .09 .16				
	.38 .27				
	.07 .11 .29				

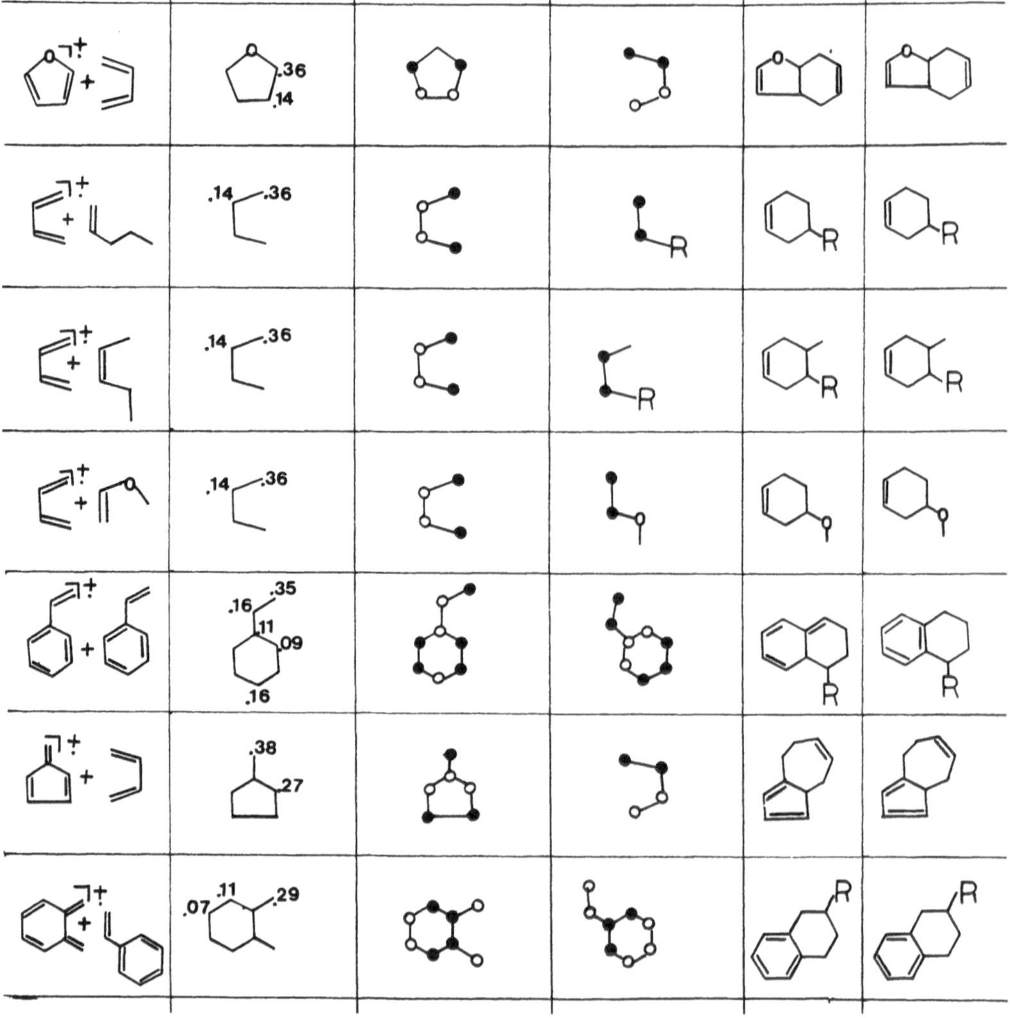

table are obtained. Note that we have not precluded shifts of hydrogen atoms (as with the styrene-ion/styrene-neutral system) to produce a more stable cycloadduct. This in no way violates the theoretical model because it occurs after adduct ion formation.

A good example of the bond-directing nature of orbital symmetry is the reaction of the furan ion with neutral 1,3-butadiene. One of the two equivalent sites of largest positive charge (+.36) is involved in sigma bond formation. Orbital overlap directs the other new sigma bond to form between the adjacent carbon (+.14), where orbital overlap is in phase, rather than across the oxygen to the other site of largest positive charge (+.36). A similar effect determines the orientation in the styrene ion-neutral system. Two of the predicted adducts based on consideration of the positive charge are shown in equations 8 and 9 while eq. 9 shows the expected adduct when orbital symmetry is also considered.

3

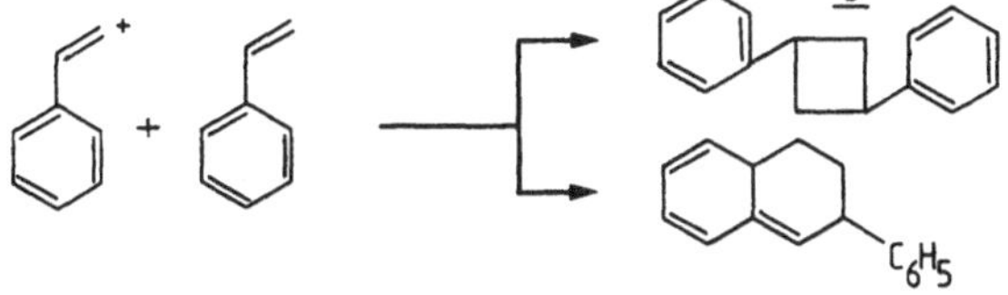

C_6H_5

(8)

(9)

Note that only one of the possible isomers (4) is shown. Thus far we have not extended this theory for predicting a preference for a regioisomer (either a or b for example), but we can use it to exclude formation of some isomeric adducts (ie. 3).

4a 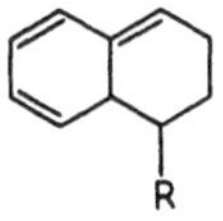 R

4b R

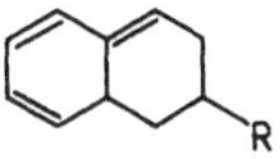

Considerable regioselectivity has been observed with neutral-neutral systems which correlates well with FMO considerations. Likewise, a preference for a particular regioisomer can be predicted for ion-molecule reactions. Ion neutral DMO's presumably interact to maximize overlap. Hence, the largest of the two reacting atomic orbitals (AO's) on each FMO overlap (in phase) while the two reacting atoms with the smaller AO's interact similarly. The new σ bonds, preferentially formed with maximum FMO overlap, lead to some regioselectivity in

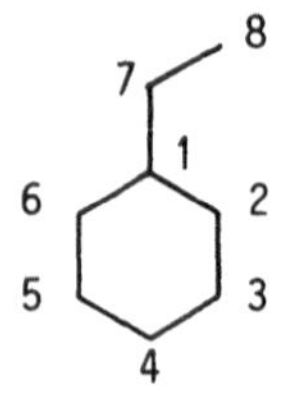

	HOMO (neutral)	LUMO (ion)
1	-0.334	-0.334
2	-0.308	0.308
3	0.130	0.130
4	0.394	-0.394
5	0.130	0.130
6	-0.308	0.308
7	0.394	-0.394
8	0.595	0.595

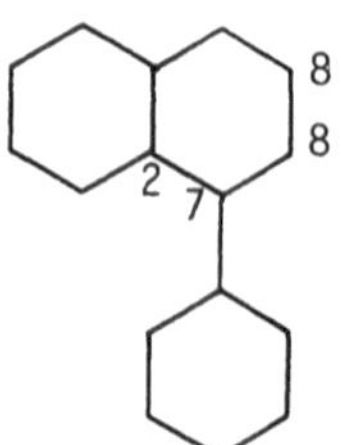

Fig. 4. Huckel Orbital Coefficients for the styrene HOMO and LUMO, and the predicted regioselectivity.

neutral-neutral systems and possibly also for ion-neutral systems. This can be illustrated for the styrene ion-molecule reaction.

The styrene ion-neutral system, with the Huckel orbital coefficients, is shown in Figure 4. Overlap of the atomic orbitals (AO) 7 and 8 on one styrene moiety with AO's 2 and 8 on the other will direct regioselectivity. Since the styrene ion could act as either the "diene" or the "dienophile" in this reaction, it is not clear whether AO's 7 and 8 of the HOMO or of the LUMO should be used. Fortunately the AO on carbon 8 has the largest coefficient in both MO's. For this reason a sigma bond between carbon 8 of the ion and carbon 8 of the neutral should form. The other sigma bond will form between atoms 2 and 7 to preferentially give the regioisomer shown, which may isomerize via a hydrogen shift to ionized 1-phenyltetralin. In fact, this is the adduct of the ion-neutral cycloaddition of styrene reported by Wilkins and Gross [27] (Recent experiments suggest that the hydrogen shift may not occur) [28]. Unfortunately, this is the only adduct thus far investigated with the appropriate asymmetry necessary to test the preference for a particular regioisomer. Furthermore, this preference will vary, decreasing as orbital energy levels become less well matched energetically. Perhaps this isomer with its narrow two electron FMO energy difference (8.8 eV) will be one of the few to exhibit such regioselectivity.

4.4 Difficulties with the Theory

The principal simplification of this theory is the exclusion of the one electron HOMO-LUMO interaction. This and, to a lesser extent, interactions of the ion HOMO with neutral lower orbitals contribute to energy changes during adduct formation. We believe this omission is justified because the one electron HOMO-LUMO interaction will generally be less important than the two electron counterpart unless E_i-E_j approaches 1/2 the value for the two electron interaction. Furthermore, interaction of the ion HOMO with lower level extra-frontier orbitals of the neutral, the second highest occupied MO for instance, should also be weak. These interactions also occur in neutral-neutral reaction systems where the electrons are all paired; however, the resulting bonding and/or antibonding orbitals produced in the cylcoaddition are only slightly perturbed by these longer range orbital interactions. These considerations and the observed experimental results are generally in accord with the assumption that neglect of these interactions is justified for most ion-neutral systems. Further refinement of this theory to include the one electron HOMO-LUMO interaction and the other odd electron interaction would probably produce a more complete theory at the risk of introducing additional complexity.

The relative importance of symmetry and charge location have not been clearly delineated thus far. It seems reasonable to expect the location of the major portion of the charge to play an important role in determining the mechanism just as charge is important in the kinetics of ion-molecule reactions. Orbital symmetry may be relegated to a role of directing formation of the adduct with preference at those charge sites (when several are present) which allow "in-phase" orbitals to interact. Perhaps symmetry of orbital overlap may play a more dominant role, especially when the charge is almost exclusively on a single carbon atom.

The exact values of (IP-EA) which allow or cause cycloaddition reactions to compete favorably with other possible mechanisms can be only experimentally determined at this point (This is also true with neutral-neutral systems). Furthermore, the required amount of orbital stabilization will vary from system to system. The chemical system for which the most data is available involves ionized 1,3-butadiene and neutral substituted ethylene. The other reactions are somewhat different and should ideally be treated separately because the required stabilization due to IP-EA may vary. Until more data are available for several different chemical systems, however, they will be treated as a group.

Finally, an experimental discrepancy has been noted. While it is known that methyl vinyl ether (MVE) and 1,3-butadiene radical cation react to form a cycloadduct (consistant with the theoretical framework), it has also been determined that MVE radical cation also reacts with neutral 1,3-butadiene in an ICR [27]. Based on orbital energies, IP-EA = 12.2 eV, this reaction should not occur when MVE carries the charge. This could be an indication that both orbital interactions should be considered for predicting cycloadditions.

However, this single discrepancy does not disprove the theoretical model, but does suggest that additional cycloaddition reactions be elucidated to further test the theory and, if necessary, alter it to more accurately reflect the nature of ion-neutral cycloadditions.

5. Conclusion

We believe that the occurrence of radical ion-neutral cycloaddition reactions is governed by three main principles. First, the charge which makes ion-neutral reactions so rapid, must be involved in new bond formation. Furthermore those orientations involving the atoms having a major fraction of the charge and also which allow the orbital symmetry of the two electron ion-LUMO/neutral-HOMO to be conserved (ie. allows the orbitals to overlap in phase) will be the preferred modes of cycloaddition. Finally, cycloaddition reactions will only be observed if orbital stabilization due to the orbital energy match (IP-EA) is sufficient. For systems similar to ionized 1,3-butadiene and ethylene, the orbital energy match should be about 11.0 eV or less.

Further refinement of the theory to include odd electron orbital interactions will add considerable complexity but may well strengthen this theoretical model when additional data on ion-neutral cycloadditions become available.

6. References

1. For a review; D.H. Russel and M.L. Gross, Lecture Notes in Chemistry, Vol. 7 (1978)
2. For example; H. Wollweber, "Diels-Alder Reactions", Georg-Thieme-Verlag, Stuttgart, Germany (1972)
3. (a) R.B. Woodward and R.H. Hoffman, "The Conservation of Orbital Symmetry", Academic Press, New York (1970);
   (b) J.S. Dewar, "The Molecular Orbital Theory of Organic Chemistry", McGraw-Hill, New York (1969);
   (c) K. Fukui and H. Fujimoto, "Mechanisms of Molecular Migrations", B.S. Thagarajan, ed., Vol. 2, pp. 118-186, Wiley, New York (1969)
4. J.O. Lay Jr. and M.L. Gross, "Theory for Predicting Structures of Intermediate Complexes of Cycloadditions", 28th Annual Conference on Mass Spectrometry and Allied Topics, ASMS, New York (1980)
5. T.A. Lehman and M.M. Bursey, "Ion Cyclotron Resonance Spectrometry", Wiley, New York, 63 (1976)
6. S.G. Lias and P. Ausloos, "Ion Molecule Reactions", American Chemical Society, Washington D.C., 64 (1975)

7. J.L. Beauchamp, "Interactions Between Ions and Molecules", P. Ausloss, ed., Vol. 6 in N.A.T.O. Advanced Study Institute Series B, (Physics), Plenum New York, (1976)

8. (a) F.W. McLafferty, "High Performance Mass Spectrometry: Chemical Applications", M.L. Gross, ed., ACS Symposia Series No. 70, American Chemical Society, Washington D.C. (1978);
   (b) R.G. Cooks, ed., "Collision Spectroscopy", Plenum Press, New York (1978);
   (c) K. Levsen and H. Schwarz, Angew. Chemie, Int. Edit. 15 (1976) 509

9. (a) K.N. Nouk, Acc. Chem. Res. 8 (1975) 361,
   (b) K.N. Houk, J. Am. Chem. Soc. 95 (1973) 4092

10. M.L. Gross, P.H. Lin, and S.J. Franklin, Anal. Chem. 44 (1972) 974

11. R. van Doorn, N.M.M. Nibbering, A.J.V. Ferrer-Correia, and K. Jennings, Org. Mass. Spectrom. 13 (1978) 729

12. (a) A.P. Bruins, K.R. Jennings and S. Evans, Int. J. Mass Spectrom. Ion Phys., 26 (1978) 395
    (b) R.S. Stradling, K.R. Jennings and S. Evans, Org. Mass Spectrom. 13 (1978) 429

13. D.H. Russel and M.L. Gross, J. Am. Chem. Soc. 102 (1980) 6279

14. K.N. Houk, J.K. George, and R.E. Duke Jr., Tetrahedron 30 (1974) 523

15. L.C. Dunn, Y.-M. Chang, K.N. Houk, J. Am. Chem. Soc. 98 (1976) 7095

16. J.H. Beynon and R.G. Cooks, Res. Dev. 22 (1976) 26

17. E.K. Chess, Ph. D. Thesis, University of Nebraska, 1982

18. For an explanation see; I. Flemming, "Chemical Reactions", John Wiley & Sons, New York, 23 (1976)

19. K. Fukui, Acc. Chem. Res. 4(1971 57

20. see ref. 9a

21. K. Fukui, Fortsch.Chem. Forsch 15 (1970) 1

22. N.D. Epiotis, Angew. Chemie, Int. Edit. Eng. 13 (1974) 751

23. T. Koopman, Physica 1 (1933) 104

24. H.M. Rosenstock, K. Draxl, B.W. Steiner and J.T. Herron, J. Phys. and Chem. Ref. Data, Vol. 6 (1977)

25. (a) K.N. Houk, J. Am. Chem. Soc. 95 (1973) 4092
(b) K.D. Jordan and P.D. Burrow, Acc. Chem. Res. 11 (1977) 341

26. M.L. Gross, D.H. Russel, R. Phenbetchara and P.H. Lin, Advances in Mass Spectrometry, 7a (1978) 129

27. C.L. Wilkins and M.L. Gross, J. Am. Chem. Soc. 93 (1971) 895

28. Unpublished results, University of Nebraska

RING IONS IN THE ION CHEMISTRY OF THIIRANE, ETHANEDIOL-1,2, ETHANEDITHIOL-1,2 AND 2-MERCAPTOETHANOL

G. Baykut*, K.-P. Wanczek and H. Hartmann
Institute of Theoretical and Physical Chemistry
University of Frankfurt, FRG

1. Introduction

The molecular ion of thiirane retains the ring structure of the neutral molecule during ionization with great probability [1,2]. The same is true for phosphirane [3]. The three-ring heterocycles C_2H_4X (X = N,O) containing the lighter elements of group V and VI react different. The oxirane molecular ion ring-opens before reaction with n-donor bases[4]. Aziridine has no ion chemistry characteristic for its structure at all [2]. The cyclopropane molecular ion [5,6] shows ion-molecule reactions indicating three equivalent carbon atoms. The tendency to form rings is quite pronounced in ion chemistry of the sulfur containing compounds 2-mercaptoethanol and ethanedithiol-1,2. The same is found for the oxygen analogue ethanediol-1,2. Among the most abundant product ions in the ion chemistries are cyclic acetals or ketals and thioacetals or thioketals, respectively.

2. Experimental

The experiments were carried out with a pulsed ICR spectrometer with a trapped ion ICR cell, as described elsewhere [1,2]. Commercially available samples were used after purification by trap-to-trap distillation.

---

* On leave from: Department of Analytical Chemistry, Faculty of Chemistry, University of Istanbul, Turkey.

Thiirane decomposes after a few days. Therefore it is stored after distillation in vacuo at liquid nitrogen temperature and only warmed up for measurements.

3. Results and Discussion

3.1 Thiirane

In the ion chemistry of thiirane several transfer reactions are observed [1,2] , they are most important for the molecular ion:

$$C_2H_4S^+ + C_2H_4S \longrightarrow C_2H_4S_2{}^+ + C_2H_4 \qquad (1)$$

$$C_2H_4S_2{}^+ + C_2H_4S \longrightarrow C_2H_4S_3{}^+ + C_2H_4 \qquad (2)$$

$$C_2H_4S_3{}^+ + C_2H_4S \longrightarrow C_2H_4S_4{}^+ + C_2H_4 \qquad (3)$$

and for the protonated molecule:

$$C_2H_5S^+ + C_2H_4S \longrightarrow C_2H_5S_2{}^+ + C_2H_4 \qquad (4)$$

$$C_2H_5S_2{}^+ + C_2H_4S \longrightarrow C_2H_5S_3{}^+ + C_2H_4 \qquad (5)$$

Study of $^{34}SC_2H_4$ shows that in reaction 1 and 4 a sulfur atom is transferred from neutral thiirane to the molecular ion ^{34}S occurs in 4.4 % natural abundance, sufficient for the study of these reactions. Transfer of a sulfur ion from the molecular ion to the neutral molecule is only a minor reaction. No further sulfur transfer occurs, even at a reaction time of 0.9 sec. This indicates a rate constant of reactions 3 and 5 at least 10 times larger than the rate constant of a next sulfur transfer step yielding $C_2H_4S_5{}^+$ and $C_2H_5S_4{}^+$, respectively.

The same types of reactions are also detected in the ion chemistry of phosphirane, in this case the PH group is transferred [3].

The protonated molecule is formed almost exclusively by the even electron fragment HCS^+:

$$HCS^+ + C_2H_4S \longrightarrow C_2H_5S^+ + CS \qquad (6)$$

and not by the molecular ion. The corresponding reaction of phosphirane is:

$$HCPH^+ + C_2H_4PH \longrightarrow C_2H_6P^+ + HCP \qquad (7)$$

The molecular ion of phosphirane also transfers a proton:

$$C_2H_4PH^+ + C_2H_4PH \longrightarrow C_2H_6P^+ + C_2H_4P \qquad (8)$$

Important for the structure elucidation of the three ions are their reactions with certain n-donor bases.
The molecular ion of thiirane transfers CH_2^+, S^+, and $C_2H_4^+$ ion to neutral ammonia:

$$C_2H_4S^+ + ND_3 \longrightarrow CH_2ND_2^+ + SCH_2D \qquad (9)$$

$$\longrightarrow SND_2^+ + C_2H_5 \qquad (10)$$

$$\longrightarrow C_2H_4ND_3^+ + S \qquad (11)$$

The question whether the molecular ions of type $C_2H_4X^+$ (X=O,S,NH,PH, CH_2) retain their cyclic structures has been studied in great detail. In the cyclopropane molecular ion the three carbon atoms are equivalent [5,6], as its CH_2^+ transfer to ammonia indicates. No such transfer is observed in the reaction of the open-chain analogue, $CH_3CHCH_2^+$, with NH_3. These results can however not be generalized. Corderman et al. [4] showed that the molecular ion of oxirane which transfers CH_2^+ has the structure $CH_2OCH_2^+$. It is not cyclic. Also for the cyclopropane molecular ion the situation seems to be more complicated. Recently, the energy surface of this ion has been calculated [7]: The lower surface shows a flat pseudorotation region. The three-membered carbon atom ring is a scalene triangle at each of the minima. Its geometry is half way between a π-complex of CH_2^+ with ethene and a trimethylene cation which is ring-closed by a long one-electron bond. In the first excited state there is a one-electron antibond between the two open ends of the ion.

Therefore the following transfer mechanism may be proposed assuming reaction of ground-state $C_3H_6^+$:

$$\dot{C}H_2\text{-}CH_2\text{-}\overset{+}{C}H_2 \longleftrightarrow \overset{+}{C}H_2\text{-}CH_2\text{-}\dot{C}H_2$$

$$+ ND_3 \longrightarrow CH_2ND_3^+ + C_2H_4 \qquad (12)$$

also indicating three equivalent carbon atoms in the ring.

The energy surfaces of thiirane and phosphirane are not known. The similarities in the ion chemistry of the three compounds C_2H_4X (X=CH_2,PH,S) may indicate similar structures of the molecular ions. Conclusions drawn from ion-molecule reactions should take into account the fact that ions may not have a structure comparable to that of neutral ground state molecules. Therefore the characteristic ion-molecule reactions should be utilized with all due caution to obtain information on ion structure.

There are five general types of transfer reactions for C_2H_4X (X=CH_2,NH, PH,O,S):

a) transfer of CH_2^+ to NH_3
b) transfer of CH_2 to NH_3^+
c) transfer to NH_3 of a molecular fragment ion containing the hetero atom,
d) transfer of the same fragment to the molecular ion,
e) transfer of CH_2 (CH_2^+) to the molecular ion (neutral molecule).

Aziridin does not undergo a single of the five reactions. Phosphirane reacts via reactions c and d:

$$C_2H_4PH^+ + NH_3 \longrightarrow HPNH_3^+ + C_2H_4 \tag{13}$$

$$C_2H_4PH + NH_3^+ \longrightarrow HPNH_3^+ + C_2H_4 \tag{14}$$

Oxirane shows reactions a, b and e [4]:

$$C_2H_4^+ + NH_3 \longrightarrow CH_2NH_3^+ + CH_2O \tag{15}$$

$$C_2H_4O + NH_3^+ \longrightarrow CH_2NH_3^+ + CH_2O \tag{16}$$

$$C_2H_4O^+ + C_2H_4O \longrightarrow C_3H_6O^+ + CH_2O \tag{17}$$

Thiirane shows reactionsa-d (reactions 9, 18, 19 and 1):

$$NH_3^+ + C_2H_4S \longrightarrow CH_2NH_2^+ + SCH_3 \tag{18}$$

$$C_2H_4S^+ + NH_3 \longrightarrow SNH_3^+ + C_2H_4 \tag{19}$$

For cyclopropane reactions a and b [6] are detected:

$$C_3H_6^+ + NH_3 \longrightarrow CH_2NH_3^+ + C_2H_4 \quad (20)$$

$$C_3H_6 + NH_3^+ \longrightarrow CH_2NH_3^+ + C_2H_4 \quad (21)$$

Also reaction e is probable.

Due to the conclusions of Corderman et al. [4] CH_2^+ transfer to certain n-donor bases like ammonia or phosphine cannot prove the structure of the ions under discussion. It should be noted that even the molecular ion of dimethyl sulfide transfers CH_2^+ to ammonia at elevated pressure [1]. Taking into account however, all the characteristic reactions the same type of reactivity is indicated for the thiirane, phosphirane and cyclopropane molecular ions. The molecular ion of thiirane transfers $C_2H_4^+$ to ammonia, reaction 11, without scrambling of the hydrogen atoms. This also indicates the cyclic structure of this ion. The structures $CH_2\text{-S-}CH_2^+$ or $CH_2\text{=}\overset{+}{S}\text{-}CH_2$ are therefore not probable.

The fragment ion $C_2H_4S^+$, the base peak in the ICR spectrum of 2-mercapto-ethanol shows the same ion chemistry as the molecular ion of thiirane. Contrary the fragment ion $C_2H_4O^+$ from 2-mercaptoethanol transfers no CH_2^+ ions to n-donor bases. The reactions of this ion, which will be discussed below indicate an acetaldehyde cation structure.

3.2.2-Mercaptoethanol, Ethanediol-1,2 and Ethanedithiol-1,2

The ion chemistries of the three compounds show a complexity increasing rapidly from ethanediol to ethanedithiol to 2-mercaptoethanol, as expected. In the ion chemistry of $HSCH_2CH_2OH$ more than 70 ion-molecule reactions are detected. They are shown schematically in Figure 1.

All three compounds show two important classes of reactions:

1. proton transfer to the neutral molecules,
2. acetal and ketal formation (c.f. Table 1).

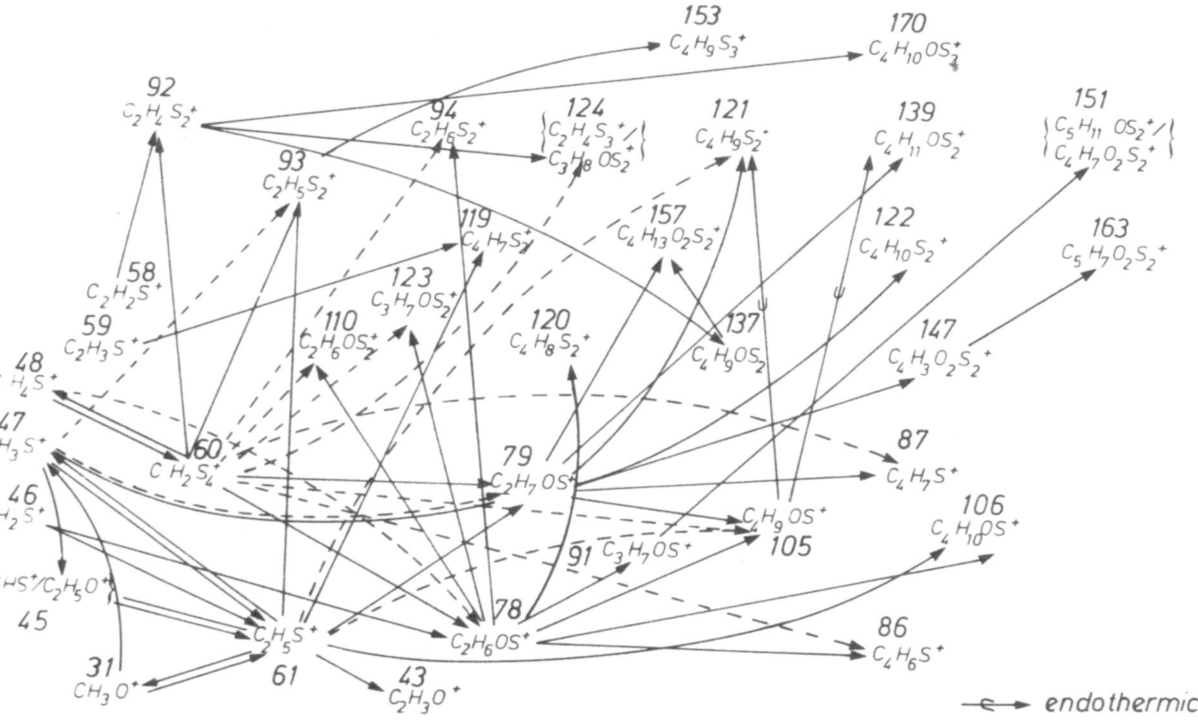

Fig. 1. Ion chemistry of 2-mercaptoethanol. If reaction in one step or a consecutive reaction are possible the one step reaction is indicated by a broken arrow.

Tab. 1 Reactions yielding cyclic products in the ion chemistry of ethanediol, ethanedithiol and 2-mercaptoethanol

product ion	$C_2H_6O_2$	no.	$C_2H_6S_2$	no.	C_2H_6OS	no.
			reacting ion			
$C_3H_6X_2^{+a}$			CH_3S^+	30		
			$C_2H_4S^+$	31		
			$C_2H_6S_2^+$	32		
$C_3H_7X_2^+$	CH_3O^+	23	CH_3S^+	33	CH_3S^+	40
	CH_5O^+	24	$C_2H_5S^+$	34	$C_2H_4S^+$	41
	$C_2H_6O_2^+$	22	$C_2H_6S_2^+$	35	$C_2H_6OS^+$	42
$C_4H_7X_2^+$	$C_2H_3O^+$	25	$C_2H_3S^+$	36	$C_2H_3S^+$	47
	$C_2H_5O^+$	26	$C_2H_6S_2^+$	37	$C_2H_5S^+$	48
	$C_2H_7O_2^+$	27				
$C_4H_9X_2^+$	$C_2H_5O^+$	28	$C_2H_5S^+$	38	$C_2H_4S^+$	43
	$C_2H_7O_2^+$	29	$C_2H_6S_2^+$	39	$C_2H_5S^+$	44,49
					$C_2H_6OS^+$	45
					$C_2H_7OS^+$	46,50
					$C_4H_9OS^+$	51

a: X = O,S

Only those reactions of the second class will be discussed here, where ring structures of the product ions are probable.

There are four types of ring structures which can be formed in the ion-molecule reactions of the three compounds:

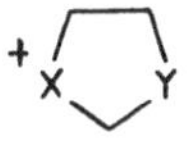

$C_3H_6XY^+$ I

$C_3H_7XY^+$ II

III a

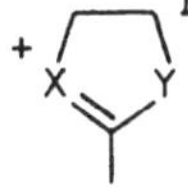

IV a

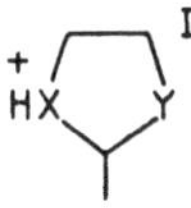

III b 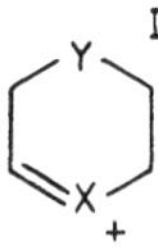 $C_4H_7XY^+$

IV b $C_4H_9XY^+$

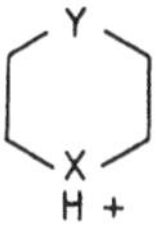

X,Y = O,S.

The product II is formed in the ion chemistry of glycol by the molecular ion and the fragment ions CH_3O^+ and CH_5O^+:

$$C_2H_6O_2^+ + C_2H_6O_2^+ \longrightarrow C_3H_7O_2^+ + (CH_5O_2) \qquad (22)$$

$$CH_3O^+ + C_2H_6O_2 \longrightarrow C_3H_7O_2^+ + H_2O \qquad (23)$$

$$CH_5O^+ + C_2H_6O_2 \longrightarrow C_3H_7O_2^+ + (H_2O + H_2) \qquad (24)$$

Like $C_3H_7O_2^+$, the product ion $C_4H_7O_2^+$ is also formed by condensation reactions:

$$C_2H_3O^+ + C_2H_6O_2 \longrightarrow C_4H_7O_2^+ + H_2O \qquad (25)$$

$$C_2H_5O^+ + C_2H_6O_2 \longrightarrow C_4H_7O_2^+ + (H_2O + H_2) \qquad (26)$$

$$C_2H_7O_2^+ + C_2H_6O_2 \longrightarrow C_4H_7O_2^+ + (2H_2O + H_2) \qquad (27)$$

A condensation of $C_2H_5O^+$ and $C_2H_7O_2^+$ with the neutral glycol molecule leads to $C_4H_9O_2^+$:

$$C_2H_5O^+ + C_2H_6O_2 \longrightarrow C_4H_9O_2^+ + H_2O \qquad (28)$$

$$C_2H_7O_2^+ + C_2H_6O_2 \longrightarrow C_4H_9O_2^+ + 2\,H_2O \qquad (29)$$

Contrary to glycol, ethanedithiol forms also the product $C_3H_6S_2^+$

$$CH_3S^+ + C_2H_6S_2 \longrightarrow C_3H_6S_2^+ + H_3S \qquad (30)$$

$$C_2H_4S^+ + C_2H_6S_2 \longrightarrow C_3H_6S_2^+ + CH_4S \qquad (31)$$

$$C_2H_6S_2^+ + C_2H_6S_2 \longrightarrow C_3H_6S_2^+ + (CH_4S + H_2S) \qquad (32)$$

The product ion $C_3H_7S_2^+$ is generated by reactions 33 - 35:

$$CH_3S^+ + C_2H_6S_2 \longrightarrow C_3H_7S_2^+ + H_2S \qquad (33)$$

$$C_2H_5S^+ + C_2H_6S_2 \longrightarrow C_3H_7S_2^+ + CH_4S \qquad (34)$$

$$C_2H_6S_2^+ + C_2H_6S_2 \longrightarrow C_3H_7S_2^+ + (CH_5S_2) \qquad (35)$$

Reaction 35 is endothermic.
The ion with mass 119, $C_4H_7S_2^+$ is formed by the following two reactions:

$$C_2H_3S^+ + C_2H_6S_2 \longrightarrow C_4H_7S_2^+ + H_2S \qquad (36)$$

$$C_2H_6S_2^+ + C_2H_6S \longrightarrow C_4H_7S_2^+ + (H_5S_2) \qquad (37)$$

A further, very intense product ion is $C_4H_9S_2^+$, m/z 121:

$$C_2H_5S^+ + C_2H_6S_2 \longrightarrow C_4H_9S_2^+ + H_2S \qquad (38)$$

$$C_2H_6S_2^+ + C_2H_6S_2 \longrightarrow C_4H_9S_2^+ + (H_3S_2) \qquad (39)$$

There are corresponding reactions in the ion chemistry of 2-mercaptoethanol:
The product $C_3H_7OS^+$ is formed by three reactions:

$$CH_3S^+ + C_2H_6OS \longrightarrow C_3H_7OS^+ + H_2S \qquad (40)$$

$$C_2H_4S^+ + C_2H_6OS \longrightarrow C_3H_7OS^+ + CH_3S \qquad (41)$$

$$C_2H_6OS^+ + C_2H_6OS \longrightarrow C_3H_7OS^+ + (CH_5OS) \qquad (42)$$

At higher pressures a product ion m/e 105, $C_4H_9OS^+$, appears with great abundance:

$$C_2H_4S^+ + C_2H_6OS \longrightarrow C_4H_9OS^+ + HS \qquad (43)$$

$$C_2H_5S^+ + C_2H_6OS \longrightarrow C_4H_9OS^+ + H_2S \qquad (44)$$

$$C_2H_6OS^+ + C_2H_6OS \longrightarrow C_4H_9OS^+ + (H_2O + HS) \qquad (45)$$

$$C_2H_7OS^+ + C_2H_6OS \longrightarrow C_4H_9OS^+ + (H_2O + H_2S) \qquad (46)$$

The ion with mass m/z 119 has the composition $C_4H_7S_2^+$ and is formed by $C_2H_3S^+$ and $C_2H_5S^+$:

$$C_2H_3S^+ + C_2H_6OS \longrightarrow C_4H_7S_2^+ + H_2O \qquad (47)$$

$$C_2H_5S^+ + C_2H_6OS \longrightarrow C_4H_7S_2^+ + H_2 + H_2O \qquad (48)$$

The product $C_4H_9S_2^+$, m/z 121, is formed by three reactions:

$$C_2H_5S^+ + C_2H_6OS \longrightarrow C_4H_9S_2^+ + H_2O \qquad (49)$$

$$C_2H_7OS^+ + C_2H_6OS \longrightarrow C_4H_9S_2^+ + 2\ H_2O \qquad (50)$$

$$C_4H_9OS^+ + C_2H_6OS \longrightarrow C_4H_9S_2^+ + (H_2 + H_2O) \qquad (51)$$

It has been shown by photoionization that the oxonium structures

$$CH_2{=}OH^+ \qquad CH_3CH{=}\overset{+}{O}H$$

are very stable structures of the ions CH_3O^+ and $C_2H_5O^+$, respectively, although several open-chained and cyclic structures may also be discussed [9-11]. These two ions are reacting via acetal formation only in glycol, not in 2-mercaptoethanol. (Reactions 23 and 28). The following mechanisms can be proposed:

$$H_2C{=}\overset{+}{O}H + \begin{array}{c}HO-CH_2\\ |\\ HO-CH_2\end{array} \longrightarrow \text{(cyclic: O–CH}_2\text{–}\overset{+}{O}\text{H, five-membered ring)} + H_2O \qquad (23)$$

$$H_3C-CH{=}\overset{+}{O}H + \begin{array}{c}HO-CH_2\\ |\\ HO-CH_2\end{array} \longrightarrow \text{(five-membered ring with O, CH–CH}_3\text{, }\overset{+}{O}\text{H)} + H_2O \qquad (28)$$

Product ions of structures II and IV a, respectively, are formed.

It is interesting that the primary ion $C_2H_4O^+$, which has most probably the structure of the acetaldehyde molecular ion [2], does not undergo ion-molecule reactions, although its abundance is high enough to allow observations of reactions. Reaction 28 of the protonated acetaldehyde is therefore protoncatalyzed.

The phenomenon of greatly increased reactivity of onium ions is quite general in the ion chemistry of the compounds discussed here, further examples are the ions $CH_2{=}XH^+$, $CH_2C{=}XH^+$, (X=O,S) and $CH_3OH_2^+$, which are also more abundant primary ions than the corresponding ions with one hydrogen atom less (CH_2X^+, $C_2H_2X^+$, X=O,S and CH_4O^+). The only exception is the ion $C_2H_4S^+$, which is not a protonated molecule because it has a different structure: It is cyclic (c.f. chapter 3.1).

The molecular ions and the protonated molecules of glycol and 2-mercapto-ethanol are both reactive, in ethanedithiol only the molecular ion is reactive. In this case the reaction mechanism is different. In the reaction of the protonated molecule ether formation must be assumed:

$$\begin{matrix} HO-CH_2 \\ | \\ H_2\overset{+}{O}-CH_2 \end{matrix} + \begin{matrix} HO-CH_2 \\ | \\ HO-CH_2 \end{matrix} \longrightarrow \text{(1,4-dioxane ring with } \overset{+}{O}H\text{)} + 2\,H_2O \qquad (29)$$

It is, therefore, probable that two different structures result for the ion m/z 89, $C_4H_9O_2^+$: protonated 2-methyl-1,3-dioxolane (reaction 28) and protonated 1,4-dioxane (reaction 29). The product ions of reactions 26 and 27 ($C_4H_7O_2^+$) have similar structures, differing only by a double bond instead of a single bond, because they contain two hydrogen atoms less than $C_4H_9O_2^+$. A further ion, $C_2H_3O^+$, reacts to $C_4H_7O_2^+$. The following mechanism may be discussed, if $C_2H_3O^+$ is assumed to have the structure of protonated ketene:

$$H_2C{=}C{=}\overset{+}{O}H + \begin{matrix} HO-CH_2 \\ | \\ HO-CH_2 \end{matrix} \longrightarrow C_4H_7O_2^+ + H_2O \qquad (25)$$

The product ion has most probably the structure IIIa. In this case, however, protonated vinylacetic acid ester should be considered as a second possible structure.

In the ion chemistry of ethanedithiol-1,2 many of the reactions corresponds to those of the oxygen analogue, mercaptals and thioethers are formed. There are, however, no reactions of the protonated molecule, also due to its low abundance. Moreover, the ion $C_3H_6X_2^{+}$ is formed only in the ion chemistry of ethanedithiol:

$H_2C=SH^+$ / (thiiranium ion, S^+) / H_2C-SH^+ | H_2C-SH $+ C_2H_6S_2 \longrightarrow$ (1,3-dithiolanium ion, S^+) $+ H_3S$ (30)

$+ CH_4S$ (31)

$+ CH_4S + H_2S$ (32)

The structure of the ion CH_3S^+ can be $CH_2{=}SH^+$ or CH_3S^+. In the literature [9,12,13] several deuterium labelling experiments are described. They indicate, that both structures can be formed initially and rearrange. The enthalpies of formation of both ions are similar. CID experiments, however, show that CH_2SH^+ is much more abundant than CH_3S^+ [14]. Our results are best understood if also the structure CH_2SH^+ is assumed.

The ion m/z 107, $C_3H_7S_2^{+}$ is probably a protonated 2-methyl-1,3-dithiolane, if formed by a reaction of $C_2H_5S^+$ (34) or a protonated 1,4-dithiane if formed by the endothermic reaction of the molecular ion (35).

In reaction 51, which is endothermic, a O atom is substituted by a sulfur atom in the ion which has structure IVb. This type of reaction is also known from the chemistry of neutral molecules. Utilizing Al_2O_3 as a catalyst reaction of 1,4-oxathiane and H_2S yield 1,4-dithiane [15].

3.3 General Discussion

The most characteristic reactions of ethanediol, 2-mercaptoethanol and ethanedithiol are the formation of protonated cyclic acetals and mercaptals. In the case of ethanediol dioxolanes or methyl dioxolanes are formed. In the ion chemistry of the thio analogue dithiolanes and methyl-dithiolanes are observed. In 2-mercaptoethanol only the mixed oxothiolane (reaction 40) is detected and not dithiolane:

$$H_2C{=}SH^+ \;+\; \begin{array}{c} HO-CH_2 \\ | \\ HO-CH_2 \end{array} \;\begin{cases} \longrightarrow \; \text{(O, }\overset{+}{S}\text{H ring)} + H_2S \\ \not\longrightarrow \; \text{(S, }\overset{+}{S}\text{H ring)} + H_2O \end{cases} \tag{40}$$

Therefore, it can be concluded, that the neutral molecule H_2X (X=O,S), which is also formed in reactions with same mechanism as reaction 40, results from the reacting ions (H_3X^+ or $C_2H_5X^+$) and not from the neutral molecules C_2H_6XY (X=O,S; Y=O,S).

Also some conclusions on the structure of the product ion $C_4H_9S^+$ from 2-mercaptoethanol can be drawn. Because no protonated dithiolane is found as product ion, the ion $C_4H_9S_2{}^+$ is a protonated dithiane formed most probably as a tertiary ion from $C_2H_7S_2{}^+$ and not directly from $C_2H_5S^+$, which should yield a methyldithiolane ion.

5. Acknowledgement

One of us (G.B.) thanks the Scientific and Technical Research Council of Turkey for a NATO fellowship.

6. References

1. G. Baykut, Dissertation, Frankfurt 1980

2. G. Baykut, K.-P. Wanczek and H. Hartmann, Dyn. Mass Spectrom. 6(1981)269.

3. Z.-C. Profous, K.-P. Wanczek and H. Hartmann, Z. Naturforsch. 30a(1975) 1470.

4. R.R. Corderman, P.R. LeBreton, S.E. Buttrill, Jr., A.D. Williamson and J.L. Beauchamp, J. Chem. Phys. 65(1976)4929.

5. M.L. Gross and F.W. McLafferty, J. Am. Chem. Soc. 93(1971)1267.

6. M.L. Gross, J. Am. Chem. Soc. 94(1972)3744.

7. J.R. Collins and G.A. Gallup, J. Am. Chem. Soc. 104(1982)1530.

8. K. Refaey and W.A. Chupka, J. Chem. Phys. 48(1968)5205

9. C. Lifshitz and Z.V. Zaretskii, The Chemistry of the Thiol Group, Vol. 1, S. Patai, Ced., Interscience, New York, 1974, p. 325.

10. J.L. Holmes, J.K. Terlouw and F.P. Lorsing, J. Phys. Chem. 80(1976)2860.

11. T.W. Shannon and F.W. McLafferty, J. Am. Chem. Soc. 88(1966)5021.

12. B.G. Keyes and A.G. Harrison, J. Am. Chem. Soc. 90(1968)5671.

13. D. Arros, R.G. Gillis, J.L. Occolowitz and J.F. Pisani, Org. Mass Spectrom. 2(1969)209.

14. J.D. Dill and F.W. McLafferty, J. Am. Chem. Soc. 100(1978)2907

15. Houben-Weyl, Methoden der Organischen Chemie, 4. Aufl. Bd. 6/4, E. Müller, Ed., G. Thieme-Verlag, Stuttgart 1966, p. 586.

KINETIC ENERGY OF FRAGMENT IONS PRODUCED IN CHARGE TRANSFER REACTIONS OF He^+ AND Ar^{++} WITH CO

R. Derai, M. Mencik, G. Mauclaire and R. Marx
Laboratoire de Résonance Electronique et Ionique
Associé au CNRS Université de Paris-Sud,
Centre d'Orsay, 91405 Orsay

1. Introduction

Determination of the amount of kinetic energy (KE) released into the products of charge transfer reactions from rare gas ions at thermal energy is often sufficient to decide whether or not a reaction goes through an energy resonant mechanism. This is true at least for all non dissociative charge transfers and for dissociative ones involving diatomic molecules.

The energy balances for such reactions are the following:

- for non dissociative charge transfers: $A^+ + BC^+ \longrightarrow BC^+ + A$

$$RE(A^+) = IE(BC^+) + KE(BC^+, A) \qquad (1)$$

- for dissociative charge transfers: $A^+ + BC \longrightarrow B^+ + C + A$

$$RE(A^+) = IE(B^+, C) + KE(B^+, C, A) \qquad (2)$$

where $RE(A^+)$ is the recombination energy of the rare gas ion, IE the internal energy of the products relative to the ground state of BC and KE the total kinetic energy release.

In non dissociative charge transfers, the KE of the product ions may only come from monentum transfer during the collision while in dissociative charge exchange it comes also partly from fragmentation of the parent molecular ion. With molecules having more than two atoms the situation is much more complex since the fragments may have some vibrational or rotational excitation.

Among the various techniques used to study ion-molecule reactions, only ICR is able to preserve translational excitation because of the low pressure and can be adapted for kinetic energy measurements. This feature has been used by Dunbar and coworkers [1] for photodissociation fragments and by us for charge transfer reactions products [2]. Some precautions are however necessary to maintain performances of the ICR cell in a large range of trapping voltage from 0.1 to 10 eV. This problem being solved, we can, by varying the trapping potential measure, for a given product ion, the velocity vector component parallel to the magnetic field. This quantity is related to the kinetic energy since in the ICR cell, the ions are formed with an isotropic distribution.

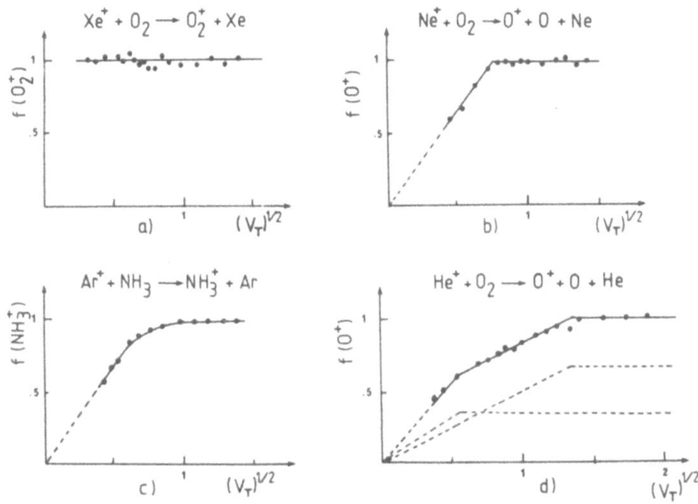

Fig. 1. Four examples of our kinetic energy measurements:
Fraction of product ions as a function of the square root of the trapping voltage. The considered ions have:
a) no kinetic energy
b) only one discrete value of kinetic energy
c) a distribution of kinetic energy
d) two discrete values of kinetic energy.

Figure 1 presents four typical examples of our KE measurements. First, in Figure 1a when the product ions have no KE, their amount is constant over the whole trapping voltage range. This is the case for the O_2^+ ions coming from the reaction of Xe^+ with O_2 (Figure 1a). Such a result indicates that no momentum transfer occurs during the collision which means that the reaction goes through a resonant mechanism in agreement with the previous results of Ajello and Laudenslager [3].

Figure 1b is obtained when all the ions have the same kinetic energy, E_k. Then, as long as the trapping voltage is higher or equal to E_k all the ions remain trapped in the cell. As soon as the trapping voltage is lower than E_k, it can be shown that, in our experimental conditions where all the ions are formed along the axis of the cell i.e., at the bottom of the trapping well, the ion current decreases linearly with the square root of the trapping potential, V_T. The KE of the ions is then deduced from the break on the curve. Figure 1b represents the fraction of O^+ ions coming from the reaction of Ne^+ with O_2 as a function of $(V_T)^{1/2}$ from which it can be deduced that the ions take away 0.65 eV [2].

If, instead of a single value, there is a broad KE distribution, a curvature is expected as shown in Figure 1c for NH_3^+ ions coming from the charge exchange Ar^+ and NH_3. In this case, the average KE is 0.65 eV and the broad KE distribution reflects a large vibrational distribution. Note that because of the square root dependence on V_T, the same KE distribution will lead to a more pronounced curvature if centered around a lower energy.

Sometimes, several breaks can be distinguished on a curve like in Figure 1d representing O^+ from ($He^+ + O_2$). This reaction produces two kinds of O^+ ions having respectively 0.3 and 1.8 eV of kinetic energy. Decomposing this curve, the relative abundances of the two kinds of ions may be determined from the heights of the plateaus [4].

Two examples of dissociative charge transfer reactions involving the same diatomic molecule CO are presented here.

2. He^+/CO

The only product ion formed in this reaction is C^+.

$$He^+ + CO \longrightarrow C^+ + O + He$$

Neither CO^+ nor O^+ were detected. The measured rate constant is 1.6×10^{-9} cm^3s^{-1} very close to the Langevin value: 1.75×10^{-9} cm^3s^{-1}. All these data are in excellent agreement with the previous results of Laudenslager and coworkers [5].

The KE of C^+ is 1.1 ± 0.2 eV as deduced from Figure 2. The energy balance for this reaction is given by equation 2 where ER (He^+) is 24.6 eV; the internal energy of the fragments may have discrete values corresponding to the various states of C^+ and O but if all three products carry away some translational energy, the total KE release cannot be determined from the KE measurement of only one species

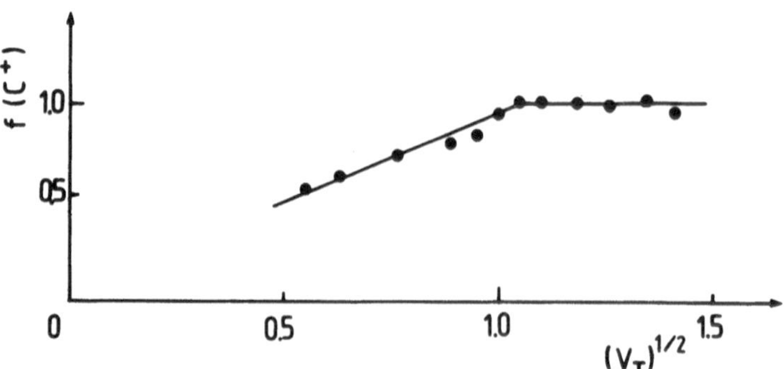

Fig. 2. Fraction of C^+ ions coming from He^+/CO as a function of the square root of the trapping voltage.

A reasonable hypothesis is then to assume that the reaction is energy resonant so that the neutral He has no KE. In this case the KE of C^+ comes only from the dissociation of CO^+ and, from momentum conservation, a value for KE (0) can be calculated: KE (0) = 0.8 $\pm$ 0.2 eV which gives a total KE of 1.9 $\pm$ 0.4 eV. By difference this leads to an internal energy for the fragments (C^++O) of 22.7 $\pm$ 0.4 eV. This value fits very well with the first dissociation limit of CO^+ into $C^+(^2P)$ + O (^3P) at 22.37 eV and the hypothesis of a near resonant charge transfer populating a CO^+ state lying around 24.6 eV and dissociating into $C^+(^2P)$ + O (^3P) is well supported by this result. Considering the potential energy curves of CO^+ [6], a possible candidate would be the $D^2\pi$ state in high vibrational levels which, according to Locht [6], could be predissociated by the $^2\Delta_r$ state leading to the observed fragments.

Our KE measurements are not sensitive enough to exclude a small contribution (less than 5%) of CO^+ dissociation into C^+ (^2P) + O (^1D) at 24.34 eV.

To conclude this first part it is worth mentioning that all the dissociative charge transfers that we have studied starting from He^+ or Ne^+ as primary ions have been shown to be compatible with an energy resonant mechanism. This is reasonable since in the energy range above 20 eV the density of states is high enough even for small molecules, to ensure a near resonance between the RE of the primary ion and an energy level of the product molecular ion.

3. Ar^{++}/CO

Reactions of doubly-charged ions with molecules have not been as much investigated as those of singly charged ions and in particular the identification of the products is not always unambiguous. At thermal energy these reactions have mainly been investigated by the groups of Howorka, Lindinger et al in Innsbruck [7], Adams and Smith in Birmingham [8] and Viggiano et al in Boulder [9] who all use high pressure techniques. These FA or SIFT experiments give the reaction rate constant and the nature of the products but no direct information upon their internal energy. Some experiments have also been performed in beam machines giving KE [10] and optical emission [11] for the products. At thermal energy, no such data are available and our purpose in this study was to obtain for doubly charged primary ions the same kind of information as for singly-charged ions.

The KE determination is specially interesting for single charge transfer reaction such as

$$A^{++} + BC \longrightarrow A^+ + (BC^+)^* \longrightarrow A^+ + B^+ + C$$

In this case, measurement of the KE carried away by A^+ allows calculation of the KE of $(BC^+)^*$ applying momentum conservation, then of the Coulombic repulsion between A^+ and $(BC^+)^*$ and finally determination of the distance r_0 at which the electron jump occurs. The DE of B^+ results from both the Coulombic repulsion and the dissociation energy of BC^+, it is therefore less useful.

Among the various possible systems we chose CO mainly because it is a heteronuclear diatomic molecule so that it is easy to distinguish the two fragments and to determine whether the reaction occurs via a single or a double charge transfer.

Concerning the rare gas ion, there are two requirements: i) the ratio of the doubly to singly charged ions has to be rather high and ii) the singly charged species have to be quickly removed from the reaction volume by cyclotron ejection. Ar^+ is a good candidate since with 90 eV electrons the Ar^{++}/Ar^+ ratio is around 15% and, because there is only one isotope, fast ejection of Ar^+ is easy.

However, a serious complication arises from the fact that there are three different states of Ar^{++}: the ground state ^3P and two metastable states ^1D and ^1S located respectively 1.74 and 4.12 eV above the ground state. The statistical relative abundances are 9/5/1. The radiative lifetimes for the metastables: 9 and 0.1s are much larger than our reaction time so that the three species have to be considered in our experimental conditions. All these data are reported in Table 1. together with the rate constants for reactions with neutral argon measured by Johnson and Biondi [12]. As generally reported for reactions of doubly-charged ions with atoms, these rate constants are very low while for reactions with molecules they are several orders of magnitude higher. As consequence even in pure argon at 2×10^{-5} torr reactions with residual molecules (residual vacuum pressure: 8×10^{-8} torr) are predominant over the reactions with argon; this prevented us to take profit of those reactions to characterize our primary Ar^{++} ions by KE measurements of the outcoming Ar^+.

Ar^{++} State	Recombination Energy (eV) ($Ar^{++} \rightarrow Ar^+\ ^2P_{3/2}$)	Relative Statistical Abundance	Radiative Lifetime(s)	k (cm^3s^{-1}) [12] $Ar^{++} + Ar \rightarrow Ar^+ + Ar^+$
^3P	27.62	9	-	4×10^{-14}
^1D	29.36	5	9	4×10^{-13}
^1S	31.74	1	0.1	6×10^{-12}

Table 1. Characterization of the different states of Ar^{++}

4. Results

The ionic products formed in the reaction are C^+ and Ar^+. No CO^+ or O^+ are detected. C^+ and Ar^+ are in equal amount indicating that the reaction proceeds via a single charge transfer and that there is only one reaction channel:

$$Ar^{++} + CO \longrightarrow Ar^+ + C^+ + O$$

This result is in contrast with double charge transfers reported by the Birmingham group for Ar^{++} with O_2, N_2 and CO_2 [8].

The measured rate constant is 1.4×10^{-9} cm^3s^{-1} very close to the calculated Langevin value 1.6×10^{-9} cm^3s^{-1}.

Kinetic energy measurement have been performed in the trapping voltage range 0.1 to 10 volt for both C^+ and Ar^+. The experimental curves are presented on Figure 3. The curve for C^+ is very different of the curve obtained for C^+ in He^+/CO (Figure 2). The shapes of the curves indicate for the two ions a very broad KE distribution, from 0.9 to 6 eV for C^+ and from 0.6 to 3.6 eV for Ar^+.

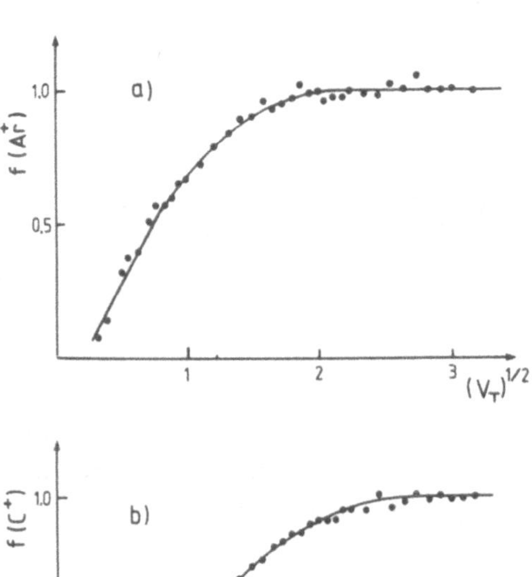

Fig. 3.
Fraction of product ions formed in Ar^{++}/CO as a function of the square root of the trapping voltage:

a) Ar^+ ions

b) C^+ ions.

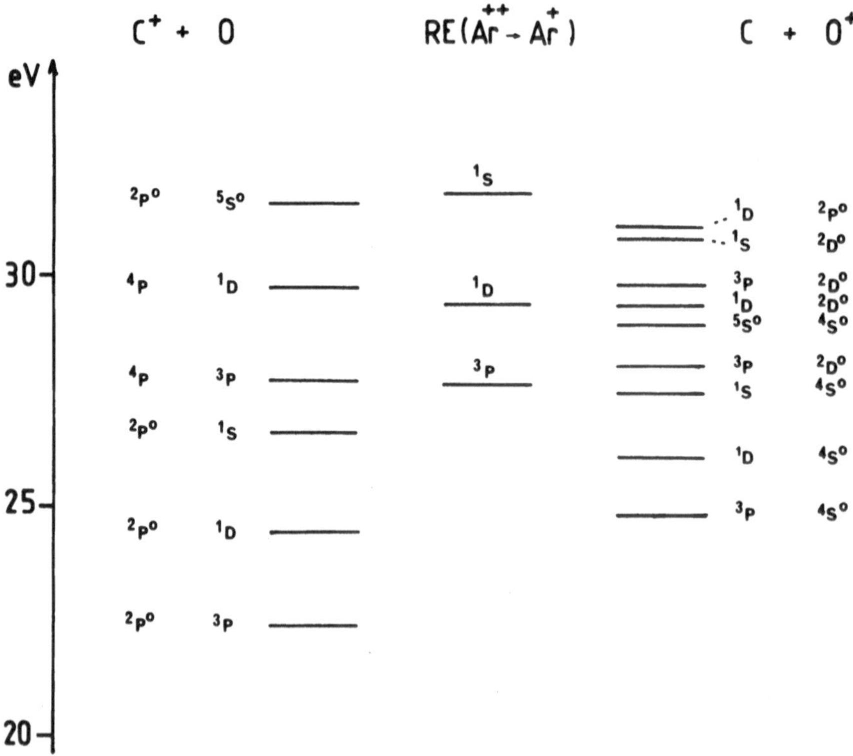

Fig. 4. Comparison between the recombination energies available from the three different states of Ar^{++} giving Ar^+ $^2P_{3/2}$ and the different accessible dissociation limits of CO^+ into C^+ and O and O^+ and C.

5. Discussion

From the recombination energy of the three states of Ar^{++} it can be seen on Figure 3 that several dissociation limits of CO^+ are accessible as well for C^+ + O as for C + O^+. However, no O^+ was observed in our experiments.

The energy balance for the reaction is

$$RE(Ar^{++} \longrightarrow Ar^+) = KE(Ar^+, C^+, O) + IE(C^+, O) \qquad (3)$$

where $RE(Ar^{++} \longrightarrow Ar^+)$ may have different values depending on the state of Ar^{++} and of the state component $^2P_{1/2}$ or $^2P_{3/2}$ of Ar^+; IE (C^+, O) corresponds to the various dissociation limits into C^+ and O relative to the ground state of CO.

The large kinetic energy distributions observed for Ar^+ as well as for C^+ is consistent with the participation of several states of the different partners of the reaction. In particular the upper values observed for KE (Ar^+) can only be explained by the occurence of

$$Ar^{++}(^1S) + CO \rightarrow Ar^+(^2P) + C^+(^2P) + O(^3P)$$

i.e, the highest state of Ar^{++} producing ground state C^+ and O.

In dissociative ionization of CO by electron impact, Locht [6] reported the formation of C^+ and O^+ ions corresponding to the various dissociation limits in the range 22 to 30 eV. In photoionization experiments, the main ion observed in the range 22-32 eV is C^+ [13,14]; the production of O^+ ion is reported only from the third dissociation limit into $O^+(^4S^0)+C(^1S)$ at 27.41 eV, but represents up to 32 eV a very low fraction of the ion current.

However, electronic states accessibly by charge transfer, photon or electron impact may be very different. Moreover, in the reaction studied here, where two of the products are positive ions, part of the available energy has to be converted into Coulombic repulsion, so that the lowest state are the most likely to be populated.

The large rate constant observed suggests a very efficient reaction mechanism which could occur preferentially via a long distance electron jump forming in a first step CO^+ in an excited state which then dissociates into C^+ and O. Since the kinetic energy resulting from Coulombic repulsion increases progressively as the charge transfer distance r_0 decreases, the internal energy available for the products decreases simultaneously. The system can thus find easily an efficient reaction channel allowing a long distance electron jump to occur. Another possible mechanism would be an intimate collision with formation of an intermediate $(ArCO^{++})^*$ exploding into Ar^+, C^+ and O but this would require a very close approach and would probably be a less efficient process.

As already mentioned, the KE of Ar^+ reflects the Coulombic repulsion between the Ar^+ and the $(CO^+)^*$ions. From the two extreme values of KE(Ar^+) we can calculate the corresponding KE of the intermediate CO^+ and then the total coulombic repulsion: $1.46 < E_r < 8.74$ eV. The internal energy of the CO^+ could then be determined as (RE($Ar^{++} \rightarrow Ar^+$) - E_r). Since we do not know which states of Ar^{++} are present in our experiments, we can just estimate that the CO^+ must have an internal energy lower than 31.74-1.46 = 30.28 eV and higher than the first dissociation limit into C^+ + O at 22.37 eV. We can also deduce a range for the intermolecular distance r_o at which the electron jump occurs. If we consider indeed that the electron jumps at the crossing point of the potential energy curves for the entrance and the exit channels, r_o can be obtained by solving the equation:

$$V_1(r_o) = V_2(r_o) - E_r$$

where

$$V_1(r) = -\frac{\alpha q^2}{2r^4} - \frac{\mathcal{D} q}{r^2}$$

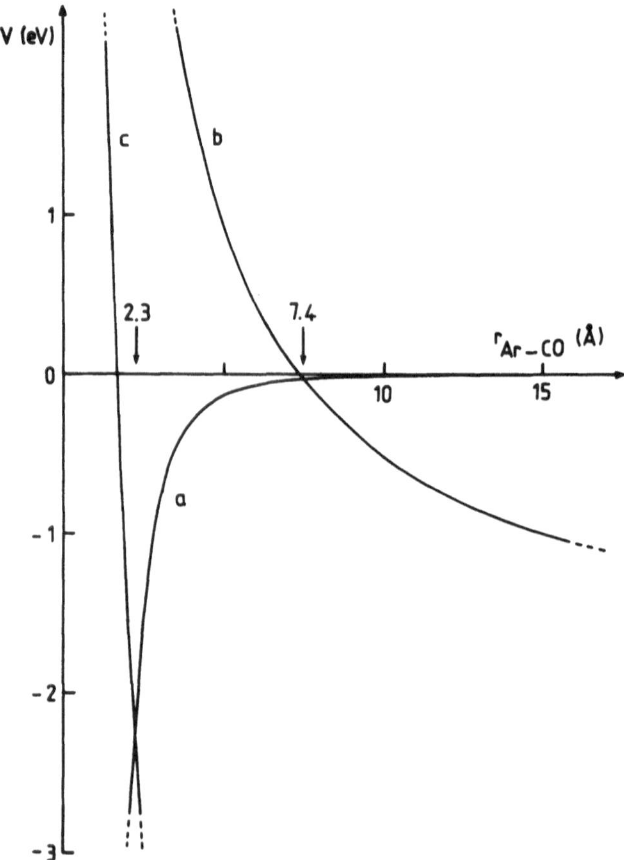

Fig. 5. Potential energy curves for the system Ar^{++}/CO

-a) Ion-induced dipole + Ion-permanent dipole potential for Ar^{++} and CO

-b) Coulombic potential for Ar^+ and CO^+ (E_r = 1.46 eV)

-c) Coulombic potential for Ar^+ and CO^+ (E_r = 8.74 eV)

is the ion-induced dipole, ion-permanent dipole potential in the entrance channel (curve a in Figure 5) (α and $\mathcal{D}$ are respectively the polarizability and the dipole of the molecule CO, q the charge of the incoming ion) and

$$V_2(r) = \frac{q'^2}{r}$$

is the Coulombic potential between the two product ions of same charge q' in the exit channel and E_r is the kinetic energy release when the ion distance is infinite. Curves b and c in Figure 5 represent the potential energy curves for the two extreme values of E_r. Intermolecular distances correponding to $2.3 < r_0 < 7.4$ Å are found in good agreement with the 3-4 Å previously reported by Hierl and Miller [10b] in Ar^{++} and Kr^{++} with N_2.

The absence of discrete breaks in the KE curves does not allow to go further in the interpretation but is not surprising in such reactions since i) the contribution of the different states of Ar^{++} cannot be separated at least in our experimental conditions; ii) at the high energies involved in this reacion, the density of states for the target molecule is large; iii) an additional broadening concerning the C^+ ion curve has to be considered: the velocity vector for a C^+ ion is indeed the resultant of the velocity vector associated to CO^+ and of that coming from the dissociation which is isotropic. This results in a very broad distribution for the C^+ velocity, v, which is still enhanced when the corresponding energies E proportional to v^2 are considered.

Yet, the measured Ar^+ can be identified by their KE as coming from the reaction: residual Ar^+ ions formed by electron impact would be near thermal. Double charge transfer which would produce both C^+ and O^+ ions is ruled out by the absence of O^+ ions. The equal abundances of C^+ and Ar^+ confirm that the reaction occurs via a single charge transfer and that the production of C^+ which may be in different states is the only channel. The large rate constant, very close to the Langevin value, evidences a very efficient mechanism probably a long distance electron jump.

6. Conclusion

ICR spectroscopy has been shown to be a very suitable tool, and until now the only one available to study kinetic energy release in products of ion-molecule reactions at thermal energy. At present, our studies have been limited to charge transfer reactions from rare gas ions to simple molecules; in non dissociative ones involving diatomic molecules, the internal energy of the products can be deduced. In more complex systems, like Ar^{++}/CO, even if the kinetic energy measurements are not sufficient, they are an important piece of the puzzle to elucidate the reaction mechanism.

7. References

1. R. Orth, R.C. Dunbar and M. Riggin, Chem. Phys., 19 (1977) 279.

2. G. Mauclaire, R. Derai, S. Fenistein and R. Marx, J. Chem. Phys. 70 (1979) 4017.

3. J.M. Ajello and J.B. Laudenslager, Chem. Phys. Lett. 44 (1976) 344.

4. G. Mauclaire, R. Derai, S. Fenistein, R. Marx and R. Johnsen, J. Chem. Phys. 70 (1979) 4023.

5. J.B. Laudenslager, M.T. Bowers and W.T. Huntress, Jr., J. Chem. Phys., 61 (1974) 4600.

6. R. Locht, Chem. Phys., 22 (1977) 13.

7. a) F. Howorka, J. Chem. Phys., 68 (1978) 804; b) H. Störi, E. Alge, H. Villinger, F. Egger and W. Lindinger, Int. J. Mass Spectrom. Ion Phys., 30 (1979) 263.

8. D. Smith, D. Grief and N.G. Adams, Int. J. Mass Spectrom. Ion Phys., 30 (1979) 271.

9. a) K.G. Spears, F.C. Fehsenfeld, M. McFarland and E.E. Ferguson, J. Chem. Phys., 56 (1972) 2562; b) A.A. Viggiano, F. Howorka, J.H. Futrell, J.A. Davidson, I. Dotan, D.L. Albritton and F.C. Fehsenfeld, J. Chem. Phys., 71 (1979) 2734.

10. a) C.Cole and P.M. Hierl, presented at the 23rd ASMS Conference, Houston, Texas (1975); b) P.M. Hierl and G. Miller, presented at the 25th ASMS Conference, Washington, DC (1977).

11. D. Neuschäfer, Ch. Ottinger and S. Zimmermann, W. Lindinger, F. Howorka and H. Störi, Int. J. Mass Spectrom. Ion Phys., 31 (1979) 345.

12. R. Johnson and M.A. Biondi, Phys. Rev. A, 20 (1979) 87.

13. M. Viard, P.M. Guyon, T. Govers, O. Dutuit, K. Ito, H. Frölich and P. Morin, unpublished results.

14. N. Nakamura, Y. Morioka, Y. IIda, H. Masuko, T. Hayashi, E. Ishiguro and M. Sasanuma, presented at the VIth international Conference on Vaccum Ultraviolet Radiation Physics, Charlottesville, Virginia (1980).

A TANDEM ICR STUDY OF THE REACTION OF N_2^+ WITH SO_2

Jean H. Futrell
Department of Chemistry
University of Utah
Salt Lake City, Utah 84112

Robert G. Orth
Department of Chemistry
Montana State University
Bozeman, Montana 59717

1. Introduction

The title reaction is an example of a growing group of reactions which exhibit very strong dependence on ion translational energy and/or internal energy of the reactant species. An extensive investigation of the charge transfer reaction between N_2^+ and SO_2 utilizing a flow drift tube has demonstrated that this reaction system has an extremely strong dependence on translational energy [1]. The rate constant reported by Dotan et al. has a miximum value of 7 x 10^{-10} cc's per molecule second at thermal energy, declines sharply to a minimum at 0.5 eV and increases at higher energy. Figure 1 illustrates this kinetic energy dependence.

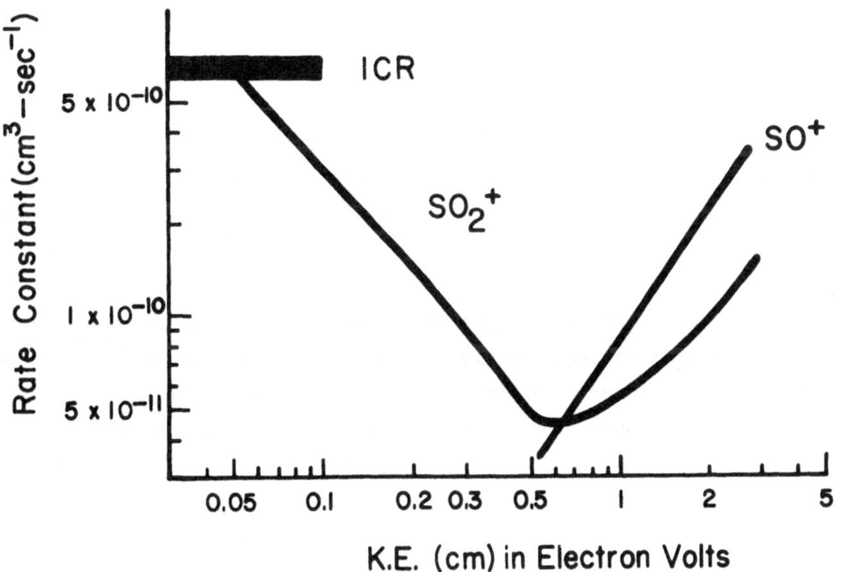

Fig. 1. Rate constant as a function of mean translational energy for the reaction of N_2^+ with SO_2 to generate SO_2^+ and SO^+ products. Value labeled ICR is from the present research; other data were extracted from Dotal, Albritton and Fehsenfeld, Ref. 1.

Also shown in this Figure is the opening of a second reaction channel, the dissociative charge transfer reaction which exhibits a threshold corresponding to the reaction endothermicity. The two principal reactions and their energetics are as follows:

$$N_2^+ + SO_2 \rightarrow SO_2^+ + N_2 + 3.28\ \text{eV} \qquad (1)$$

$$N_2^+ + SO_2 \rightarrow SO^+ + O + N_2 - 0.42\ \text{eV} \qquad (2)$$

The reaction energetics quoted refer to the v = 0 level of ground electronic states of reactants and products.

A particular advantage of the flow drift tube experiment is that this technique ensures ground state reactants; the reactant ions undergo tens of thousands of collisions with bath gas molecules prior to entering the reaction zone. For this reason the reaction displayed in Figure 1 involves only the ground state $X^3\Sigma_u$, v = 0 level of N_2^+ in reactions 1 and 2 . Thus this experiment isolates the effect of ion translational energy on both the reaction rate and the branching ratio.

As shown in the figure, reactions 1 and 2 are extremely sensitive to ion kinetic energy, both in terms of the absolute rate constants which are observed for these reactions and the branching ratios between the dissociative and non-dissociative electron transfer reaction. It is apparent the translational energy is utilized very efficiently to drive the endothermic dissociative charge transfer reaction 2 . The relatively low endothermicity of reaction 2 and the relatively large vibrational energy spacing for ground state N_2^+ both suggest that this reaction may be equally sensitive to internal energy in the reactant ion. It appeared therefore, that a study of internal effects with the Utah tandem ICR would be strongly complementary to the translational energy study of Dotan, Albritton and Fehsenfeld [1].

Figure 2 summarizes the relevant energy levels for the SO_2^+ and N_2^+ systems. As illustrated in this figure, reaction 2 becomes exothermic for N_2^+ ions which have vibrational energy corresponding to $v \geq 2$ (or for $v \geq 1$ if the data from Dibeler and Liston are used. See footnote 2 for discussion of thermochemical data.) jf it is assumed that vibrational energy is effective in driving reaction 2 . The vibrational reaction of N_2^+ may be investigated by observing the branching ratio between the two products SO^+ and SO_2^+. An analysis of the vibrational population of the ground state N_2^+reactant ions and the effectiveness of collisional relaxation of excited N_2^+ by collision with N_2 will be presented in this paper.

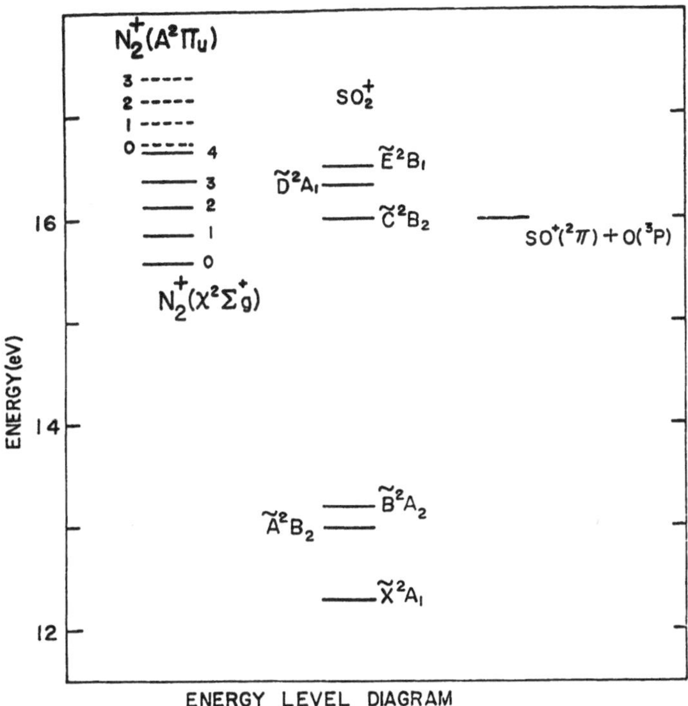

Fig. 2. Energy level diagram for the N_2^+/SO_2 charge transfer system. See Footnote 2 for references to original data and a brief discussion.

This paper is logically divided into two sections. Since the tandem ICR instrument has not been described in a previous edition of the Lecture Notes in Chemistry series and since it has been suggested that this particular instrument may transmit a mixture of low energy and high energy ions, the first will describe the instrument and some tests which were carried out to characterize the kinetic energy distribution of reactant ions. Our conclusion is that only low velocity ions are presented as reactants such that the apparatus effectively isolates the effects of internal energy from translational energy in the investigation of reactions 1 and 2 . We then describe the experimental procedures used and present a discussion of the results obtained in our study of the title reaction.

2. General Description of the Spectrometer

The first stage mass spectrometer is a 180^0 magnetic deflection instrument of 5.7 centimeters radius which injects a mass analyzed beam of reactant ions into a two-section ICR cell. Figure 3 is a block diagram of the overall instrument. The vacuum canister indicated in the figure is between the pole pieces of a 12-inch Varian magnet.

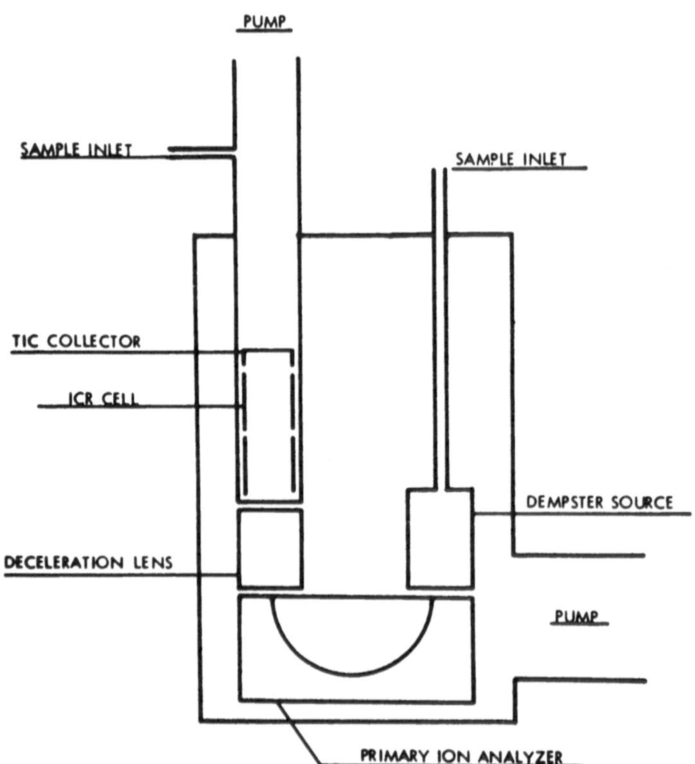

Fig. 3. Block diagram of the Tandem ICR spectrometer. Main vacuum canister fits within the pole pieces of a Varian 12-inch electromagnet.

The basic optical properties of the 180^{0} direction focusing mass spectrometer were described by Dempster and are summarized in most books on mass spectrometry. Our design is copied relatively closely from the Consolidated 21-103B mass spectrometer geometry. The ion source is an Isotron® salvaged from such a machine and the ion acceleration and deceleration optics are geometrically located in a similar manner to the "metastable suppressor" version of that machine. A detailed but schematic representation is presented in Figure 4. The key features of the design are the split repellers, the location of the electron beam midway between the repeller plates and the source exit slit, symmetrical placement of the focus plates and a matching pair of deceleration slits, and mechanical offset of the object slit and mass resolving slit of the spectrometer from the mid-lines of the ion source radius. It is also necessary to apply asymmetric focus potentials and deceleration potentials to optimize the ion transmission through the object slit and the ICR cell inlet slit, respectively, thereby ensuring that the ions execute a trajectory which is symmetrical about a plane parallel to the magnetic field located halfway between the object slit and mass resolving slit of the spectrometer. This is the fundamental property of the Dempster geometry which is required for the present application.

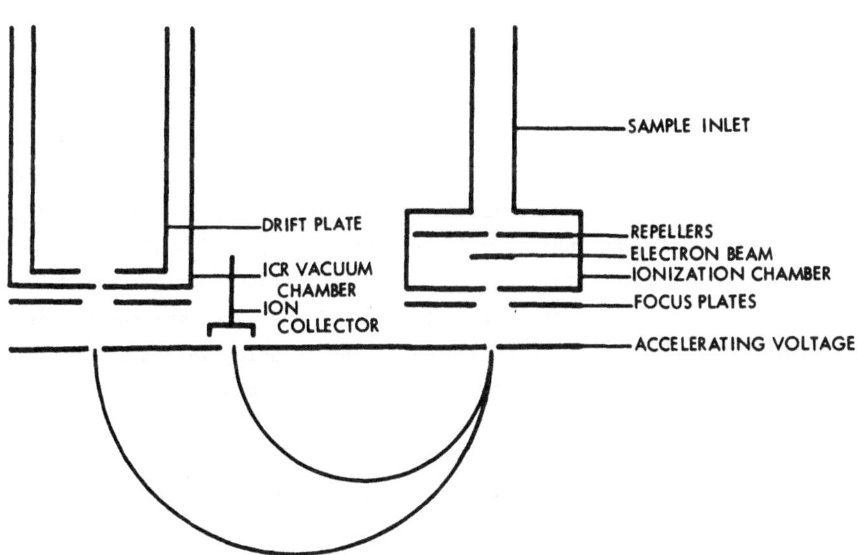

Fig. 4. Detail (schematic) view of the tandem ICR optics showing split-repeller ion source, acceleration and deceleration focus plates, shaped drift plates and auxiliary ion collector.

The ICR vacuum chamber is held at ground potential while the ion source potential is variable by ± 20 V with respect to ground. The repeller voltages are controlled independently to provide the field gradient necessary to extract ions from the source. At low ion energies quite asymmetric potentials are required so that the complex cycloidal motion of ions generated in the electron beam will intersect the ion exit slit and allow ions to be extracted by the weak penetration field of the acceleration slits. At higher energies the repeller potentials become asymptotically equal to each other; above 10 volts per centimeter extraction fields, the ion paths are accurately calculable from simple electrostatic theory. An important feature of the ion source is that the electron beam be much wider than the ion exit slit to accommodate the cycloidal paths of ions moving under the influence of the E X B force exerted on ions formed at rest in the plane of the electron beam.

The electron beam enters the source through a slit 0.010 cm by 0.150 cm, the plane of which is located midway between the split repellers and the ion exit slit. The repeller to ion exit slit distance is 0.25 cm and the ion exit slit 0.25 cm x 0.150 cm. The system is evacuated by a 700 liter per second oil diffusion pump and differential pumping isolation between the ion source and ICR cell is about 10^4. Consequently ion source pressures as high as .05 torr may be used without significantly increasing the pressure inside the ICR vacuum chamber. The volume of the ion source is about 1 cubic centimeter.

Ions leaving the source are accelerated by a simple asymmetric immersion lens to kinetic energies of a few kilovolts. A metal cage shields the high energy region from the grounded vacuum canister. Those ions with a radius of curvature of 5.7 cm pass through the mass resolving slit of the spectrometer, are decelerated by a lens identical to the acceleration lens and injected into the ICR cell, which is at ground potential. An auxiliary ion collector at a radius of 3.5 cm serves as an independent mass analyzer for scanning the mass spectrum generated by the Dempster ion source independent of the ICR cell. A quasi-thermal energy ion beam is transmetted by the final deceleration lens and injected into the ICR cell where the mean trajectory of the ions is defined by an equipotential line of the ICR cell.

The final slit is acually a "slot" of .035 cm width and 0.110 cm depth. This serves as an energy filter for the device such that it will transmit only those ions whose cyclotron radii are less than .017 cm. Collectively these special characteristics of the Dempster mass spectrometer and deceleration lens provide a means for injecting very low energy ions into an ICR cell using a simple and rugged mass analyzer and deceleration lens system. The ICR cell functions as both the reaction chamber and as an ICR mass analyzer. The two section ICR cell permits the measurement of reaction time using previously developed methods. Because no slits separate the collision chamber in the final mass analyzer, product ion detection is independent of reaction kinematics. The marginal oscillator detector signal may be directly related to the number density of ions generated in the ICR collision chamber/detector. Alternatively the ion ejection method may be used to measure ion signals by noting the decline in the total ion current signal registered by an electrometer amplifier. A comparison of the two methods provides an absolute calibration of the marginal oscillator. The cross section of the ICR cell used in this instrument is shown in Figure 5. Two sections, each having dimensions of 7/8"x5/8"x2" are terminated by a total ion current collector. The cell was formed by rolling 0.010" Nichrome V® into a rectangular open-ended box as shown in Figure 5. The two opposing sides of the box serve as the trapping plates whose fields constrain the ions from drifting out of the cell in the direction parallel to the magnetic field. Drift plates are formed by spot-welding 0-80 nuts to strips of 0.010" thick Nichrome V®; 0-80 rod isolated from the rectangular box section by ceramic spacers is used to attach the driftplates to the box, which also serves as a mounting frame for the assembly. The principal advantages of this method of cell construction are simplicity of assembly and disassembly, low capacitance between the sections and physical rigidity, accompanied by compactness of the overall configuration.

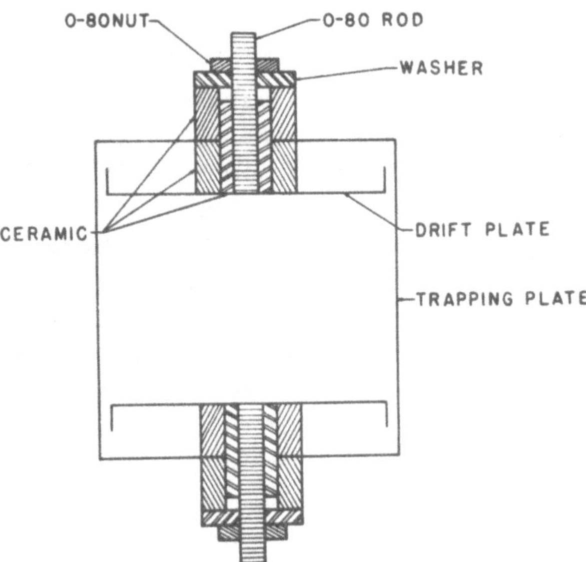

Fig. 5. Cross-sectional view of ICR drift cell section showing method of construction and monitoring of plates within the differentially-pumped Reaction/Analyzer region of the Tandem ICR.

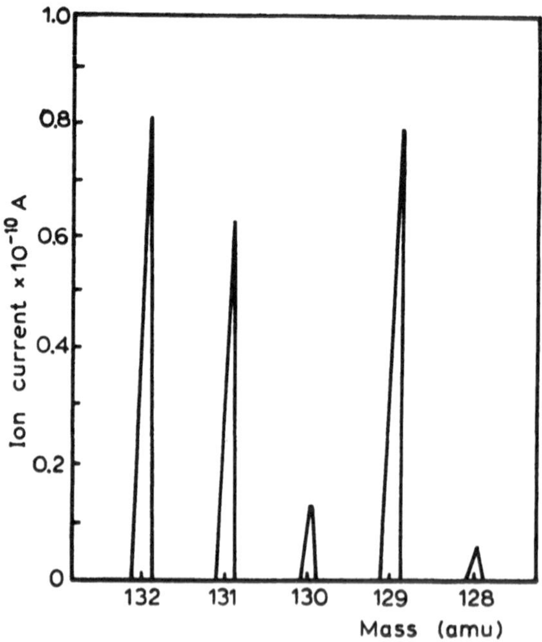

Fig. 6. Mass spectrum of Xe^+ isotopes measured with the total ion current collector as the magnetic field is scanned. Note the sharp triangular shape of the peaks, indicative that the detector slit is much narrower than the source object slit. A nominal resolving power of $m/\Delta m > 500$ (10 % valley) is deduced from the scan. See discussion of anomalously high resolution in text.

The mass resolution of the first stage analyzer is demonstrated in Figure 6. This is a plot of the xenon isotopes monitored by the total ion current collector as the magnetic field is swept with the accelerating voltage held constant at 1 kilovolt. The resolution illustrated in this figure $m/\Delta m$ is approximately 500, much greater than the theoretical resolution defined by the radius of the magnetic field and the slit widths of the object slit (0.009 cm) and the mass resolving slit (0.035 cm). This geometry corresponds to a theoretical mass resolution $m/\Delta m$ = 130 which should be just sufficient to resolve the xenon isotopes.

The anomalously high resolution of the first stage Dempster mass spectrometer may be understood in terms of the action of the deceleration lens. The requirement that ions cycloid through the slit which is long in the beam dimension creates a virtual slit which is much narrower than its geometrical dimensions. A trivial calculation based on the observed resolution suggests that the effective width is of the order of .002 cm. (The same phenomenon of electrical narrowing of a slit by creation of a saddlepoint in the ion path was used in metastable suppressor versions of the Consolidated 21-103B series of mass spectrometers.) The triangular shape of the mass peaks and the anomalously high resolution are also indicative of the the effectiveness of the energy filter principles described below.

The transmission of ions through the deceleration lens as a function of ion source voltage at constant repeller potentials of +2 and -1 volt is illustrated in Figure 7. A sharp maximum in ion transmission at an ionization chamber voltage near zero followed by no detectable transmitted current for about one volt and an increased transmission with energy at higher voltage is further evidence for the energy filtering action of the slot lens. The sharp peak at low energy represents the transmission of ions executing cycloidal motion constrained by the width of the slot with cyclotron radii less than 0.017 cm. Higher energy ions strike the edges of the final element of the deceleration lens (entrance aperture for the ICR cell) and are neutralized at higher energies. Ions with a kinetic energy greater than a few eV pass though the slot in one loop.

The width of the transmission window at low energy may be interpreted in different ways. One interpretation is that the width simply represents the range of ion kinetic energies transmitted by the deceleration lens. This would imply that ions have nearly thermal energies and a half-width at half-maximum of about 0.1 eV. A second interpretation is that the energy transmission width of the deceleration lens is extremely narrow (of the order of .03 eV) and that the width of the low energy peak of Figure 7 is determined almost entirely by the spread

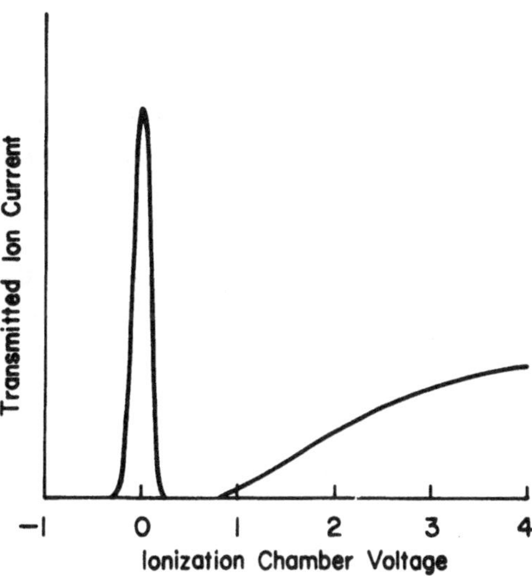

Fig. 7. Ion transmission as a function of ion kinetic energy. The sharp peak near zero shows the filtering action of the slot lens.

of ion kinetic energies extracted from the ion source under these operating conditions. In either case it is clear that this thick slit or "slot" lens element serves as an absolute energy filter which establishes a rather low upper limit to ion kinetic energy. Recognizing that ion motion under the less than ideal conditions of ion acceleration and deceleration presented by this rather simple lens system is too complex for detailed analysis, we take the conservative view that the described operation and the low energy transmission maximum of Figure 7 are evidence that quasi-thermal kinetic energy reactant ions enter the ICR cell with an energy distribution bounded by an upper limit of the order of 0.15 eV.

In characterizing the performance of the tandem ICR we have carried out a number of experiments which were designed to provide independent information on ion kinetic energy. One example is the reaction of CO^+ with CH_4,

$$CO^+ + CH_4 = CH_4^+ + CO \tag{3a}$$

$$CO^+ + CH_4 = CH_3^+ + H + CO \tag{3b}$$

Reaction 3a is exothermic by 1.3 eV while 3b is endothermic by 0.22 eV for ground state reactant ions. At low ion source pressure (and high repeller fields) both reaction channels were observed with the branching ratio $CH_3^+/CH_4^+ = 0.12$. However, on increasing the pressure of carbon monoxide in the first stage ion source to 0.05 torr, the branching ratio CH_3^+/CH_4^+ declined to 0.01. Consequently, the population of the endothermic channel 3b reflects the population of internally excited CO^+ under collision-free conditions in the ion source. When the ions are collisionally relaxed, the endothermic channel closes. If we assume that translational energy is also effective in driving this reaction we may interpret the closure of the endothermic channel with increasing ion source pressure as evidence that the translational energy of reactant ions is less than 0.5 eV (laboratory frame).

The reaction of C^+ with OCS was recently investigated by the present authors [3]. The principal product channel

$$C^+ + OCS \rightarrow CS^+ + CO \tag{4}$$

is exothermic by about 4 eV, a significant fraction of which appears as translational energy of the products. The method of Orth, Dunbar and Riggin [4] was used to demonstrate that the mean kinetic energy of the CS^+ product was 1.14 ± 0.1 eV. The same trapping voltage scan method was used to establish an upper limit to the kinetic energy of C^+ as 0.4 eV. Applying the same method in the presently reported study of N_2^+ reactions with SO_2 confirmed the energy of N_2 to be less than 0.2 eV.

The rate constant and branching ratio experiments for reactions 1 and 2 also provide experimental evidence that the Utah tandem mass spectrometer provides a means of selecting translational energies for the reactant ions which limits ion kinetic energy to rather low values. The reaction energetics have already been discussed and the closure of the endothermic channel with increasing nitrogen pressure in the Dempster ion source provides support for the hypothesis that ion kinetic energies are less than 0.15 eV, analogous to the information presented for reaction 3 . In addition, because of the very steep dependence of the rate constant on ion kinetic energy, the numerical value for the rate constant for this charge provides strong evidence that our ion energy is less than 0.1 eV. The tandem ICR rate constant entered in Figure 1 is in almost exact agreement with the thermal energy value for the rate constant for this reaction obtained in the flow drift tube experiment [1]. Consideration of the probable experimental errors in both sets of measurements place an upper limit of 0.2 eV on ion translation energy.

In summary, the measurements we have carried out to characterize the properties of the tandem ICR provide both physical and chemical evidence that the kinetic energy of ions passing through the deceleration lens is significantly less than 0.05 eV. Considerations of ion motion and semi-quantitative interpretations of the action of the final "slot" lens element suggest an upper limit for ion energies of the order of 0.15 eV. We therefore conclude that this apparatus is well suited for the study of internal energy effects on the rate constants and branching ratios of ion-molecule reactions.

3. Experimental Procedures

For the investigation of the reactions of N_2^+ with SO_2 the reactant ions were generated in the ion source at several electron energies; eg., 16.8 eV, 17.1 eV, 21 eV and 50 eV. The overall rate constant was measured by observing the decrease in the N_2^+ resonance signal as the SO_2 pressure in the ICR cell was increased. ICR power absorption was utilized in the present experiments because of its intrinsically broader dynamic range and superior sensitivity over ion current measurements. The magnetic field was fixed while the frequency of the resonance oscillator detector was varied. Since the sensitivity of the oscillator depends upon its frequency it was necessary to calibrate the sensitivity of the detector using total ion current measurements to calibrate the oscillator signal. Total rate constants were measured from the decline in the N_2^+ resonance signal while the branching ratio was measured by comparing the intensities of SO_2^+ and SO^+ using the previously calibrated marginal oscillator detector.

From the application of first order kinetics the total rate constant is calculated from the relationship $\ln(N_2^+)_0 = kn(SO_2)t$, where $(N_2^+)_0$ and N_2^+ are the number densities of N_2^+ in the ICR cell at zero SO_2 pressure and pressure of the measurement, respectively. $n(SO_2)$ is the number density of SO_2 molecules in the cell at pressure P, and t is the reaction time ions spend transitting the ICR cell. Reaction time t, was measured by a pulsing technique, while the pressure of SO_2 was determined using a calibrated ionization gauge attached to the cell chamber.

The pressure in the ion source is measured by calibrating the ion gauge in the pumping line external to the source as indicated in Figure 3. The calibration was achieved using the ion-molecule reactions of D_2^+ with D_2 and CH_4^+ with CH_4 as reference standards. The auxiliary detector of Figure 4 was used to measure the mass spectrum of ions exiting the source and thereby monitor the extent of reaction. Voltage-scanning of the mass spectrum, the technique commonly used for Dempster type mass spectrometers, was used to develop the mass spectrum.

Both of these reactions have rate constants which approximate the Langevin collision rate constants for these systems. Consequently the ion current ratios D_3^+/D_2 and CH_5^+/CH_4^+ provide a quantitative measure of the product source pressure and resonance time. No independent measures of the number density and reaction time was possible; however, the product of these factors is the relevant parameter for discussing collisional relaxation of ions. The ratio between the ionization gauge reading in the pump line and the (nt) product deduced from the measurement of the extent of reaction with the auxiliary electrometer detector provides a calibration factor. A small correction to the ionization gauge reading for the relative response of the gases used in this experiment compared to deuterium and methane, respectively, provides a calibration constant to relate the gauge reading to the effective nitrogen pressure within the source. The estimated error in the reported ion source pressure and the average number of ion source collisions to extraction is estimated to be $\pm$ 20%.

4. Results and Discussion

The total rate constant for the reaction of N_2^+ with SO_2 as a function of N_2 pressure in the source of the Dempster mass spectrometer is shown in Table 1. Under the conditions used in this experiment, the total rate constant was found to be essentially independent of the electron energy used for ionization of the nitrogen molecules. The rate constants were the same within experimental error for the electron energies fo 16.8 eV, 17.1 eV and 22 eV. As is clear from Table 1, the total rate constant is also independent of the N_2 pressure in the source.

Also reported in Table 1 is the rate constant for the reaction of N_2^+ generated in a 10/1 mixture of argon with nitrogen. This gas mixture filters out all the excited N_2^+, leaving a pure preparation of $X^2\Sigma_g^+$, v = 0 ions. Considering the possible ion source reactions,

$$Ar^+ + N_2 \rightarrow N_2^+(X^2\,\Sigma_g^+,\ v = 1) \qquad (5a)$$

$$Ar^+ + N_2 \rightarrow N_2^+(X^2\Sigma_g^+,\ v = 0) \qquad (5b)$$

$$N_2^+(X^2\Sigma_g^+,\ v = 0) + Ar \rightarrow Ar^+ + N_2 \qquad (5c)$$

$$N_2^{+*}(X^2\Sigma_g^+,\ v \geq 0;\ A,\ B) + Ar \rightarrow Ar^+ + N_2 \qquad (5d)$$

Lindinger et al. [5] have recently shown that $k_{5a} = 1 \times 10^{-11} cm^3\ mol^{-1}sec^{-1}$, $k_{5b} < 10^{-12} cm^3\ mol^{-1}sec^{-1}$, $k_{5c} = 1.7 \times 10^{-13} cm^3\ mol^{-1}sec^{-1}$ and $k_{5d} = 4 \times 10^{-10}$ $cm^3\ mol^{-1}sec^{-1}$. Accordingly, for this experiment the Ar^+ ions are essentially unreactive with nitrogen, while the ground electronic state, v = 0, ions of N_2^+ are similarly unreactive with Ar. However, excited N_2^+ are essentially quantitatively removed in reaction (5d), leaving a pure preparation of $N_2^+(X^2\Sigma_g^+,\ v = 0)$ reactant ions. As shown in Table 1, the rate constant is the same, within our ex- experimental error, as that measured for the mixed distribution of ground and excited state N_2^+ generated by electron impact.

Table 1. Total Rate Constant for the Reaction of N_2^+ with SO_2

N_2 pressure in the ion source (torr)	$k(cm^3 molecule^{-1} sec^{-1})$
2.9×10^{-2} torr[(a)]	7.0×10^{-10}
2.5×10^{-1} torr	6.8×10^{-10}
5.5×10^{-2} torr	6.4×10^{-10}
3.0×10^{-2} torr	7.4×10^{-10}

[(a)] Reactant ions generated in a 10:1 mixture of $Ar:N_2$ at a total pressure of 0.3 torr, according to the reactions sequence:

$$Ar + e \rightarrow Ar^+ + 2e$$
$$N_2 + e \rightarrow N_2^+ + 2e$$
$$\rightarrow N_2^{+*} + 2e$$

followed by reaction 5a-5c.

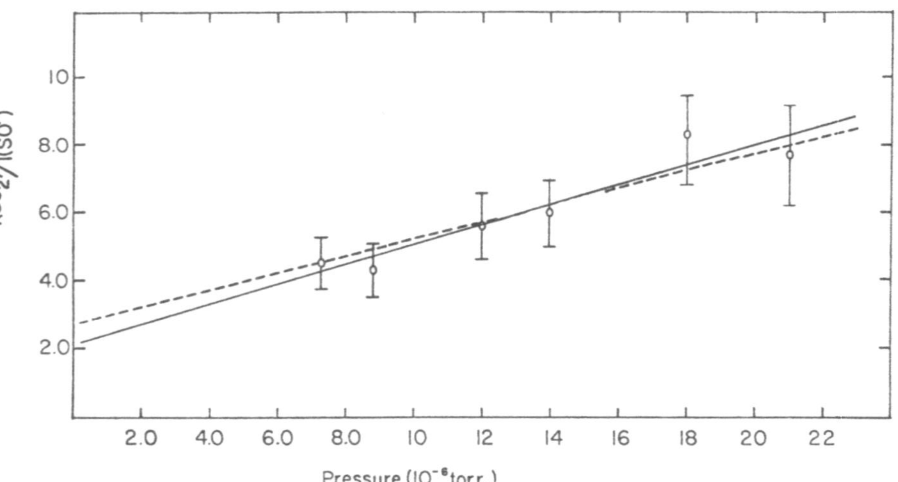

Fig. 8. Branching ratio SO_2^+/SO^+ as a function of SO_2 pressure in the ICR cell. Solid line is a least-squares fit of equation E4, while the dashed line is a least-squares fit of equation E7. See text for discussion.

The total rate constant for the reaction of N_2^+ with SO_2 reported in Table 1 is about one fifth the collision rate constant given by the "average dipole orientaion theory" (ADO) [6]. As shown in Figure 1, our rate constant is in excellent agreement with the value obtained in a flow-drift tube at the same center-of-mass energy of 0.1 eV [1]. The agreement of our experimental value for the rate constant with the flow-drift tube study is gratifying in view of the very strong kinetic energy dependence of the reaction rate.

It should be noted, however, that the velocity distributions are somewhat different for the two types of experiments at the same mean relative energy. In the flow-drift tube experiments the ion kinetic energy distribution is determined by the ion-bath gas collisions under the influence of a linear electric field. It is assumed, that the ion kinetic energy is given by the Wannier expression :

$$KE_{ion} = \frac{1}{2} m_i v_d^2 + \frac{1}{2} m_b v_d^2 + \frac{3}{2} kT \qquad \text{(E1)}$$

where m_i and m_b are the masses of the ion and bath gases, respectively, v_d is the (measured) drift velocity, and $\frac{3}{2}$ kT is the thermal energy of the ion. Further, the center-of-mass kinetic energy is given by:

$$KE_{cm} \doteq \frac{1}{2}\mu(v_i^2 + v_n^2) \qquad \text{(E2)}$$

where μ is the reduced mass of the reacting pair, and v_i and v_n are the ion and neutral velocities, respectively. This contrasts with the velocity distribution in the tandem-ICR experiments in which the beam defining slits serve as an energy filter, limiting the ion energy to $\leq$ 0.15 eV. Equation E2 leads to a center-of-mass energy of 0.1 eV for this reaction system. However, in neither experiment--flow dirft or tandem-ICR--is the distribution of energies found to be truly Maxwellian.

Figure 8 shows the ratio of SO_2^+ to SO^+ as a function of SO_2 pressure in the ICR cell. The ratio increases with increasing SO_2 pressure, suggesting that at least part of the SO^+ is generated via a predissociation mechanism. Assuming that all of the SO^+ derives from a predissociation mechanims, the quenching of this predissociation channel can be discussed in terms of the following simplified reaction scheme:

<u>Ion Source Reactions</u>

$$e + N_2 \rightarrow N_2^+ + 2e \qquad \text{(6a)}$$

$$e + N_2 \rightarrow N_2^{+*} + 2e \qquad \text{(6b)}$$

Collision Chamber Reactions

$$N_2^+ + SO_2 \xrightarrow{k_{1a}} SO_2^+ + N_2 \qquad (1a)$$

$$N_2^{+*} + SO_2 \xrightarrow{k_{1b}} SO_2^{+*} + N_2 \qquad (1b)$$

$$SO_2^{+*} \xrightarrow{k_7} SO^+ + O \qquad (7)$$

$$SO_2^{+*} + SO_2 \xrightarrow{k_8} SO_2 + SO_2^* \qquad (8)$$

Applying the steady state treatment to the concentration of SO_2^{+*} leads to the following relationship for the product ratio:

$$\frac{(SO_2^+)}{(SO^+)} = \frac{k_{1a}k_7(N_2^+) + [k_{1a}k_8(N_2^+) + k_{1b}k_8(N_2^{+*})]\,(SO_2)}{k_{1b}k_7(N_2^{+*})} \qquad (E3)$$

Further, as shown in Table 1 and discussed earlier, the overall rate of reaction of N_2^+ with SO_2 is independent, within experimental error, of electron energy, N_2 pressure in the source, and mode of formation. Therefore equation E3 can be simplified by substituting $k_{1a} = k_{1b} = k_1$. The resultant expression is as follows:

$$\frac{(SO_2^+)}{(SO^+)} = \frac{(N_2^+)}{(N_2^{+*})} + \frac{k_8}{k_7}\,\frac{[(N_2^+) + (N_2^{+*})]\,(SO_2)}{(N_2^{+*})} \qquad (E4)$$

The line shown in Figure 8 is a least squares fit of this equation to the experimental data for 17 eV electron ionization of N_2 at low ion source pressure (near collision-free conditions). The intercept for this line is 2.1 (S.D. = 0.8) indicating that the N_2^{+*} reactant ion responsible for generating the predissociative state(s) of SO_2^{+*} constitutes 22-42% of the total ion beam under these conditions.

This is a surprisingly large number in view of the demonstrated fact that SO^+ is not generated from thermally relaxed (v = 0) N_2^+ ions. Accordingly the simplified kinetics scheme was examined to ascertain the implications of the steady-state assumption. This assumption requires that the rate of SO_2^{+*} generation equals its loss rate and that the "induction period" to build up the steady-state level of SO_2^{+*} be small compared to the ICR cell residence time. The validity of these assumptions can be checked by calculation, provided the individual rate constants are known. However, only ratios of the rate constants k_8/k_7 are deduced directly; there is no independent measure of the lifetime of the predissociative state, $\tau_{SO_2^{+*}} = k_7^{-1}$.

As an alternative approach, it can be postulated that the lifetime of SO_2^{+*} is sufficiently long (~1 ms.) that the detected signal includes SO_2^+ and SO_2^{+*}. Therefore, the concentration of SO_2^{+*} should be included in the kinetic analysis, i.e.,

$$(SO_2^{+*}) = \frac{k_1(SO_2)e^{-k_1(SO_2)t}}{k_7(1-e^{-k_1(SO_2)t})} \qquad (E5)$$

By expanding the exponential to second order we obtain

$$(SO_2^{+*}) = \frac{1}{k_7t\,[1 + \frac{1}{2}k_1(SO_2)t]} \qquad (E6)$$

The ratio of the total (detected) SO_2^+ signal is therefore,

$$\frac{(SO_2^+) + (SO_2^{+*})}{(SO^+)} = \frac{(N_2^+)}{(N_2^{+*})} + \frac{1}{k_7t\,[1 + \frac{1}{2}k_1(SO_2)t]} + \frac{k_8[(N_2^+) + (N_2^{+*})]\,(SO_2)}{k_7(N_2^{+*})} \qquad (E7)$$

This expression is an intrinsically more satisfactory description of the overall reaction kinetics.

In Figure 8 a non-linear least squares fit for equation (E7) to the experimental results is indicated by the dashed curve. It is clear that experimental errors do not allow us to distinguish between equation (E4) and equation (E7). Nevertheless, the coefficients, $(N_2^+)/N_2^{+*})$, k_7 and k_8 obtained from the non-linear fit to (E7) are physically reasonable. The values $(N_2^+/N_2^{+*}) = 2.2$ (S.D. = 0.4), $k_7 = 1.62 \times 10^3\ sec^{-1}$ and $k_8 = 1.39 \times 10^{-9}\ cm^3/mol^3 sec$ are obtained. If the quenching of SO_2^{+*} is efficient, then the expected value for k_8 would be the collision rate or some fraction thereof. Using the method of Barker and Ridge [7], we calculate the rate constant for SO_2^+/SO_2 collisions to be $1.5 \times 10^{-9}\ cm^3/mol\text{-}sec$, comparing favorably with our value for k_8, $1.4 \times 10^{-9}\ cm^3/mol^{-1}sec^{-1}$. This indicates that the quenching process approaches unit collision efficiency. The other internal consistency requirement for the de-excitation mechanism is that the lifetime of SO_2^{+*} should be of the same order as the ICR cell residence time. This implies a value for k_7 in the range of $10^3\ sec^{-1}$, which is indeed comparable to the value of $1.5 \times 10^3\ sec^{-1}$ obtained in this kinetic analysis.

The value for $(N_2^+)/(N_2^{+*})$ of 2.2 ± 0.4 indicates that a substantial fraction of the N_2^+ reactant ion beam is excited nitrogen ions capable of driving the predissociative charge-transfer reaction. As shown in Figure 2, the reaction is nominally exothermic only for $X^2\Sigma_g^+$ ($v \geq 2$) N_2^+. However, the population of excited states fulfilling this prescription indicated by the kinetic ananlysis given above (eg., 26-36% N_2^{+*}) cannot be justified by Franck-Condon considerations.

Table 2 summarizes the Franck-Condon factors for primary excitation processes in the N_2^+ system [8]. Assuming that Franck-Condon factors for the ionization of nitrogen with 22 eV photons may be used to estimate the distribution from low energy electron ionization, we conlude that v = 0 and v = 1 levels of the X $^2\Sigma_g^+$ state of N_2^+ are the major species generated in our experiment. Direct population of v = 2 and higher levels is negligible. Data for the A $^2\Pi_u$ and B $^2\Sigma_u^+$ excited states of N_2^+ are also included in Table 2. For the 16.5 eV nominal electron energy used in the present experiments, the N_2^+ formed by electron impact consists of a mixture of X $^2\Sigma_g^+$ and A $^2\Pi_u$ states. Maier [9] reported the

Table 2. Differential Photoionization Cross Sections and Franck-Condon Factors for the Ionization of N_2 with 21.22 eV Radiation [(a)]

Transition	v	Franck-Condon Factor	Observed Relative Intensity
$N_2^+X^2\Sigma_g^+$	0	0.90	100
	1	0.09	6.9
	2	0.006	0.3
$N_2^+A^2\Pi_u$	0	0.24	87.1
	1	0.31	100
	2	0.23	76.3
	3	0.12	43.5
	4	0.06	19.1
	5	0.02	7.1
	6	0.01	2.5
$N_2^+B^2\Sigma_u^+$	0	0.89	100
	1	0.11	9.8

[(a)] Reference 8, Tables 7 and 8, p. 200

relative population X:A:B states of N_2^+ formed by 19.2 eV electron impact on N_2 was 0.769:211:0.106. Lindinger et al. have shown that the distribution of excited states of N_2^+ is nearly independent of electron energy a few eV above threshold [5]. Consequently we assume that the X and A states are formed in the same ratio in our work and estimate that a maximum of 22% of the reactant N_2^+ are generated in the A $^2\Pi_u$ electronic state.

The lifetime of vibrational levels 1-5 of the A $^2\Sigma_u$ states have been determined in luminescence studies of N_2^+ emission [10]. They range from 13.9 x 10^{-6} sec for v' = 1 to 9.1 x 10^{-6} sec for v' = 5. Clearly the lifetimes of these states substantially exceed the source residence time (ca. 1.4 μsec) of our experiments. Consequently at low source pressure the reactant ion beam injected into the ICR reaction cell of the tandem instrument includes any $A^2\Pi_u$ states of N_2^+ formed by electron impact. However, the lifetimes are also very much shorter than the residence time of ions in the ICR cell (approximately 1 millisecond). Therefore, these electronically excited states would decay by luminescence emission to form vibrationally excited ground state $(X^2\Sigma_g^+)$ states prior to reacting with SO_2.

From this information we conclude that the states formed by electron impact are a mixture of ground state $N_2^+X^2\Sigma_g^+$ (v = 0,1) and excited state $N_2^+A^2\Pi_u(v_i)$. At low ion source pressure (collision free conditions) this mixture of states is injected into the collision chamber. Since the reaction time is more than an order of magnitude greater than the radiative lifetimes of the $A^2\Pi_u$ state ions, they decay to a vibrational distribution of ground state $N_2^+X^2\Sigma_g^+$ (v_i) prior to reaction. The population of vibrational states can be estimated by Franck-Condon analysis. The distribution of vibrational levels of the X and A states initially formed is convoluted with the Franck-Condon factors for A → X emission to give the final distribution appropriate to our experimental conditions.

Utilizing the Franck-Condon factors deduced by Maier from a study of the luminescence decay of $N_2^+[A^2\Pi_u(v_i)] \rightarrow N_2^+[X^2\Sigma_g^+(v_j)]$ [9], the data of Table 2, and the assumption discussed above that the initial ratio of $N_2^+(X^2\Sigma_g^+)/N_2^+(A^2\Pi_u)$ is 0.78/0.22, the population of vibrational states of the reactant ion beam $N_2^+(X^2\Sigma_g^+)$ v=0:v=1:v=2 is estimated to be 0.82:0.13:0.05. It is clear from this result and the kinetic analysis given above that <u>all</u> the vibrational states above v = 0 must be invoked as reactant ions capable of generating predissociative SO_2^{+*}. Consequently excited N_2^{+*} is identified as N_2^+ $X^2\Sigma_g^+(v \geq 1)$.

Indeed, it is difficult to account for the total amount of N_2^{+*} apparently present in the reactant ion beam even with the v = 1 level included in our reaction scheme. The Franck-Condon analysis described above leads to an estimated excited state population of 18% N_2^{+*}, while the kinetics treatment estimates the population of N_2^{+*} as 32 ± 6%. However, this value is in excellent agreement with the estimate by Lindinger et al. [5] that electron impact generates 30-40% excited N_2^+. Because of approximations in both the kinetics scheme and the estimation of the vibrational level population from spectroscopic data, this discrepancy is probably not significant. In particular, the use of Franck-Condon factors obtained from short-wavelength photoionization studies to approximate the distribution from low energy electron impact is suspect. We suggest that this is the major source of the discrepancy.

Recent photoelectron spectroscopy [11], luminescence [2], and threshold photoelectron-photoion coincidence [13] studies of SO_2 permit an indentification of SO_2^{+*} with the third band of the photoelectron spectrum. This band is an overlapping mixture of $\tilde{C}\ ^2B_2$ and $\tilde{D}\ ^2A_1$ states of SO_2^+. On energetic grounds it is tempting to associate the $^2B_2$ state with this transition, i.e.,

$$N_2^+(X^2\Sigma_g^+,\ v = 1) + SO_2(\tilde{X}\ ^1A_1) \longrightarrow \tag{10}$$

$$N_2(^1\Sigma_g^+) + SO_2^+(\tilde{C}\ ^2B_2,\ 000) - 0.14\ \text{eV}$$

possibly followed by

$$SO_2^+\ (\tilde{C}\ ^2B_2) \rightarrow SO^+(X^2\Pi,\ v = 0) + O(^3P) + 0.0\ \text{eV} \tag{11}$$

However, Meisels et al. have shown that the dissociation threshold is below that for forming $SO_2^+(\tilde{C}\ ^2B_2,\ 000)$, at 15.930 eV.

Both the PES [1] and TPE-CPI [12] studies howed that SO^+ was formed with excess kinetic energy. The method of Orth, Dunbar and Riggin [4] was therefore used to compare the kinetic energy distribution of SO_2^+ under out experimental conditions. In contrast with the other studies, our experiments show that the SO^+ product possesses kinetic energy $0 < KE < 0.2$ eV. Thus, it appears that the charge transfer reaction selectively populates a different state (or states) from those excited directly by photon impact. This result is consistent with our deduction that the predissociative state populated in our experiments has such an extraordinarily long lifetime.

It is impossible, of course, to identify this extraordinarily long-lived predissociating state by our kinetics measurements. One cannot choose among the possiblities mentioned above, since the nature of the barrier to dissociation is unknown at this time. Unusual difficulty in intermodal energy transfer and/or curve crossing are likely possibilities. In any case, it appears that some states of SO_2^+ dissociate promptly into energetic SO^+ and O; these states are distinctly different from the extraordinarily long-lived state (or states) studied in the present experiments which gives near thermal energy products and has a lifetime of the order of milliseconds.

The deduction from the above kinetic analysis that $N_2^+(X^2\Sigma_g^+, v = 1)$ reacts is somewhat surprising, since this reaction (equation 15 is nominally endothermic by 0.14 eV. A recent PI study places the endothermicity at 0.07 eV; see reference 2 for discussion. As discussed earlier, the nominal reaction ion kinetic energy is $\leq$ 0.15 eV. Consequently an efficient mechanism for utilizing both translational energy and vibrational energy to drive the reaction must be invoked. One possibility is the strongly attractive potential between N_2^+ and SO_2 relative to that between the SO_2^+ and N_2 products. Consequently the incoming trajectory accelerates the reactants, while the products are retarded less strongly. This potential-to-kinetic energy conversion process would clearly augment the available energy present in translation and/or internal energy of the reactants.

We have also investigated the collisional relaxation of excited N_2^{+*} in the ion source. Figure 9 illustrates the dependence of the product ion ratio $(SO^+)/(SO_2^+)$ as a function of the average number of collisions between N_2^+ and N_2 in the ion source. The average number of collisions that N_2^+ undergoes, $\bar{n}$, is given

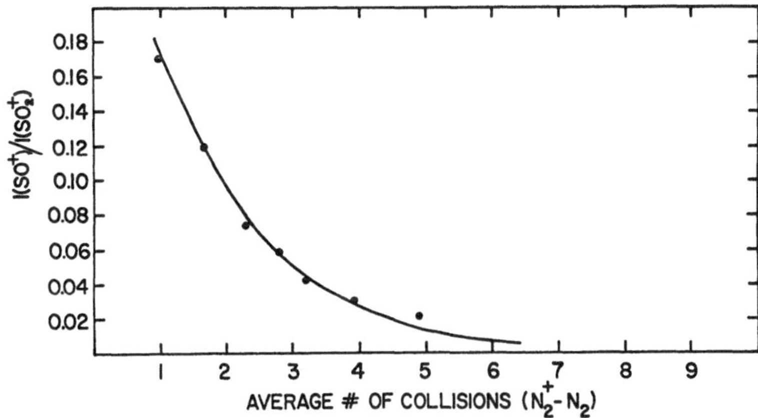

Fig. 9. Branching ratio SO_2^+/SO^+ as a function of N_2 pressure in the Dempster spectrometer ion source, expressed as the average number of N_2^+/N_2 collisions. Data are consistent with the interpretation that 2 collisions are sufficient to relax excited N_2^+ to the v = 0 level of the ground electronic state.

by $\tau k_c(N_2)$ where τ is the source residence time, k_c the collsision rate, and (N_2) the number densitiy of N_2 within the source. As can be seen in Figure 9, the ratio of $I_{SO^+}/I_{SO_2^+}$ decreases as the number of N_2^{+*}-N_2 collisions increases. After five collisions, less than 2% of the products is SO^+, while for an average collision number of one, more than 14% of the products was SO^+. This suggests that the excited N_2^{+*} state is de-excited after suffering from one to two collisions with N_2.

This high efficiency for the relaxation of N_2^{+*} on collision with N_2 can be further supported by considering the distribution in actual number of collisions corresponding to a given average collision number. As discussed elsewhere [14], the probablitity of n collisions, P(n) is given by,

$$P(n) = \frac{(\bar{n})^n}{n!} e^{-\bar{n}} \tag{E8}$$

where $\bar{n}$ is the average number of collisions experienced during the source residence time. Table 3 tabulates the distributions corresponding to $\bar{n}$ = 1, 2 and 3.

Table 3. Distribution of the Number of Ion Neutral Collsions, n.
$(P(n) = \frac{(\bar{n})^n}{n!} e^{-n})$ (a)

$\bar{n}$ \ n	0	1	2	3	4	5	6	7	8
1	0.37	0.37	0.18	0.06	0.02				
2	0.14	0.27	0.27	0.19	0.09	0.04	0.01		
3	0.05	0.15	0.22	0.22	0.17	0.10	0.05	0.02	0.01

(a) This equation is derived in reference 14.

It is instructive to consider Figure 9 in light of the distribution shown in Table 3. The zero-collision intercept implies a ratio of $(SO^+)/(SO_2^+)$ of 0.25, which is proportional to the ratio of $(N_2^{+*})/(N_2^+)$ in equation 14 for the collision chamber pressure of this series of experiments. At an average collision number of unity the ratio has decreased to 0.17, a decrease of 32%. In comparison with Table 3, 26% of the ions have undergone two or more collisions. Similarly at an average collision number of two, the ratio has decreased from the intercept value by 64%, while the number of ions having undergone two or more collisions is 59%. At an average collision number of 3, the ratio had decreased by 80%, while 80% have also undergone two or more collisions. Thus the data are consistent with the conclusion that about two collisions of N_2^{+*} with N_2 are sufficient to relax the ion to the lower energy state N_2^+ incapable of driving the dissociative charge-transfer reaction.

4. Summary and Conclusion

From the above discussion we may summarize the detailed reaction mechanism as follows:

Source Reactions

$$N_2 + e \rightarrow N_2^+ (X^2\Sigma_g^+ , v = 0) + 2e \quad (14)$$

$$N_2 + e \rightarrow N_2^+ (X^2\Sigma_g^+ , v = 1) + 2e \quad (15)$$

$$N_2 + e \rightarrow N_2^+ (A^2\Pi_u , v = 0,1,2) + 2e \quad (16)$$

$$N_2^+ (A^2\Pi_u) + nN_2 \rightarrow N_2^+ (X^2\Sigma_g^+ , v = 0) + xN_2^* + (n - x) N_2 \quad (17)$$

$$N_2^+ (X^2\Sigma_g^+ , v = 1) + nN_2 \rightarrow N_2^+ (X^2\Sigma_g^+ , v = 0) + xN_2^* + (n - x) N_2 \quad (18)$$

ICR Cell Reactions

$$N_2^+ (A^2\Pi_u) \rightarrow N_2^+ (X^2\Sigma_g^+ , v \geq 1) + h\nu \quad (19)$$

$$N_2^+ (X^2\Sigma_g^+ , v = 0) + SO_2 \rightarrow SO_2^+ + N_2 \quad (20)$$

$$N_2^+ (X^2\Sigma_g^+ , v \geq 1) + SO_2 \rightarrow SO_2^+ + N_2 \quad (21)$$

$$SO_2^{+*} \rightarrow SO_2 + O \quad (22)$$

$$SO_2^{+*} + SO_2 \rightarrow SO_2^+ + SO_2^* \quad (23)$$

Although condensed and somewhat simplified, this reaction scheme is consistent with the experimental facts. As discussed above, considerations of Franck-Condon factors and ionization cross sections, coupled with the pragmatic requirement for a stable ion beam of adequate intensity, results in a mixture of $X^2\Sigma_g^+$ and $A\ ^2\Pi_u$

in the primary beam. The composition depends upon source pressure; at the highest ion source pressure used the N_2^+ ion beam consists almost entirely of ground state $X^2\Sigma_g^+$ ions with insufficient vibrational energy (i.e., only ground state, v = 0) to drive the dissociative charge-transfer reaction. These facts are summarized in equations 14 - 18 .

The well-established radiative transition $A^2\Pi_u \rightarrow X^2\Sigma_g^+$ is represented by reaction 19. The radiative lifetime of various $A^2\Pi_u$ states is about an order-of-magnitude longer than the source-analyzer residence time and about an order-of-magnitude shorter than the reaction cell residence time. Hence, for low extent of reacion (single-collision conditions) reaction 19 is essentially complete before charge-transfer reactions 20 and 21 occur. Finally, Figure 8 demonstrates that much, if not all, of the dissociative charge-transfer proceeds via a predissociation mechanism (reaction 22) which can be collisionally quenched (reaction 23).

Noteworthy in this study is the extremely efficient quenching of both the N_2^{+*} reactant ions (by N_2) and SO_2^{+*} product ions (by SO_2). These energy relaxation processes are much more efficient than have been observed for many other ion-molecule systems. It is plausible to speculate that near-resonant-charge- exchange mechanisms are responsible for this result. This is especially likely for the quenching of the C state of SO_2^+ since the quenching rate constant is of the order of the gas kinetic rate constant.

5. References

1. I. Dotan, D.L. Albritton and F.C. Fehsenfeld, J. Chem. Phys. 64 (1976) 4334.

2. The energetics given in the text assume that the reactants and the products are in their ground state. The ionization potential of SO was taken from J.M. Dyke, L. Golob, N. Jonathan, A. Morris, M. Okuda and D.J. Smith, J. Chem. Soc. Faraday Trans. II 70 (1973) 1809. Thermochemical data used in constructing Figure 2 were taken from J.L. Franklin, J.G. Dillard, H.M. Rosenstock, J.T. Herron, K. Draxl and F.H. Field, Ionization Potentials, Appearance Potentials, and Heat of Formation of Gaseous Positive Ions (U.S. Government Printing Office, Washington, D.C., 1969) NSRDS-NBS 26, and D.R. Stull and H.H. Prophet, JANAF Thermochemical Tables, 2nd Ed. (U.S. Government Printing Office, Washington, D.C., 1971) NSRDS-NBS 37. It should be noted, however, that an earlier photoionization study by V.H. Dibeler and S.K. Liston, J. Chem. Phys. 49 (1968) 482 reports the threshold for SO^+ formation as 15.81 $\pm$ 0.02 eV. Further, Meisels et al. (reference 21) report a threshold of 15.93 eV. These lead to endothermicities of 0.21 and 0.35 eV, respectively, for reaction 2 .

3. R. Orth, J.H. Futrell and K. Jex, Int. J. Mass Spectrom. Ion Phys. (to be published).

4. R. Orth, R.C. Dunbar and M. Riggin, Chem. Phys. 19 (1977) 279.

5. W. Lindinger, F. Howorka, P. Lukac, S. Kuhn, H. Villinger, E. Alge and H. Ramler, Phys. Rev.

6. T. Su and M.T. Bowers, Int. J. Mass Spectrom. Ion Phys., 12 (1973) 347.

7. R.A. Barker and D.P. Ridge, J. Chem. Phys. 64 (1976) 4411.

8. J.W. Rabalais, Principles of Photoelectron Spectroscopy, New York, John Wiley (1978), Tables 7 and 8, p. 200.

9. W.B. Maier II and R.F. Holland, J. Chem. Phys. 59 (1973) 4501.

10. J.R. Peterson and J.T. Moseley, J. Chem. Phys. 58 (1973) 172.

11. B. Brehm, J.H.D. Eland, R. Frey and A. Kustler, Int. J. Mass Spectrom. Ion Phys. 12 (1973) 197.

12. K.T. Wu and A.J. Yencha, Can. J. Phys. 55 (1977) 767.

13. M.J. Weiss, T.C. Hsieh and G.G. Meisels, J. Chem. Phys. 71 (1979) 567.

14. A. Fiaux, D.L. Smith and J.H. Futrell, Int. J. Mass Spectrom. and Ion Phys. 20 (1976) 223.

INTERNAL ENERGY DEPENDENCE OF THE REACTION OF NH_3^+ WITH H_2O; A TANDEM ICR STUDY

P.R. Kemper and M.T. Bowers
Department of Chemistry, University of California
Santa Barbara, California 93106

1. Introduction

The Tandem ICR (TICR) is a concept with a great deal of appeal. The concept originated with Smith and Futrell[1] at Utah who constructed the first instrument. The TICR constists of an ion source, a 180^0 Dempster magnetic mass selector and a differentially pumped ICR cell. All three are positioned between the pole camps of a 12" electromagnet. The ions are formed in the source, either by direct electron impact or electron impact followed by chemical reaction. They are then mass analyzed and injected through an entrance slit into the ICR cell where their subsequent reactions are studied. The advantages over a conventional ICR are obvious: reaction complexity is greatly reduced with a single primary ion; reactions of product ions may be easily studied; ions may be formed and thermalized under high pressure source conditions and reacted under low pressure ICR conditions (ions such as Ar_2^+ may be studied this way) and lastly ions with different amounts of internal energy may be formed in the source using charge transfer reactions permitting a determination of absolute reaction rate constants and branching rations and their dependence on internal reaction energy. In this paper we describe briefly the tandem instrument at UCSB and preliminary results of a study of a reaction as a function of ion internal energy.

2. The Tandem Instrument

The Santa Barbara Tandem consists of tow differentially pumped chambers, one containing the ion source and the Dempster mass filter and the other containing the ICR cell. The source chamber vacuum is maintained with a 1500 l/sec Turbo Torr pump, providing about 800 l/sec of pumping at the source. The ICR is pumped with a 1-inch diffusion pump and cold trap.

With the ICR entrance slit normally used a pressure differential of ca. 2×10^5 can be maintained between the source and ICR. The source is surrounded with a cooling passage allowing operation at low temperatures which results in increased gas density at a given maximum pressure. Usually compressed air is used to maintain a source temperature of ca 40^0. A split repeller/extraction plate combination is used to focus the ions on the exit slit. This is necessary due to the strongly curved motion of the ions in the source due to the magnetic field. After leaving the exit slit the ions pass through a plit "steering" lens and are accelerated to between 1 and 4 KV by the high voltage plate. After passing through the high voltage slit the ions enter a field free region. The magnetic field is set to focus the proper mass on the exit slit of the field free region. After exiting, the ions are decelerated through another steering lens to a small kinetic energy and pass through the ICR entrance slit into the cell. The mass resolution is ca. 150. An intermediate focus/current collector is used to tune the source voltages. The mechanical layout and ion lenses are shown in Figures 1 and 2.

An obvious question about the TICR concerns the energy of the ions entering the ICR cell. Smith and Futrell [1] used diagnostic product distributions to probe the reactant ion kinetic energies and suggested they were less than 0.1 eV. Recent flowing afterflow work by Howorka et.al. [2] brought this into question. A study of the energy dependence for the product distribution in the $N^+ + O_2$ reaction (Figure 3) indicated the Utah TICR ion energies were ca 0.6 eV (C.M.).

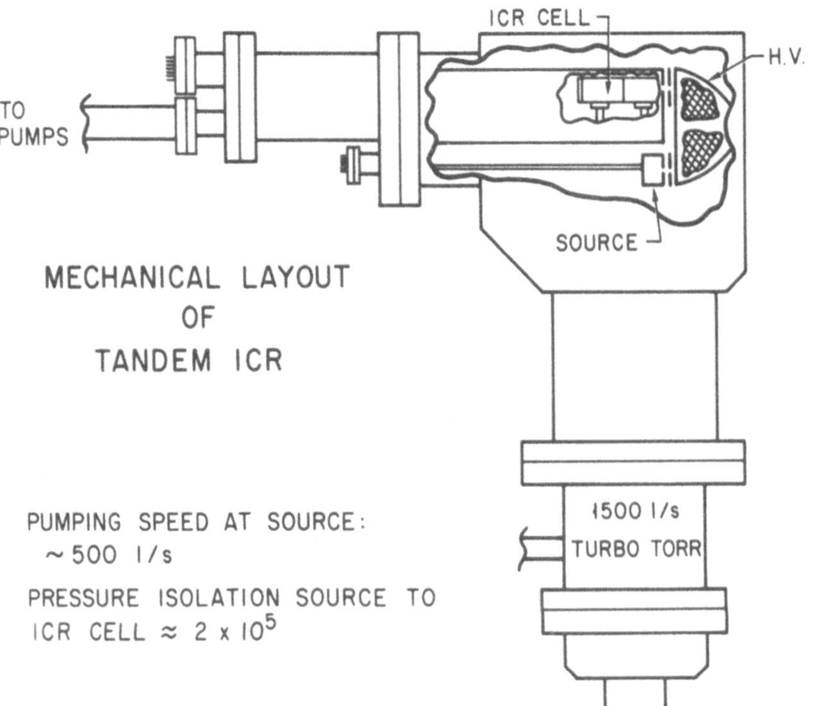

Fig. 1. Mechanical layout of the Santa Barbara Tandem ICR.

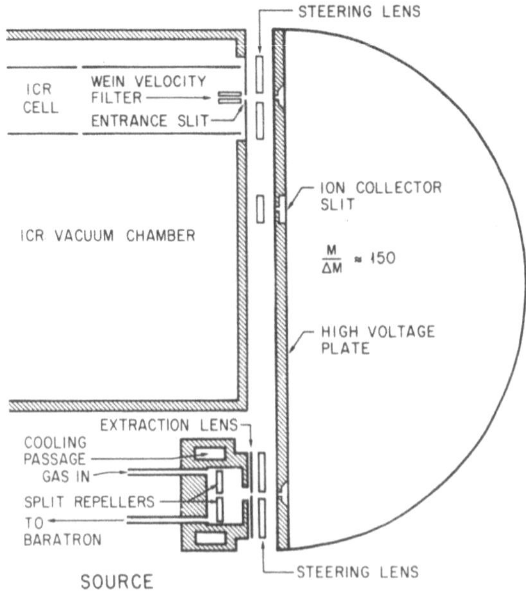

Fig.2. Ion source and lenses of UCSC Tandem ICR.

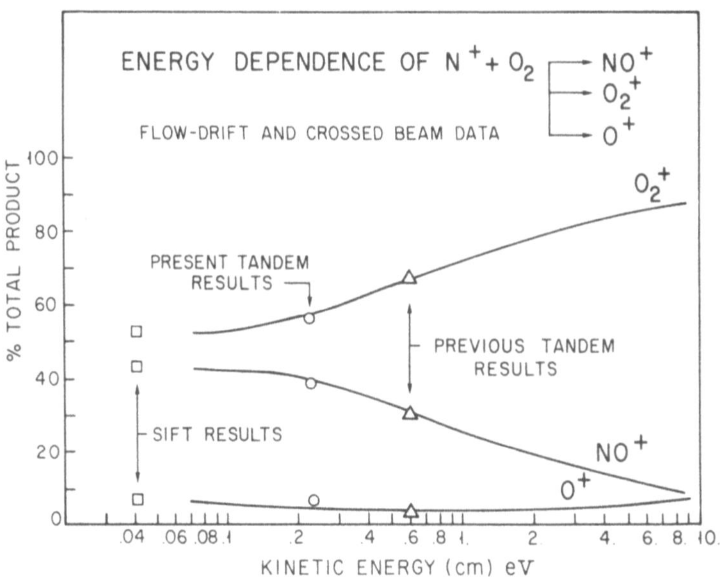

Fig. 3. Energy dependence of the $N^+ + O_2$ reaction product distribution.

The experiment was repeated on the Santa Barbara TICR (using a nearly identical lens structure) with the same results, indicating that translationally excited ions were entering the ICR cell. In an effort to reduce this effect, a Wien Velocity Filter (shown in Figure 2) was placed directly after the ICR entrance slit.

The filter consists of two plates .015" apart and 0.1" long. Voltages on the plates establish an electric field E and only ions with a velocity equal to E/B (where B is the magnetic field) pass through in a straight line. Ions with larger or smaller velocities are deflected and hit one of the plates. The spacing of the plates and the electric and magnetic fields establish a bandwidth of ion velocities which can enter the cell. With the filter in place $N^+ + O_2$ product distribution was remeasured and the results indicated ion energies of ca 0.2 eV were present (Figure 3). While this is a large improvement, it is not the final solution and efforts are underway to reduce the ion energies to less than 0.1 eV.

3. Internal Energy in NH_3^+

We have investigated the average amount of internal energy present in NH_3^+ ions using the product distribution of the diagnostic reaction

$$(NH_3^+)^* + H_2O \longrightarrow \begin{cases} H_3O^+ + NH_2 + 0.47\ \text{eV} \\ NH_4^+ + OH \ - 0.23\ \text{e} \end{cases} \qquad (1)$$

Anicich et al. [3] have shown that the relative amounts of H_3O^+ and NH_4^+ products from reaction 1 are a strong function of the NH_3^+ internal energy. Thus we expected to see large changes in the product distribution if NH_3^+ could be formed with different amounts of internal energy before it underwent reaction 1. The actual tandem experiment was done as follows: NH_3^+ ions were formed in the ion source by charge transfer from a variety of different ions (Xe^+, Kr^+, Ar^+, N_2^+, O_2^+, CO^+ and CO_2^+). The mass-selected NH_3^+ ions were then injected into the ICR cell where the reaction with H_2O occured and the product distribution was measured. Figure 4 presents data obtained using the charge transfer ions Ar^+, Xe^+ and CO_2^+. Clearly the three different charge transfer ions give very different results. The change in $\%NH_4^+$ with charge transfer gas pressure is due to the change from electron impact to chemical ionization conditions. Note also that the equilibrium $\%NH_4^+$ is reached at much lower pressure with Ar^+ and CO_2^+, presumably due to different charge

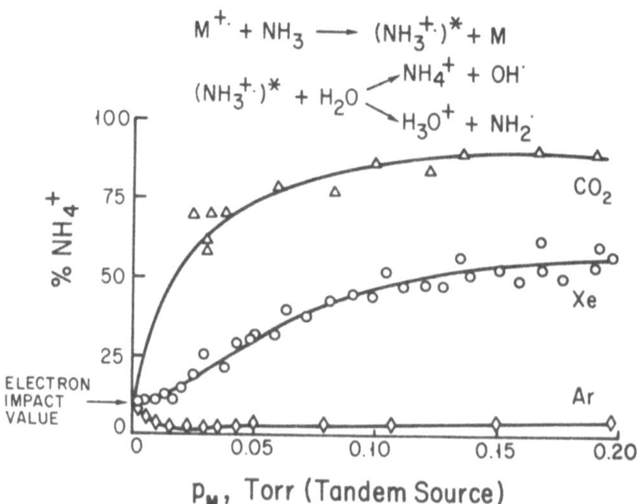

Fig. 4. Variation of NH_3^+ + H_2O product distribution with source gas pressure.

transfer rate constants (CO_2^+ = 1.8 x 10^{-9} cm^3/s , Ar^+ = 1.6 x 10^{-9} cm^3/s , Xe^+ = 0.6 x 10^{-9} cm^3/s).

Similar product distribution data is obtained with the other charge transfer ions. A corresponding trend is also found in the total rate of reaction. The rate increases from ca 4 x 10^{-10} cm^3/s with CO_2^+ to 7.5 x 10^{-10} cm^3/s with Ar^+.

It is clear from these results that different charge transfer reactions yield NH_3^+ with differing amounts of internal energy and that these differences in internal energy cause changes in the product distribution of reaction 1. This suggests that the product distribution can be used to measure the NH_3^+ internal energy. To do so, the scale must be calibrated using charge transfer ions which produce NH_3^+ with known amounts of internal energy. Mauclaire, Derai and Marx [4a] have measured the kinetic energy released, E_t (NH_3^+ - X), in the charge transfer reactions between the rare gas ions Xe^+, Kr^+ and Ar^+. The NH_3^+ internal energy is then determined from the energy balance equation.

$$E_{int}(NH_3^+) = R.E.(X^+) - IP(NH_3) - E_t(NH_3^+ - X) - E_{int}(X). \qquad (2)$$

E_{int} (X) is of course zero for X = Ar, Kr, Xe. Thus, by measuring the resulting distributions from reaction 1 when NH_3^+ is formed from Xe^+, Kr^+ and Ar^+, a semiquantitative scale of % NH_4^+ product vs NH_3^+ internal energy can be constructed.

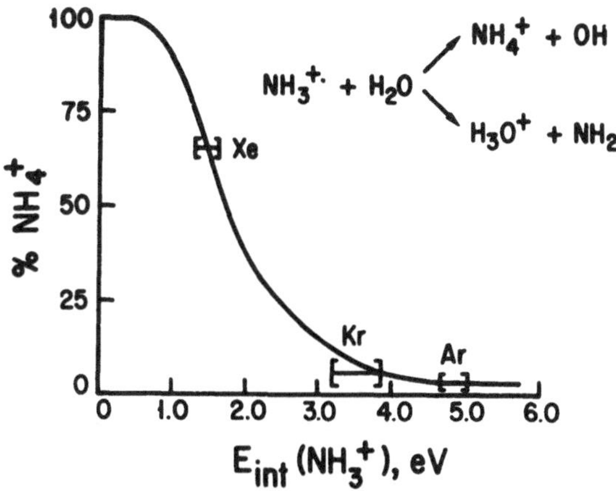

Fig. 5. NH_3^+ + H_2O product distribution vs. NH_3^+ internal energy.

Actually four calibration points are available since the reaction to form H_3O^+ product is 0.47 eV endothermic and NH_3^+ with less than 0.47 eV internal energy produces 100% NH_4^+ product [5,6]. This curve of % NH_4^+ product vs NH_3^+ internal energy is shown in Figure 5.

There are several questions that must be answered before these results can be accepted as correct. First, it will be noted on Figure 5 that the NH_3^+ internal energy resulting from Xe^+ charge transfer is shown as 1.5 $\pm$ 0.1 eV, a small uncertainty. There are however, two electronic states in Xe^+, $^2P_{1/2}$ and $^2P_{3/2}$, separated by 1.3 eV. In assigning the NH_3^+ internal energy the effect of higher energy state ($^2P_{1/2}$) has been neglected. This is justified for two reasons: The population of the $^2P_{3/2}$ state is twice that of the $^2P_{1/2}$ (due to the 2J + 1 degeneracy factor); and the rate of charge transfer [7] of the $^2P_{3/2}$ is 4.6 times that of the $^2P_{1/2}$. Thus the amount of NH_3^+ formed by the Xe^+ $^2P_{1/2}$ state is ca 10% of the total amount. The difference in ratio between states in Ar^+ and Kr^+ is not as great and the contribution from both states must be included. For this reason, much larger uncertainty in the NH_3^+ internal energy is shown in Figure 5 for the Ar^+ and Kr^+ points.

A second question concerns the effect of excess translational energy in the NH_3^+ ions entering the ICR cell. As indicated before this is probably 0.2 to 0.3 eV. If this energy is constant, it will cause a shift in the curve, assuming that translational energy and internal energy will affect reaction 1 similarly. The magnitude of the translational energy effect cannot be greater than a few tenth on an eV since product distributions very near 100% NH_4^+ have been measured.

The third and most important problem is that of possible collisional deactivation of the NH_3^+ in the ion source. As the charge transfer reagent gas pressure is increased not only does the fraction of NH_3^+ produced by charge transfer increase but also the number of collisions between the product NH_3^+ and the neutral gas. For residence times between 1 and 5 μsec (typical in our source) and a pressure of 0.1 Torr the average number of collisions is between 3 and 15 (assuming a collision rate of ca. 1 x 10^{-9} cm^3/s). This number of collisions could have a substantial effect on the average internal energy of the NH_3^+ leaving the source. The Ar^+ and Kr^+ data do, in fact, show some indication of collisional deactivation. The %NH_4^+ obtained drops from ca. 15% with no charge transfer gas to ca. 2% (Ar^+) and 4% (Kr^+) at 2-3 x 10^{-2} torr and then rises to ca. 8% (Ar^+ and Kr^+) by 2 x 10^{-1} torr. This is a small change in these cases, but in other systems it could be substantial. However, the presence of a plateau in the product distribution vs pressure data argues strongly against a large effect. Substantial collisional deactivation would cause the %NH_4^+ to increase monotonically toward 100% as the source pressure is increased. Since, in fact a plateau is attained, collisional deactivation in these experiments appears to be small. We are currently attempting to device experiments in which collisional deactivation is not a problem but, for now, based on the small observed effects in Ar^+ and Kr^+ and the plateau in the other data we shall assume the magnitude of the error is smaller than the experimental uncertainty already present.

Within the limiations discussed above the product distribution vs internal energy curve for reaction 1 has been determined. We may now use this curve to determine the NH_3^+ internal energies that correspond to a given product distribution. Thus, we measure the %NH_4^+ obtained with N_2^+, CO^+, CO_2^+ and O_2^+ charge transfer ions and use the product distribution curve in Figure 5 to determine the internal energy each of these ions deposits in the NH_3^+ ion. This is shown in Figure 6 and summarized in Table 1.

Table 1. $M^+ + NH_3 \rightarrow NH_3^+ + M$

	Reaction Energy R.E.(M^+)-I.P.(NH_3)	$E_{int}(NH_3^+)$	$E_t(NH_3^+ - M)^c$
Ar^a	5.61/5.79	4.86-5.02	0.87
N_2	5.4	3.35	
Kr^a	3.85/4.52	3.22/3.89	0.63
CO	3.8	2.2	
CO_2	3.6	0.9	
Xe^b	1.94	1.5	0.41
O_2	1.87	1.3	

a) for $^2P_{3/2}$ and $^2P_{1/2}$ states respectively

b) for $^2P_{3/2}$ only

c) Data from Mauclaire, R. Derai and R. Marx (private communication).

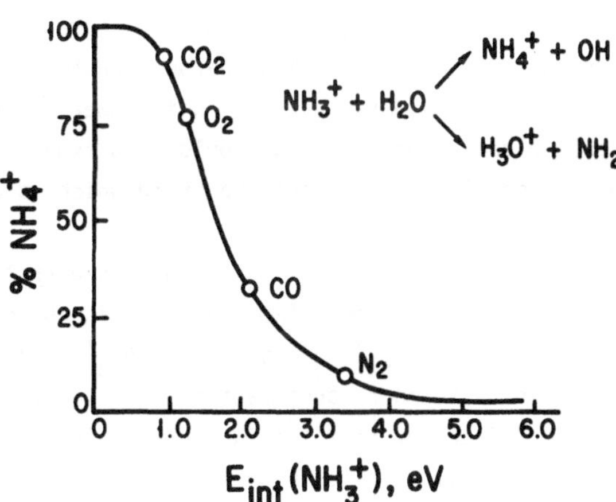

Fig. 6. Determination of NH_3^+ internal energy produced from charge transfer from N_2^+, O_2^+, CO^+, CO_2^+.

There are a number of interesting points to be made from this data. First it is clear that the energy available for reaction R.E.(X^+)-I.P.(NH_3) is not always a reliable indication of the resulting NH_3^+ internal energy. The charge transfer reaction of Kr^+, CO^+ and CO_2^+ all provide ca. 3.8 eV of available energy, yet the resulting NH_3^+ internal energies range from 0.9 eV (CO_2^+) to 2.2 eV (CO^+) to 3.2 eV (Kr^+). In contrast to the Kr, CO, CO_2 data is that of Xe^+/O_2^+. Here, both M^+ ions provide ca 1.9 eV to the reaction, and both produce NH_3^+ with 1.3 to 1.5 eV of internal energy - identical, within the accuracy of the experiment. A possible explanation for this difference may lie in the availability of a vertical ionization process with Xe^+ and O_2^+ (due to a large NH_3^+ Frank-Condon manifold at these energies) and the low probability for a vertical process with the reaction energies provided by Kr^+, CO^+ and CO_2^+ [8]. Such a vertical ionization with Xe^+ and O_2^+ could occur at larger internuclear distance. The difference between these ions would thus be minimized and the energies of the reaction (which are very similar for these ions) would be emphasized. The result might well be that both Xe^+ and O_2^+ deposit the same internal energy in NH_3^+. Since the photoelectron spectrum of NH_3 shows no transitions in the energy range provided by Kr^+, CO^+ or CO_2^+ (ca 3.8 eV above threshold), a vertical ionization is not probable [8]. Thus these ions must undergo a strongly coupled collision with the NH_3 in order for charge transfer to occur. This permits the partitioning of the available reaction energy among the neutral internal modes of molecule X, NH_3^+ internal energy, and NH_3^+ - X translational energy. Under these circumstances we would expect (for a given energy of reaction) that a charge transfer ion with a large number of internal modes would produce NH_3^+ with lower internal energy than an ion with a smaller number of modes. We would also expect the resulting neutral X to be more excited. This is in fact the trend which is observed in the Kr^+, CO^+, CO_2^+ series.

The results presented so far are in their own right. However they open the door to experiments which may be even more important. It is obvious from the energy balance equation 2 that if $E_{int}(NH_3^+)$ and $E_{trans}(NH_3^+ - M)$ are both determined then $E_{int}(M)$ can be found by substraction [9]. This would allow the complete determination of energy partitioning in a series of charge transfer reactions, a very useful result. These kinetic energy release experiment are currently underway as a collaborative effort with Dr. Rose Marx's laboratory at the Université de Paris-Sud in Orsay [10].

4. Ackowledgement

The support fo the National Science Foundation under grant CHE80-20464 is gratefully acknowledged. We also wish to thank Drs. Mauclaire, Derai and Marx for communicating results prior to publication and for many helpful discussions.

5. References

1. D. Smith and J. Futrell, Int. J. Mass. Spectrom. and Ion Phys. 14 (1974) 171.

2. F. Howorka, I. Dotan, F.C. Fehsenfeld and D.L. Albritton, J. Chem. Phys. 73 (1980) 758.

3. V.G. Anicich, J.K. Kim and W.T. Huntress, Int. J. Mass Spectrom. Ion Phys. 25 (1977) 433.

4. a) G. Mauclaire, R. Derai and R. Marx, to be published; b) See, for example, G. Mauclaire, R. Derai, S. Fenistein and R. Marx, J. Chem. Phys. 70 (1979) 4017; R. Derai et al., Chem. Phys. 44 (1979) 65; R. Marx et al., J. Chem. Phys., 76 (1979) 1077.

5. H.M. Rosenstock, K. Draxl, B.W. Steiner and J.T. Heron, J. Phys. Chem. Ref. Data 6 (1977) Supp No. 1.

6. D.H. Aue and M.T. Bowers, Chapter in "Gas Phase Ion Chemistry", Vol. II, M.T. Bowers (ed.), Academic Press, N.Y. (1979), pp. 2-52.

7. N.G. Adams, D. Smith and E. Alge, J. Phys. B. 13 (1980) 3235.

8. D.W. Turner, C. Baker, A.D. Baker and C.R. Brundle, "Molecular Photoelectron Spectroscopy", Wiley-Interscience, New York (1970).

9. For molecular charge transfer ions care must be taken to generate them with little or no internal energy. For our apparatus symmetric charge exchange between X^+ and X before reaction with NH_3 is a very efficient deactivation mechanism.

PRECISION DETERMINATION OF CYCLOTRON FREQUENCIES OF FREE ELECTRONS AND IONS

G. Gräff
Institut für Physik
Universität Mainz, West Germany

1. Introduction

Within the last two decades the electrodynamical storage of electrons and ions developed into an experimental method of great versatility. That this method is now being used in so many different fields of physics and chemistry results primarily from the long storage times which nowadays can be achieved. Under ultrahigh vacuum conditions and in sufficiently strong electromagnetic fields the particles can easily be trapped for hours or even days. This really long storage time offers the possibility of studying reactions of very slow rate to the chemist and of precision measurement of photon-ion interactions to the physicist. The accuracy of photon-ion interaction measurement is finally limited by Heisenberg's uncertainty relation. Therefore long interaction times correspond to narrow line widths. An excellent example is the determination of the hyperfine structure of stored Barium ions [1] . The transition frequency is about 10 GHz, the absolute line width achieved in this experiment was a few Hz only. Therefore the fractional line width is of the order of one part in 10^{10} opening the introduction of this method as a future frequency or time standard. In an analogous fashion electrons were trapped to measure the anomalous part of their magnetic moment to one part in 10^{8}, now the best known elementary particle poperty at all [2] . Last not least atomic masses have been measured to high accuracy.

2. Ion traps

There are several possibilities to trap ions and electrons. However, for reasons to be discussed later the Penning configuration is mostly used (Figure 1). The Penning configuration consists of an electrostatic quadrupole with superimposed homogeneous magnetic field. The quadrupole electrodes are formed by a hyperbolic ring and two hyperbolic end caps. The dimensions of this kind of trap varies between mm and cm. If the rest gas pressure is kept below 10^{-10} torr and the magnetic field strength above 1 T, ions can be trapped for hours.

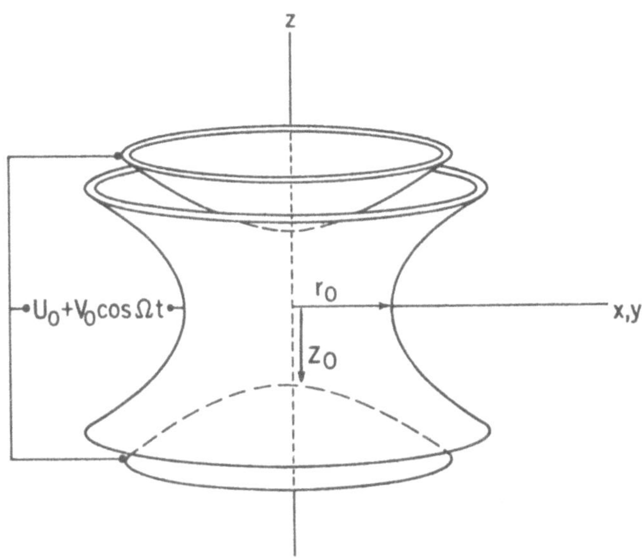

Fig. 1. The electrostatic quadrupole trap.

One advantage of the Penning configuration is the fact that the Schrödinger equation of this potential can be solved exactly. The solution shows that the ion behaves like being exposed to the potential of a three dimensional oscillator. Neglecting magnetic moment interactions the ion energy is given by

$$W = \omega_z\,(n_z + \tfrac{1}{2}) + \omega_+\,(n_+ + \tfrac{1}{2}) - \omega_-\,(n_- + \tfrac{1}{2}) \qquad n = 0,1,2,........$$

with $\omega_z = (eU_0/mR_0^2)^{1/2}$

$$\omega_+ = \omega_{c/2} + \omega_0$$

$$\omega_- = \omega_{c/2} - \omega_0$$

with $\omega_0 = (\omega_c^2/4 - \omega_z^{2}/2)^{1/2}$

In the limit of vanishing electric field strength we find $\omega_+(U_0 \to o) = \omega_c = (e/m)\,B$. These equations demonstrate another advantage of this electrodynamic configuration: the fundamental orbital frequency ω_+ of the ions is not broadened by the presence of the electric trapping field but shifted to a lower value by ω_-. Therefore by measuring both frequencies ω_+ and ω_-, the cyclotron frequency can be determined. However, inhomogenieties of the radio frequency field and imperfections of the trapping potential lead generally to a coupling of the different degrees of freedom. Therefore combinations of the three fundamental oscillations can be observed in a single transition, in particular also the direct cyclotron frequency ω_c.

3. Trapped Particle Detection

Trapped particles are usually detected by observing the image voltage which is induced in the electrodes by the oscillatory motion of the charges. The detection sensitivity could be significantly improved by the introduction of resonance circuits with high Q-values and phase sensitive amplification combined with low temperature technique. By this means Dehmelt and collaborators succeeded in detecting one single electron [3] .

A single trapped ion was also observed (Toscheck el al) [4] . In this experiment a single Ba^+ ion was continuously excited by two lasers and the resulting fluorescence light ovserved by a microscope. However, the application of this method is still restricted to specific ions. Sometimes it is easier to extract the ions out of the trap and count them by a channel plate detector.

4. Measurement of Fundamental Oscillatory Frequencies

The measurement of the oscillatory frequencies of the trapped ions is usually performed by induction of the relevant frequencies. This leads to an increase of kinetic energy and thereby to an increase of the detection signal. Due to the strong coupling between the different degrees of freedom the energy is rapidly transferred from the motion perpendicular to the magnetic field into that motion which is parallel to the magnetic field. Generally the increase of kinetic energy can therefore be observed in both directions. If the cyclotron frequency of electrons is induced, energy electrons gain energy which is transferred to their axial motion. Thus the signal induced in the end caps is increased.

5. Determination of Ion Masses

Both ω_+ and ω_z depend on the mass m of the ion and can therefore be used to determine its mass. The axial frequency is appropriate only for low precision measurement since the effective electric field strength depends usually on contact potentials, thermal electromotoric forces, temperature dependent dimensions, etc. which are difficult to control. In addition the truncation of the three electrodes gives rise to higher order terms of the basically parabolic potential. To compensate for this effect van Dyck et al [5] introduced two additional electrodes. Thus they were able to reduce the contribution of the higher order terms of the potential by a factor hundred. The resulting relative line width of ω_z was smaller than a few parts in 10^7.

Contrary to the axial frequency the cyclotron frequency is ideally suited to determine masses, since it is no problem to stabilise the magnetic field strength using superconducting magnets. The cyclotron frequency can be induced as a single transition at frequency $\omega_c = \omega_+ + \omega_-$ or alternatively as a sum of two different transitions at the fundamental frequencies ω_+ and ω_-, respectively. The drift frequency ω_- can be determined to an accuracy of about one part in 10^4. This imposes an uncertainty of approximately a few parts in 10^8 on the determination of ω_c of an ion.

The most important contributions to the line width stem from possible misalignments of the electrodes relative to the magnetic field axis, from magnetic field inhomogenieties, from space charge effects, and finally from the finite interaction time between radio frequency fields and the charge of the ions. Space charges generally shift and broaden the cyclotron frequency. Since the signal to noise ratio is proportional to the number of stored ions, high sensitivity of the charge monitoring system is of great importance.

Three different methods of detecting cyclotron frequencies of ions are known. The observation of the image current in the electrodes which increases if the charges are coherently accelerated was mentioned already. The most sensitive method however, has been developed by Dehmelt and his collaborators [2]: In this method a slightly inhomogeneous magnetic field is superimposed. Since the circular ion motion corresponds to a rotational magnetic moment

$$\mu = (2n_+ + 1)\mu_B$$

this inhomogeneous field acts as an additional force along the z direction. Consequently the axial oscillation frequency ω_z is changed proportional to μ and therefore to the quantum number n_+. The absolut value of the inhomogeneity of the additional magnetic field is chosen such that a change of the rotational magnetic momont with $\Delta n_+ = 1$ results in an axial frequency shift of 2 Hz. The axial frequency is then continously monitored to this precision.

The line width of the axial frequency is less than a few Hz. Therefore it is possible to observe a shift of 2 Hz and thus a change of the cyclotron orbit quantum number from $n_+ = 0$ to $n_+ = 1$ of a single electron! Since the magnetic field also gives rise to a force parallel to the z direction, Dehmelt suceeded in determining not only the cyclotron frequency but also the anomalous magnetic moment of the electron with an accuracy of a few parts in 10^8.

There is still a third possibility to detect the cyclotron frequency which has recently been used for a precise determination of the electron/proton mass ratio [6]. Figure 2 shows the experimental arrangement. In this experiment a superconducting magnet is used providing a field of 6 T. To ensure a sufficient homogeneity over the trap volume a series of coils is mounted on the vacuum

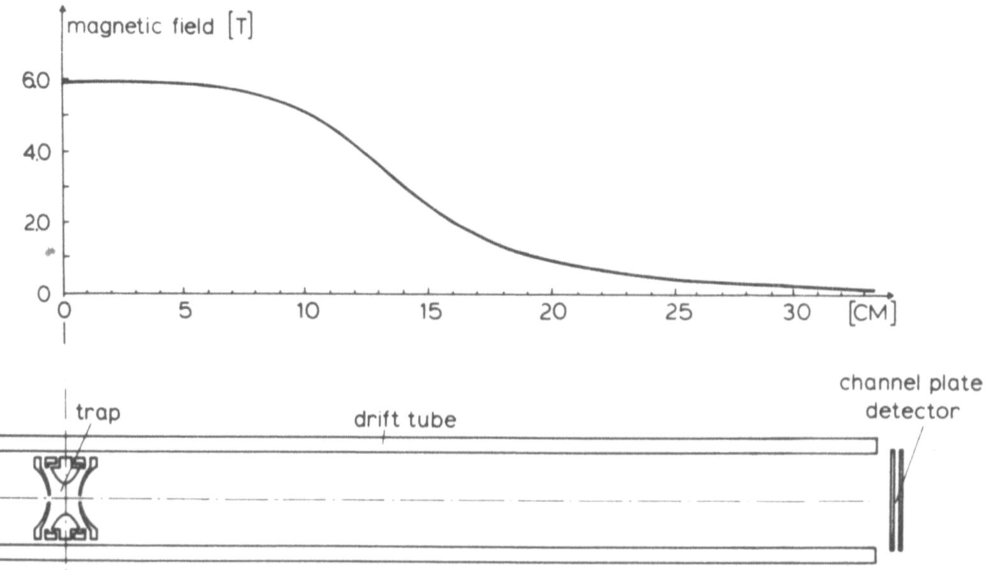

Fig. 2. The experimental arrangement of the time of flight method.

enclosure. They allow a systematic correction of linear field gradients in three directions and of the quadratic term along the axis of symmetry. Outside the trap the magnetic field falls off rapidly. At one side of the quadrupole trap, at a distance of 40 cm and near to the magnetic field axis, a heated tungsten wire serves as electron source. The primary electrons may penetrate through the entrance hole of the end cap and produce protons inside the trap by ionisation of the residual gas. Or they impinge on the interior surface of the trap producing secondary electrons some of which are trapped. Due to synchrotron radiation in the strong magnetic field the electrons loose their initial energy of a few eV and cool down to thermal energies of about 30 meV within a second. The trap is installed in a 40 cm long copper drift tube of 30 cm diameter. In order to count the number of stored particles the ions are ejected out of the trap along the axis of symmetry. The ions drift through the tube until they finally hit a channel plate detector 40 cm far from the trap at the end of the drift tube.

To obtain a vacuum better than 10^{-9} torr the complete apparatus is baked out. The trap and tube can be cooled down to liquid helium temperatures.

To detect the cyclotron frequency the following method is used: After trapping a few protons (electrons) the potential is kept constant for one second during which the protons cool down. Since the protons are generated by electron bombardment of the residual gas, other ions of heavier masses are also produced and stored inside the trap. These ions contribute, of course, to the space charge and broaden the cyclotron frequency. Therefore the trap was cleared of all other ions in the following way: One second after the ion creation the trap voltage was linearly decreased down to 1 V. Simultaneously, a rf field is applied, the frequency of which is chosen so that all ions heavier than protons experience their axial resonant frequency ω_z. The amplitude of this rf field is high enough to garantee the ejection of all unwanted ions.

Then the cyclotron frequency is applied for about 500 ms, and finally the trap is cleared by a linear voltage sweep superimposed by a sequence of ejection pulses which define the starting time of the ejected particles. The ions leave the trap and drift along the axis of symmetry until they hit the channel plate detector. With the same series of ejection pulses a sequence of contiguous gates is generated opening fast scalers. Thus a time of flight spectrum of the ions is built up.

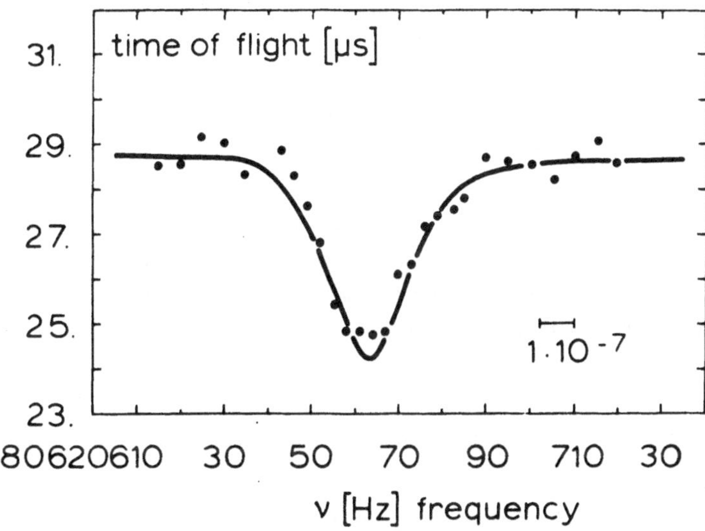

Fig. 3. The proton cyclotron frequency determined by the time of flight method.

The time of flight reflects the initial kinetic energy of the particles along the magnetic field axis, but also their transverse energy incorporated in the cyclotron motion. The transverse energy corresponds to a rotational magnetic moment, and in the inhomogeneous magnetic field the particles experience an acceleration proportional to their rotational magnetic moment. When the cyclotron frequency is induced the trapped ions gain energy in their transverse degree of freedom. Thus the cyclotron orbits of the particles and consequently their rotational magnetic moment increases. When these particles are expelled out of the trap they experience an additional acceleration in the inhomogeneous part of the magnetic field. The axial energy increases, leading to a corresponding decrease of the time of flight which is observed. This idea was proposed originally by Bloch [7] and used in different experiments. Figure 3 shows a typical result. The relative linewidth is about 2 parts in 10^7. The corresponding cyclotron frequency of electrons is broadened and shifted by the relativistic mass increase of the electron.

6. Summary

Ion cyclotron resonance spectrometry has become a useful tool for studying ion chemistry in the gas phase and for precise mass determination, too. In this new technique ion storage plays a dominant role. It has been demonstrated, that the Penning configuration (electrostatic quadrupole field with superimposed homogeneous magnetic field) has several advantages, since the solution of the Schrödinger equation is known for this potential. The cyclotron frequencies are not broadened by the electric trapping field. At least for light ions the relative line width of the cyclotron frequency may be as small as a few parts in 10^7. Further improvements and extension of the method to heavier masses seem possible.

7. References

1. W. Becker, R. Blatt and G. Werth, Europhysics Conference Abstracts - ECAP vol. 5a, part 5, p. 208

2. R. S. van Dyck, Jr., P. B. Swinberg and H. G. Dehmelt, Phys. Rev. Lett. 38 (1977) 310

3. D. J. Wineland, P. Ekstrom and H. G. Dehmelt, Phys. Rev. Lett. 31 (1973) 1279

4. P. E. Toscheck and W. Neuhauser, Physikal. Blätter, 36, nr. 7 (1980)

5. R. S. van Dyck, Jr., D. J. Wineland, P. Ekstrom and H. G. Dehmelt, Appl. Phys. Lett. 28 (1976) 446

6. G. Gräff, H. Kalinowsky and J. Traut, Z. Phys. A 297 (1980) 35

7. F. Bloch, Physica 19 (1953) 821

TOWARD A FREQUENCY SCANNING MARGINAL OSCILLATOR

Paul R. Kemper and Michael T. Bowers
Department of Chemistry, University of California, Santa Barbara, CA 93106

1. Introduction

The advantages of a frequency scannable detector in ICR spectrometry have long been obvious. The ability to operate at constant magnetic field allows uniform trapping efficiency in trapped ion experiments and, in drift cell work, eliminates differential effects due to changing ion density, drift times and extents of reaction. Experiments where one ion is continually ejected are possible as well. That a great need exists for this type of detector is obvious from the tremendous interest in Bridge Circuit Detectors (BCD) which exists at present. A BCD suitable for ICR work was first presented by Wobschall [1]. McIver has recently developed a solid state version, [2,3] and other workers have followed [4]. Throughout this development, the possibilities of a frequency scannable Marginal Oscillator (MO) have been ignored. Recent work in our laboratory and others [1, 4, 5] indicate, however, that the sensitivity of the MO surpasses that of the BCD by a significant factor. While not universally applicable, the scanning MO appears to be the detector of choice in many experiments. We present here a summary of the requirements a scanning MO must fulfill, the basic approaches we have taken to satisfy them, and finally a short derivation of the relative sensitivities of Bridge Circuit and Marginal Oscillator detectors. A complete description of the scanning MO will be submitted for publication elsewhere.

2. Requirements

Five requirements must be met to permit frequency scanned operation: (1) a continous frequency range from ca 100 KHz to 1 MHz; (2) a means of calibrating MO sensitivity at different frequencies; (3) an oscillation level control circuit to maintain a constant level as the frequency is scanned; (4) a means of generation a x axis drive voltage proportional to frequency; and (5) a drive mechanism for sweeping the frequency. The approaches taken to satisfy these requirements are outlined below.

A continously adjustable frequency range is easily accomplished using several inductors for different ranges and a varialbe capacitor to tune within a given range. Theoretical consideration indicate that greater sensitivity is obtained with higher inductance to capacitance ratios (L/C) in the tuned circuit. Thus, large tuning capacitors are to be avoided. This factor must be weighed against the greater ease of scanning with fewer ranges. The present design has four ranges covering 80 KHz to $\sim$1500 KHz (L = 5, 1.25, 0.33 and 0.08 mHy; C = 50 $\rightarrow$500 pf).

The sensitivity as a function of frequency is measured using a Q-Spoiler standard signal. This method is accurate to within a few percent over small frequency ranges and to within 20% for a factor of 10 change in frequency [6].

Some form of oscillation level control circuit (OLC) is needed to both maintain oscillation and prevent high levels leading to excess ion heating. Our design senses the rectified AC voltage after the first stage of amplification, smooths it and combines it with a preset reference voltage to produce a control voltage. The control voltage is then used to vary the gain of the second amplifier stage. In the vacuum tube circiut used, the signal controls the plate voltage on the second amplifier tube. The method works well from 100 KHz to $\sim$800 KHz. Above 1 MHz the capacitive roll off in the first amplifier results in a decreased input to the OLC leading to an increase in actual oscillation level. This effect is small (a factor of two increase by 1500 KHz) and a solution is currently being sought.

The x axis drive voltage is generated using a frequency to voltage converter (F$\rightarrow$V). The MO frequency counter output is shaped and used by the F$\rightarrow$V to generate a train of pulses at the same frequency. The pulse train is than filtered to give a DC level proportional to frequency. Frequency/output voltage ranges from 500 Hz/volt to 50 KHz/volt are selectable. A zero offset control allows on scale operation from 0 Hz to 2 MHz. Thus, it is possible to scan from 1000 KHz to 1005 KHz with a 0.0 to 10.0 volt output.

The actual scan is done with a variable speed, reversible DC motor coupled to the tuning capacitor. The motor is electrically and vibrationally isolated from the MO. The connection to the tuning capacitor is via a teflon clutch and a 10:1 worm gear reducer. A 20:1 reducer would probably be useful for scanning at higher frequencies.

The five modifications discussed above allow for reliable, frequency scanned Marginal Oscillator detection. An obvious question arises as to the relative performance of the Scanning Marginal Oscillator and the Bridge Circuit Detector. One of the main considerations is relative sensitivity. A comparison can be made as follows.

In both the MO and BCD, the ion power absorbtion (PA) is the property which is used for detection. Since ions in resonance absorb power from the irradiating field and increase their translational energy, they are functioning as a shunt resistance across the cell capacitance. Thus, both the MO and BCD measure a change in resistance across the cell. It follows intuitively and theoretically [1] that the sensitivity of the detector (here defined as the fractional change in detector voltage caused by one ion) is governed by

$$\text{Sensitivity} = \frac{\Delta V}{V} \propto \frac{Z_{detector}}{Z_{ion}} \tag{1}$$

where $Z_{detector}$ is the impedance of the detector and Z_{ion} is the resistance/ion. Thus, since a small change in a large resistance is being detected, a large detector impedance is needed. The impedance of the MO tank circuit has been shown many times to be [7, 8]

$$Z_{MO} = Q\omega L = Q/\omega C \tag{2}$$

since

$$\omega^2 = 1/LC \tag{3}$$

where Q is a dimensionless "quality" factor, L is the inductance of tank circuit, and C is the total capacitance across tank circuit. In the BCD the impedance is governed by the capacitance due to the cell, cables, balance capacitor and preamp input capacitance.
The sum of these is denoted C' and

$$Z_{BCD} = 1/\omega C' \tag{4}$$

An ion in resonance, in the absence of collisions, absorbs power according to

$$\text{PA/ion} = \frac{q^2 V^2 t}{4m\ell^2} \tag{5}$$

where q is the electron charge, V the peak irradiating field, t the time in resonance, m the ion mass and ℓ the spacing of cell plates. Since the power dissipated in a resistor is given by V^2/R it follows that

$$Z_{ion} = 4\,m\,\ell^2/q^2 t. \tag{6}$$

Then the sensitivities of the MO and BCD may be written:

$$\frac{\Delta V}{V_{MO}} = \frac{Q}{\omega C} \times \frac{q^2 t}{4m\ell^2} \quad (7a)$$

$$\frac{\Delta V}{\Delta V_{BCD}} = \frac{1}{\omega C'} \times \frac{q^2 t}{4m\ell^2} \quad (7b)$$

Equation 7b has been derived rigorously as well [1,3,9] . These expressions are nearly identical and show the relative sensitivities to depend on the factors Q/C (for the MO) and 1/C' (BCD). The quality factor Q in MO tank circuits is typically equal to or greater than 100 and the total capacitance C varies from ca. 150 pf to 600 pf. The ratio of Q/C thus varies in the range of ca. .6 to .15 pf^{-1}. The capacitance C' in the BCD has been given values from 100 pf[9]to 30 pf[10].Thus, 1/C' is in the range from 0.01 to 0.03 pf^{-1}. This analysis indicates the Marginal Oscillator Detector to be more sensitive than the Bridge by a factor between 5 and 60. Wobschall [1] reports a value of C' = 3 pf obtained by driving the cable shields to null their capacitance. This seems unrealistically low and, as McIver [9,10] has pointed out, very low values of C' will lead to non-linear response.

There are both other disadvantages and advantages to the BCD and the MO. In some BCD designs there appears to be difficulty in maintaining the proper phase shift and level in the reference signal when a large frequency range is swept. Also, although theory [3,8] predicts BCD sensitivity to be independent of frequency, when a Q-spoiler sensitivity calibration was done on our Bridge Detector variations of $\pm$ 30% were observed between 100 KHz and 1 MHz. To our knowledge, this is the only such measurement that has been made on the frequency depenence of the BCD detector sensitivity.

Despite the above, there are a number of applications where only a Bridge Circuit can be used. Computer controlled experiments, rapid frequency scans, signal averaging etc. are difficult or impossible with a Marginal Oscillator Detector. The limitation in sensitivity can be overcome in some applications (especially analytical applications) by observing the image currents created by the excited ions in a Bridge Detector [2,3,8] . This greatly increases the available signal but also introduces the possibility of differential damping effects (e.g. different ion collision rates) leading to uncertainties in actual ion concentrations.

In summary, the Marginal Oscillator Detector is not the dinosaur that it is sometimes made out to be. The scanning MO has significant advantages over Bridge Detectors in many applications and represents a very large improvement over fixed frequency MOs.

3. Acknowledgement

The support of the National Science Foundation under grants CHE 77-15449 and CHE 80-20464 is gratefully acknowledged.

4. References

1. D. Wobschall, Rev. Sci. Instrum. 36 (1965) 466.
2. R.L. Hunter, Ph.D. Thesis, University of California at Irvine, 1979.
3. R.T. McIver, R.L. Hunter, E.B. Ledford, M.J. Locke and T.J. Francl, Int. J. Mass. Spectrom. and Ion Phys. (in press).
4. D. Ridge
   a) 2nd International ICR Meeting in Mainz;
   b) Proceedings of the 29th Annual ASMS Meeting, Minneapolis, Minn., 1981.
5. M.B. Comisarow, Proceedings of the 2nd International ICR Meeting in Mainz.
6. P.R. Kemper and M.T. Bowers, Rev. Sci. Instrum. 48 (1977) 1477.
7. A. Warnick, L.R. Anders and T.E. Sharp, Rev. Sci. Instrum. 45 (1974) 929.
8. R.T. McIver, Rev. Sci. Instrum. 44 (1973) 1071.
9. R.T. McIver, E.B. Ledford and R.L. Hunter, J. Chem. Phys. 72 (1980) 2535.
10. R.L. Hunter and R.T. McIver, Second Conference on Ion Cyclotron Resonance, Mainz, Germany, March 1981.

AN FTICR SPECTROMETER - DESIGN PHILOSOPHY AND PRACTICAL REALISATION

J.H.J. Dawson
Laboratory of Organic Chemistry, University of Amsterdam,
Nieuwe Achtergracht 129, 1018 WS Amsterdam, The Netherlands

1. Introduction

Many papers have appeared in the literature describing the theoretical background to Fourier Transform Ion Cyclotron Resonance Mass Spectrometry (FTICR) and presenting block diagrams for such spectrometers [1-6]. This paper will endeavour to draw attention to some of the practical difficulties which must be solved by constructors, and to describe in detail how the most serious problems have been solved in the building of the Amsterdam FTICR instrument.

A designer working from scratch and with a free hand will probably choose a superconducting magnet [7]. This will enable him to have a larger cell than has been customary hitherto, but it will also involve him in working at higher frequencies. Although the designs presented here are all based upon the idea of 3 MHz as the upper signal frequency limit, they could probably be extended, with care, up to 5 MHz which would correspond to m/z 15 at 49 kG (4.9 T). The principal consideratations in choosing the computer will be the resolution required in normal wideband spectra, which will determine the size of the data block, and the length of time which it will take to do the transform on that block. Roughly, the size of the data bank will need to be four times the required resolution, for example, 16K for a resolution of 4,000. The instrument described here used the very cost-effective "MINC" package from Digital Equipment Corporation. This is a PDP11 family computer, built in LSI with 30K of memory and two floppy disc units.

Having decided that the maximum signal frequency to be handled should be 3 MHz it follows that the ADC must take samples at a rate over 6 MHz. The question then arises as to how this numerical fusillade can be got into memory. I do not like to say that general purpose computers will never have such fast memory (with direct memory access), but it seems certain that for several years to come some form of special dedicated fast hardware memory will be needed. The designer has two choices, either the extra fast memory will be only a buffer between the ADC and the computer's general memory, or the extra fast memory will be made larger so that it can itself perform the accumulation of successive transients. In this design it takes about 250 ms to add each transient from the 8 MHz, 16K byte simple buffer store into memory, which means that the operation costs 1 minute for every 256 transients accumulated. Now comes the major practical problem. If transient signals from a

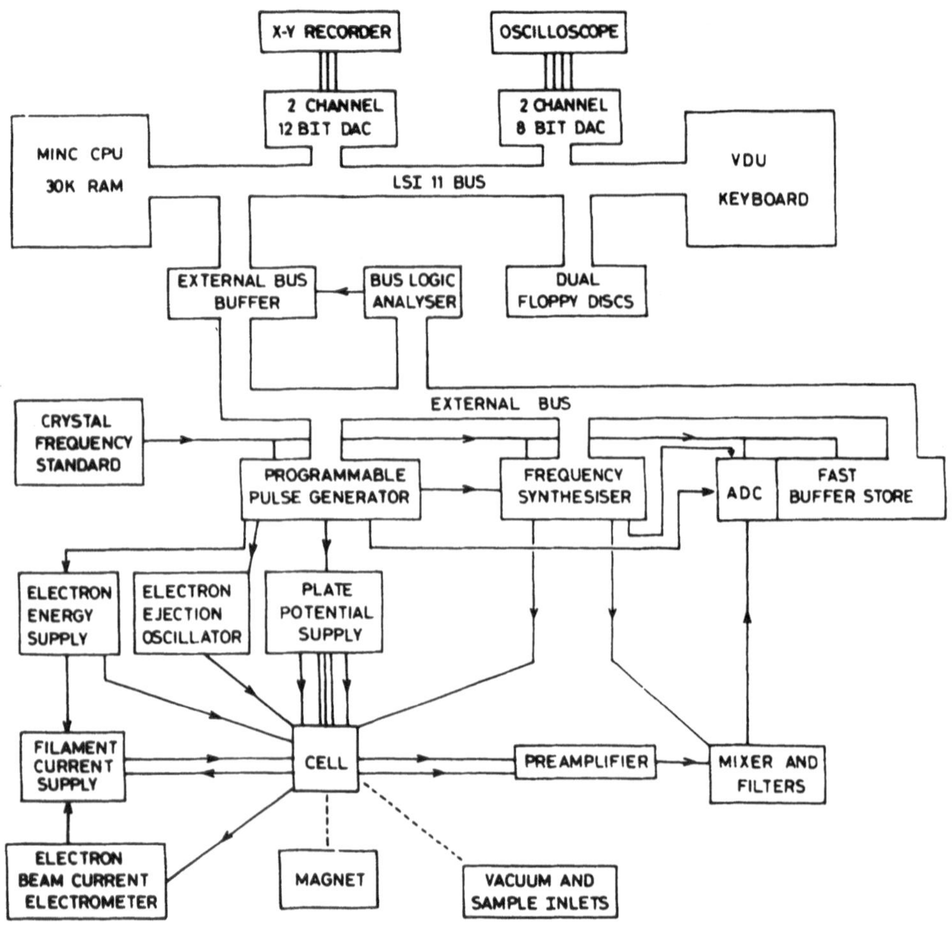
AMSTERDAM FTICR INSTRUMENT
X-Y RECORDER
OSCILLOSCOPE
2 CHANNEL 12 BIT DAC
2 CHANNEL 8 BIT DAC
MINC CPU 30K RAM
LSI 11 BUS
VDU KEYBOARD
EXTERNAL BUS BUFFER
BUS LOGIC ANALYSER
DUAL FLOPPY DISCS
EXTERNAL BUS
CRYSTAL FREQUENCY STANDARD
PROGRAMMABLE PULSE GENERATOR
FREQUENCY SYNTHESISER
ADC
FAST BUFFER STORE
ELECTRON ENERGY SUPPLY
ELECTRON EJECTION OSCILLATOR
PLATE POTENTIAL SUPPLY
FILAMENT CURRENT SUPPLY
CELL
PREAMPLIFIER
MIXER AND FILTERS
ELECTRON BEAM CURRENT ELECTROMETER
MAGNET
VACUUM AND SAMPLE INLETS

Fig. 1.

series of chemistry cycles are to be accumulated so as to enhance the signal to noise ratio, then the relationship in time and phase between ion excitation and the running of the ADC must be absolutely reproducible. It follows that no part of this process may be under real time software control unless the computer and its programs are themselves capable of fulfilling the time reproducibility requirement. It will be a rare and specialised computer that can completely fulfill all the necessary conditions.

In this design, shown in outline in Figure 1, it was decided that the whole operation of the chemistry cycle should be carried out by a dedicated hardware system, programmed by the computer, but executing all the instructions independently. Once the hardware system has received the START command it carries out all the operations, including double/multiple resonance experiments without any further intervention from the computer; indeed the computer is effectively locked out until the Fast Buffer Store has been filled with the new transient.

The extent to which the experimental parameters will be under computer control or open to interrogation by the computer, will be a matter of personal choice and convenience. Since this instrument grew out of an existing drift cell instrument it was most convenient to continue to use the existing independent analogue filament, electron energy and magnet controls.

2. Ion Excitation

Unfortunately the pulse excitation technique used in FTNMR [9] is completely unsuitable for the wideband excitation required for FTICR. The ions in the cell must be excited electrostatically by the application of some reproducible radio frequency field covering the required spectral bandwidth. It would seem that the necessary field might perhaps be generated in one of three ways:

1) By a linearly swept burst of radio frequency
2) By summing the outputs of a number of "oscillators", each one tuned to the frequency of an ion of interest
3) By digital to analogue conversion of the data produced by a reverse Fourier transform of a block spectrum.

The frequency sweep [10] method which has been used so far does have the disadvantage that one end of the spectrum is excited before the other, but that is only going to be a serious objection when the sample pressure is higher than normal. There is also a considerable redundancy in the sense that most of the sweep is wasted in exciting the voids between mass numbers. The second method (and perhaps the third) has the advantage that it can more easily excite only ions of interest, but the practical difficulty lies in building all the independent (?)

"oscillators", ensuring that they are all "on tune" and that they all start with the same phase relationship in each chemistry cycle. At the moment this seems like an electronics nightmare, but for applications like isotope ratio analysis it might have a future. The reverse Fourier transfer method has the practical disadvantage that an exceedingly large data pool will be needed if a high frequency is to be generated for more than a few milliseconds. However, if a limit of say 2 ms is placed on the duration of the FR burst then it becomes clear that all that is required is an inverted ADC/Fast Buffer Sore, viz a Fast Buffer Store/DAC plus filter combination. The fast store could as suggested be preloaded with the output of a reverse FT, or if it was (for example) decided always to excite one frequency decade the data could be put once and for all in a PROM/DAC unit. This has the convenience that by changing the clock frequency (outputing rate) of the PROM/DAC (and by changing to an appropriate output filter) the decade could be moved up or down the RF spectrum.

However, let us return from speculation to the technique which has always been used [10]: a sweepable frequency synthesizer. Unfortunately, the demand on consecutive rapid sweeps more or less rules out all the commonly employed methods of frequency synthesis. Even a phase locked loop VCO could not be asked to stay in a jitter-free lock under the exceedingly fast scan conditions required for FTICR. That technique might however be successful if the VCO were operated at VHF and its output mixed down. Thankfully, there is one presently exotic technique which fits the bill perfectly: the method known as Direct Digital Synthesis.

Despite its practical complexity, the operating principal of a Direct Digital Synthesizer is probably easier to explain than that of any other kind of frequency synthesizer. It is like a computer which is programmed to calculate incrementing angular values of a sine wave function and to output them via a digital to analogue converter (DAC) at a fixed rate. For example, a 100 Hz sine wave would be generated by moving in steps of 36° every millisecond. Even if the program fetched new sine values from a data table rather than calculating them from first principals every time, the computation time would limit the maximum output frequency to about 50 kHz. To get the output frequency up into the RF band the computation time must be drastically reduced: a dedicated hardware processor has to be built. In the design presented here the computation time is reduced to 125 ns, different output frequencies being obtained by varying the angular step size. Since the output function is generated numerically it is reproducible.

In fact the synthesizing section of such synthesizer is not particularly complicated, the real trouble comes in ensuring that it receives and executes its (scan) instructions consistently. To overcome these difficulties a complete unit was designed, incorporating scanning, timing, synthesizing, attenuating, routing

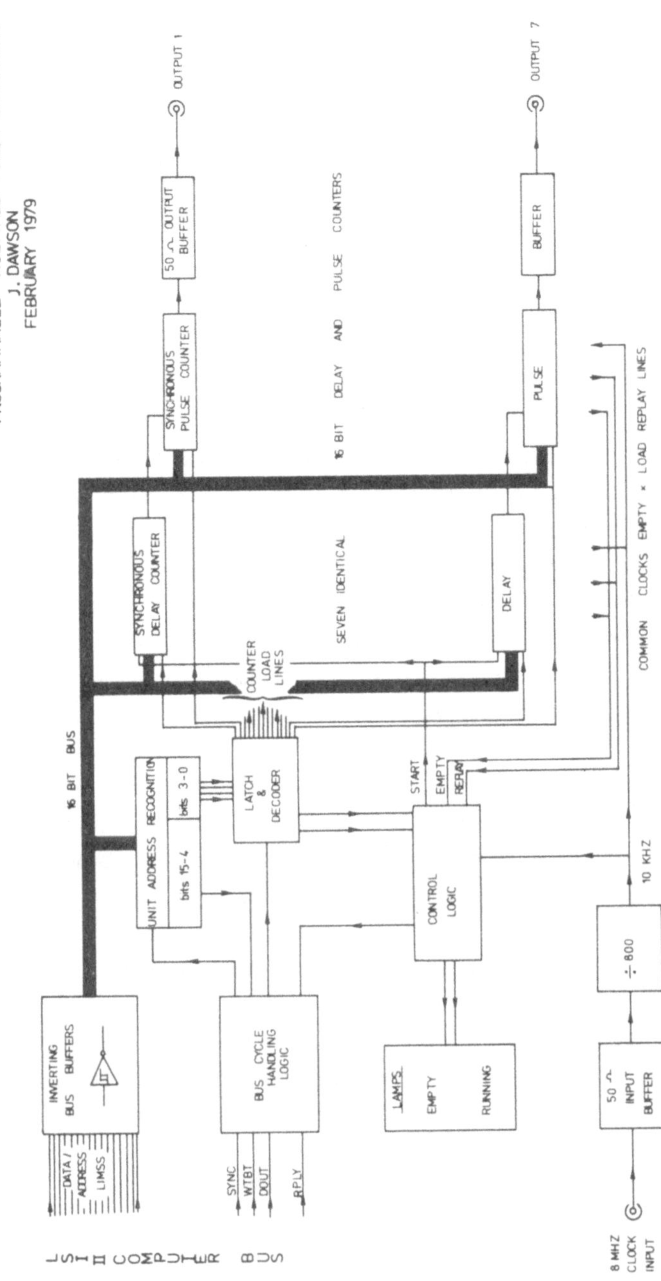

PROGRAMMABLE MULTIPLE PULSE GENERATOR
J. DAWSON
FEBRUARY 1979
OUTPUT 1
OUTPUT 7
50 Ω OUTPUT BUFFER
BUFFER
SYNCHRONOUS PULSE COUNTER
PULSE
SYNCHRONOUS DELAY COUNTER
DELAY
16 BIT DELAY AND PULSE COUNTERS
SEVEN IDENTICAL
COUNTER LOAD LINES
COMMON CLOCKS EMPTY × LOAD REPLAY LINES
16 BIT BUS
UNIT ADDRESS RECOGNITION
bits 15-4
bits 3-0
LATCH & DECODER
START
EMPTY
REPLAY
CONTROL LOGIC
10 KHZ
÷ 800
50 Ω INPUT BUFFER
8 MHZ CLOCK INPUT
LAMPS
EMPTY
RUNNING
BUS CYCLE HANDLING LOGIC
INVERTING BUS BUFFERS
DATA / ADDRESS
SYNC
WTBT
DOUT
RPLY
LSI II COMPUTER BUS

Figure 2

and control logic. The unit has enough programmable RAM to enable it to execute all the operations required for a single chemistry cycle on cue from the pulse generator.

Finally, it should be stated that synchronous operation of the Pulse Generator, Frequency Synthesizer and ADC/Fast Buffer Store is ensured by providing them all with 8 MHz clock pulses from a common temperature controlled crystal frequency standard.

There now follow the four main sections of this paper which describe the Pulse Generator, the Frequency Synthesizer, the Preamplifier and the ADC/Fast Buffer Store.

3. Programmable Pulse Generator

The pulse generator illustrated in Figure 2 is rather basic. The design is based on the great practical convenience of accepting time quantisation in 100 μs steps (10 KHz clocking) so that a system of presettable 16 bit synchronous down counters (SN74LS191) will be able to handle count times up to 6.5535 seconds. If this range is considered inadequate more bits could be added, but it could more easily be extended by dedicating the different sections and driving different counters at different clock frequencies. If such dedication is considered undesirable one could accept the increased complexity of implementing a series of exponent mantissa counters. Dedication would permit a further simplification in that the pulse counters from sections required only to deliver a trigger pulse could be eliminated completely. The only essential point to bear in mind in any detailed design is that all the output pulses should be truly synchronised to the high frequency clock (8 MHz), which means that numerous internal resynchronisations should be made and all race conditions elimimated.

In the process of counting down, the programmed information is lost. It hardly seems worthwhile to save the time required for reprogramming before each chemistry cycle by providing the unit with an extra layer of memory. That might not be true if the unit was not to be "on" the computer's parallel bus. The wired-or EMPTY line, which only goes Hi when all the counters have counted down to zero, enables a security lock to be introduced to protect the system against the unit being reprogrammed at an inappropriate moment. The unit does not respond or reply to the start command from the computer if all the counters are empty, nor does it accept or reply to new data after being started until it has become empty again. Thus, if the computer attempts to reprogram the unit whilst a chemistry cycle is still in progress, a bus time out error and program crash will ensue. This feature is most effective in bringing to light some classes of program and and operator errors.

Figure 3

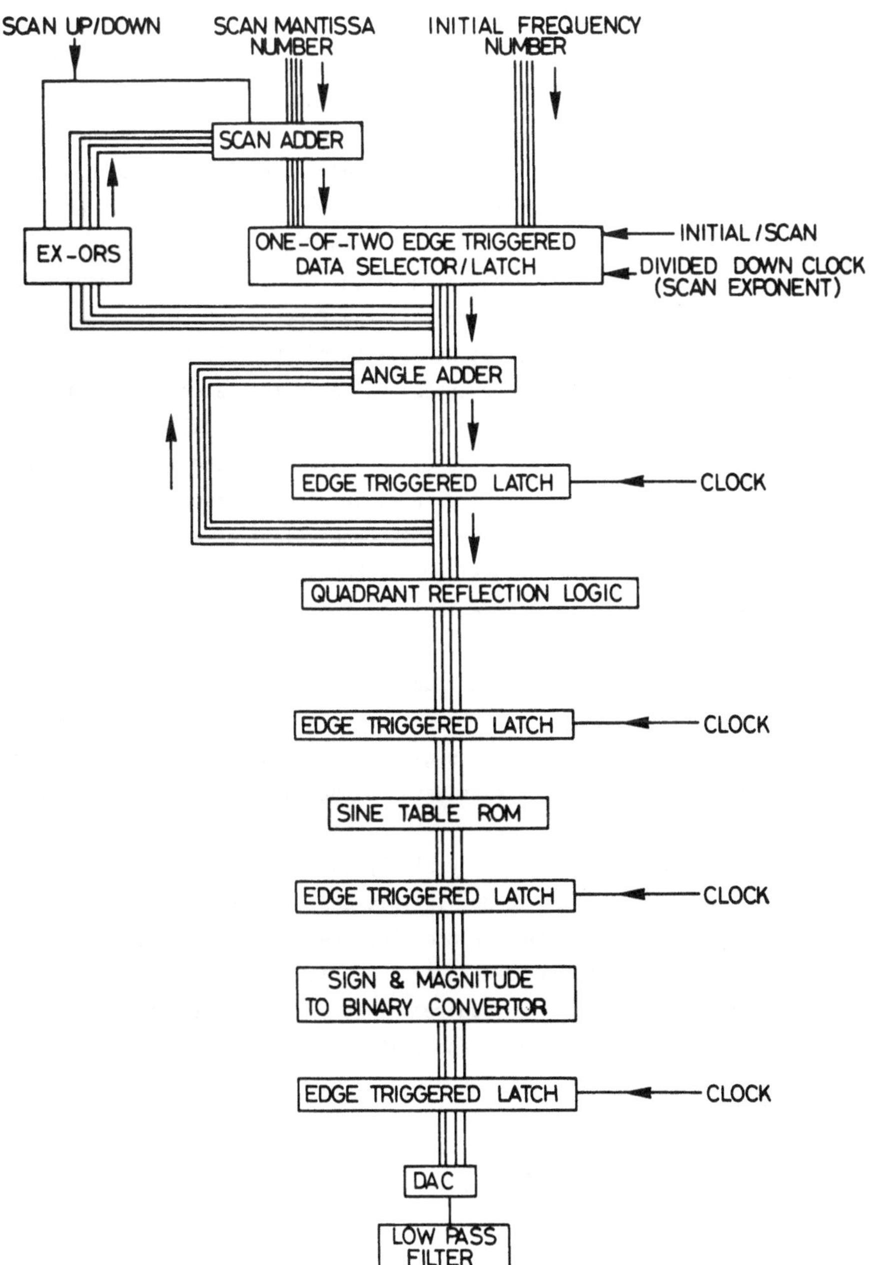

SCAN UP/DOWN
SCAN MANTISSA NUMBER
INITIAL FREQUENCY NUMBER
SCAN ADDER
EX-ORS
ONE-OF-TWO EDGE TRIGGERED DATA SELECTOR/LATCH
INITIAL/SCAN
DIVIDED DOWN CLOCK (SCAN EXPONENT)
ANGLE ADDER
EDGE TRIGGERED LATCH
CLOCK
QUADRANT REFLECTION LOGIC
EDGE TRIGGERED LATCH
CLOCK
SINE TABLE ROM
EDGE TRIGGERED LATCH
CLOCK
SIGN & MAGNITUDE TO BINARY CONVERTOR
EDGE TRIGGERED LATCH
CLOCK
DAC
LOW PASS FILTER

4. Frequency Synthesizer

Both in design and in construction the Frequency Synthesizer circuits are divided into five logic boards plus the post-DAC RF sections. The latter are built in a shielded honeycomb subchassis.

To understand the evolution of the total design it is first necessary to study Figure 3 which sets out the basic design of the synthesizing and scanning logic. Imagine for a moment that the scanning logic is over-ridden and that the 16 bit unsigned binary initial frequency number is fed directly to one port of the angle adder. Although this number is referred to as a frequency it is really an angle, and must be thought of as such here. Its most significant bit (MSB) corresponds to 90°. The output of the edge triggered latch which follows the angle adder may be imagined as being initially clear, so that at the first clock pulse it will change to be the inital frequency number. Let us call this angle Θ. Since the angle adder always adds this angle to the previous value of the latch output it follows that the latch output will step forward by Θ at every clock pulse. The adder and latch are 17 bits wide so that multiples of 360° are discarded as and when the advancing angle $n\Theta$ exceeds them. All the remaining logic between this point and the digital to analogue converter (DAC) is the equivalent of a big 360° sine look up table. The operation of the synthesizer is thus simplicity itself: the angle adder keeps advancing through the sine table so that successive numerical values of a sine wave are generated and ouput via the DAC. The actual output frequency is determined by the angular step size Θ and the clock rate, which in this design is fixed at 8 MHz. For example, a 1 MHz output is generated by stepping through the sine table by 45° every 125 ns. This gives the stepped analogue output shown in Figure 4 which is smoothed to a sine wave by the sharp low pass filter which immediately follows the DAC.

Frequency scanning may be achieved by slowly incrementing or decrementing the number which is fed to the angle adder. In this design the initial frequency is selected and fed to the angle adder for one clock period only. Thereafter this number is circulated through the scan adder where it is periodically modified unless a fixed frequency output is required. That brings us back to the question of how are the Scan Exponent/Mantissa and Initial Frequency Number, which must ultimately be derived from the computer, to be presented to the synthesizing logic at the appropriate moments. It is clear that the intermediacy of a small read/write memory will be a great convenience, and much of the ensueing design arose from the availability of 16x4 bit chips to fulfil this function. It was decided to take four to these chips (6561 Schottky RAMs) to make up a 16 word memory arranged to hold eight pairs of Scan and Initial numbers.

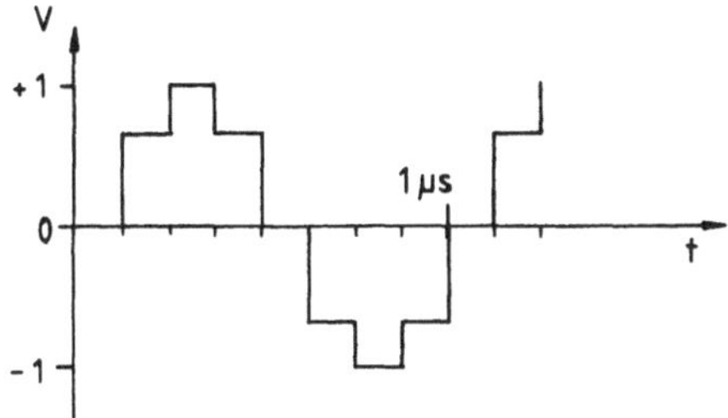

Fig. 4.

For the further understanding of the design it now becomes essential to grasp the fact that the eight initial frequency numbers are stored in even RAM addresses (ie., Q_A of ic F of Figure 8 Lo) with their Scan instructions contained in the succeeding odd RAM addresses (Q_A Hi). Since the synthesizing logic requires the initial frequency number only once, the RAM address normally rests odd, going even for single clock periods only when a new inital frequency is required. Implicit in all of this is the idea that the unit will be pre-programmed by the computer with eight different frequency bursts, which it will execute sequentially, returning the RAM to its computer controlled write mode only after the eight and final "segment" has run.

The above begs several questions, in particular, how is the unit to know when to start and stop any particular segment. To provide the answer to these and other questions the unit is provided with a second 16 word RAM called RAM II to distinguish it from the above mentioned RAM called RAM I. Both RAMs share the same address lines. Even addresses in RAM II contain a control word, and odd addresses contain, in exponent mantissa form, an instruction as to the length of time for which that particular segment must run.

The control word contains information for each segment about the direction of the scan, which external trigger pulse is to be used to initiate the running of that segment, the initial phase angle, the destination of the ouput and its amplitude. Seen from the computer these control words occur at the octal bus addresses 1741x2 and are defined as follows:

bit 15	0	= Scan Up, 1 = Scan Down	
bits 14-20	000	= Trigger Pulse input (Status)	1
	001	= ditto	2
	010	= ditto	3
	011	= ditto	4
	1xx	= Computer controlled (infinite Run) Status.	

bits 11-8 are latched but unused (may be used as flags within the software).

bits 7,6	00 =	Initial Phase Angle	0°
	01 =	ditto	90°
	10 =	ditto	180°
	11 =	ditto	270°

However, note that the 90° component of a phase change can only occur for a segment which does not "run on" immediately from a previous segment.

bit 5 1 = Output to Mixer (or ADC rate) ON, 0 = OFF
bit 4 1 = Output to Cell ON, 0 = OFF

bits 3-0 Control the ouput amplitude in 3 dB steps from 111 = Full Output down to 0000 = -45 dB.

At this point it is necessary to point out that the unit must know whether it is supposed to be listening to the computer so as to have its RAMs programmed, or whether it is supposed to be looking into the RAMs and waiting for the relevant trigger pulse or otherwise busying itself. This Either/Or problem is solved for the unit by a special command which it receives from the computer via the bus address 174076. Upon receipt of the datum 1 at this address the unit goes into what is called the BUSY state. It blocks further writing by the computer into the RAMs, giving RAM address control over to ic F of Figure 8, which is cleared and set into its counting mode. The initial frequency number and control word for the first segment are fetched from RAM, and the RAM address is incremented. The relevant trigger pulse input latch will be inspected, and if, as is probable, it has not yet been set by a pulse from the pulse generator, the unit will grind to a halt. The RAM address is now odd. When the relevant trigger pulse is received segment 1 will run until the time instruction contained in RAM II has been counted down. $\overline{\text{END OF COUNT}}$ is generated, the RAM address goes even for one clock period and the control word for the next segment is fetched. Immediately, or in due course, that will be run. However, at the conclusion of running of the 8th segment the BUSY state is revoked and the unit goes back to its programmable mode. Note that segments must run sequentially, although the Status numbers may be arranged in any ordering which is compatible with this requirement. The BUSY state can always be aborted by the computer by sending the datum 2 to the control address 174076. Two other commands are also recognised, 3 and 4, which are used by the computer to Start and Stop segments which have been given a control word containing the infinite run Status. Infinite runs can also be initiated by substatus trigger pulses. Note that the unit must of course be sitting at the relevant segment address at the time that the Start command is received. The infinite run facility was originally introduced for test purposes, but has proved to solve difficulties which would otherwise have arisen in writing the software for variable rate ADC experiments etc. The complete set of paramenter required from the computer to define each of the eight segments are as follows:

1. Initial Frequency - Bus Address 1741x0
   A 16 bit unsigned binary number of which the least significant bit is 2^{-14} MHz, ie 61.03515625 Hz.
2. Control Word - Bus Address 1741x2, as defined above.
3. Scan Rate - Bus Address 1741x4
   Bits 7-0 (the eight least significant bits) define the scan mantissa "A" and bits 12-8 define the scan exponent "N". The scan rate is then $Ax2^{-N}x2^{-16}x10^{6}$ KHz/ms.
   The mantissa has the full range of 0-255, but the exponent is limited to 0-23. The range goes from 1.8 Hz/s to 3.9 GHz/s.
4. Segment Time - Bus Address 1741x6
   Again in exponent "N" (bits 15-12) and mantissa "A" (bits 11-0) form.

$$\text{Segment Time} = Ax2^{N}\ \mu s$$

   Both exponent and mantissa have their full numerical range, giving a total range from 1 μs to 134 s. Unwanted (null) segments should be given a scan time of 1 μs; giving them 0 μs does not actually cause a crash, but it does result in a certain degree of internal inelegance.

The somewhat abstract parameters for each segment are rearranged and presented to the operator via the VDU in the form of eight pages as shown in Figure 5.

Figure 5. Set Up Frequency Synthesizer Segment No. 6

```
ACTUALLY USED VALUES
      FROM MASS  10.000 amu.
        TO MASS  86.011 amu.
      TIME        1.499 ms.
      (Scan rate = 1083.374 KHz/ms)
      RF LEVEL  42 dB.
      STATUS     2
      ROUTE      1
      PHASE      0°

Type C to change this segment;
Type S to exit;
Type X to exit and clear all subsequent segments;
Type N to clear this segment;
Push RETURN to go to next segment.
```

If the operator changes one of the variables on a page the new set is converted to the four parameters words. The time value is always rounded down and a preference is exercised to make the scan (if any) go in a large number of steps of 61 Hz or less rather than in a smaller number of larger steps. These four approximated parameters are written in computer memory and immediately re-read, decoded and displayed in the standard format. If the operator does not like the way in which approximations have occurred he can try a new set of entries.

The synthesizer requires that segments must run sequentially. Since ion excitation is usually placed in segment 6 it follows that if any of the prior five segments are not required for ion ejection (etc) they must be programmed to run in advance of segment 6, but ineffectually. Such "null" segments are therefore made to run at zero frequency for 1 μs with the attenuator and routing firmly shut. If required they can be a null segment, but be given an extended period of time so that they act as additional time delays to one or more subsequent segments.

The fact that frequency scanning is performed by the synthesizer itself means of course that it is available just as much for ion ejection segments as for the ion excitation segment(s). Thus whole bunches of ions may be expelled by a wide sweep or by a combination of single and swept frequencies. In fact it has become quite common in our laboratory to use a narrow sweep (say ± 0.05 amu) to expell

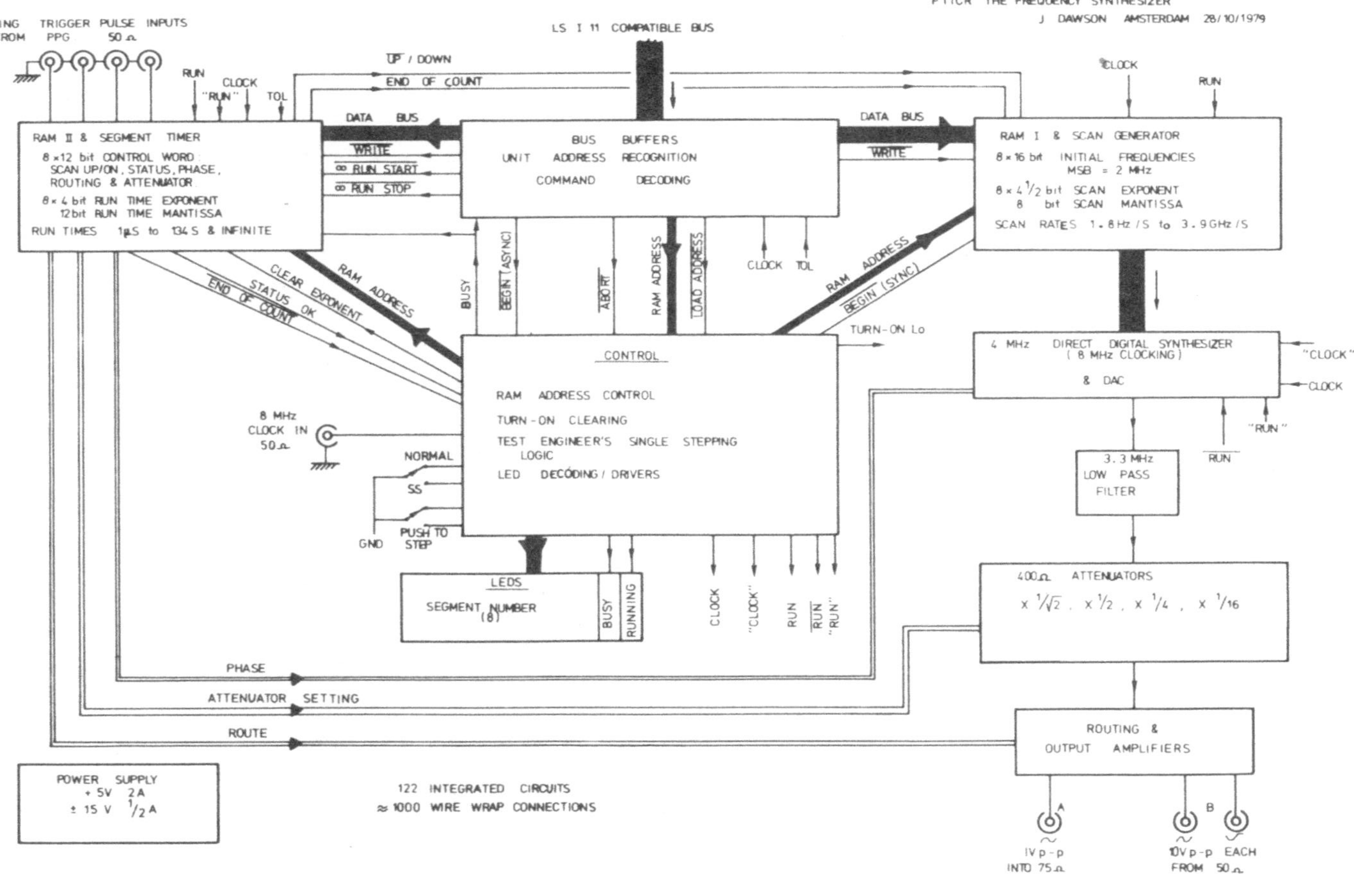
FTICR THE FREQUENCY SYNTHESIZER
J DAWSON AMSTERDAM 28/10/1979
TIMING TRIGGER PULSE INPUTS
FROM PPG 50 Ω
LS I 11 COMPATIBLE BUS
RUN
CLOCK
"RUN"
TOL
UP / DOWN
END OF COUNT
DATA BUS
WRITE
∞ RUN START
∞ RUN STOP
RAM II & SEGMENT TIMER
8 × 12 bit CONTROL WORD
SCAN UP/ON, STATUS, PHASE,
ROUTING & ATTENUATOR
8 × 4 bit RUN TIME EXPONENT
12 bit RUN TIME MANTISSA
RUN TIMES 1μS to 134 S & INFINITE
BUS BUFFERS
UNIT ADDRESS RECOGNITION
COMMAND DECODING
RAM I & SCAN GENERATOR
8 × 16 bit INITIAL FREQUENCIES
MSB = 2 MHz
8 × 4 1/2 bit SCAN EXPONENT
8 bit SCAN MANTISSA
SCAN RATES 1.8 Hz/S to 3.9 GHz/S
BUSY
BEGIN (ASYNC)
ABORT
RAM ADDRESS
LOAD ADDRESS
CLOCK
TOL
RAM ADDRESS
BEGIN (SYNC)
TURN-ON Lo
CLEAR EXPONENT
STATUS OK
END OF COUNT
RAM ADDRESS
CONTROL
RAM ADDRESS CONTROL
TURN-ON CLEARING
TEST ENGINEER'S SINGLE STEPPING
LOGIC
LED DECODING / DRIVERS
8 MHz
CLOCK IN
50 Ω
NORMAL
SS
PUSH TO
STEP
GND
LEDS
SEGMENT NUMBER
(8)
BUSY
RUNNING
CLOCK
"CLOCK"
RUN
RUN
"RUN"
4 MHz DIRECT DIGITAL SYNTHESIZER
(8 MHz CLOCKING)
& DAC
"CLOCK"
CLOCK
"RUN"
RUN
3.3 MHz
LOW PASS
FILTER
400 Ω ATTENUATORS
x 1/√2 , x 1/2 , x 1/4 , x 1/16
PHASE
ATTENUATOR SETTING
ROUTE
ROUTING &
OUTPUT AMPLIFIERS
A
B
1V p-p
INTO 75 Ω
10V p-p EACH
FROM 50 Ω
POWER SUPPLY
+ 5V 2A
± 15 V 1/2 A
122 INTEGRATED CIRCUITS
≈ 1000 WIRE WRAP CONNECTIONS

Figure 6

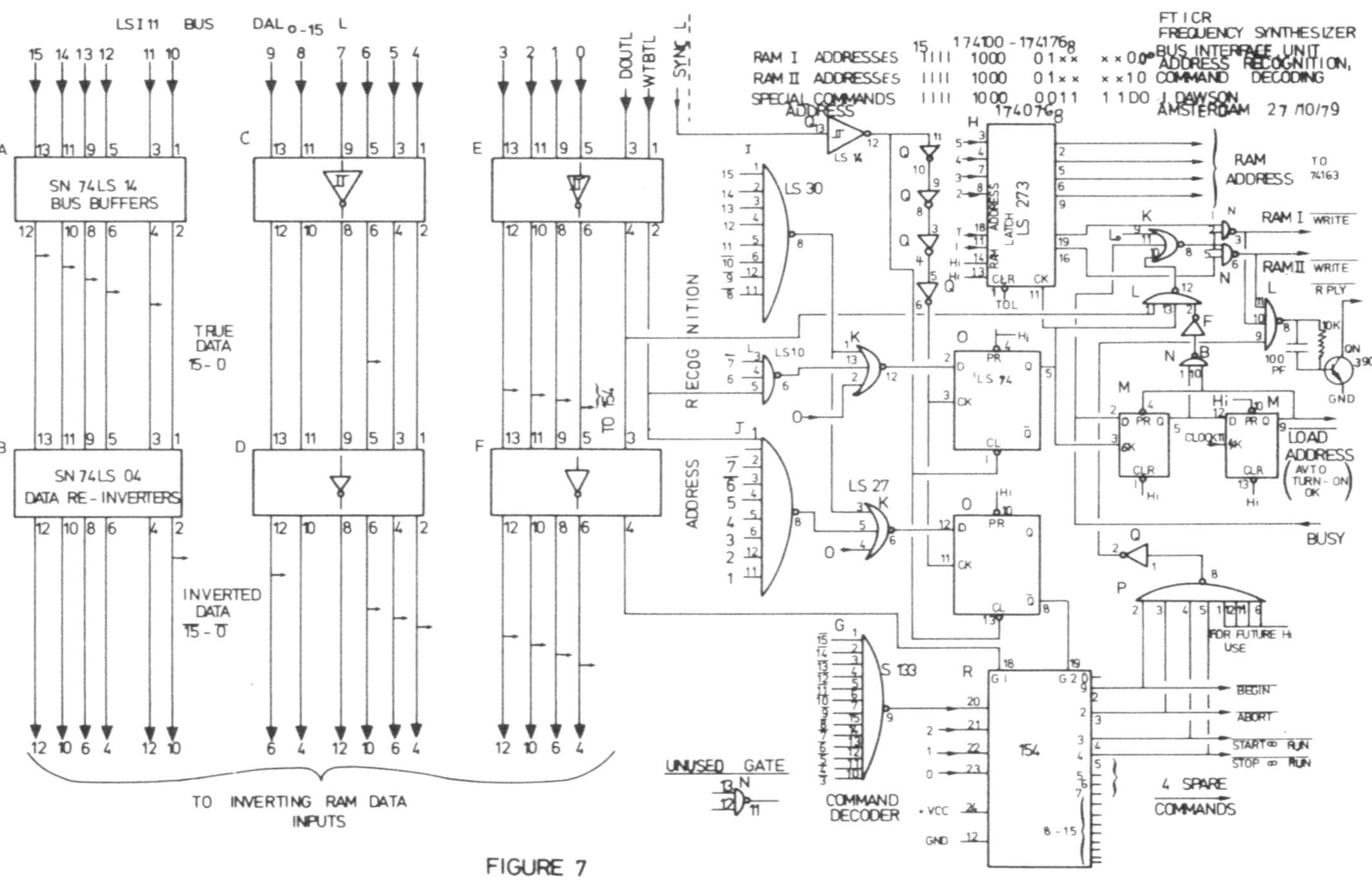
FTICR
FREQUENCY SYNTHESIZER
BUS INTERFACE UNIT
ADDRESS RECOGNITION,
COMMAND DECODING
J DAWSON
AMSTERDAM 27/10/79
RAM I ADDRESSES
RAM II ADDRESSES
SPECIAL COMMANDS
LSI 11 BUS DAL 0-15 L
SN 74LS 14
BUS BUFFERS
SN 74LS 04
DATA RE-INVERTERS
TRUE DATA
15-0
INVERTED DATA
$\overline{15-0}$
TO INVERTING RAM DATA INPUTS
RECOGNITION
ADDRESS
RAM ADDRESS
TO 74163
RAM I WRITE
RAM II WRITE
RPLY
LOAD ADDRESS
AUTO TURN-ON OK
BUSY
FOR FUTURE USE
BEGIN
ABORT
START RUN
STOP RUN
4 SPARE COMMANDS
COMMAND DECODER
UNUSED GATE

FIGURE 7

single ions so as to avoid the possibility of being slightly out of tune.

There now follows the detailed circuit description, taken figure by figure. Figure 6 shows the interconnections.

5. Bus Interface (See Figure 7)

The computer bus for which the unit is designed is logically equivalent to the Digital Equipment Corporation LSI 11 bus, but is built in normal power open collector ttl. An "intelligent" two way bus buffer connects this external bus to the computer's internal high power bus. The external bus serves the four peripheral units of Figure 1 in a daisy chain with a resistive termimation.

Inputs from the bus are buffered by low power Schottky Schmitt trigger gates A,C,E,Q_{12}. The bus is actually in inverted logic (1 = active Lo) so another set of inverters B,D,F are introduced because the RAMs used are themselves inverting. Gate I output goes Lo when a bus address occurs in the range 174000 to 174377. Gates I,L_6, K_{12} are specific for the RAM addresses 1741xx (even only) and gates I,J,K_6 are specific for the control address 174076. Latches O_5 and O_8 are clocked by the bus address valid synchronization pulse ($\overline{\text{SYNC}}$) to catch these conditions, latch H catching simultaneously the individual RAM address concerned. Note that the address information contained in H can only set the RAM address when it has been passed synchronously through chip F of Figure 8, which can occur only when the unit is not BUSY. Chips M and N_8 ensure that no attempt is made to write into RAM unless the new RAM address has first settled. The unit responds only to DATO cycles of the LSI 11 bus. After the address has been placed on the bus and signalled, it is removed and the data to be output is placed on the bus. The bus signal $\overline{\text{DOUT}}$ asserts that the data is now valid and the peripheral unit is called upon to accept the data and to signal that it has done so by asserting the return bus signal $\overline{\text{RPLY}}$. Generation of $\overline{\text{RPLY}}$ is controlled by gate L_8 which searches for the conditions which indicate that the unit has actually responded to the attempted DATO bus cycle. $\overline{\text{RPLY}}$ will not be generated if an invalid command is sent to the control address, or if RAM reprogramming is attempted whilst the unit is BUSY.

If the interface has to be redesigned for a different computer bus it should be noted that the $\overline{\text{LOAD ADDRESS}}$ line emanating from it must be synchronous with the (8 MHz) clock. If required, four further commands can easily be extracted from chip R with further links to the input of chip P.

6. Control Board (see Figure 8)

The circuitry of the control board is acually rather simple. It provides the necessary combinational logic to link together the control signals supplied or required by the other sections of the unit. It is responsible for:

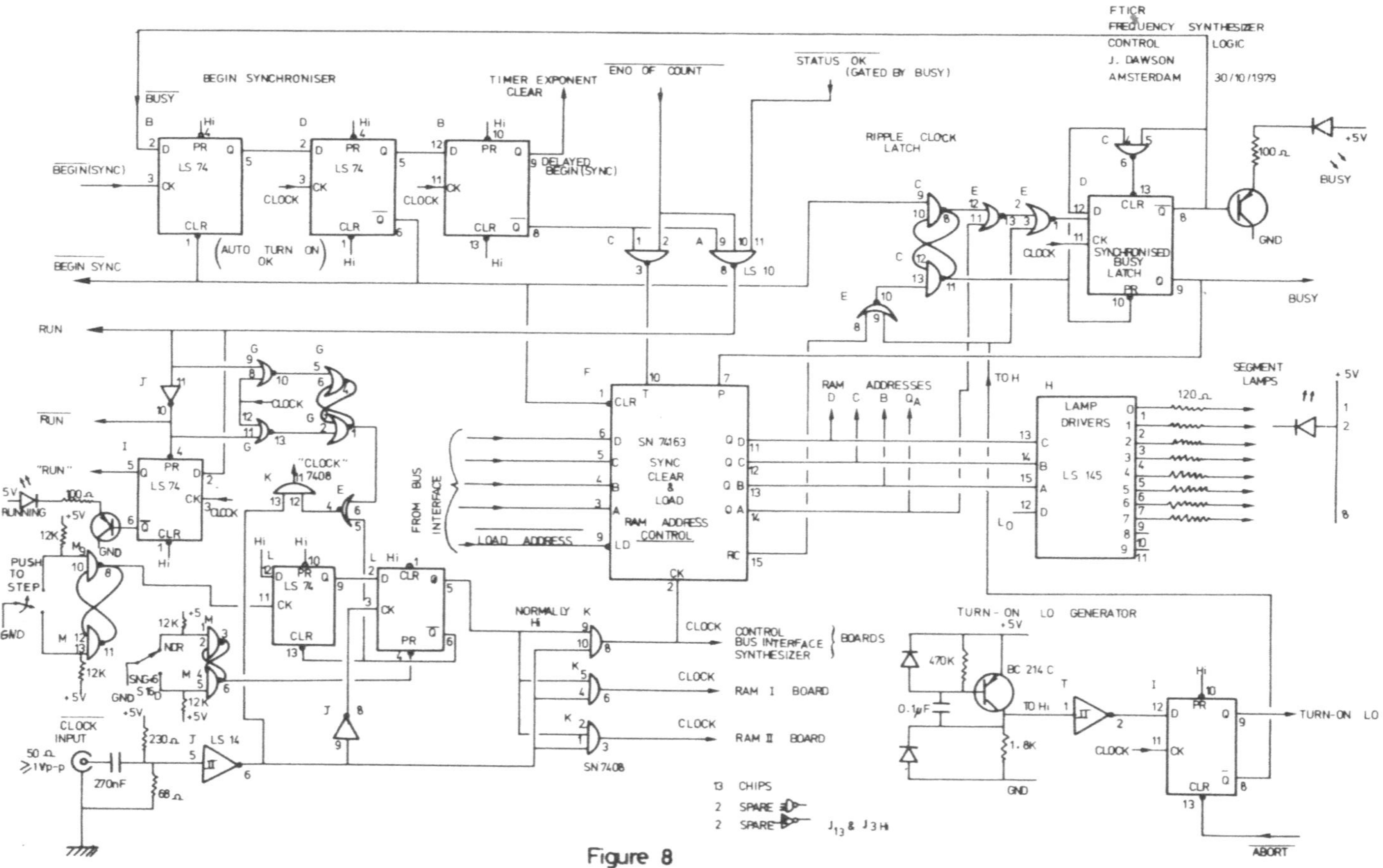
FTICR
FREQUENCY SYNTHESIZER
CONTROL LOGIC
J. DAWSON
AMSTERDAM 30/10/1979
BEGIN SYNCHRONISER
TIMER EXPONENT CLEAR
END OF COUNT
STATUS OK (GATED BY BUSY)
RIPPLE CLOCK LATCH
SYNCHRONISED BUSY LATCH
BUSY
BEGIN(SYNC)
BEGIN SYNC
DELAYED BEGIN(SYNC)
(AUTO TURN ON OK)
RUN
"RUN"
RUNNING
PUSH TO STEP
CLOCK INPUT
SN 74163
SYNC CLEAR & LOAD
RAM ADDRESS CONTROL
FROM BUS INTERFACE
LOAD ADDRESS
RAM ADDRESSES
LAMP DRIVERS
LS 145
SEGMENT LAMPS
CONTROL BUS INTERFACE SYNTHESIZER BOARDS
RAM I BOARD
RAM II BOARD
SN 7408
TURN-ON LO GENERATOR
BC 214 C
TURN-ON LO
ABORT
13 CHIPS
2 SPARE
2 SPARE

Figure 8

1. asserting BUSY when the unit is waiting to run or is running a segment,
2. determining the RAM address when in the BUSY state,
3. giving RAM address control to the computer when the unit is to be generated,
4. generating RUN, "RUN" and "CLOCK" when the next segment is to be generated,
5. ensuring that when the unit is first turned on it will settle in a controlled way into the non-BUSY state,
6. distributing CLOCK pulses to other sections, and providing the test engineer with the facility to single step the unit.

The control board receives three instructions from the bus interface: $\overline{\text{BEGIN}}_{async}$, $\overline{\text{ABORT}}$ and $\overline{\text{LOAD RAM ADDRESS}}$. $\overline{\text{ABORT}}$ simply induces a synchronously terminated pseudo turn-on condition. The asynchronous $\overline{\text{BEGIN}}$ command is latched if the unit was not already BUSY and is then made synchronous. The command then makes the unit BUSY, clears the RAM address from chip F (Q_{DCBA}), and sets the RAM I data selector chips C,F,I,L, R to receive an initial frequency. A delayed version of the synchronous BEGIN signal then makes the unit run through from RAM address 0000 to 0001 as if it had happened at the end of a previously running segment.

The control board acts upon two other instructions $\overline{\text{END OF COUNT}}$ and $\overline{\text{STATUS OK}}$ which it receives synchronously from the RAM II board. $\overline{\text{The END OF COUNT}}$ signal always lasts for two clock periods and so if the unit is BUSY it enables chip F to count through two RAM addresses (odd → even → odd). The new segment will run if and when $\overline{\text{STATUS OK}}$ is asserted. "RUN" is like RUN but is extended in time by one clock period and "CLOCK" consists of a train of clock pulses which lasts for the duration of RUN but which is delayed by one whole clock period.

When chip F counts up to the final RAM address 1111 the ripple clock latch $C_{8,11}$ is set so that the next time F_{14} goes Lo (as if to run the first segment again) gate E_{13} allows the BUSY latch to be cleared.

A timing diagram for the main control functions has been compiled in Figure 13 which shows how they behave at BEGIN, at a normal start and stop, at an infinite run start and stop, at a run-on between two segments and at the final stop.

7. RAM II (See Figure 9)

As explained in the introduction, RAM II contains for each segment the control word and a word defining the time for which the segment must run. Since the control word defines a number of parameters it is best to see the RAM II board as consisting of three subsections:

1. The exponent/mantissa segment timer counter
2. The status trigger pulse input logic
3. the feed-forward delay for the attenuator controls etc.

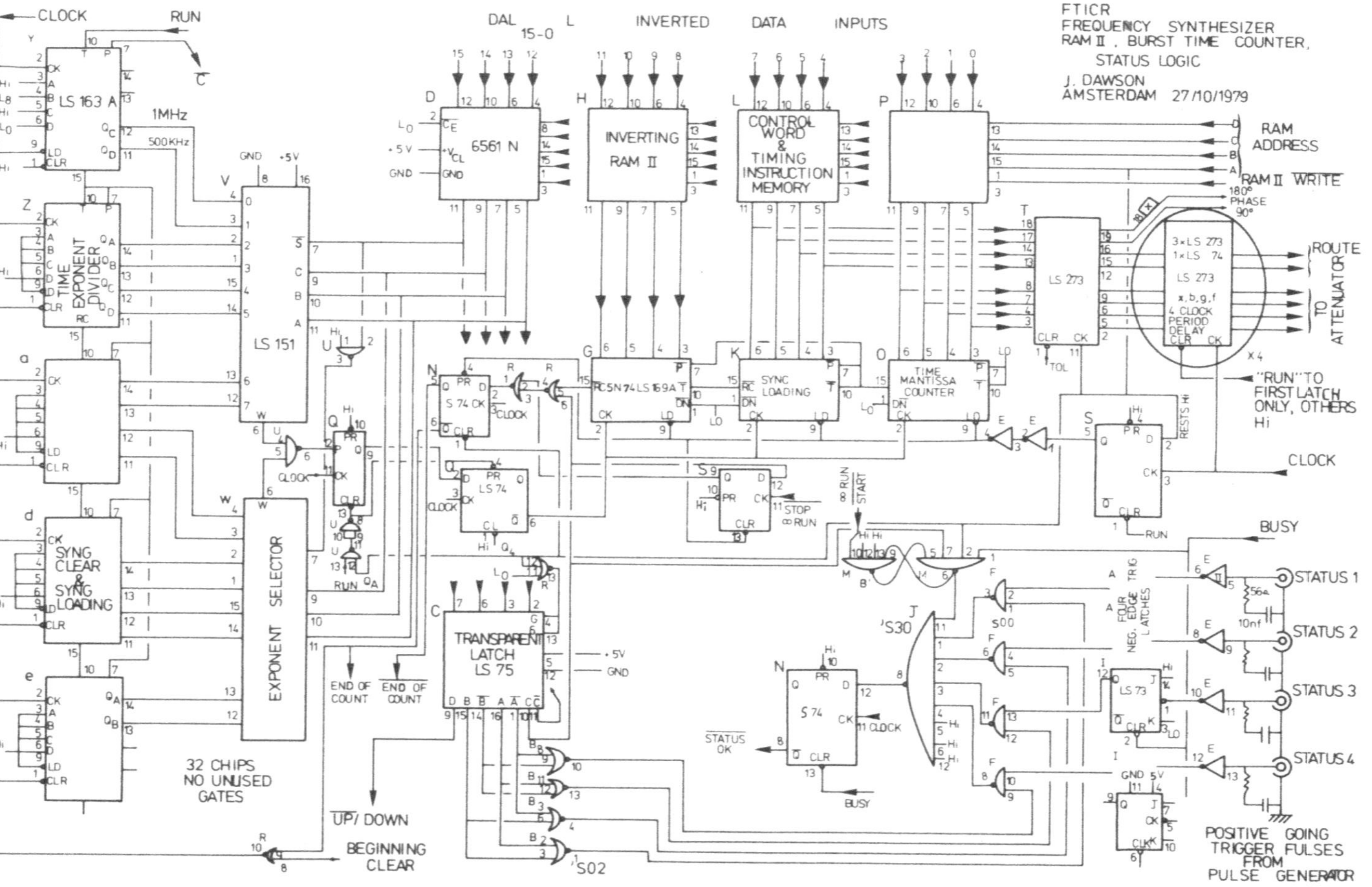
FTICR
FREQUENCY SYNTHESIZER
RAM II , BURST TIME COUNTER,
STATUS LOGIC
J. DAWSON
AMSTERDAM 27/10/1979
CLOCK
RUN
DAL 15-0
INVERTED
DATA
INPUTS
LS 163 A
1MHz
500KHz
TIME EXPONENT DIVIDER
SYNG CLEAR & SYNG LOADING
LS 151
EXPONENT SELECTOR
6561 N
INVERTING RAM II
CONTROL WORD & TIMING INSTRUCTION MEMORY
LS 273
3×LS 273
1×LS 74
x,b,g,f
4 CLOCK PERIOD DELAY
RAM ADDRESS
RAM II WRITE
180° PHASE 90°
ROUTE
TO ATTENUATOR
"RUN" TO FIRST LATCH ONLY, OTHERS Hi
SN74LS169A
SYNC LOADING
TIME MANTISSA COUNTER
S 74
LS 74
STOP ∞ RUN
RUN START
CLOCK
BUSY
STATUS 1
STATUS 2
STATUS 3
STATUS 4
FOUR NEG. EDGE TRIG. LATCHES
LS 73
'S30
S 74
STATUS OK
TRANSPARENT LATCH LS 75
END OF COUNT
END OF COUNT
UP/ DOWN
BEGINNING CLEAR
32 CHIPS NO UNUSED GATES
'S02
POSITIVE GOING TRIGGER FULSES FROM PULSE GENERATOR

Figure 9

At the end of the previous segment the control board drives the RAM address even for one clock period. During this time the four bit latch C goes transparent so that the data in the four most significant bits of the control word may be acted upon immediately. The scan up/down control bit is fed directly to the scanning logic board and the status bits are rapidly decoded by the normal power Schottky chips B, F and J so that if the relevant trigger pulse latch $A_{9,12}$ or $I_{9,12}$ has already been set the $\overline{\text{STATUS OK}}$ latch N_8 will be ready to register the fact at the clock pulse when the RAM address goes odd again. At this clock edge latch T catches the lower eight bits of the control word, and latch C goes opaque. Because of delays in the synthesizing logic the routing and attenuator control bits need to be delayed too, and this is handled by latches X, b, g and f which are shown in detail in Figure 12. When there is a discontinuity between two segments (i.e. when the unit temporarily stops running) latch X gets cleared so that the attenuator and routing switches shut down the output stages.

If infinite run status has been selected by setting bit 14 of the control word, latch $M_{6,8}$ is enabled and gate R_4 renders irrevelant the condition of the mantissa time counter. The infinite run latch M will be set upon receipt of the next datum 3 at the bus control adress 174076. Receipt of the infinite run stop command (datum 4) is latched by S_9 which then feeds a pseudo time mantissa count down pulse to the $\overline{\text{END OF COUNT}}$ latch N_5 via gate R_1. Latch M is cleared when the control word for the next segment is fetched, and all the trigger input latches are held clear when the unit is not BUSY. Infinite runs can also be started by setting the appropriate substatus trigger latch.

The upper four bits defining the time exponent control the data selector chips V and W so as to select one of sixteen clocking frequencies for the mantissa counter which are derived from the 8 MHz clock frequency by the binary divider chain Y,Z,a,d,e. Prior to the running of any new segment this exponent divider is set to the count state 5 to compensate for the fact that five clock period delays will occur during the subsequent synchronous operation of Q_9,Q_6,N_5, the mantissa counter itself and chip F of the control board.

The load line for the mantissa counter is derived by delaying the LSB of the RAM address in chip S_5 with further slight delay in gates E_2,E_4 to allow for delay in chip Q_6. Q_6 and the gates U_{11},U_8 are required to provide the clock pulse which is required for synchronous loading of the '169A down counter chips G,K,O.

8. <u>RAM I, Scan Generator</u> (see Figure 10a,b)

The elements outlined in Figure 3 may easily be indentified: the 20 bit scan adder B,E,H,K,Q; the Ex-or gates J,P; and the one-of-two edge triggered data

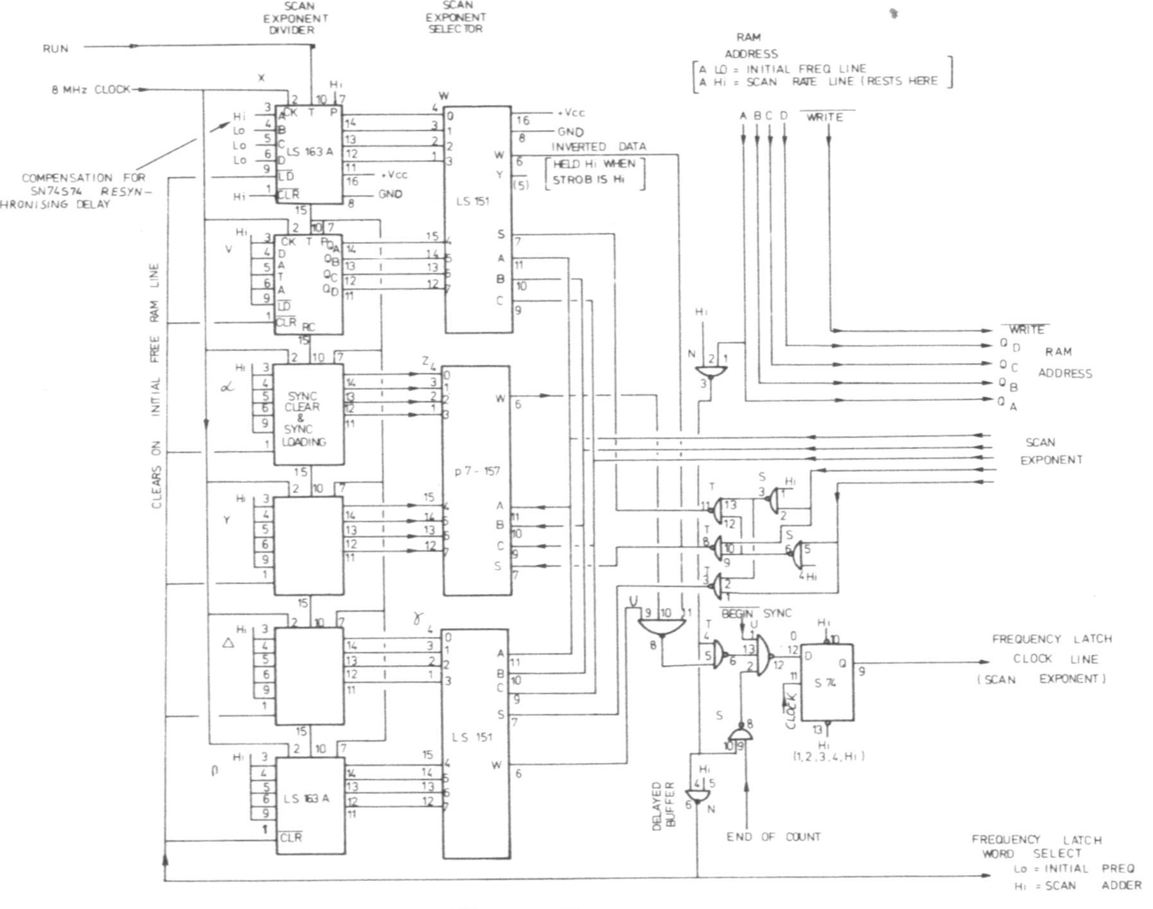
SCAN EXPONENT DIVIDER
SCAN EXPONENT SELECTOR
RUN
8 MHz CLOCK
COMPENSATION FOR SN74S74 RESYN-CHRONISING DELAY
RAM ADDRESS
A LO = INITIAL FREQ LINE
A Hi = SCAN RATE LINE (RESTS HERE)
INVERTED DATA
HELD Hi WHEN STROB IS Hi
LS 163 A
LS 151
SYNC CLEAR & SYNC LOADING
p 7 - 157
CLEARS ON INITIAL FREE RAM LINE
WRITE
RAM ADDRESS
SCAN EXPONENT
BEGIN SYNC
S 74
FREQUENCY LATCH CLOCK LINE (SCAN EXPONENT)
DELAYED BUFFER
END OF COUNT
FREQUENCY LATCH WORD SELECT
Lo = INITIAL PREQ
Hi = SCAN ADDER

Figure 10a

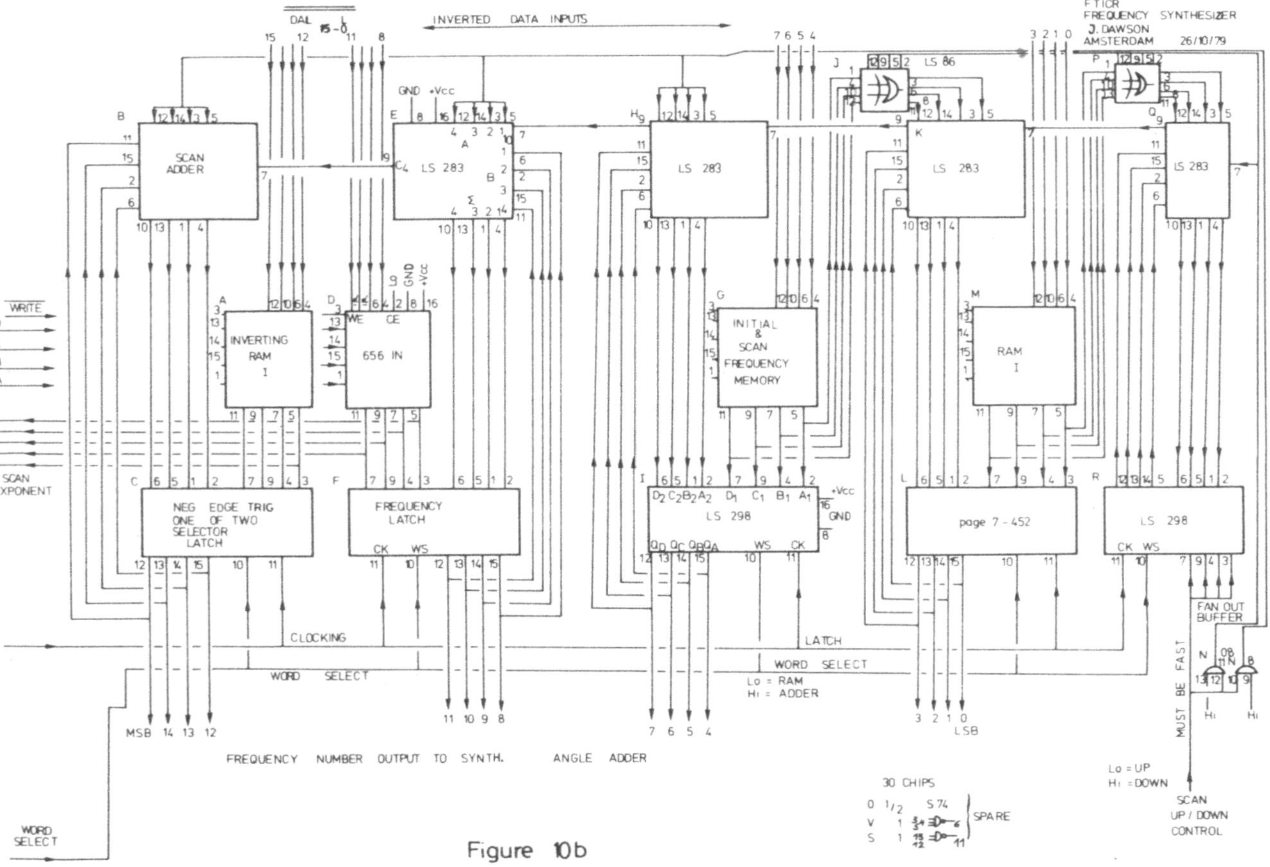
FTICR
FREQUENCY SYNTHESIZER
J. DAWSON
AMSTERDAM 26/10/79
INVERTED DATA INPUTS
DAL 15-0
SCAN ADDER
LS 283
INVERTING RAM I
656 IN
INITIAL & SCAN FREQUENCY MEMORY
RAM I
LS 86
NEG EDGE TRIG ONE OF TWO SELECTOR LATCH
FREQUENCY LATCH
LS 298
page 7 - 452
FAN OUT BUFFER
WRITE
SCAN EXPONENT
CLOCKING
WORD SELECT
LATCH
Lo = RAM
Hi = ADDER
MSB
LSB
MUST BE FAST
FREQUENCY NUMBER OUTPUT TO SYNTH.
ANGLE ADDER
30 CHIPS
SPARE
Lo = UP
Hi = DOWN
SCAN UP / DOWN CONTROL

Figure 10b

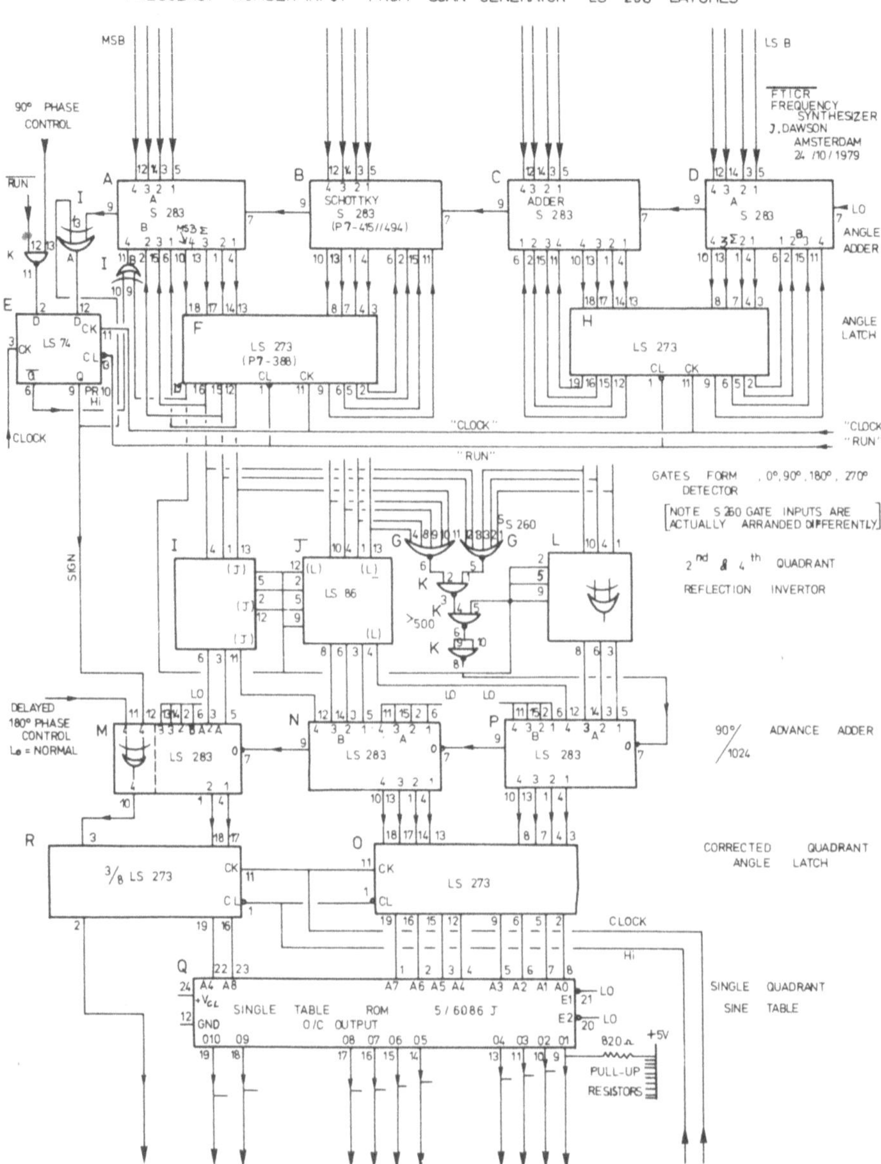
FREQUENCY NUMBER INPUT FROM SCAN GENERATOR 'LS 298 LATCHES
MSB
LS B
FTICR FREQUENCY SYNTHESIZER
J. DAWSON
AMSTERDAM
24 /10 / 1979
90° PHASE CONTROL
RUN
S 283
SCHOTTKY
S 283
(P7-415//494)
ADDER
S 283
S 283
LO
ANGLE ADDER
LS 74
LS 273
(P7-388)
LS 273
ANGLE LATCH
CLOCK
"CLOCK"
"CLOCK"
"RUN"
"RUN"
GATES FORM , 0°, 90°, 180°, 270° DETECTOR
NOTE S 260 GATE INPUTS ARE ACTUALLY ARRANGED DIFFERENTLY
S 260
SIGN
LS 86
>500
2nd & 4th QUADRANT
REFLECTION INVERTOR
DELAYED 180° PHASE CONTROL
Lo = NORMAL
LS 283
LS 283
LS 283
90°/1024
ADVANCE ADDER
3/8 LS 273
LS 273
CORRECTED QUADRANT ANGLE LATCH
CLOCK
Hi
SINGLE TABLE ROM 5 / 6086 J
O/C OUTPUT
GND
820 Ω
+5V
PULL-UP RESISTORS
SINGLE QUADRANT SINE TABLE

FIGURE 11a

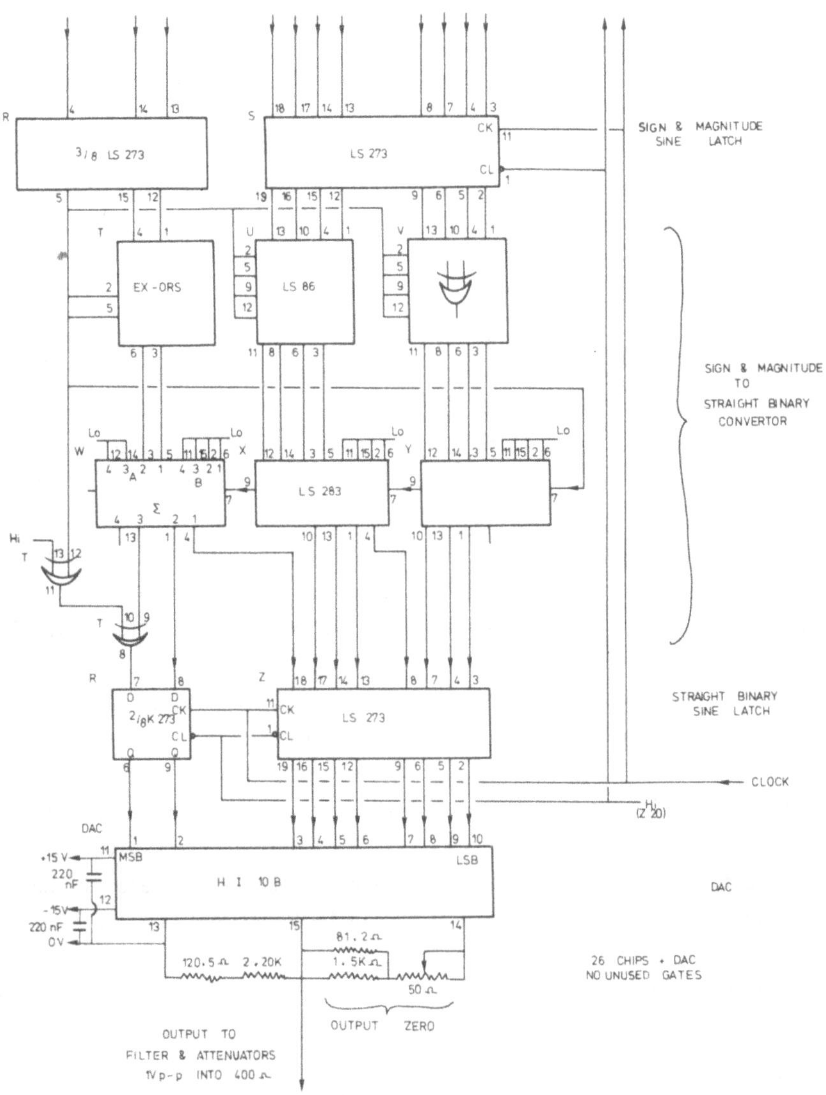
3/8 LS 273
LS 273
SIGN & MAGNITUDE SINE LATCH
EX-ORS
LS 86
LS 283
SIGN & MAGNITUDE TO STRAIGHT BINARY CONVERTOR
2/8 K 273
LS 273
STRAIGHT BINARY SINE LATCH
CLOCK
DAC
H I 10 B
MSB
LSB
+15 V
-15V
0V
220 nF
120.5 Ω
2.20K
81.2 Ω
1.5K Ω
50 Ω
OUTPUT ZERO
26 CHIPS + DAC
NO UNUSED GATES
OUTPUT TO
FILTER & ATTENUATORS
1Vp-p INTO 400 Ω

Figure 11b

selector/frequency latch C,F,I,L,R. Whenever the RAM address goes even gates N_3 and N_6 drive the data selectors into the mode where at their next negative going clock edge they will latch the data provided by the RAMs, viz. the new initial frequency. In order that there can be such a negative going edge it is first necessary to ensure that the clocking line is Hi; this is achieved by $\overline{\text{BEGIN}}_{sync}$ or by END OF COUNT, the logic of S_8, U_{12} and the fast normal power Schottky flip flop O_9. It is then the subsequent operation of these gates plus T_6 that ensures the negative transition at the moment that the RAM address begins to go odd again.

Thereafter the 'LS298 chips are clocked at a rate determined by the scan exponent and derived from the divider chain X,V,α,Y,Δ,β. Although the upper five bits of the scan rate word are used to define the scan exponent they are decoded only so as to provide 24 different exponent rates, not 32. The exponent divider chain is always set to count state 1 rather than clear before the running of any new segment because there always occurs another delay of one clock period in chip O_9.

When the RAM address is odd the 'LS298 chips accept data not from the RAMs but from the scan adder. Note that to avoid astronomic scan rates the scan adder has been widened to 20 bits, with the scan mantissa shifted four bits right. Every time the 'LS298s receive another clocking pulse their output is incremented or decremented by the scan mantissa. Incrementing occurs when the scan control line (derived from chip C of Figure 9) is Lo so that the Ex-ors are not inverting and the adder simply adds. Decrementing (scan down in frequency) requires that the adder should substract - this being achieved by the usual method of complementing (binary inversion) the full 20 bit mantissa and adding 1 (at Q_7, the carry input). (This is best understood by noting that scanning down with a zero mantissa will indeed result in there being no scan.) The initial frequency number is sort of "rounded up" ($R_{7,9,4,3}$) prior to a scan down, for reasons that are now obscure and were probably mistaken.

9. The Synthesizer (see Figure 11a,b)

As explained in relation to Figure 3, the output of the 17 bit angle latch E_9,F,H advances at every "CLOCK" pulse by the angle which it receives from the frequency latches C,F,I and L of Figure 10b. The ex-or gate I_{11} functions as a half adder for the 180° bit. Whenever the unit stops running the angle latch is cleared (by "RUN"), but when segments run without interruption the second segment takes its initial phase angle from the end of the first segment - there is no discontinuity in the waveform. Gates K_{11},I_8 and the flip flop E_6 can advance the initial phase angle by 90° if there has been a break since the running of the last segment (i.e. if RUN has gone Hi). Once this 90° phase shift has been introduced it can only be removed at a discontinuity. Note that it really is true that if the first of a set of consecutive segments is given a 90° phase advance, then so are the subsequent segments of the set.

Now come the complications that arise from the fact that the Sine Table read only memory (ROM), in common with its paperback antecedents, only contains directly the angles between 0° and just less than 90°. Being a 10 bit chip it divides the first quadrant up into $\frac{90°}{1024}$ quanta. To find the sine value of an angle in the second quadrant (similarly for the fourth) one must address the ROM at 180° minus the angle of interest. In binary this operation amounts to inverting the angle, adding 1, and discarding the carry bit. This operation is performed by the ex-ors I,J,L and the adders M,N,P. There remains the problem that the ROM does not contain 90° itself, but this is easily overcome by using the normal power Schottky gates G, K_3 and K_6 to specifically detect for the angles 90° and 270° and to withhold the "add 1" from the inversion operation. The result is that the ROM treats these two angles as if they were the same as the biggest angle which it does in fact contain. The error is absolutely negligible.

There can be no overflow into the most significant section of the adder chip M, so it is safe to use M_{10} as an ex-or gate to provide the 180° phase advancement control. Synchronising latches R, O and S are placed around the ROM because it is rather slow (access time 100 ns) and must be allowed one whole clock period in which to operate. If the synthesizer is to be clocked faster so as to give higher output frequencies it might be necessary to find a faster ROM, or to multiplex two ROMs.

After the ROM there comes the further difficulty that the sine value latched by $R_{5,15,12}$,S is in sign and magnitude form, whereas ultra high speed DACs tend to require a straight or offset binary input. However, it is simple to effect code conversion by the now familiar compliment (ex-ors $T_{6,3}$U,V) and add 1 (adders W,X,Y) process. Note that the overflow from the adder W_{13} must be added to the inverted sign bit in T_8 if a numerical mistake is not to occur at 180° itself. The output from the code converters is made synchronous by the latches $R_{6,9}$,Z so that with the very fast DAC used deglitching is unnecessary. The DAC output is adjusted by the external resistors so as to be equivalent to a current source of $\pm$ 2.5 mA in parallel with 400 ohms.

10. Analogue Stages (see Figure 12)

Although the ouput from the DAC does contain the desired output frequency as indicated in Figure 4, it also contains the high frequency components to be expected from such a stepped waveform and various image frequencies [11]. These are all removed quite effectively by the sharp elliptic function filter [12]

LOW PASS FILTER
CUT OFF FREQUENCY 3.3 MHz
MAX PASSBAND ATTENUATION 1db
MIN STOPBAND ATTENUATION 50db
MIN STOPBAND FREQUENCY 4.65 MHz
FIRST STOPBAND NODE 4.81 MHz
SECOND STOPBAND NODE 7.78 MHz
FROM DAC
18.6µH
14.5µH
26.9pF
75.5pF
233 pF
288 pF
197 pF
1Vp-p
ATTENUATORS
R1
R2
780 Ω
421 Ω
180 Ω
43 Ω
68 Ω
130 Ω
228 Ω
330 Ω
2.7K Ω
SD 5000 QUAD SWITCH
"ON" RESISTANCE 30 Ω
OUTPUT TO MIXER AND/OR ADC RATE
1V p-p INTO 75 Ω
SD 5000 OUTPUT ROUTE SWITCHING
2Vp-p
+15V
-15V
BC 184
BC 182 B
BC 239 B
BC 184
BC 214
2N 2219A
2N 2905 A
10mA QUIESCENT
200 Ω OUTPUT ZERO
8.1mA
OUTPUT TO CELL 10 V p-p FROM 50 Ω
OUTPUT TO CELL
SD 5000 DRIVE
3,14 (+15V)
(-15V) 6,11 SUBSTRATE
ttl LOGIC INPUT
(Lo) (ATTENUATE)
BC 212
BC 182
FROM CHIP "T" OF RAM II TIMING BOARD
DAL
ATTENUATOR
ROUTE
180°
DELAYING LATCHES
LS 273
LS 74
CLOCK
TO SYNTHESIZER CHIP M
ROUTE: MIXER CELL
"RUN"
CLOCK

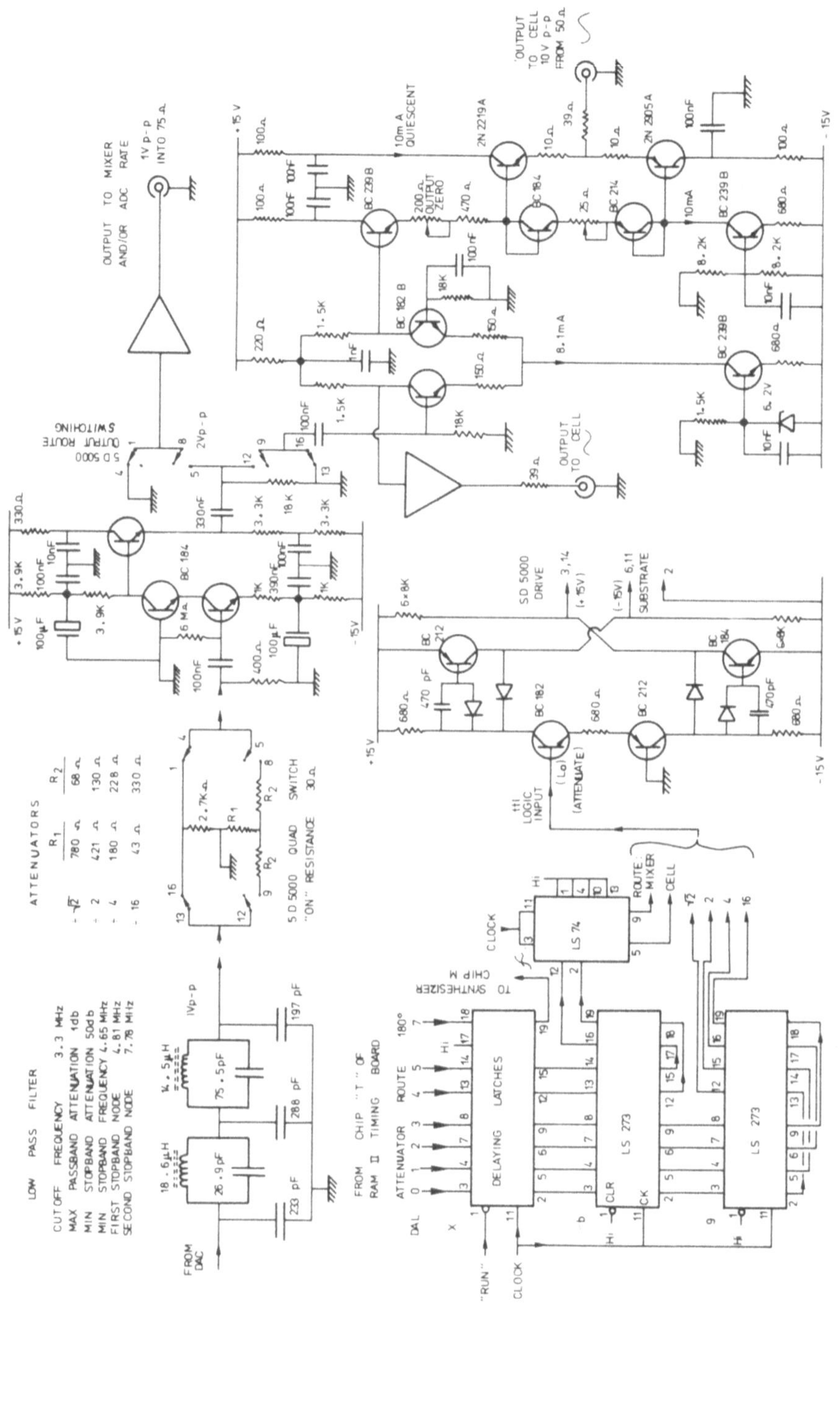

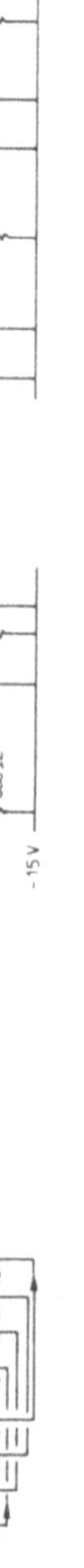

Figure 12

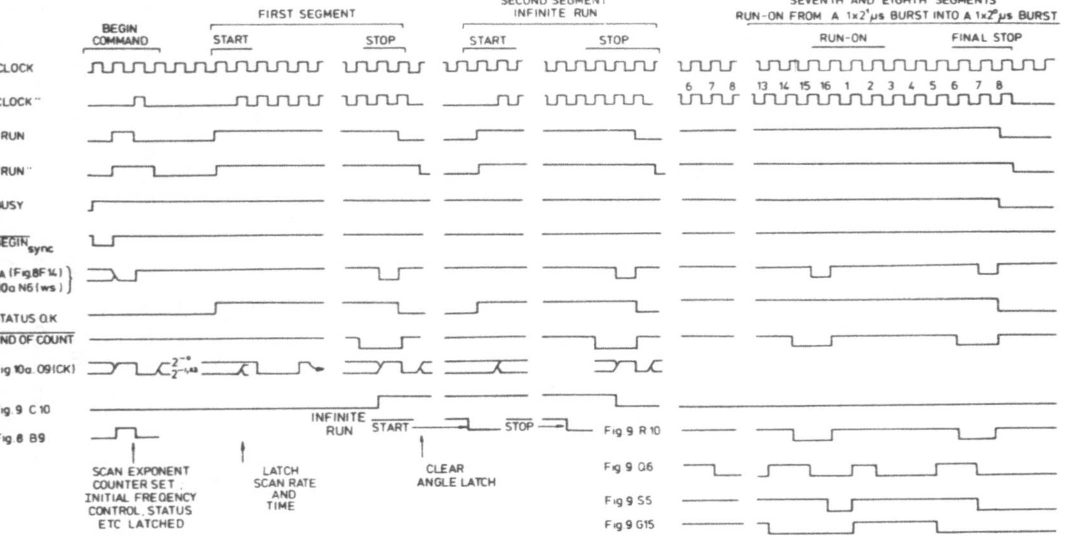

BEGIN COMMAND
FIRST SEGMENT
START
STOP
SECOND SEGMENT
INFINITE RUN
START
STOP
SEVENTH AND EIGHTH SEGMENTS
RUN-ON FROM A 1x2¹μs BURST INTO A 1x2⁰μs BURST
RUN-ON
FINAL STOP
CLOCK
"CLOCK"
6 7 8 13 14 15 16 1 2 3 4 5 6 7 8
RUN
"RUN"
BUSY
BEGIN sync
Q_A (Fig 8 F14)
Fig 10a N6 (ws)
STATUS O.K
END OF COUNT
Fig 10a O9 (CK)
Fig 9 C 10
Fig 8 B9
INFINITE RUN
START
STOP
Fig 9 R 10
Fig 9 Q6
Fig 9 S5
Fig 9 G15
SCAN EXPONENT COUNTER SET. INITIAL FREQENCY CONTROL, STATUS ETC LATCHED
LATCH SCAN RATE AND TIME
CLEAR ANGLE LATCH

Figure 13

which gives 50 db of attenuation to the 4.7 MHz image associated with the nominal maximum output frequency of 3.3 MHz. The filter and attenuators are all designed for a characteristic impedance of 400 ohms. Each section of the attenuator uses an SD5000 quad D-MOS analogue switch. Since these switches each have an ON resistance of 30 ohms the "non attenuating" positions do in fact attenuate and require the 2.7 Kohm resistors to maintain the constant impedance. The amplifier which terminates the attenuator chain has a gain of 4 to overcome this loss and to bring the signal up to 2 Vp-p (max). That amplifier and all subsequent stages are AC coupled. The terminating amplifier drives another SD5000 chip which turns each output amplifier On or Off. The ouput amplifier for the ICR cell consists of amplifying phase splitter and two identical low output impedance drivers. The original idea was that these antiphase outputs would drive the cell directly, but in the event they have been coupled to the primary of a (pulse) transformer with a centre tapped secondary (1:2 ct) which drives the cell with a maximum differential signal of 40 Vp-p. Restrospectively it seems that a definite commitment could have been made to transformer coupling to the cell, with consequent simplification of the output amplifier.

11. Preamplifier (cf. Figure 14)

The requirements for the signal preamplifier shown in Figure 14 were that its bandwidth should extend up to 3 MHz and that at its maximum gain the theoretical input noise should be amplified up to about the nominal rated output voltage of 1 Vp-p into 75 ohms. The actual design was arrived at after deciding that in practice front end noise would most probably be minimised by minimising the signal carrying component count rather than by elaboration. After the front end, it did not appear that any discrete component design could improve upon the performance offered by the NE592 differential video amplifier ic.

Junction gate fets are used in the front end for the sake of their relative hardiness; the type chosen are a readily available plastic encapsulated high mutual conductance RF type. If it is thought essential (for ultra high resolution work) to remove the small imbalance of the input potentials caused by mismatched gate leakage currents, this may be done with supplementary coupled input resistors, or by allowing for independent external plate potential supplies. The two current sources must each be adjusted on the bench to deliver 5 mA. Independent current sourcing of the AC coupled input long tailed pair allows for variable plate potentials whilst preserving the utmost simplicity at the critical input ports. To conform to the simplistic philosophy a conventional rather than a complementary cascode configuration is employed. The price to be paid

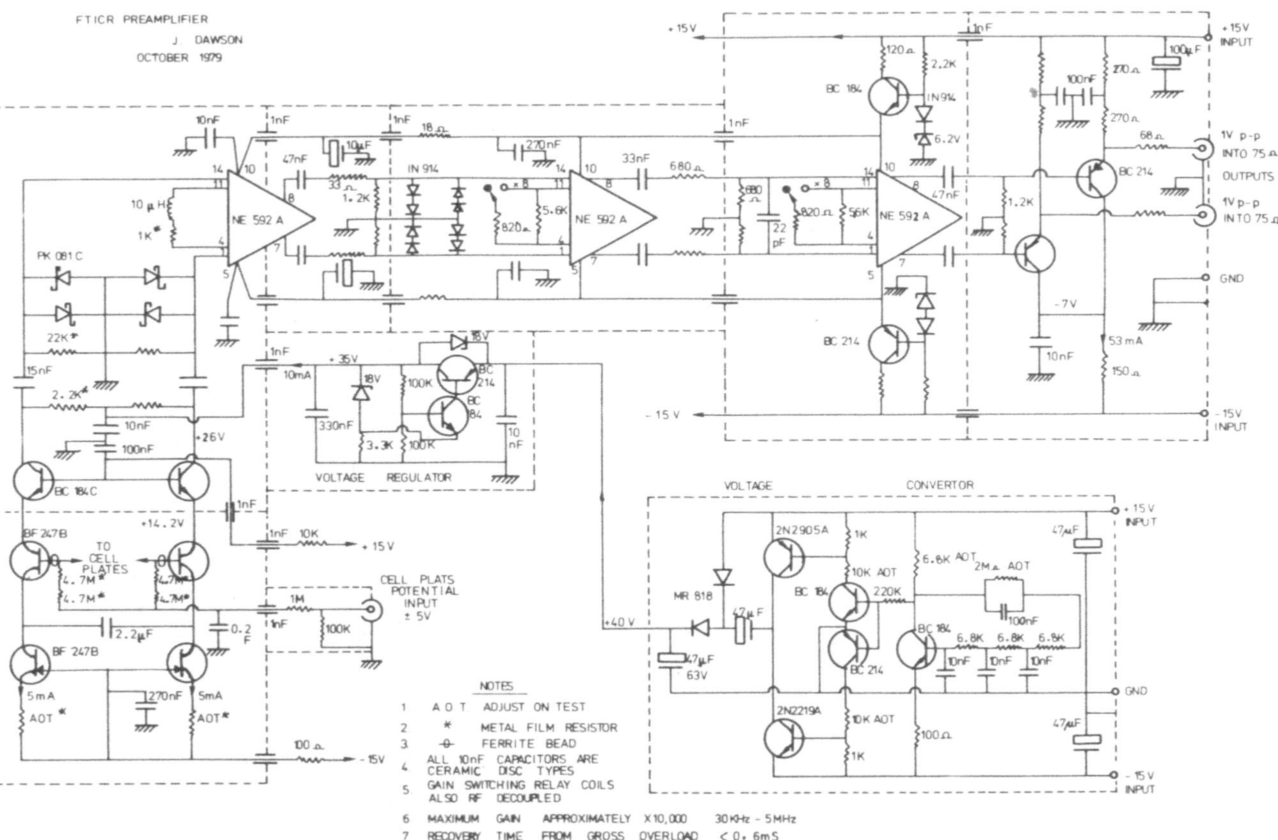
FTICR PREAMPLIFIER
J. DAWSON
OCTOBER 1979
+15V INPUT
1V p-p INTO 75 Ω
OUTPUTS
1V p-p INTO 75 Ω
GND
-15V INPUT
NE 592 A
BC 184
BC 214
IN 914
VOLTAGE REGULATOR
VOLTAGE CONVERTOR
2N2905A
2N2219A
MR 818
BF 247B
PK 081 C
TO CELL PLATES
CELL PLATS POTENTIAL INPUT ± 5V
+40 V
+35V
+26V
+14.2V
NOTES
1 A O T ADJUST ON TEST
2 * METAL FILM RESISTOR
3 FERRITE BEAD
4 ALL 10nF CAPACITORS ARE CERAMIC DISC TYPES
5 GAIN SWITCHING RELAY COILS ALSO RF DECOUPLED
6 MAXIMUM GAIN APPROXIMATELY X 10,000 30 KHz - 5MHz
7 RECOVERY TIME FROM GROSS OVERLOAD < 0.6mS

Figure 14

is that a + 35 V supply is required, but this is derived from the more convenient ±15 V supplies via an inverter and an inverted configuration output referenced series regulator which is ideal in this application.

The input transistors themselves should be quite tolerant of the input overloads which will be caused by the quench pulse and the ion excitation burst, but the video amplifier ics require protection. In the front end where the signal is at relatively high impedance (2 Kohm) ultra low capacitance Schottky diodes are used to clamp the signal, but after the second stage ordinary silicon diodes suffice. The overall gain of the preamplifier is varied by relay switching of the input coupling resistors of the third and fourth stages.

Each section of the circuit was constructed in a seperate compartment of a honeycomb chassis, built so that the front end compartment sealed directly onto the ends of two 5 mm internal diameter pipes which lead down into the vacuum system and which serve both to mount the cell and to screen each of the taut 0.1 mm diamater wires joining the cell receiver plates to the input transistors. (Vacuum seals at the cell ends) to stop UHF instability it was necessary to place ferrite beads on each input gate lead, and to avoid (an albeit small amount of) pickup from the inverter it was ultimately necessary to locate it remotely. If it had been better located in an isolated subsection of the honeycomb that might not have been necessary.

The measured bandwidth is substantially flat from 30 KHz to 5 MHz. No second harmonic distortion has ever been detected*, but in spectra containing a single intense ion a small (<3%) third harmonic peak is often seen. Although not specifically designed to have a fast recovery from input overload (ion excitation pickup) the preamplifier is in fact always adequately recovered within 0.6 ms. Changing the coupling capacitors to extend the bandwidth to lower frequencies might spoil this.

12. ADC and Fast Buffer Store

Fortunately for FTICR the state of integrated circuit technology has now progressed to the point where a number of manufacturers are producing fast ADCs to satisfy the demands of the video market. The TRW TDC 1007J 30 Megasamples/sec 8 bit ADC was chosen for this design because it can be obtained on a subassembly

---

*It could be that it is present, but is very effectively removed by the PAL-ing operation which causes alternate transients to be excited with inverted phase and then substracted into memory.

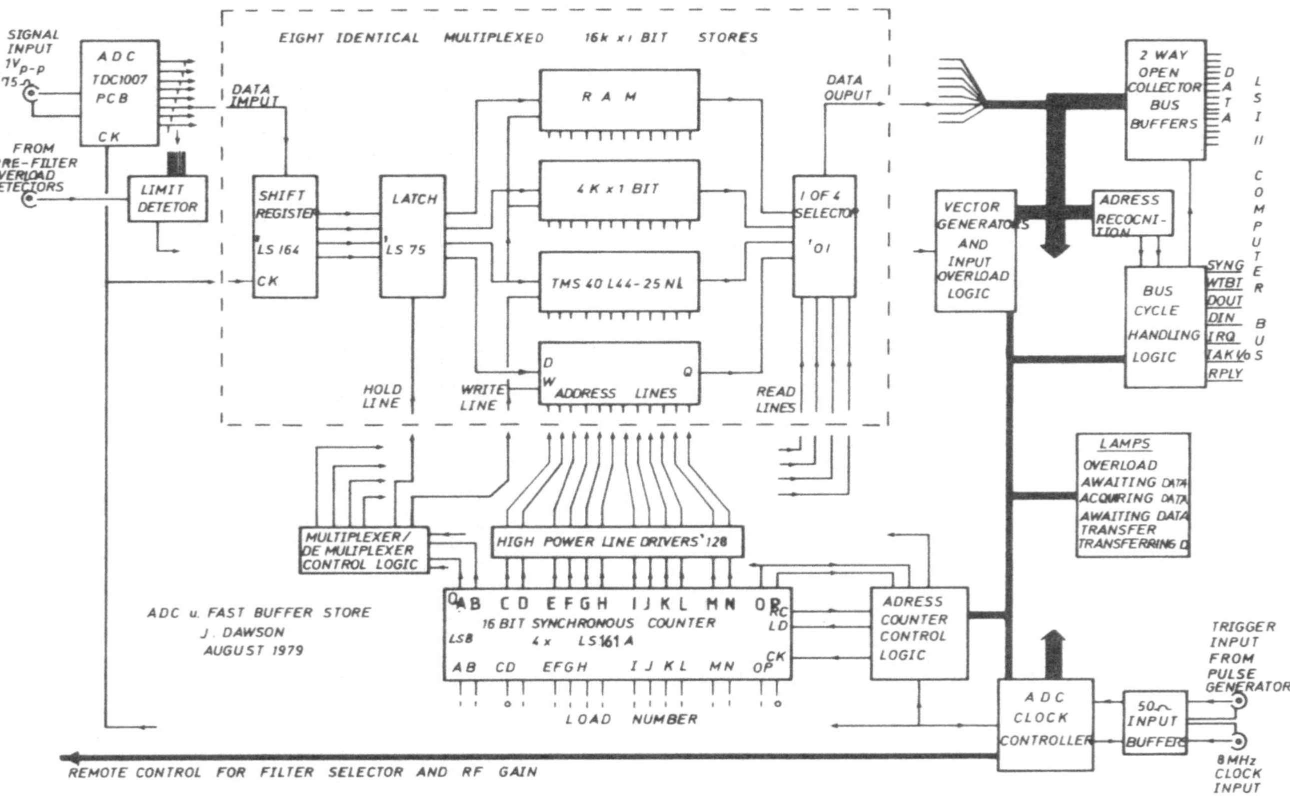
SIGNAL
INPUT
1V p-p
75 Ω
ADC
TDC1007
PCB
CK
FROM
PRE-FILTER
OVERLOAD
DETECTORS
LIMIT
DETETOR
EIGHT IDENTICAL MULTIPLEXED 16k x 1 BIT STORES
DATA
IMPUT
SHIFT
REGISTER
LS 164
CK
LATCH
LS 75
RAM
4 K x 1 BIT
TMS 40 L44-25 NL
D
W
Q
ADDRESS LINES
HOLD
LINE
WRITE
LINE
READ
LINES
1 OF 4
SELECTOR
'01
DATA
OUPUT
2 WAY
OPEN
COLLECTOR
BUS
BUFFERS
DATA
LSI II COMPUTER BUS
ADRESS
RECOGNI-
TION
VECTOR
GENERATORS
AND
INPUT
OVERLOAD
LOGIC
BUS
CYCLE
HANDLING
LOGIC
SYNC
WTBT
DOUT
DIN
IRQ
IAKI/OS
RPLY
LAMPS
OVERLOAD
AWAITING DATA
ACQUIRING DATA
AWAITING DATA
TRANSFER
TRANSFERRING D
MULTIPLEXER/
DE MULTIPLEXER
CONTROL LOGIC
HIGH POWER LINE DRIVERS '128
AB CD EFGH IJKL MN OP
16 BIT SYNCHRONOUS COUNTER
4 x LS161 A
LSB
RC
LD
CK
AB CD EFGH IJKL MN OP
LOAD NUMBER
ADRESS
COUNTER
CONTROL
LOGIC
ADC u. FAST BUFFER STORE
J. DAWSON
AUGUST 1979
ADC
CLOCK
CONTROLLER
50 Ω
INPUT
BUFFER
TRIGGER
INPUT
FROM
PULSE
GENERATOR
8 MHz
CLOCK
INPUT
REMOTE CONTROL FOR FILTER SELECTOR AND RF GAIN

Figure 15

board with input and supply buffers at little extra coast, and because it can be used without an additional external sample and hold circuit when the input signal does not include frequency components in excess of 7 MHz. This ADC consists of 256 parallel analogue comparators, each fed by the signal and by a tap from a reference voltage resistive devider. The parallel outputs of all this comparators feed a priority encoder network to produce an 8 bit parallel binary coded output. Having decided upon the maximum ADC sampling rate (in this design 8 MHz) and having chosen the ADC accordingly, design of the Fast Buffer Store becomes to a large extent a matter of searching through manufacturers' catalogues to find the most suitable memory chips currently available. For a one-off construction there seems little point in struggling with the extra design effort required to save a little money by using dynamic RAMs. The design outlined in Figure 15 is based upon the decision to build up the 16K byte memory from 32 4Kx1 bit static RAMs. This has the great convenience that the required speed of 8 MHz can be obtained by sectioning each of the 8 16Kx1 bit stores (one for each bit from the ADC) into a 4Kx4 bit stores which with suitable input multiplexing need only operate at 2 MHz. The multiplexing of each of these 4Kx4 bit stores to a data input line is carried by two ordinary ttl chips - a serial in, parallel out shift register, and a 4 bit latch which catches four outputs from the shift register once in every four clock periods. These four latest data values are presented to each 4Kx4 bit memory array for the best part of 500 ns whilst the shift registers go on to acquire the next four sample values from the ADC. There is no speed problem in readout from the store - this requires only some 1 of 4 line data selectors as the demultiplexer. By adopting this design configuration each of the eight identical multiplexed 16Kx1 bit stores becomes a small board of 7 chips.

The rest of the buffer store circuitry is largely concerned with incrementing the active RAM address in step with the ADC and multiplexer during data acquisition and in step with the computer and demultiplexer during data transfer to the computer. At the heart of this is a 16 bit fully synchronous counter which as well as supplying the RAM address also signals to control logic when it has just filled the last memory position, and again when it is reading out from the last memory position. In retrospect it would have been wise to have given the computer a control command with which to reset the counter, and it would have opened up another method for avoiding data overflow during transient accumulation if the computer could have reversed the count direction during data transfer so that software could direct "un-accumulation" of the partially accumulated transient. This hazzard can however be prevented by periodic software controlled look ahead operations on the accumulated signal.

The output of the ADC is monitored during data acquisition to search for the extreme data values of -128 or +127 which are regarded as indicating clipping of the signal. If such a condition occurs, or if one of the pre-filter analogue overload detectors is triggered during the data acquisition period, then a latch in the vector generator logic is set (it is un-set at the end of data transfer). When the store has just acquired the 16Kth data value the unit sends an interrupt request to the computer. When this is acknowledged the overflow latch is inspected and a different vector sent to the computer if it is set. This gives the operator the choice of asking the program to discard overloaded transients, or to accumulate them regardless.

For speed, the data transfer process takes place under a software infinite loop. To terminate this at the appropriate moment the Buffer Store sends a second interrupt request the moment that it sees that the computer is initiating the bus cycle to fetch the 16Kth data value. The vector sent when this interrupt is acknowledged causes the program to jump out of the infinite loop and go on with the next chemistry cycle, or whatever other operation is required. After that the Buffer Store resets the address counter to its initial state and awaits receipt of the next trigger pulse from the pulse generator. The rate at which the ADC and store run is determined by a control word sent from the computer. The rates run in multiples of 2 from 62.5 KHz up to 8 MHz. The control word also contains some bits used to control the preamplifier gain etc and these together with information about the required filter setting are passed on to the analogue signal handling units.

13. Acknowledgements

The author wishes to thank André Noest and Frans Pinkse for many helpful discussion concerning the design of the digital electronics discussed in this paper. Drs. J.C. Kleingeld assisted in the preparation of the manuscript.

All the work was carried out in co-operation with Professor N.M.M. Nibbering and was financed by the University of Amsterdam and the Netherlands Organisation for Pure Research - SON/ZWO.

14. References

1. M.B. Comisarow and A.G. Marshall, J. Chem. Phys. 62 (1975) 293.

2. M.B. Comisarow in "Transform Techniques in Chemistry" (P.R. Griffiths, Ed.), Plenum Press, New York, 1978, p. 257.

3. M.B. Comisarow, J. Chem. Phys. 69 (1978) 4097.

4. A.G. Marshall, Anal. Chem. 51 (1979) 1710.

5. A.G. Marshall, M.B. Comisarow and G. Parisod, J. Chem. Phys. 71 (1979) 4434.

6. A.G. Marshall and D.C. Roe, J. Chem. Phys. 73 (1980) 1581.

7. M. Alleman, Hp. Kellerhals and K.-P. Wanczek, Chem. Phys. Letters 75 (1980) 328.

8. J.H.J. Dawson, A.J. Noest and N.M.M. Nibbering, Int. J. Mass Spectrom. Ion Phys. 29 (1979) 205.

9. T.C. Farrar in "Transform Techniques in Chemistry" (P.R. Griffiths, Ed.), Plenum Press, New York, 1978, p. 199.

10. M.B. Comisarow and A.G. Marshall, Chem. Phys. Letters 26 (1974) 489.

11. IEEE Trans., Audio Electroacoust. 1971 AU-19, pages 48-56.

12. The filter was designed from data contained in Table A4-4, Appendix 4, page 146 of "Simplified Modern Filter Design" by Philip R. Geffe, published by Iliffe Books Ltd., London.

A MICROCOMPUTER-BASED FOURIER TRANSFORM ION CYCLOTRON RESONANCE MASS SPECTROMETRIC DETECTION SYSTEM

R. J. Doyle, Jr., T. J. Buckley and R. Eyler[a]
Department of Chemistry, University of Florida, Gainesville, Fl 32611

1. Introduction

Since its development less than ten years ago, the technique of Fourier transform ion cyclotron resonance (FTICR) mass spectrometry [1] has been applied in a number of laboratories and has shown great promise for advances in several fields of analytical and physical chemistry. Developments in FTICR theory [2-5] enhancements of the technique based in part on theoretical concepts [6-9] and some limited applications [9-13] have appeared. All of the recently published work has involved FTICR spectrometers utilizing relatively expensive minicomputers developed originally for use in Fourier transform nuclear magnetic resonance spectrometry.

Despite the numerous studies referenced above, there has been a paucity of instrumental detail in papers published to date, with two simplified block diagrams [9,14] and one slightly expanded block diagram [15] the only published information in this respect. This paper reports in some detail the FTICR development which has been ongoing in our laboratory for several years. Both block diagrams and detailed schematics of various subsystems are given. As contrasted with most other FTICR users, we have relied upon two relatively low cost microcomputers for FTICR experimental control and limited data processing. Section 2 of this paper discusses the overall block diagram of our system and the analog signal detection electronics. Section 3 deals with some important aspects of the digital signal averaging and processing, the two microcomputer systems, and requisite timing considerations for correct synchronization of digital data acquisition and manipulation components.

---

[a] Camille and Henry Dreyfus Foundation Teacher-Scholar, 1978-82

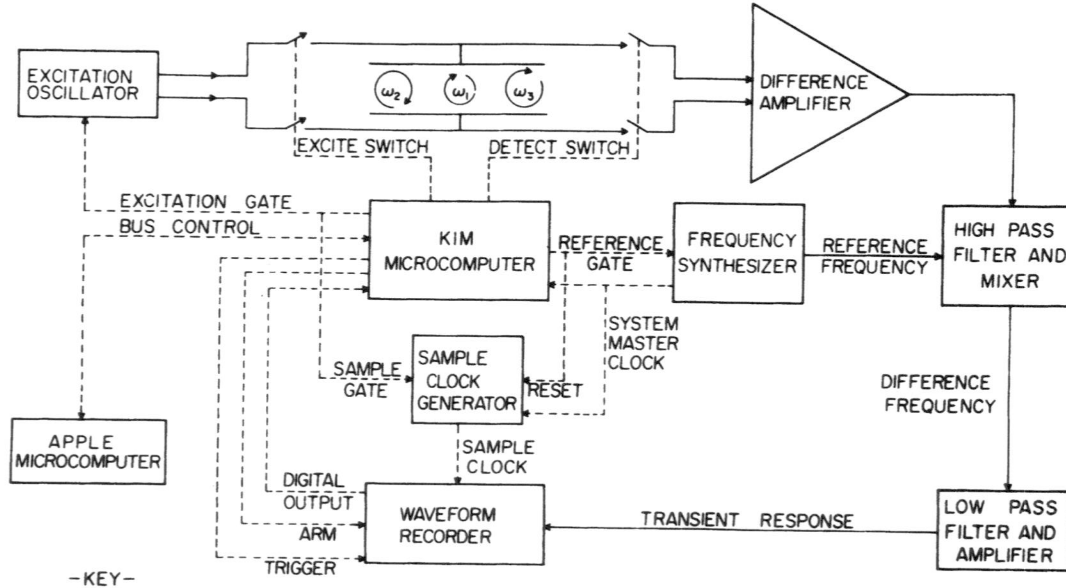

Fig. 1. Block diagram of microcomputer-based Fourier transform ion cyclotron resonance mass spectrometric detection system.

In Section 4 we present representative FTICR mass spectra and discuss general system performance, advantages, disadvantages and, finally, some recent and planned improvements.

2. General Comments and Analog Circuitry

2.1. Overall System Block Diagram

Figure 1 shows a block diagram of our microcomputer-based FTICR excitation and detection system. Our initial development has centered around the "spectral segment extraction" mode of FTICR [16] , in which the ion transient response is mixed with a reference oscillator, and the resulting (lower frequency) differency signal is digitized and subjected to discrete Fourier transformation. This allows analysis of limited ion mass ranges, but can lead to higher resolution than "wide-band" FTICR in which a wide range of ion masses are detected at the same time. The reader is referred to any of the earlier references in this paper (esp. 1b) for a detailed description of the FTICR method. The electronics shown in Figure 1 are in addition to those normally used for formation, trapping and removal of ions in the pulsed mode of ICR mass spectrometric operation [17] . The FTICR system represented in Figure 1 does, however, replace the gated marginal oscillator detector [18] and the "postdetection system" described in Section III of reference 17.

The excitation oscillator, frequency synthesizer, and waveform recorder shown in Figure 1 are commercial instruments [19] and will not be discussed further except with respect ot some timing pulse considerations. The KIM-I and Apple II microcomputers and sample clock generator will be discussed in Section 3. The remainder of this section is devoted to a discussion of the excite and detect switches, difference amplifier, which collectively comprise the analog detection electronics of our system.

2.2. Excite and Detect Switches

Figure 2 shows the F E T switches and associated driving electronics used to block or pass the excitation voltage or the ion transient response signal. The excitation oscillator (a multigenerator with differential output) is used to apply an alternating potential to the upper and lower plates of the ICR cell. A voltage ramp applied to the voltage controlled frequency (VCF) input of the excitation oscillator by a second function generator leads to gated frequency sweeping for these experiments.

An arrangement of Motorola 2N5555 junction FET's is used to prevent saturation of the detection electronics during excitation and to prevent loading of the upper and lower cell plates by the low output impedance of the excitation oscillator during collection of the transient response signal.

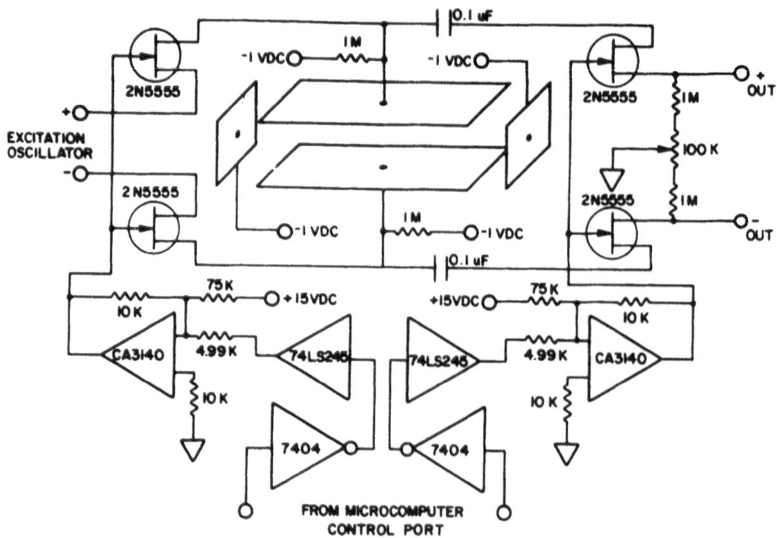

Fig. 2. FT ICR cell (trapping plates not shown) with excite and detect switching circuitry.

These devices have a maximum "on" resistance of 150 ohms, an "off" resistance on the order of 10^{12} ohms, and a maximum turn-on time of 10 nanoseconds. The transistors are driven, in pairs, by two RCA 3140 operational amplifiers in the summing-inverter configuration [20] . Resistor values were chosen to provide the necessary on-off potentials of -2.2 and -8.0 V respectively. Pulses from the microcomputer control port are inverted by a 7404 inverter and driven by a 74LS245 line driver.

2.3. Difference Amplifier and High-Pass Filter

Figure 3 shows the difference amplifier and high-pass filter circuitry. A Burr-Brown 3622J differential input instrumentation amplifier was selected for pre-amplification. Shown in Figure 3 with a gain of 100, its bandwidth is 2 MHz for ± 3 db flatness and the common mode rejection ratio is over 60 db. The input impedance is 10^{11} ohms with 3 pF input capacitance. High pass filtering is accomplished with a Burr-Brown 3508J operational amplifier in the multiple feedback filter configuration. The filter, as shown in Figure 3 has a bandwidth of over 20 MHz at a gain of 4 with a low frequency cutoff of 50 kHz.

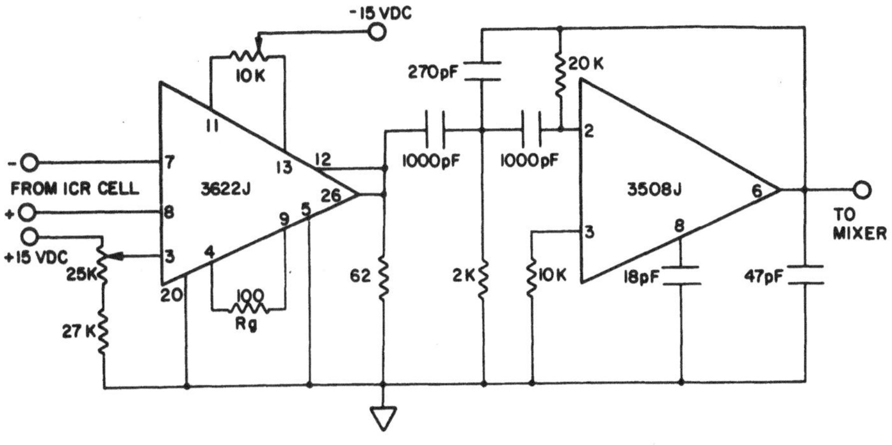

Fig. 3. Differential pre-amplifier and high-pass filter used in FT·ICR detection electronics.

2.4. Mixer

Figure 4 shows the mixer circuitry utilized in our detection electronics. Isolating the high pass filter from the mixer is a Burr-Brown 3608J operational amplifier in the inverting voltage follower configuration compensated for stability. Mixing is performed by a Motorola MC1496 balanced modulator-demodulator set up as a doubly balanced mixer with broadband input and output networks. The unit is operated at ± 15 V D.C. Mixer gain is dependent upon the reference oscillator level. A programmable frequency synthesizer [19] provides the reference frequency to the mixer.

2.5. Low-Pass Filter and Amplifier

The low-pass filter and amplifier stage of the detection electronics is shown in Figure 5. Final filtering is performed by Frequency Devices Model 745PB-4 programmable, four pole, Butterworth, low-pass filter. External switches control the corner frequencies and any combination of 3.1, 6.4, 12.8 and 25.6 kHz may be selected. Final amplification is accomplished with an Analog Devices AD521 instrumentation amplifier. Gain is externally controlled by adjusting the 50 kohm

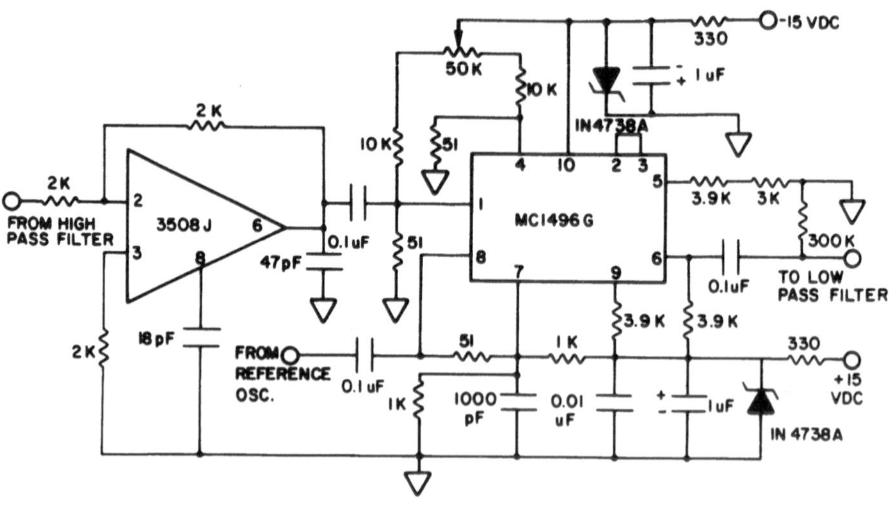

Fig. 4. Inverting voltage follower and mixer stage of FT ICR detection electronics

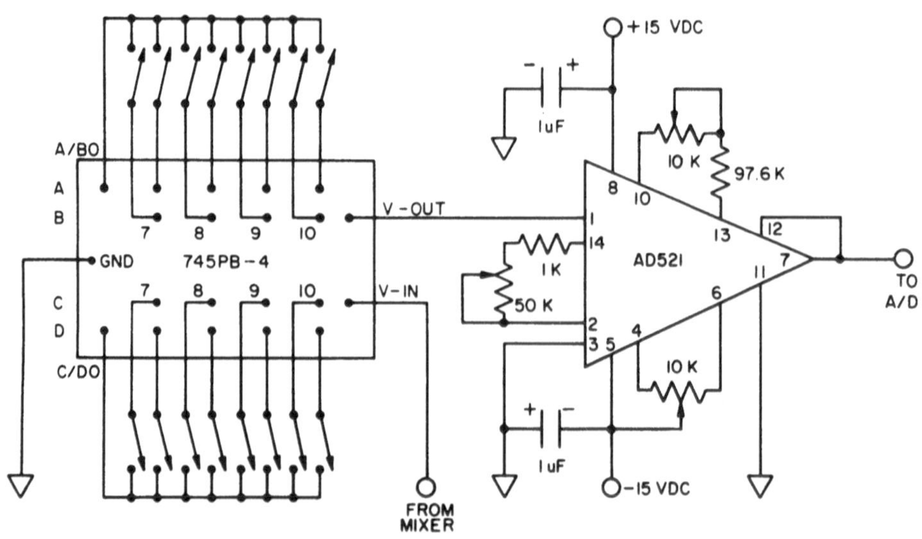

Fig. 5. Low-pass filter and amplifier used in FT ICR detection electronics.

potentiometer located between pins 2 and 14. Gain may be varied from 2 to 100 in this configuration.

3. Digital Electronics

3.1. General Discussion

The analog electronics described in the previous section are controlled by the KIM-1 [21] microcomputer shown in Figure 1. The primary responsibility of this computer is to perform routine tasks of timing and data processing necessary to obtain a transient response from the system. The Apple II microcomputer carries out further processing and handling of the acquired transient response signals.

Four digital signals which control the analog electronics in Figure 1 are provided by the KIM-1 microcomputer. Two of these activate the excite and detect switches connected to the ICR cell. The inverters shown in Figure 2 provide the proper logic polarity such that the switches are opened by a logical zero and closed by a logical one. The excitation oscillator is started on its preset frequency sweep by placing a logical one on the excitation gate line for the duration of the excitation period. The frequency synthesizer produces its reference frequency upon receiving a reference gate pulse. The reference frequency starts with a phase angle of zero degrees and lasts for the length of the reference gate pulse.

The ion transient response signal from the analog detection electronics is digitized by a waveform recorder [19] under computer control. An arm pulse prepares the recorder for data acquisition while a trigger pulse initiates the process. Two thousand fortyeight 8-bit precision points fill the recorder's memory at a rate determined by an external sample clock. The data is transferred to the microcomputer's memory via a digital interface under software control.

The signal-to-noise ratio of the digitized transient is enhanced by signal averaging. However, this requires a very reproducible timing sequence for each transient recorded. The necessary synchronization for the experiment is accomplished by a 1 MHz system master clock provided by the frequency synthesizer's internal clock. This signal is supplied directly to the KIM-1 microcomputer's phase zero input from which all computer timing is derived. The slower sample clock rate used to control the waveform recorder is derived from the master clock by a simple divide-by-N circuit [22] . The inclusion of a sample gate and reset pulse insures the phase coherence of all critical clock signals.

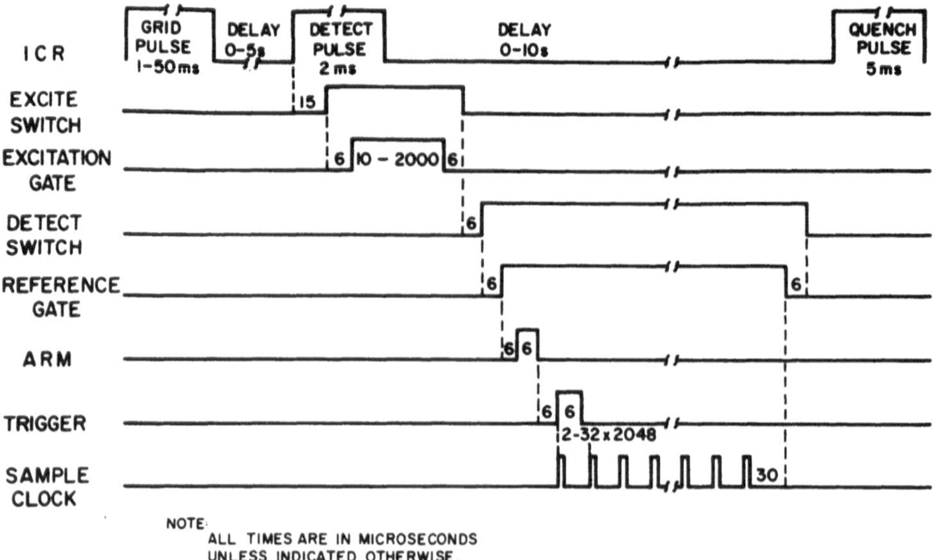

Fig. 6. Timing pulse sequence used in FT ICR experiments.

3.2. Timing Pulse Sequence

The pulse sequence used in our FTICR experiments is shown in Figure 6. The experiment begins with the ICR control electronics which generate a pulse sequence (see Ref. 17 for details). First a grid, which normally repels the electrons from a filament, is pulsed positive with respect to the filament for a period of time. During this time electrons enter the cell and ions are generated by electron impact ionization. Next a variable delay period may be introduced to allow ion-molecule reactions or thermalizing ion-neutral collisions to take place, if desired. A detect pulse is then sent to the KIM-1 microcomputer to start the transient response detection scheme.

Upon receiving the leading edge of the detect pulse the microcomputer first services the interrupt request by taking a few microseconds to perform some housekeeping operations. Next the excite switch is closed, thus connecting the output of the excitation oscillator to the upper and lower plates of the ion trapping cell. Then the excitation oscillator is activated. During this period the ions absorb energy from the oscillating electric field between the upper and lower cell plates. The excitation gate is then turned off, followed by the opening of the excite switch.

After translational excitation of the ions has been achieved detection can be started. This is done by closing the detect switch which connects the cell to the transient response detection electronics. Next the reference oscillator is turned on by the reference gate. During this period the reference frequency is mixed with the dignal from the cell and a difference frequency is produced. Data acquisition is started by an arm and a trigger pulse sent to the waveform recorder. The difference frequency is then digitized at a selectable clock rate until 2048 sample points have been taken. Next the reference gate and detect switch are turned off.

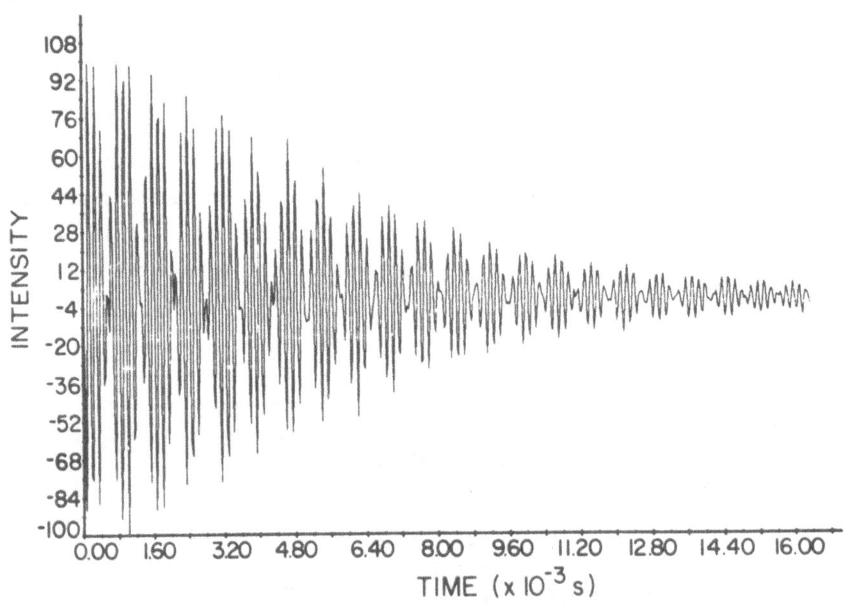

Fig. 7. 512 point transient response of m-bromotoluene at 3.6 x 10^{-6} torr with a 30 ms grid pulse followed by a 90 ms reaction time. A 1 V peak-to-peak excitation sweep from 99.6 to 114.9 kHz for a duration of 400 μs was used. The transient was mixed with a 100 kHz reference frequency and digitized at a rate of 32 μs per point. 256 passes were averaged. The magnetic field was 1.2100 Tesla.

The transient response is then transferred from the waveform recorder to the KIM-1 microcomputer where it is signal averaged and stored in memory. Finally a quench pulse, which removes all ions from the cell, is sent from the ICR control electronics. This sequence is repeated until the desired number of passes have been taken for signal averaging.

An example of a signal averaged transient response obtained by this system is shown in Figure 7. The superposition of ion cyclotron frequencies characteristic of the ions present and excited in the ICR cell, is exponentially damped due to collisional dephasing processes and ion-molecule reaction losses.

3.3. Dual Microcomputer System

Following accumulation in the KIM-1 microcomputer memory, the transient response is then subjected to further processing such as display for visual inspection, storage for later use, or discrete Fourier transformation to obtain the various ion frequencies and intensities. A block diagram of the computer arrangement we employ is shown in Figure 8. The two microcomputers used in this experiment both use a 6502 8-bit microprocessor as their central processing unit. The limited number of control lines necessary in this experiment and the 8-bit data from the waveform recorder make an 8-bit microprocessor well suited for this application.

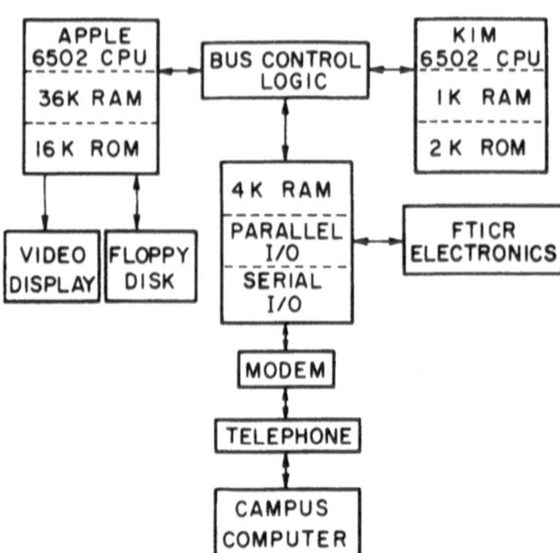

Fig. 8. Block diagram of dual microcomputer system used in FT ICR experiments.

The Apple II microcomputer is responsible for interaction with the user, program development (for both itself and the KIM-1), graphics, storage of programs and data on disk, and Fourier transformation. The read only memory (ROM) contains a system monitor, assembler, and disassembler as well as integer and floating-point BASIC interpreters. The 36 K of random access memory (RAM) allows sufficient room for progress and data in memory. A black and white TV monitor acts as a video display terminal while a floppy disk serves as a mass storage device. The KIM-1 microcomputer performs the routine tasks needed to run the FTICR experiment. These include system initialization, pulse generation, signal averaging, and system status display. All of its instructions come directly from the Apple II. Communication between the two microcomputers is accomplished through a common S-100 bus and its control logic. The Apple II microcomputer, being the source of commands determines use of the S-100 bus and halts instruction execution of the KIM-1 when the bus is desired. All data and programs which pass between the two microcomputers are placed in the common 4K memory block on the S-100 bus. The FTICR electronics and waveform recorder are controlled by several parallel input-output (I/O) ports. A serial I/O port serves as a communication channel to the campus computer. All components on the S-100 bus can be used by either processor.

The serial I/O port operates at 300 baud and sends its signal over a telephone line to the campus computer through an acoustic modulator-demodulator (MODEM). Once the transient response has been transferred to the campus computer additional data treatment is performed. This includes storage, Fourier transformation [23] , and obtaining hard copy spectra from the Gould digital plotter. Previously the initial Fourier transformation was accomplished by the Apple II microcomputer using a fast Fourier transform (FFT) algorithm written in BASIC. A 512 point transform of 16-bit averaged data required approximately six minutes computation time. A faster FFT routine written in 6205 assembly language has recently been completed [24]. This routine performs a 512 point transform on 16-bit data in 14 seconds, thus greatly enhancing the speed of data analysis. More involved transforms with apodization and various degrees of zero filling are currently performed on the campus computer.

4. Results and Discussion

Sample spectra obtained by our system are shown in Figures 9 and 10. These transforms were obtained on our campus computer using three orders

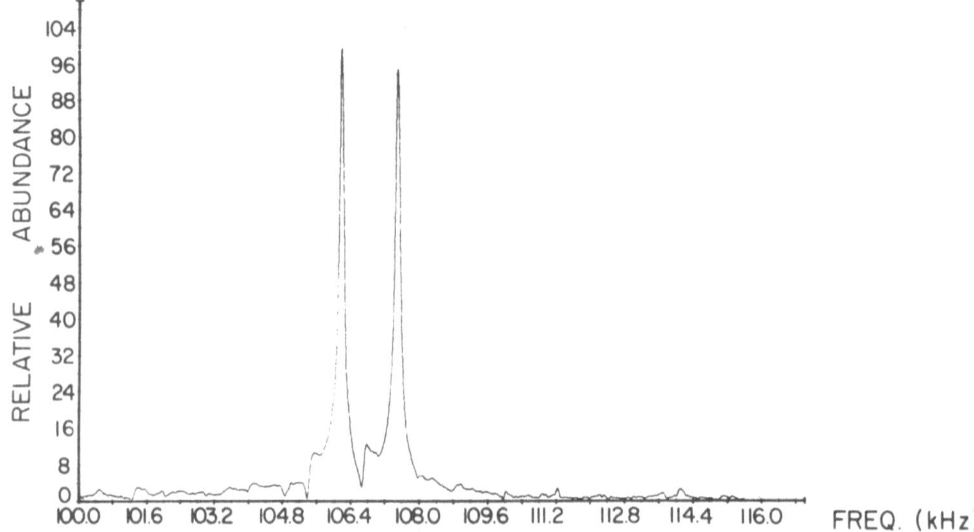

Fig. 9. Fourier transform of transient response shown in Fig. 7 from m-bromotoluene showing predominant peaks at m/e 172 and 170 (left to right). Trapezoidal apodization and 3 orders of zero-filling were used. Resolution $(f/\Delta f_{1/2})$ is 971 after zero-filling.

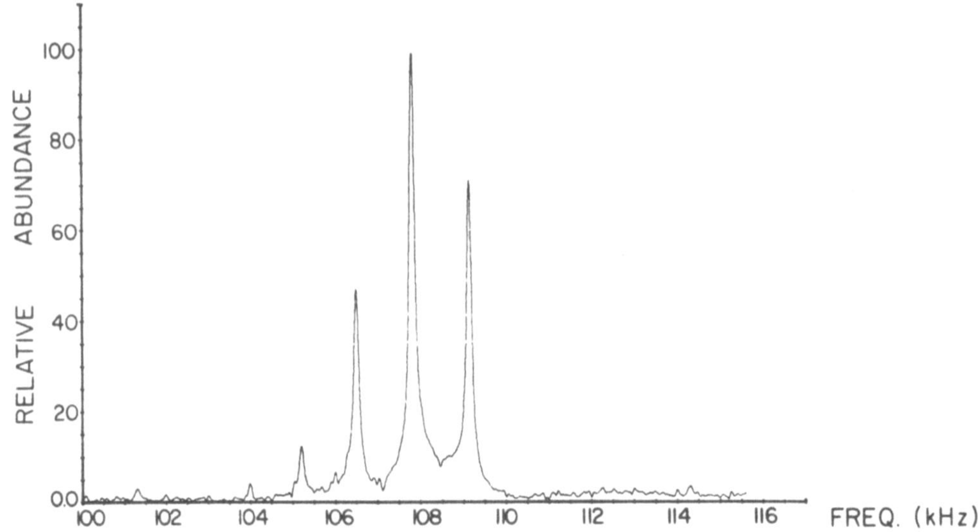

Fig. 10. Fourier transform of transient response of tetrachloroethene at a 8.3 x 10^{-7} torr showing peaks at m/e 172,170,168,166, and 164 (left to right). Trapezoidal apodization and 3 orders of zero-filling were used. The 512 point transient was digitized at a rate of 32 µs per point and signal averaged 256 times. A 1 V peak-to-peak excitation sweep from 102.5 to 115.0 kHz for a duration of 600 µs was used. The transient response was mixed with a reference signal of 100 kHz frequency. The grid pulse was 26 ms long followed by a 74 ms reaction time. Magnetic field strength was 1.1846 Tesla. Resolution $(f/\Delta f_{1/2})$ is 678 after zero-filling.

of zero filling [8]and trapezoidal apodization of 512 point transient collected by the system described in Section 2 and 3. The plots were produced by the Gould electrostatic plotter associated with our campus computer.

The peaks shown for m-bromotoluene and tetrachloroethene are separated by two mass units due to the halogen isotope masses. The resolution of these spectra is improved when compared to that which we obtain on our conventional marginal oscillator detector. The intensity ratios in Figure 10 are quite close to those predicted (1:12:54:108:81) from the ca. 3:1 natural abundance ratio between ^{35}Cl and ^{37}Cl. A slight intensity reversal is seen, however, in Figure 9, since the peak at ca. 107.5 kHz (m/e 170) should be slightly larger (50.5:49.5) than that at ca. 106 kHz (m/e 172) due to the natural abundance ratio between ^{79}Br and ^{81}Br. The spectra shown could be improved by acquiring more data points from the transient response, but this would require that the transient response extend for a longer time period. For the experiments which produces Figures 9 and 10, thelength of the transient response waslimited by the relatively high pressure used.

As mentioned in Section 2.2., in our initial experiments we used an analog multigenerator for ion excitation. This device in some respects did not generate excitation sweeps of sufficient quality for this experiment. Frequency drifts over time and trigger jitter caused slight phase differences between sweeps. These problems negated the effect of signal averaging over long periods of time. This problem has now been corrected by using the frequency synthesizer for both the excitation sweep and reference frequency, as is currently done by other FTICR groups [9,15] .

Another drawback with this system is that it was designed for use as a second detection system for our existing pulsed ICR electronics. The cell geometry dictated by laser experiments which we currently perform requires the use of switches for cell isolation. Various other electrical connections which supply trapping potentials for the cell introduce extraneous noise. A vacuum and electonics system optimized for the sole use of the Fourier transform detector and utilizing a cubic ICR cell 25 has also recently been developed in our laboratory.

The main advantages of this system are its relatively low costs and the flexiblity afforded by its dual microcomputer configuration. We believe the above mentioned developments and modifications will produce a system very well suited for the needs of our research efforts, and one which is of definite interest of others entering the FTICR area.

5. Acknowledgements

Design and troubleshooting advice as well as the loan of various items of equipment by the Department of Chemistry Electronics Shop is gratefully acknowledged. Portions of this work were supported by the University of Florida Division of Sponsored Research.

6. References

1. a) M. B. Comisarow and A. G. Marshall, Chem. Phys. Lett. 25 (1974) 282.
   b) M. B. Comisarow in "Transform Techniques in Chemistry", ed. P. R. Griffiths, Plenum Press, New York, 1978, ch. 10.
2. M. B. Comisarow and A. G. Marshall, J. Chem. Phys. 64 (1964)110.
3. M. B. Comisarow, J. Chem. Phys. 69 (1978) 4097.
4. A. G. Marshall, M. B. Comisarow and G. Parisod, J. Chem. Phys. 71 (1979) 4434.
5. A. G. Marshall and D. C. Roe, J. Chem. Phys. 73 (1980) 1581.
6. A. G. Marshall, Anal. Chem. 51 (1979) 1710.
7. A. G. Marshall, Chem. Phys. Lett. 63 (1979) 515.
8. M. B. Comisarow and J. D. Melka, Anal. Chem. 51 (1979) 2198.
9. a) E. B. Ledford, Jr., S. Ghaderi, R. L. White, R. B. Spender, P. S. Kulkarni, C. L. Wilkins and M. L. Gross, Anal. Chem. 52 (1980) 463.
   b) R. L. White, E. B. Ledford, Jr., S. Ghaderi, C. L. Wilkins and M. L. Gross, Anal. Chem. 52 (1980) 1525.
10. M. B. Comisarow, V. Grassi and G. Parisod, Chem. Phys. Lett. 57 (1978) 413.
11. G. Parisod and M. B. Comisarow, Chem. Phys. Lett. 62 (1979) 303.
12. G. Parisod and T. Gäumann, Chimia, 34 (1980)271.
13. a) E. B. Ledford, Jr., R. L. White, S. Ghaderi, C. L. Wilkins and M. L. Gross, Anal. Chem. 52 (1980) 2450.
    b) E. B. Ledford, Jr., S. Ghaderi, C. L. Wilkins and M. L. Gross, Adv. Mass Spectrom 8 (1980) 1707.

14. a) M. B. Comisarow, Adv. Mass Spectrom. 7 (1978) 1042.
b) M. B. Comisarow, Adv. Mass Spectrom. 8 (1980) 1698.

15. M. B. Comisarow, G. Parisod and V. Grassi, Proc. 26th Ann. Conf. Mass Spectrom. Allied Topics, paper WP12, p. 633, St. Louis, 1978.

16. M. B. Comisarow and A. G. Marshall, J. Chem. Phys. 62 (1975) 293.

17. R. J. Dugan, L. N. Morgenthaler, R. O. Daubach and J. R. Eyler, Rev. Sci. Instrum. 50 (1979) 291.

18. R. T. McIver, Jr., Rev. Sci. Instrum. 44 (1973) 1071.

19. Excitation oscillator: Exact Electronics Inc., Hillsboro OR 97123, Model 124B Multigenerator. Frequency Synthesizer: Rockland Systems Corp., West Nyack, NY 10994, Model 5100. Waveform Recorder: Biomation Div. of Gould, Inc., Cupertino, CA 95014, Model 805.

20. W. G. Jung, "IC Op-Amp Cookbook, H. W. Sans & Co., Inc., Indianapolis, IN, 1974, p. 335.

21. KIM-1 Microcomputer: MOS Technology, Valley Forge Corporate Center, 950 Rittenhouse Road, Norristown, PA 19401. Apple II Microcomputer: Apple Computer Co., 10260 Bandly Drive, Cupertino, CA 95014.

22. D. Lancester "TTL Cookbook" H. W. Sans & Co., Indianapolis, IN, 1974, p. 214.

23. "Harwell Subroutine Library" compiled by M. J. Hopper, Theoretical Physics Division, U.K.A.E.A. Research, Harwell, UK 1973.

24. D. C. Grande, T. J. Buckley and J. R. Eyler, presented at the 181st Natl. Am. Chem. Soc. Meeting, Atlanta, Georgia, 1981.

25. M. B. Comisarow, Int. J. Mass Spectrom. Ion Phys. 37 (19819 251.

FT ICR SPECTROMETRY WITH A SUPERCONDUCTING MAGNET

M. Allemann and Hp. Kellerhals
Spectrospin AG, Fällanden, Switzerland,
and K. P. Wanczek
Institute of Theoretical and Physical Chemistry,
University of Frankfurt, FRG

1. Introduction

Since ICR spectrometry [1] was introduced in 1965 [2] great progress has been made in instrumentation. Among the most important new methods were: 1. Introduction of pulsed spectrometry with a trapped - ion analyzer cell [3] and 2. Fourier transform ICR spectrometry [4]. The early instruments utilized a magnetic field scan to obtain a mass spectrum. This has several disadvantages: The scan is slow, trapping efficiency and sensitivity are changing with magnetic field strength. The strength of the magnetic field is not well defined and its homogeneity is not high. Therefore a frequency scan at constant magnetic field is preferable. Once operating at constant magnetic field the use of superconducting high field magnets yields substantial improvements:

1. Very high resolution,
2. Large mass range
3. Very long trapping times
4. Capability of simple mass scale calibration stable for extended periods of time utilizing the great field stability of the superconducting magnet.

2. Small-band and Broad-band Spectra

The spectrometer is equipped with a Bruker BZH-200 WB/A superconducting magnet with vertical room temperature bore of 8.9 cm diameter, operated at 4.7 T [5]. Relative field variation is less than 10^{-5} over the active volume with a set of superconducting shim coils activated. The $(2.9\ \text{cm})^3$ cubic ICR cell [6] is located in the center of the homogenous field region in a non-magnetic stainless steel vacuum system. The cell is constructed of non-magnetic materials [7]. The vacuum system is equipped with a Turbomolecular pump (Pfeiffer

TPU 270) and a DUO 012A roughing pump (Balzers, Liechtenstein). The turbomolecular pump has to be shielded against the magnetic field. Pressures better than 5 x 10^{-9} Torr are obtained. If lower pressures are needed, a titanium sublimation pump (Ti-Ball, Varian Ass., Palo Alto, Calif.) inserted in a liquid nitrogen trap can be activated. For pressure measurement an ionization gauge is used.

Fourier transformation of the transient signal is performed with a Bruker computer Aspect 2000. This computer also controls the operation of the spectrometer. A typical pulse sequence consists of quenching, electron beam, excitation and detection pulses.

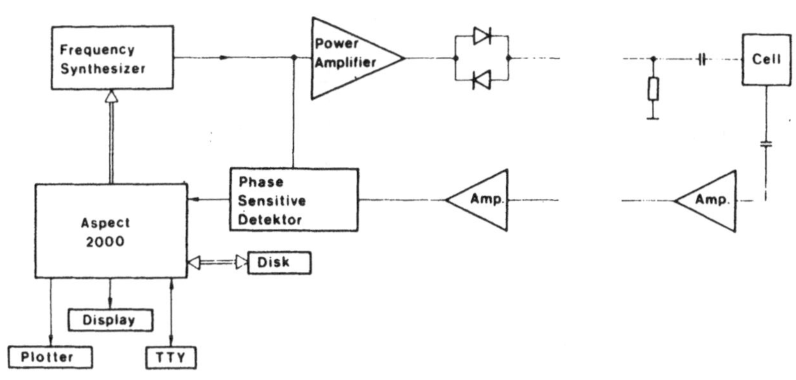

Fig. 1. Block diagram of the narrow-band electronics of the spectrometer (from [5]).

The spectrometer can be operated in a small band and a broad band mode. A block diagram of the small band electronics is shown in Figure 1. The ions are excited by a pulse of constant frequency of approximately 0.2 ms duration and 30 V amplitude (p-p), generated by a frequency synthesizer and an amplifier. The transient signal generated by the coherently gyrating ions is amplified, mixed in a phase sensitive detector and finally directly digitized and Fourier transformed by the computer.

The arrangement of the cell in the magnetic field and of transmitter and receiver is shown in Figure 2. It is the same for broad-band and narrow-band mode of the instrument.

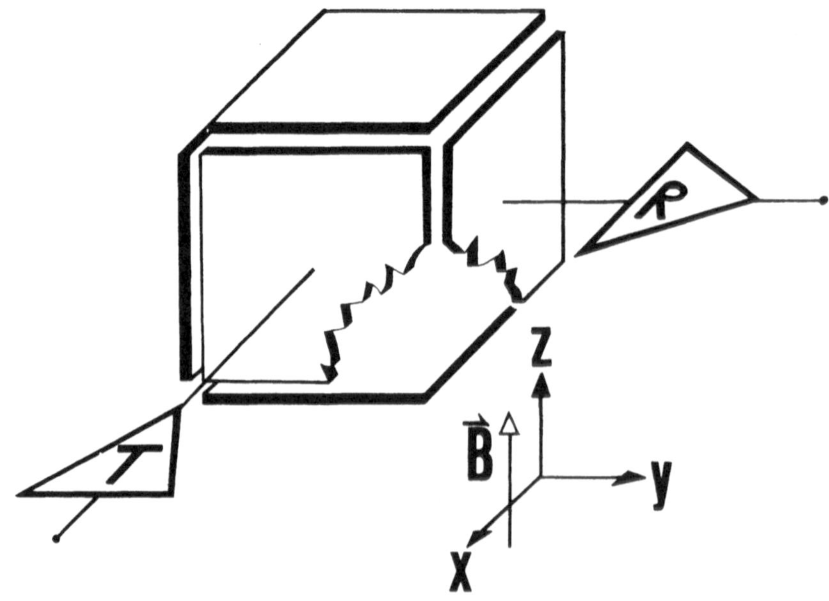

Fig. 2. Arrangement of the ICR cell, the transmitter (T), the receiver (R) and the coordinate system.

Figure 3 shows a very high resolution spectrum of the molecular ion of tetrachloroethane, m/z 166 at a sample pressure of 2×10^{-8} Torr. 20 sweeps were accumulated each for 6.8 sec. The resolution is $m/\Delta m = \omega/\Delta\omega = 1.5 \times 10^{6}$ obtained from the full width at half height $\Delta\omega = 0.29$ Hz of the peak.

In Figure 4 the first portion of the transient signal is shown after mixing and filtering.

Trapping of positive or negative ions is achived by changing the sign of trapping voltages in the ICR cell. It is therefore possible to record spectra of positive and negative ions almost simultaneously, only a few tenth of a second apart, inversing the polarity of the voltages by the computer and adding the stored spectra of positive and negative ions. This is shown for $^{35}Cl^{+}$ and $^{35}Cl^{-}$ ions in Figure 5. The distance between the two peaks is two

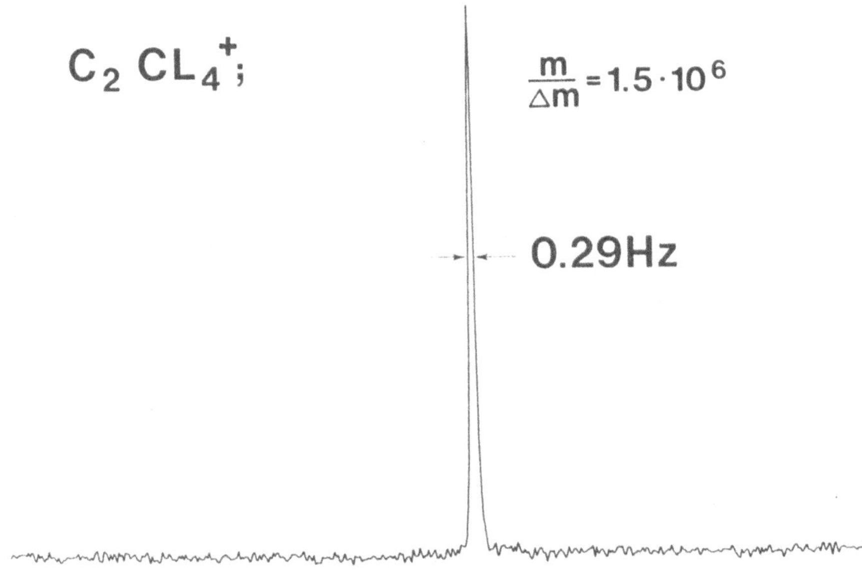

Fig. 3. Very high resolution spectrum of $C_2Cl_4^+$, m/z 166 (from [5]).

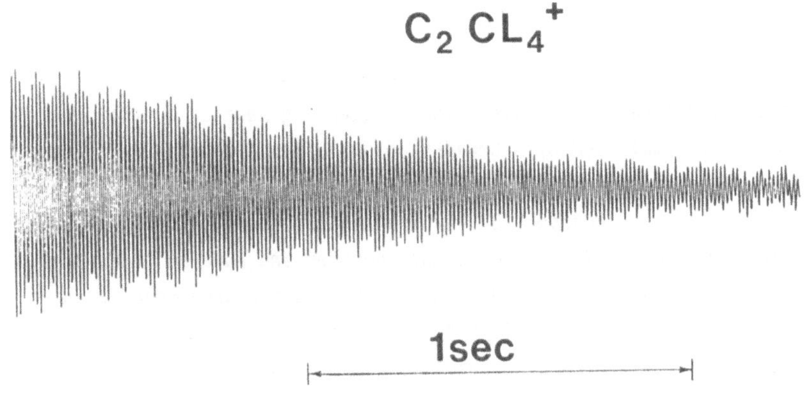

Fig. 4. First part of the transient of $C_2Cl_4^+$ signal shown in Figure 3 (from [5]).

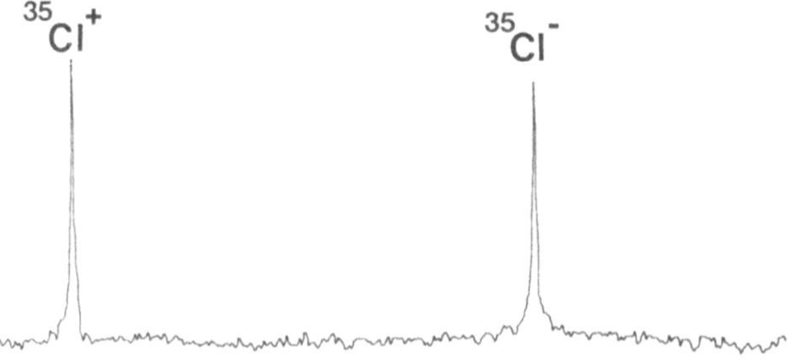

Fig. 5. Spectrum of $^{35}Cl^{+}$ and $^{35}Cl^{-}$ at $p = 10^{-8}$ Torr.

electron masses. The deviation from the exact value of the electron mass, which may be calculated from this spectrum, is approximately 5 %. This may be due mainly to two reasons: 1. Inversing the polarity of the cell potentials was done without calibration. 2. No electron ejection was applied, yielding different total charge densities in the cell at positive and negative ion modes.

In the broad band (c.f. Figure 6) mode a frequency sweep is used, corresponding to the mass range studied. The bandwidth of transmitter and receiver is 60 KHz to 5 MHz corresponding to a mass range from m/z 14 to 1,200 at the magnetic field strength utilized. The frequency generated by the frequency synthesizer is amplified by a broad-band amplifier. The signal is composed of the resonance frequencies of all the ions in the ICR cell. It is digitized by a fast transient recorder (Bruker Transi-Store BC 104). The complete transients are then transferred to the computer and Fourier transformed. In Figure 6 the broad-band frequency spectra of tetrachloroethene is shown in the frequency range 200 to 1600 kHz. It was one of the first broad-band spectra obtained with the described instrument. The resolution is at least unit resolution.

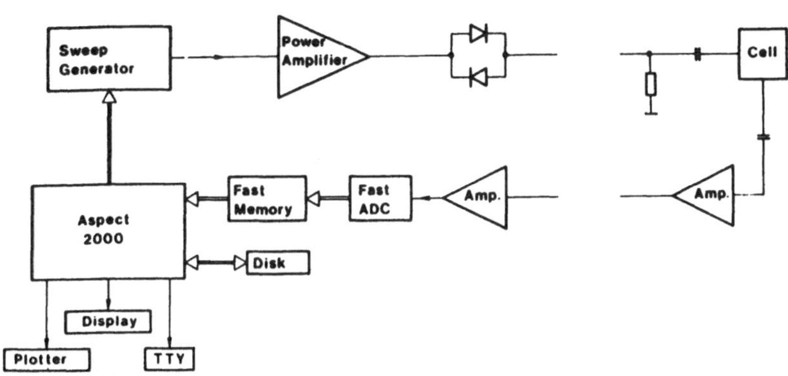

Fig. 6. Block diagram of the broad-band spectrometer electronics (from [5]).

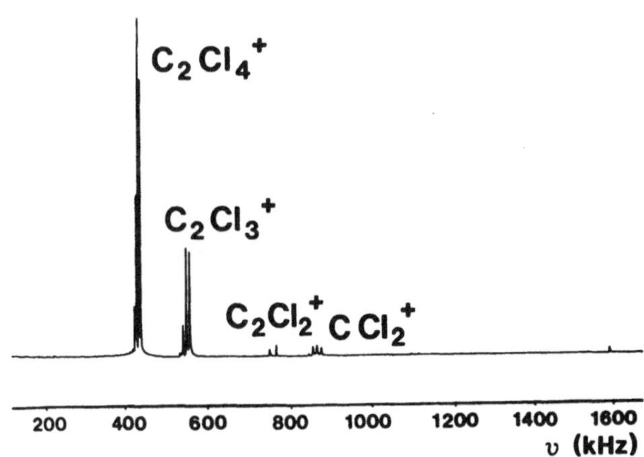

Fig. 7. Broad-band ICR spectrum of tetrachloroethylene (from[5]).

3. Sidebands [8]

With the experimental arrangement shown in Figure 2 the sidebands due to the coupling of cyclotron and drift motions of the ions can be detected. These sidebands are well-known. In Penning ion traps [9,10] side bands at frequencies $\omega_c \pm n\omega_d$ (n = 1,2) have been found, where ω_c is the cyclotron frequency and ω_d the drift frequency of the ions. In the ICR spectrometry diffuse sidebands have been observed in drift cells [11]. Narrow sidebands were found for the first time in ICR with cylindric cells [12,13]. An ICR cell is comparable to a Penning trap. In both cases homogeneous magnetic and inhomogeneous electric fields are used to trap the ions. The hyperbolically shaped electrodes of the Penning trap generate a quadrupolar electrical field if a dc voltage is applied. For the dc electric field in the ICR trapped-ion cell a quadrupole approximation has been used successfully to describe ion motion near the cell center [14]. N_2^+ spectra with sidebands obtained in this study

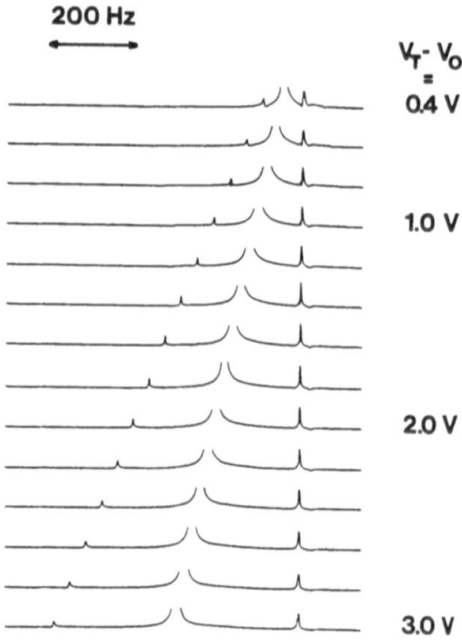

Fig. 8. N_2^+ ICR peak with sidebands at different voltages V_t-V_o (from [8]).

are shown in Figure 8. The frequency difference between main peak and sidebands depends linearly on $V_t - V_o$, where V_o is the trapping potential and V_t the potential applied to the other four cell plates. It is interesting, that the position of the sideband at higher frequency does not depend on $V_t - V_o$.

The intensity of the sidebands is normally less than 5 % of the intensity of the main peak. This intensity ratio depends on the position of the entrance hole of the electron beam in the cell [12] and several further parameters and is not fully understood.

The resolution in Figure 8 is $m/\Delta m = \omega/\Delta\omega = 1.5 \times 10^6$. The CO^+ peak has a frequency of approximately 1 kHz higher than the N_2^+ peak and is out of range. Part of the transient signal of the N_2^+ peak is shown in Figure 9. Modulation due to the side bands is very pronounced.

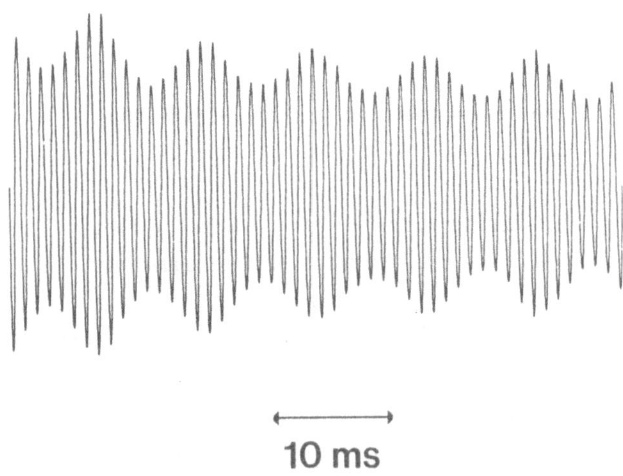

Fig. 9. Portion of transient of N_2^+ ions shown in Figure 7 modulated by the sidebands (from [8]).

4. Mass scale calibration

With the three-dimensional quadrupole approximation for the electric field [8,15] it can be shown that the high frequency side band has the frequency

$$\omega_r = \frac{zB}{m} \tag{1}$$

with good approximation for cubic cells. Its position corresponds to the undisturbed cyclotron frequency.

The three-dimensional quadrupole electric potential in the ICR cell is

$$V(x,y,z) = \frac{1}{2}(V_t + V_o) + (V_t - V_o)\,[-\alpha(x/a)^2 - \lambda(y/a)^2 + \beta(z/a)^2 - \gamma] \tag{2}$$

where α, λ, β and γ are constants depending on the geometry of the cell. For the cell employed in this study we obtain:

$$\alpha = 1.574; \quad \beta = 2.999; \quad \lambda = 1.425$$

(This cell is not exactly cubic) a is the distance between the two cell plates in x direction (cf. Figure 2).

For a cell with the same dimensions in x and y directions

$$\alpha = \lambda \tag{3}$$

After separating the trapping motion of the ions in z direction, one obtains solutions for ω_c and ω_d, the cyclotron and drift frequencies:

$$\omega_c = (\omega_{eff}^2 + \omega_t^2)^{1/2} \tag{4}$$

with

$$\omega_t^2 = \frac{2\beta z}{ma^2}(V_t - V_o) \tag{5}$$

where ω_{eff} is the effective cyclotron frequency, which is measured for the main peak and

$$\omega_d = \frac{2(\alpha\lambda)^{1/2}}{a^2 B}(V_t - V_o) \tag{6}$$

and for the cell utilized

$$\omega_d = K\,(V_t - V_o) = 88.9\,(V_t - V_o) \tag{7}$$

The two frequencies ω_{eff} and ω_d yield a receiver signal U(t)

$$U(t) = U_o \sin(\omega_{eff} t) \; [1 + \varepsilon \sin \omega_d t] \qquad (8)$$

which gives after Fourier transformation a carrier frequency ω_{eff} with sidebands $\omega_{eff} \pm \omega_d$.

For the sideband at higher frequency ω_r, already mentioned, one obtaines with the aid of eq. 5 and 6 and expanding the square root:

$$\begin{aligned} \omega_r &= \omega_c + \frac{V_t - V_o}{a^2 B} \; [2(\alpha\lambda)^{1/2} - \beta] \\ &= \omega_c + \omega_{cor} \end{aligned} \qquad (9)$$

ω_{cor} is much smaller than ω_c and one has in good approximation

$$\omega_r = \omega_c = \frac{zB}{m} \qquad (10)$$

for all ICR cells with square cross sections perpendical to the magnetic field.

To compare the theoretical prediction with experiment the constant K (cf. eq. 7) is determined from the distance of the sidebands from the main peak $2\pi\Delta\nu = \Delta\omega = \omega_r - \omega_{eff} = \omega_{eff} - \omega_\ell$ for the ions H_2O^+, N_2^+ and $C_2Cl_4^+$ to 90.4, 88.8 and 89.4 Hz/V, respectively. Agreement with the theoretical value is good. $\Delta\omega$ does not depend on the mass of the ion, as is predicted by the theory. This also rules out the possibility that the sidebands may be caused by the trapping motion , because ω_t is proportional to $m^{-1/2}$.

If one wants to calculate the m/z ratio of ions with eq. 10, the magnetic field strength has to be determined at the site of the ICR cell. Therefore an NMR probe was introduced in the vacuum chamber removing the top flange. Corrections, suggested by Pendleburg [16] was included. One obtaines B = 4.695957 T. This value may be slightly erroneous, because the conditions of measurement are not exactly same in ICR experiments. Furthermore the filament current may cause a small disturbance of the magnetic field. Therefore a difference between experimental and calculated m/q values of approximately 60 ppm in the lower mass range and 70 ppm in the higher mass range is obtained (cf. Table 1, first three columns).

Table 1. Literature values of m/z for several ions and (a) values calculated with eq. 10 and (b) with the parametrized equation from ICR spectra (from [8]).

Lit. value (amu)	Exp. mass (a) (amu)	Error (ppm)	Exp. mass (b) (amu)	Error (ppm)
18.01002	18.01113	-61.5	18.01003	-0.7
28.00560	28.00734	-61.9	28.00562	-0.6
46.96831	46.97127	-62.9	46.96834	-0.6
48.96535	48.96836	-61.4	48.96530	1.0
81.93716	81.94254	-65.6	81.93728	-1.4
83.93421	83.93959	-64.1	83.93419	0.2
93.93716	93.94319	-64.2	93.93711	0.6
95.93421	95.94023	-62.8	95.93401	2.1
97.93125	97.93799	-68.9	97.93162	-3.8
128.90602	128.91446	-65.5	128.90587	1.2
130.90306	130.91166	-65.7	130.90293	1.0
132.90010	132.90928	-69.1	132.90039	-2.2
163.87487	163.88558	-65.3	163.87436	3.1
165.87191	165.88300	-66.8	165.87162	1.8
167.86896	167.88084	-70.8	167.86930	-2.1
169.86600	169.87807	-71.0	169.86638	-2.2

Much better agreement between calculated and experimental masses is obtained when ω_{cor} and B are choosen as free parameters, determined from experiment by a least-squares fit. The agreement is now excellent, Table 1, columns 1, 4 and 5. The average error is 1.5 ppm. Once carried out, the mass scale calibration can be utilized for several days, due to the great stability of the superconducting magnet.

The sidebands are relatively small compared with the main peak. Therefore intensity problems may arise when the method described is employed for the mass determination of small peaks. The intensity ratio can be improved using Penning traps or cylindrical ICR cells [12,13]. It is not difficult to differentiate sidebands from small ICR peaks utilizing the fact that the high frequency sideband does not depend on V_t-V_o as do the ICR peaks.

5. References

1. For a recent review cf. K.-P. Wanczek, Dyn. Mass Spectrom. 6(1981)14.
2. Chem. Eng. News 43(1965)155.
3. R.T. McIver, Jr., Rev. Sci. Instrum. 41(1970)555.
4. M.B. Comisarow and A.G. Marshall, Chem. Phys. Lett. 25(1974)282.
5. M. Allemann, Hp. Kellerhals and K.-P. Wanczek, Chem. Lett. 75(1980)328.
6. M.B. Comisarow, Int. J. Mass. Spectrom. Ion Phys. 37(1981)251.
7. K.-P. Wanczek, Z. Naturforsch. 30a(1975)329.
8. M. Allemann, Hp. Kellerhals and K.-P. Wanczek, Chem. Phys. Lett. 84(1981)547.
9. H.G. Dehmelt and F.L. Walls, Phys. Rev. Lett. 21(1968)927.
10. G. Gräff, J. Kalinowsky and J. Traut, Z. Physik A297(1980)35.
11. M. Riggin, Intern. J. Mass Spectrom. Ion Phys. 22(1976)35.
12. S.H. Lee, Dissertation, Frankfurt (1980).
13. S.H. Lee, K.-P. Wanczek and H. Hartmann, Adv. Mass. Spectrom. 8(1980)1645.
14. T.E. Sharp, J.R. Eyler and E. Li., Int. J. Mass Spectrom. Ion Phys. 9(1972)421.
15. M. Allemann,Dissertation, University Bremen, 1982.
16. J.M. Pendleburg, Rev. Sci. Instrum. 50(1979)535.

ANALYTICAL FOURIER TRANSFORM MASS SPECTROMETRY

Michael L. Gross
Department of Chemistry
University of Nebraska
Lincoln, NE 68588

Charles L. Wilkins
Department of Chemistry
University of California
Riverside, CA 92521

This is a survey concerning the current state of Fourier Transform mass spectrometry (FT-MS) for chemical analysis. Four general analytical applications are considered: (1) High resolution measurements, (2) Precise and accurate mass measurement, (3) Chemical ionization spectra, and (4) Gas chromatography/mass spectrometry. Most work to date has been devoted to feasibility studies with the objective of assessing the technique's capabilities. Accordingly, this chapter should be considered a progress report, rather than a definitive description of long-established analytical FT-MS practises.

1. Introduction

Fourier Transform mass spectrometry was first demonstrated by Comisarow and Marshall [1] at the University of British Columbia. The underlying phenomenon upon which the technique is based is the motion of charged particles in a magnetic field An ion stored in a uniform magnetic field undergoes circular motion with a frequency which is dependent on the charge-to-mass ratio (z/m) of the particle (equation 1), where ω is

$$\omega = zB/m \tag{1}$$

the angular frequency and B is the strength of the magnetic field. For a singly charged ion of mass 100 daltons in a magnetic field of 1.0 tesla, the frequency will be about 153.000 cycles per second or 153 kHz.

The circular motion can be excited, and the ion trajectory increased to larger radii by applying an alternating electromagnetic field of the same frequency. Ions absorb power from the external alternating field. In early ion cyclotron resonance (ICR) spectrometers resonances were detected by measuring power absorption using a marginal oscillator for excitation of ions at a single mass. Mass spectra were obtained by a slow scan of the magnetic field using a constant excitation frequency bringing ions of one mass after another into resonance sequentially. A scan from m/z 30 to m/z 300 required 10 - 20 minutes.

In Fourier Transform mass spectrometry, all ions are put into resonance by a rapid scan of the excitation (about 1 millisecond is required) at a constant magnetic field strength. The excitation is nearly simultaneous for all mass ions rather than sequential as in slow scanning ICR spectrometry. Excitation is admitted to the cell on either the front of back plate (called the transmitter plate) of a cubic cell (see Figure 1).

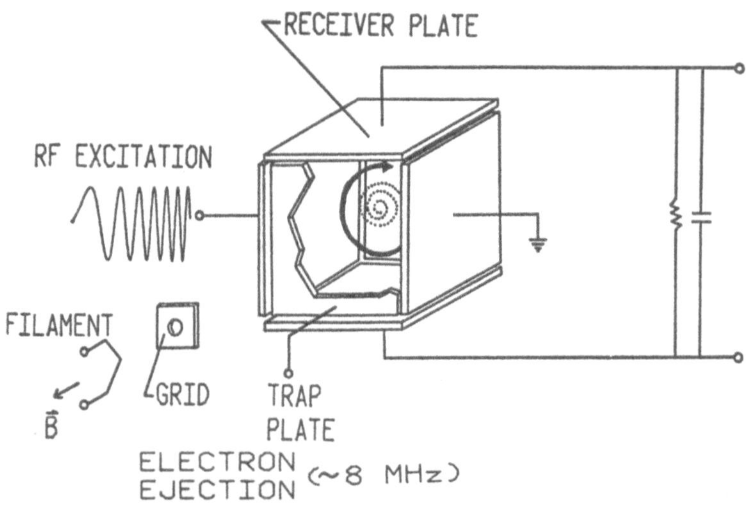

Fig. 1. Cubic analyzer cell. (Used with permission from E.B. Ledford, Jr., et al., Anal. Chem. 52 (1980) 485.

A common cell design uses a cubic 2.54 x 2.54 x 2.54 cm^3 stainless steel box located in a high vacuum chamber situated between the poles of an electromagnet. The magnetic field lines are perpendicular to the plane of the figure. As the ions absorb power during the excitation, their radius of trajectory is increased (depicted by the broken line in Figure 1) until the excitation is completed. At that time, they will remain in the larger radius orbital (depicted by the solid line) until their motion is interrupted by collisions with background gas. Ion loss from the cell is prevented by applying a small ($\sim$1V) voltage (of the same polarity as the ions under study) to the side plates.

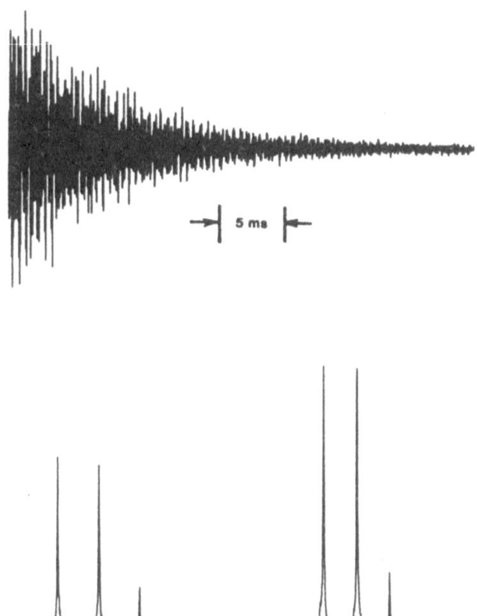

Fig. 2. Time domain transient and FT mass spectrum of 1,1,1,2-tetrachloroethene. (Used with permission from E.B. Ledford, Jr., et. al., Anal. Chem. 52 (1980) 485).

Detection of the ions is truly simultaneous and is based on the phenomenon of image currents. The cycloiding ions induce small alternating currents to flow in the circuit consisting of the receiver plates shunt resistor and capacitor (see Figure 1). The induced image currents give rise, across the resistor, to alternating voltages of the same frequencies as the excited ions. These signals are a composite of all ion frequencies and appear as a transient decay in the time domain as shown in Figure 2. The signal decays because the coherent motion of the ions is disrupted by collisions with background gas as mentioned earlier. Fourier transformation of the time domain signal gives a frequency domain output which is a mass spectrum (a display of all the frequencies and amplitudes contributing to the time domain ion decay). Such a spectrum is displayed on the bottom of Figure 2.

Of course, for ions to be observed they must first be formed. In the electron impact mode, an electron beam is pulsed on for about 5-10 milliseconds. Electrons enter the cell and ionize a small fraction (say 0-1%) of the background gas (N) according to equation 2.

$$N + e = N^{+} + 2e \tag{2}$$

Other methods of sample ionization will be discussed later under the topic of Chemical Ionization.

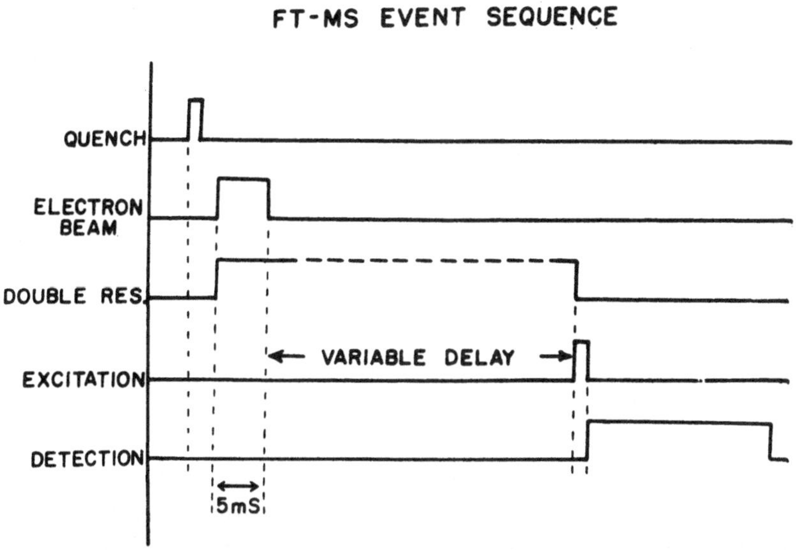

Fig. 3. Typical sequence of events for the FT-MS experiment. (Used with permission from Ghaderi, et al., Anal. Chem. 53 (1981) 428.

The entire sequence of events leading to the acquisition of one mass spectrum is depicted in Figure 3. First, a general pulse is applied to the cell by applying a high potential (~15V) to one of the trapping plates. This drives out any ions remaining from an earlier experiment. The electron beam is then turned on for 5-10 milliseconds to prepare a new set of sample ions. Often it is desireable to remove certain ions from the cell in order to determine their fate by comparison with an experiment in which they are present. This is done by admitting a high amplitude excitation signal at the frequency corresponding to the desired mass during the double resonance pulse. After a suitable delay, ions are excited by turning on the excitation "chirp" (rapidly scanned excitation frequency) for about 1 millisecond. This is followed by turning on the detection circuitry.

If no variable delay is used (ions are excited immediately after their formation), the entire sequence of events takes about 30 milliseconds, resulting in a full, low resolution, electron impact mass spectrum. This "fast scanning" can permit the analyst to aquire as many as 33 mass spectra per second.

By using a variable delay, the analyst can observe reactions of ions with background neutral molecules (ion-molecule reactions) in a time-resolved way.

High resolution mass spectra are obtained in a narrow band scan by exciting a single ion and observing its transient decay for a long period (100 milliseconds to 5 seconds). (The relationship between transient decay times and mass resolution will be discussed in the following section.) This experiment is done in a mixer mode to prevent overloading the computer memory during a long term experiment. In this mode, the ion decay signal is multiplied by a reference of nearly the same frequency. The difference frequency is extracted and digitized at a lower rate than the original frequency.

2. High Mass Resolution

Using an instrument constructed in our laboratory, we have obtained resolution up to 760,000 for the molecular ion of benzene at m/z 78 (see Figure 4a) [2]. This result was obtained using a benzene pressure of ca. 5×10^{-9} torr at a magnetic field strength of 1.2T. In a related experiment, the m/z 84 doublet from a mixture of benzene-d_6 and cyclohexane (see Figure 4b) at a total pressure of 2×10^{-8} torr was resolved. Other relevent experimental parameters for the latter measurement are: instantaneous source emission current, 300 nA; beam duration, 5 msec; electron energy, 70 eV; rf level, 24V p-p applied to the cell; frequency excitation window, 219.920 to 220.060 kHz, rf Chirp rate, 2 Hz/microsecond; trap voltage, 0.25V; quench pulse applied between successive scans, 100 signal averaged (time domain) scans; 1 msec delay between electron beam turn off and excitation scan start. Conditions, with the exception of the frequency window chosen and pressure, were the same for the benzene measurement.

This result stimulated our interest in the relationship between signal-to-noise ratio (S/N) and resolution in the FT-MS experiment. We have shown that it is possible to simultaneously increase mass resolution and S/N, a feature unique to the FT instruments compared to conventional mass spectrometers.

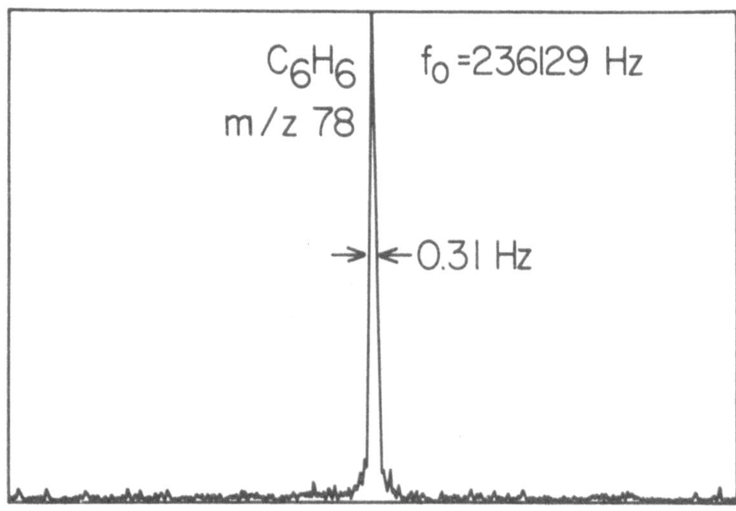

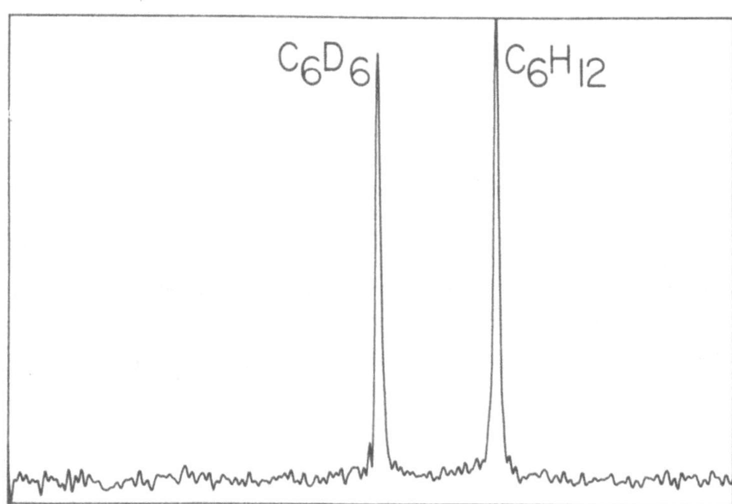

Fig. 4. (a) Ultra-high-resolution (not magnitude mode) spectrum of benzene at m/z 78, obtained from a signal with τ of 1.0 s at a magnet field strength of 1.2 T. Resolution is 760,000 (FWHH definition). (b) Resolved benzene-d_6/cyclohexane m/z 84 doublet, absorption mode, at 1.2 T. Resolution is 220,000 (FWHH definition). (Used with permission from R.L. White, et al. Anal. Chem. 52 (1980) 1525.

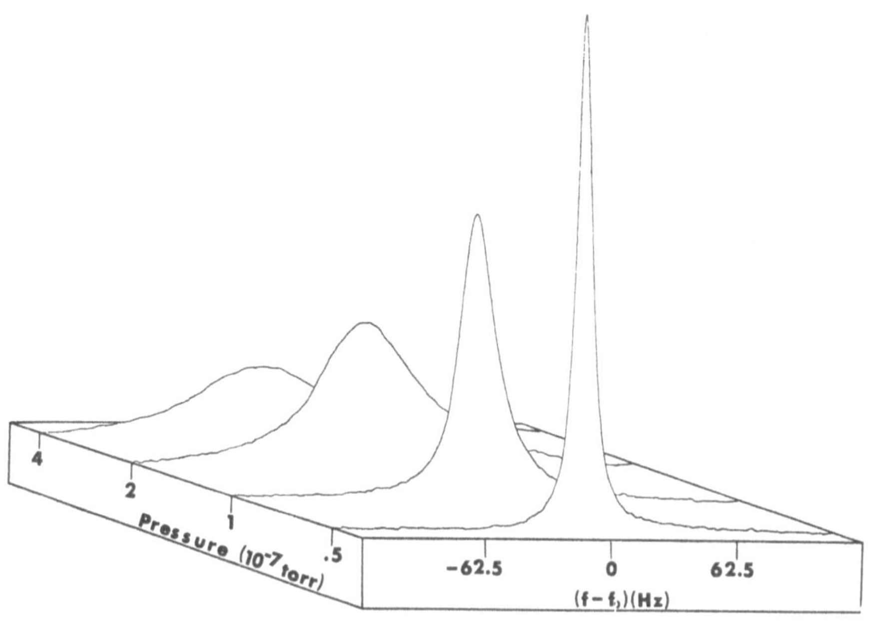

Fig. 5. Simultaneous increase in resolution and signal-to-noise ratio with decreasing pressure for benzene at m/z 78. As benzene pressure was varied, electron beam emission current was adjusted to keep the number of ions in the cell constant for each spectrum. Spectra are absorption mode, rather than magnitude mode displays. (Used with permission from R.L. White et al., Anal. Chem. 52 (1980) 1525.

We may examine the relationship between signal-to-noise and resolution by deriving an expression for the latter. The frequency resolution (full-width-at-half-height definition) assuming the entire transient signal is observed, is found to be

$$m/\Delta m = \pi f_0 \tau = zB\tau/2m \tag{3}$$

in which B is magnetic field strength and z/m is charge to mass ratio of ions. Therefore, the resolution increases linearly with magnetic field strength and varies inversely with mass. Furthermore, the resolution increases in direct proportion to τ, the time constant for decay. As the value of this parameter is increased, say by lowering the pressure in the analyzer cell, signal-to-noise ratio and resolution increase together if other parameters are held constant. Alternatively stated, the area under the peak remains constant as τ is changed and other parameters are unchanged. This phenomenon is illustrated in Figure 5.

The essential trade-off is resolution vs. mass range rather than resolution vs. signal-to-noise, as in other forms of mass spectrometry. The mass range trade-off is inherent in mixer-mode operation of the FT-MS.

We believe the unique relationship between resolution and signal-to-noise in FT-MS will prove important in analytical applications of the method, particularly in gas chromatographic FT-MS experiments.

3. Precise and Accurate Mass Measurement

It is well known that the capability to measure exact masses of chemical substances is extremely important for both structure proofs of new compounds and for highly specific trace analysis. These measurements, which can lead to unambiguous assignments of elemental compositions, have been made traditionally using complicated double-focusing mass spectrometers. One of the principal limitations of the high resolution mode is that it is extemely difficult to acquire high resolution data across a wide mass range in short times (e.g. the several second duration of a capillary column gas chromatographic peak) in either a scanning or a peak switch mode. Nonetheless, rapid peak switching to obtain high resolution measurements of ions widely separated on the mass scale would be invaluable in conjuction with high resolution gas chromatographic separation of mixture components.

As a first step, we have made a preliminary investigation of the mass measurement capabilities of Fourier Transform mass spectrometry [3]. The result of our study is a practival method for rapidly obtaining exact masses of sufficient accuracy and precision for determination of elemental compositions by FT-MS.

Consideration of equation 1 reveals that a very simple relationship exists between mass and frequency in FT-MS. Unfortunately, that relationship is not sufficiently accurate because it does not take into account the fact that the trapping fields perturb the ion frequencies. We need to measure frequency accurately and then relate frequency to mass in order to make accurate mass measurements. We have derived the necessary relationship for a cubic cell, and the results have been published elsewhere [3]. It is sufficient to state here that the calibration equation is of a quadratic form (equation 4) in which m_1 is the mass of an ion,

$$m_1^2 f_1^2 = a_o - b_o m_1 + c_o m_1^2 \qquad (4)$$

f_1 is its cyclotron frequency, and a_o, b_o, and c_o are calibration constants.

The derived mass calibration procedure was tested against the six major ions in the spectrum of 1,1,1,2-tetrachloroethane, which provides six intense calibration peaks in a useful region of the mass spectrum (see Figure 2). Mass accuracy was evaluated over various mass ranges using the quadratic mass calibration formula.

We find that the parabolic calibration procedure of equation 4 produces mass measurements of sufficient accuracy and precision to permit elemental composition assignments over a mass range of at least 18 amu, centered at m/z 126. Frequency measurement precision for the six major ions in the spectrum of 1,1,1,2-tetrachloroethane, (standard deviation over ten successive Fourier Transform mass spectra) was typically 0.3 Hz and in no case exceeded 1.5 Hz. The quantity $(m_1 f_1)^2$ was fitted to the parabolic form in equation 4 and the masses then back-calculated (see Table 1 for results) using the empirically determined values of a_0, b_0, and c_0. Over an 18 amu range, the parabolic calibration procedure gives mass errors of less than 1 millimass unit.

In practice, we have found the mass accuracy shown in Table 1 is routinely obtained over a broad range of operating conditions including various values of electron beam emission currents.

We have also calibrated successfully over a wider mass range extending from m/z 78 to nearly m/z 240. Mass measurement errors were less than 6 ppm for the masses assigned.

Table 1. Typical Results of a Parabolic Mass Calibration over the Major Ions in the Mass Spectrum of 1,1,1,2-Tetrachloroethane

		error[b]	
m_i	calculated masses[a]	amu	ppm
116.9066	116.9064 ± 0.0003	-0.0002	(1.7)
118.9036	118.9041 ± 0.0003	+0.0005	(4.2)
120.9007	120.8999 ± 0.0008	-0.0008	(6.6)
140.9222	130.9226 ± 0.0004	+0.0004	(3.1)
142.9192	132.9188 ± 0.0004	-0.0004	(3.0)
134.9163	134.9164 ± 0.0006	+0.0001	(0.7)

[a]Average of ten determinations made on ten successive mass spectra.
[b]Tolerances shown represent single standard deviations over ten trials.

It is well known that high mass resolution is not necessarily needed for exact measurements, so long as mass spectral peaks are not composed of multiplets of isobaric ion species. The measurements discussed here were made at mass resolution of ca. 1000 (full width at half height definition). For analytical purposes, it should be possible to perform exact mass measurements on wideband, low resolution ICR mass spectra, then switch to high resolution multiple ion monitoring mode to resolve potential multiplet peaks as a check. Since this operation would involve only frequency switching and pulse width adjustments, it could be performed rapidly and automatically under computer control. We reemphasize that FT-MS is capable of much higher mass resolution than 1000. The use of higher resolving power is expected to further increase the precision, and possibly the accuracy, of mass measurements by FT-MS.

4. Chemical Ionization

Sample ionization can also be accomplished by ion-molecule reactions. The most common mode is proton transfer. The neutral sample molecules are protonated by reagent ion in a gas-phase acid-base reaction. The advantage of this method of ionization is that molecules which give weak or no molecular ions by electron impact can be protonated to give intense $M + H^+$ ions where M is the neutral molecule. The second advantage of chemical ionization is that fragmentation of the $M + H^+$ ion can be readily controlled by choosing a "gentle" protonating reagent.

Chemical ionization is now an established method in conventional mass spectrometry and is available on most commercial instruments. It is well known that the sources of these spectrometers must be operated at a relatively high reagent gas pressure (0.1-1.0 torr) and sample partial pressures (10^{-5}-10^{-4} torr) for significant conversion of neutrals into ions. The high pressure is needed to compensate for the short residence time of the ions in the source ($\sim 10^{-4}$ seconds).

Chemical ionization can be conducted using FT-MS at low reagent gas pressure (10^{-6} torr) and even lower sample pressure ($\sim 10^{-8}$ torr) because the ions can be trapped for long periods of time (up to 1 second and more). This lower pressure CI should give, in principle, higher sensitivity (lower detection limits) than electron impact because sufficient time can be provided during the variable delay to permit quantitative transfer of ionization from reagent (ionized at higher pressure) to sample (present at lower pressure). This advantage has not been achieved in CI using conventional mass spectrometers.

We have recently demonstrated the capabilities of low pressure CI for a wide range of compounds and reagent gases [4]. Furthermore, we have described some alternative ways of obtaining CI spectra in the FT-MS mode. Some highlights of that work will be presented here.

The first approach for obtaining spectra in the CI mode is to introduce an appropriate delay time between the ionization and excitation, thus allowing ions and neutral molecules to react before mass analysis. The reagent gas is introduced to approximately 10^{-6} torr and the sample can be evaporated from a direct probe. Selection of the delay time is determined by the time required to establish a high concentration of reagent ions. For example, in methane at 10^{-6} torr, the concentration of the reagent ions CH_5^+ and $C_2H_5^+$ is constant after about 100 milliseconds. Formation of H_3O^+ in water at 2×10^{-6} torr is complete within 10 milliseconds. These reagent ions are now available for proton transfer. A comparison of the EI and CI spectra for the abused drug "Angel Dust" is shown in Figure 6. The CI spectrum originates from protonation of the sample by NH_4^+ followed by a minor amount of decomposition of $M + H^+$.

In absence of reagent gas, CI can take place by reaction of the fragment ions of the sample and the sample neutral molecules. We term this method "self CI". This procedure is well suited for molecules containing large hydrocarbon moieties such as fatty acids, fatty acid esters, pheromones, and other long chain aliphatics containing one or more functional groups. The EI spectra of these molecules contain many closed-shell ions which are potential proton transfer reagents. After a suitable delay, the ion current borne by these fragments is transferred to $M + H^+$ permitting higher sensitivity detection of the sample using a single ion of high structural significance. An example is given by the self CI spectrum of methyl stearate at 1×10^{-7} torr (Figure 7).

A third method is "Quench-Off CI". In this mode, we forego the quench pulse (see Figure 3) and allow ions to accumulate in the cell. The ion concentration grows until a steady state is reached which represents a balance of ion production by electron impact and ions loss by drift from the cell.

The principle advantage of "Quench-Off CI" is the capability to produce spectra at a more rapid rate than the two procedures described above. In some cases, the extent of protonation produced in a quench-off CI experiment using a delay of approximately 10 milliseconds is equivalent to that produced in a quench-on mode using a delay of 200 milliseconds. Thus, data acquistion can be 20 times more rapid in the former mode.

We have also demonstrated that CI spectra can be obtained for negative ions, for radical cation reagent ions, and for deuterium exchange to count exchangeable hydrocarbons [4]. The unique double resonance capability of FT-MS permits us to study the mechanisms of formation of the various ions in addition to applying the chemistry for analysis.

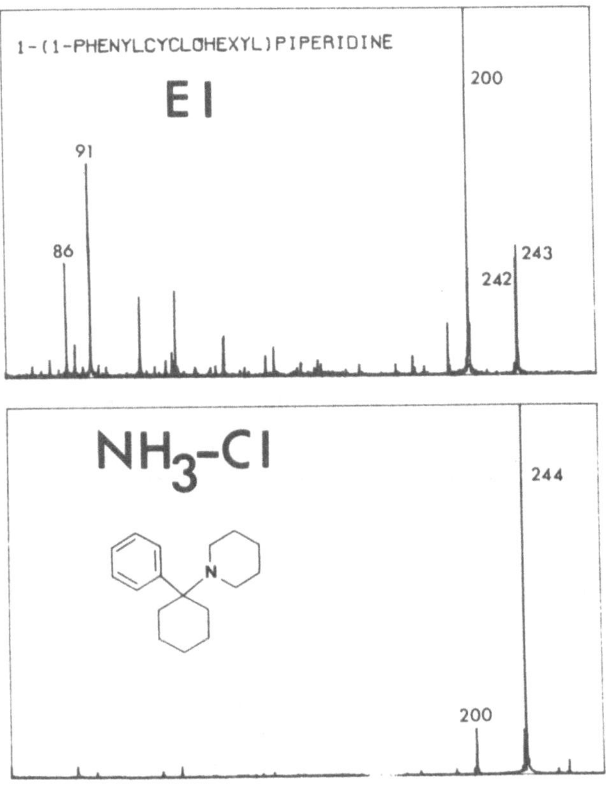

Fig. 6. EI and ammonia CI mass spectra of 1-(1-phenylcyclohexyl)-piperidine (PCP or "Angel Dust") over identical bandwidths. (Used with permission from S. Ghaderi, et al., Anal. Chem. 53 (1981) 428.

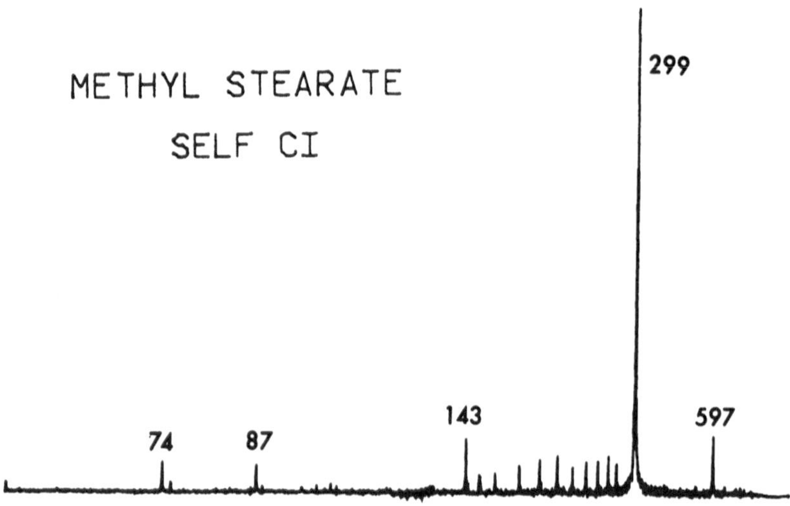

Fig. 7. Self-CI spectrum of methyl stearate at 1 x 10^{-7} torr. (Used with permission from S. Ghaderi et al.m Anal. Chem. 53 (1981) 428.

5. Gas Chromatography/Fourier-Transform Mass Spectrometry

An important potential analytical application requires interfacing a gas chromatograph to an FT-MS for the purpose of obtaining high resolution mass scans of mixture components emerging from a capillary column. This requires: (a) the interface of a GC to an FT-MS, (b) fast scanning to obtain full spectra of components from a capillary column, (c) sufficient pumping speed at the cell to permit high resolution measurements over narrow mass ranges, (d) sufficient sensitivity for high resolution single ion monitoring for analysis of single components, (e) developments of peak switching capability to permit multiple ion monitoring to increase the certainty of identifying specific chemical substances, (f) evaluation of the capability to make precise and accurate mass measurements in a multiple ion monitoring mode, and (g) development of the capacity to obtain full spectra at medium resolving power (6,000-10,000) and to assign accurate masses to the peaks in the spectra of components emerging from a capillary column.

We have recently developed an interface for GC and the FT-MS instruments and have demonstrated that the fast scanning capabilities of the FT-MS are sufficient to obtain full low resolution mass spectra [5].

The FT-MS instrument used in that study was specially designed with a high gas conductance vacuum chamber. This allows the mass analyzer to operate at low pressure, which is necessary for high mass resolution. The forward end of the vacuum chamber, which contains a cubic trapped ion analyzer cell (plate separation 0.0254 m), has a nominally rectangular shape. The width (internal 0.0667 m) is sufficient to permit insertion into the 0.0762 m air gap of a Varian V-7300 0.305 m electromagnet. That volume expands into a 0.152 m i.d. stainless steel tube which is outfitted with a 0.152 m gate valve. The tube terminates at 500ℓ/second turbomolecular pump which is backed by a fast mechanical pump.

The pumping speed at the cell is 360 ℓ/s which was determined by calculation and verified experimentally by introducing a known flow rate of helium and observing the pressure. A schematic of the vacuum can and separator is given in Figure 8.

The electromagnet is equipped with one pair of 0.305 m cylindrical ring-shim pole caps with iron-cobalt ring for high homogeneity. Its maximum field strength is 1.37T at the center of the air gap, and it was operated at 1.2T for these experiments.

The vacuum system is interfaced to a Perkin-Elmer Sigma II gas chromatograph containing an OV 101 15.2 m x 0.5 mm SCOT capillary column at 150°C (isothermal). The interface consisted of either a direct coupled transfer line (0.5 mm i.d.,

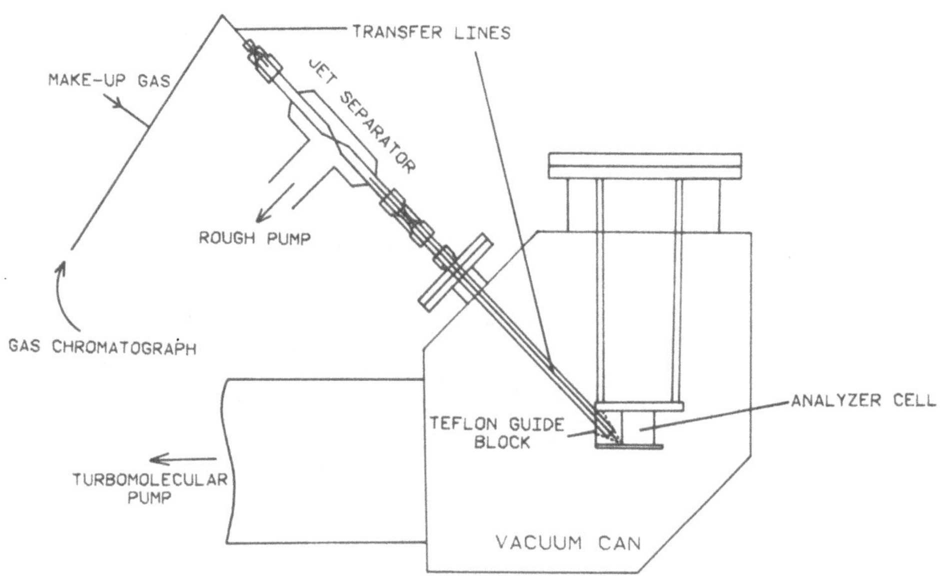

Fig. 8. Schematic diagramm of the capillary gas chromatograph-FTMS interface.

0.6mm o.d. gas-lined stainless steel obtained from Scientific Glass and Engineering) or a standard glass jet separator coupled to the transfer line. Both were held at 200°C. When the separator was used, 20 mℓ/minute of make-up helium gas was introduced prior to the jet separator.

Full low resolution spectra were obtained at a rate of 0.5 seconds per spectrum for a 1:1:1 mixture of benzene, toluene, and xylene (see Figure 9). This serves as a demonstration of the fast scanning ability of the combined system.

The separator design and pumping speed are sufficient to permit high resolution single ion monitoring over narrow mass bands. Using the jet separator, we have been able to observe the m/z 78 (molecular ion of benzene) with a resolution of 23,000 (full width at half height definition). Two components at m/z 106: xylene (m/z 106.0782) and benzaldehyde (m/z 106.0419) were separated easily with a mass resolution of 9,000. This experiment was done as a demonstration of useful resolution by admitting benzaldehyde via a liquid inlet and the xylene via the capillary column.

Since the spectral bandwidth selection in FT-MS involves setting the frequency sweep of an rf sine wave excitation, it should be possible to alter both the size and position of a mass observation window by changing a few computer words.

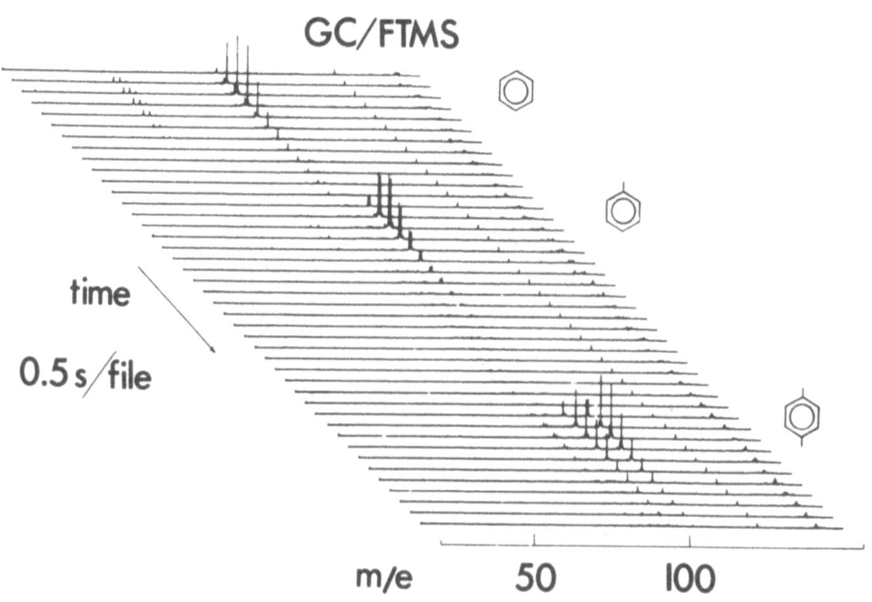

Fig. 9. GC-FT/MS spectra of a mixture of equal amounts of benzene, toluene and xylene. Twenty-five signal averaged scans were taken per file. (Used with permission from E.B. Ledford, Jr., et al., Anal. Chem. 52 (1980) 2450.

This could be easily done within microseconds under computer control. Switching of this type would allow for acquisition of high-resolution narrow mass range data for two or more widely separated masses during the elution of a single chromatographic peak. This has been demonstrated for a mixture of benzene and bromobenzene. Both molecular ions (m/z 78 and m/z 156) were monitored sequentially at a mass resolution of 20,000 (at m/z 78) at a rate of 3.5 Hertz.

We believe a similar experiment could be done in which full low resolution spectra were acquired in one half cycle and a single ion were monitored at high resolution in the other half cycle. This remains to be demonstrated.

We have made some preliminary measurements of sensitivity by injecting decreasing quantities of material onto the GC and using the FT-MS in a single ion monitoring mode. In this way, we have established a detection limit of about 1 nanogram of material (see Figure 10). We anticipate that better detection limits will be achieved by making improvements in the design of the detection circuitry and of the interface of the transfer line to the FT-MS cell. It also may be possible to improve the detection limit by employing new cell designs in lieu of the present cubic one.

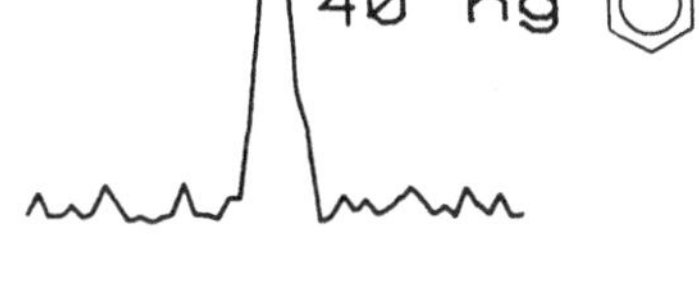

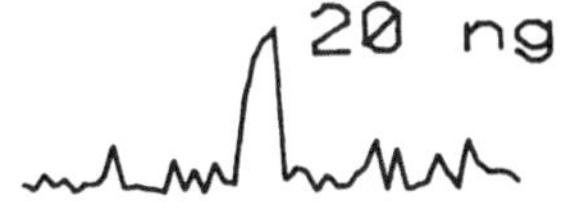

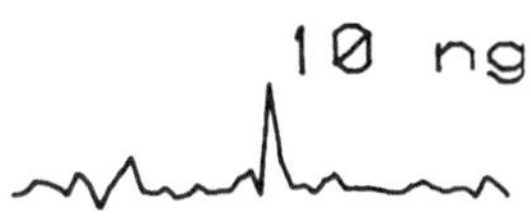

Fig. 10. GC/FTMS sensitivity as estimated for benzene molecular ion at m/z 78.

In summary, it has been verified that high resolution measurements can be made routinely using FT-MS, and that precise and accurate mass measurements are possible using this new form of mass spectrometry. Furthermore, a wide variety of chemical ionization options can be used in FT-MS. Finally, the important goal of interfacing an FT-MS with a GC has been achived, and the capabilities for multiple ion monitoring have been demonstrated. It is expected that future work will directed at improving on these capabilities and will result in application of FT-MS to important problems in chemical analysis.

6. Acknowledgement

This work was supported by the National Science Foundation (Grants CHE-77-03964 and CHE-80-18245), the United States Environmental Protection Agency (Grant R807251010), and the Gulf Research Foundation.

7. References

1. M. B. Comisarow and A. G. Marshall, J. Chem. Phys. 64 (1976) 110.
2. R. L. White, E. B. Ledford, Jr., S. Ghaderi, C. L. Wilkins and M. L. Gross, Anal. Chem. 52 (1980) 1525.
3. E. B. Ledford, Jr., S. Ghaderi, R. L. White, R. B. Spencer, P. S. Kulkarni, C. L. Wilkins and M. L. Gross, Anal. Chem. 52 (1980) 463.
4. S. Ghaderi, P. S. Kukarni, E. B. Ledford, Jr., C. L. Wilkins and M. L. Gross, Anal. Chem. 52 (1980) 428
5. E. B. Ledford, Jr., R. L. White, S. Ghaderi, C. L. Wilkins and M. L. Gross, Anal. Chem. 52 (1980) 2450.

THERMOCHEMICAL INFORMATION FROM ION-MOLECULE RATE CONSTANTS

Sharon G. Lias
National Bureau of Standards
Washington, D. C. 20234

1. Introduction

In the study of ion-molecule reactions, there has long been practice of inferring exothermicity from the fact that a reaction is observed to occur, or, on occasion, endothermicity from the non-occurrence of a particular reaction. This "bracketing" technique has, for example, been used to establish relative proton affinities by ascertaining whether or not the reaction:

$$MH^+ + B \rightleftarrows BH^+ + M \tag{1}$$

occurs preferentially from left to right or from the right to the left. Since the advent of the measurement of equilibrium constants for a bimolecular ion-molecule reactions in 1973 [1], most quantitative thermochemical information about ion-molecule reactions is derived from such measurements. Nevertheless, there are situations in which equilibrium constant determinations can not be made; this happens, for example, when one of the neutral bases is a free radical, or when a fast competing process precludes the establishement of an equilibrium.

This chapter presents a brief description of recent work, primarily in the author's laboratory, in which rate constant measurements are used in these situations in which equilibrium constants can not be measured, to derive quantitative or semi-quantitative information about the thermochemistry of ion-molecule reactions. One result of such measurements has been the establishment of the position in the gas phase proton affinity scale of serveral possible primary standards for the absolute value of that scale. A summary of values for the proton affinity scale based on recent work is included.

2. The Measurements of Rate Constants of Corresponding Exothermic and Endothermic Reactions

In the event that the occurrence of some competing process interferes with the establishment of the equilibrium:

$$A^+ + B \underset{k_r}{\overset{k_f}{\rightleftarrows}} C^+ + D \tag{2}$$

it may be possible to derive a value for the equilibrium constant by individually measuring k_f and k_r, if possible under conditions such that the competing process does not occur. For example, the measurement of equilibrium constant for the reaction:

$$sec\text{-}C_3H_7^+ + M \rightleftarrows MH^+ + CH_3CH{=}CH_2 \qquad (3)$$

is difficult because sec-$C_3H_7^+$ reacts with propylene. However, this problem can be overcome by measuring k_f in systems in which sec-$C_3H_7^+$ is generated from some process other than reaction of MH^+ with $CH_3CH{=}CH_2$. In a recent study [2] of reactions 3 , sec-$C_3H_7^+$ ions were generated in propane in the presence of various bases, M, for the determination of k_f. Corresponding values of k_r were measured in mixtures of M with $CH_3CH{=}CH_2$, but the back reaction, as well as the reaction of sec-$C_3H_7^+$ with propylene, was prevented from occurring by the continuous ejection of sec-$C_3H_7^+$ from the system. The results obtained for a number of bases, M, are given in Table 1. It is seen that for many bases, the probability of a reactive (sec-$C_3H_7^+$ + M) collision is approximately unity. However, as one proceeds down the proton affinity scale in choice of M, the efficiency of this reaction begins to drop off as it approaches the thermoneutral point. As this reaction approaches thermo-neutrality, the reverse reaction begins to occur with a measurable reaction rate constant, which increases with increasing exothermicity of this reaction.

Estimates of the equilibrium constant for proton transfer equilibria involving CH_3OH, CH_3NO_2, AsH_3 and HCOOH are obtained from the ratios of the rate constants, $k(3)_f$ and $k(3)_r$. A comparison of the relative values of $\Delta G°$ cited in the table with the values derived from the ratios of the measured rate constants shows that the internal consistency is good, ± 0.1 kcal/mol. From the estimates of $\Delta G°(3)$ listed in the Table, it is possible to obtain estimates of $\Delta H°$ for reaction 3 , and thereby to establish the position of propylene in the proton affinity scale. Recent photoionization [3] and photoion-photoelectron coincidence [4,5] studies have given values for the heat of formation [6] of sec-$C_3H_7^+$ at 300 K as 190.8±1.0 kcal/mol [4], 191.1±0.7 kcal/mol [5], or 191.8±0.4 kcal/mol [3]. Taking a value of 191.1 kcal/mol for this heat of formation, the absolute proton affinity of propylene is 179.4 kacl/mol. Table 2 shows the position of propylene relativ to various selected bases in the proton affinity scale, and the absolute values of the protom affinities of these bases using the proton affinity of propylene as a primary standard. This Table is discussed more fully below.

3. Estimation of Thermochemical Information from Reaction Efficiencies as a Function of ΔH° or ΔG° of Reaction

In Table 1, we showed a number of rate constants for proton transfer reactions from sec-$C_3H_7^+$ to polar bases in which the observed reaction efficiencies, k/Z (where Z is the corresponding ion-molecule collision rate constant [7]) diminish as the proton transfer reaction approaches thermoneutrality. This is a widely observed trend for ion-molecule rate constants; examples of such trends have been

Table 1. Rate Constants at 350 K and Thermodynamic Parameters for the Reaction:

$$sec\text{-}C_3H_7^+ + M \underset{B}{\overset{A}{\rightleftarrows}} MH^+ + CH_3CH{=}CH_2$$

M	Relative ΔG° *kcal/mol*	$k_A \times 10^{10}$ cm^3/molec-s	k_A/Z_A^a	$k_B \times 10^{10}$ cm^3/molec-s	k_B/Z_B^a	ΔG°= $-RT\ln k_A/k_B$ *kcal/mol*
$(i\text{-}C_3H_7)_2O$	-23.5	17.7±1.3	1.0			
$(C_2H_5)_2CO$	-18.8	21.9±1.0	0.94			
$(C_2H_5)_2O$	-17.8	14.3±0.8	0.90			
$C_2H_5COOCH_3$	-17.5	16.7±0.4	0.95			
$c\text{-}C_4H_8O$	-17.0	16.7±0.6	0.95			
CH_3COCH_3	-14.7	19.7±0.6	0.83			
$i\text{-}C_3H_7CHO$	-10.8	23.4±0.8	1.0			
C_2H_5CHO	-7.9	20.7±1.0	0.91			
$HCOOCH_3$	-7.5	15.8±1.0	0.94			
PH_3	-6.2	-5.2±0.5	0.43			
CH_3SH	-5.8	11.4±0.9	0.73			
CH_3CHO	-4.9	16.0±0.4	0.72			
CH_3OH	-1.0	7.1±0.4	0.42	1.6±0.5	0.11	-1.0
$CH_3CH{=}CH_2$	0.0					
CH_3NO_2	+0.2	3.4±0.3	0.14	4.8±0.3	0.41	+0.24
AsH_3	+1.0	0.97±0.2	0.097	3.5±0.2	0.31	+0.9
HCOOH	+1.4	1.1±0.3	0.077	8.5±0.5	0.68	+1.4

Results from reference 2.

a. Collision rate constant estimated from formulations in reference 7.

Table 2. Absolute Values for the Proton Affinity Scale Based on $CH_3CH=CH_2$, *iso*-$C_4H_9^+$, and $C_6H_5CH_2$ Proton Affinities as Primary Standard.

M	Ab Initio[e]	Propylene (kcal/mol)	Isobutene (kcal/mol)	Benzyl Radical (kcal/mol)
NH_3	205.6	202.9±1.5[a] 202.2±1.5[b] 202.0±1.5[c] 201.7±1.5[d]	205.7±0.8[a] 205.0±0.8[b] 204.8±0.8[c] 204.5±0.8[d]	203.5±2.0[a] 202.8±2.0[b] 202.6±2.0[c] 202.3±2.0[d]
$C_6H_5CH_2$[f]		198.4±1.5	201.2±0.8	199.0±2.0[g]
iso-C_4H_8[h]		192.4±1.5	195.2±0.8[i] (194.9)[j]	193.0±2.0
PH_3[a,d]	188.9	186.8±1.5	189.6±0.8	187.4±2.0
$CH_3CH=CH_2$[k]		179.4±1.5[l]	182.2±0.8	180.0±2.0
H_2S	171.7	169.8±1.5[c] 172.2±1.5[a]	172.6±0.8[c] 175.0±0.8[a]	170.4±2.0[c] 172.8±2.0[a]
H_2O	168.8	166.9±1.5[c] 168.3±1.5[a]	169.7±0.8[c] 171.1±0.8[a]	167.5±2.0[c] 168.9±2.0[a]

a. From the relative gas phase basicity scale reported in J. F. Wolf, R. H. Staley, I. Koppel, M. Taagepera, R. J. McIver, Jr., J. S. Beauchamp, and R. W. Taft, J. Am. Chem. Soc. 99 (1977) 5417 . Temperature corrections discussed in referenc 2 are incorporated.

b. From the unpublished results cited in D. H. Aue and M. T. Bowers, Chapter 9 in "Gas Phase Ion Chemistry"(M. T. Bowers, Editor) Academic Press, New York, 1979.

c. From the relative gas phase basicity scale reported in the thesis of Y. K. Lau, University of Alberta, 1979.

d. From the relative gas phase basicity scale reported in reference 2.

e. D. S. Marynick, K. Scanlon, R. A. Eades, and D. A. Dixon, J. Am. Chem. Soc., submitted for publication.

f. Related to the proton affinity scale through experiments from reference 11, give here in Table 3.

Footnotes to Table 2 (Continued)

g. Absolute proton affinity calculated taking $\Delta H_f(C_6H_5CH_2)$ = 48.7±1.0 kcal/mol (W. Tsang, Int. J. Chem. Kinetics 10 (1978) 41) and $\Delta H_f(C_6H_5CH_3^+)$=215.39±0.30 kcal/mol (reference 13 and references cited therein). If one were to chose the value for $\Delta H_f(C_6H_5CH_2)$ = 47.8±1.5 kcal/mol (M. Rossi and D. M. Golden, J. Am. Chem. Soc. 101 (1971) 1230) the absolute proton affinity of the benzyl radical would be increased by 0.9 kcal/mol.

h. Related to the proton affinity scale through experiments reported by Wolf et al (see footnote *a*), by Y. K. Lau (see footnote *c*) and by P. Ausloos and S. G. Lias in reference 16.

i. Absolute proton affinity calculated taking ΔH_f(*iso*-C_4H_8) = -4.26±0.15 kcal/mol (J. B. Pedley and J. Rylance, "N.P.L. Computer Analysed Thermochemical Data: Organic and Organometallic Compounds", University of Sussex, 1977) and ΔH_f (*tert*-$C_4H_9^+$) = 166.2±0.8 kcal/mol (reference 3).

j. Value obtained for the absolute proton affinity of isobutene based on a heat of formation of *tert*-$C_4H_9^+$ of 166.5 kcal/mol derived from ΔH_f(*tert*-C_4H_9) = 12±1 kcal/mol (W. Tsang, Int. J. Chem. Kinetics 10 (1978) 821) and an ionization potential for this radical of 6.70 eV (F. A. Houle and J. L. Beauchamp, J. Am. Chem. Soc. 101 (1979) 4067). More recent determinations of the heat of formation of this radical give values of 9.4±1.0 kcal/mol (A. L. Castelhano, P. R. Marriott, and D. Griller, J. Am. Chem. Soc. 103 (1981) 4262) or 10.5±1.0 kcal/mol (C. E. Canosa and R. M. Marshall, Int. J. Chem. Kinetics 13 (1981) 303). A value for the ionization potential of *tert*-C_4H_9 of 6.58±0.01 eV has also been reported (J. Dyke, N. Jonathan, E. Lee, A. Morris, and M. Winter, Phys. Scr. 16 (1977) 197). Substitution of any of these alternate values would lead to a lower heat of formation of *tert*-$C_4H_9^+$ and, therefore, a higher predicted proton affinity of isobutene. It should be remembered, however, that the adiabatic ionization potential corresponds to the difference in the heat of formation of the ion and the radical at zero degrees Kelvin, and that a simple addition of this number to the 300 K heat of formation of the radical does not necessarily lead to a correct value for the 300 K heat of formation of the ion.

k. Related to the proton affinity scale through experiments reported in reference 2, and shown here in Table 1.

l. Absolute value of the proton affinity calculated taking $\Delta H_f(CH_3CH{=}CH_2)$ = +4.88± 0.16 kcal/mol (J. B. Pedley and J. Rylance, "N.P.L. Computer Analysed Thermochemical Data: Organic and Organometallic Compounds", University of Sussex, 1977) and ΔH_f(*sec*-$C_3H_7^+$) = 191.1±0.7 kcal/mol (reference 5). Other recent determinations of $\Delta H_f(C_3H_7^+)$ give values of 190.8±1.0 kcal/mol (reference 4) and 191.8±0.4 kcal/mol (reference 3).

reported for hydride transfer reactions [8], proton transfer reactions [9], and charge transfer reactions [10]. In a recent study [11], advantage was taken of this behavior in order to estimate the position of the benzyl radical in the proton affinity scale. Rate constants were measured for the reactions:

$$C_6H_5CH_3^+ + M \rightarrow MH^+ + C_6H_5CH_2 \qquad (4)$$

From the results, given in Table 3, it appears that the $\Delta G°$ of reaction 4 is approximately zero when M is diethyl ketone. Since the heat of formation of the benzyl radical has been established as 48.7±1.0 kcal/mol [12], and the heat of formation of $C_3H_5CH_3^+$ is known to be 215.39±0.30 kcal/mol [13], the proton affinit of the benzyl radical (and thus, of diethyl ketone) is 199.0±1.3 kcal/mol.

Table 3. Rate Constants and Reaction Efficiencies for the Reaction: $C_6H_5CH_3^+ + M \rightarrow MH^+ + C_6H_5CH_2$

M	k x 10^{10} cm^3/molec-s[a]	k/Z[b]	Relative Gas Basicity[c] *kcal/mol*
$(sec\text{-}C_4H_9)_2O$	8.1	0.57	-9.1
$(iso\text{-}C_3H_7)_2O$	6.4	0.46	-6.3
$(tert\text{-}C_4H_9)OCH_3$	6.7	0.51	-2.3
$(n\text{-}C_3H_7)_2O$	4.9	0.35	-2.1
$(C_2H_5)_2CO$	2.3	0.13	-1.2
$(iso\text{-}C_3H_7)COCH_3$	1.5	0.10	-1.2
$CH_3COOC_2H_5$	0.47	0.032	-0.3
$(C_2H_5)_2O$	0.27	0.021	0.0

a. 335 K. Results from reference 11.

b. Collision rate constants estimated from formulations in reference 7.

c. 335 K. From the gas phase basicity scales reported in references 2, 11, and 20.

Table 2 shows the position of the benzyl radical relative to various selected bases in the proton affinity scale, and absolute values for the proton affinity scale based on the benzyl radical as a primary standard. Table 2 is discussed below.

4. Thermochemical Information from the Temperature Dependence of the Rate Constants of Endothermic Reactions

The rate constant of a chemical reaction for which the reactants must traverse an energy barrier is given by:

$$k = \frac{kT}{h} e^{-\Delta H^{+}/RT} e^{\Delta S^{+}/R} \qquad (5)$$

where $\Delta H^{\dagger}$ and $\Delta S^{\dagger}$ are the heat and entropy of activation, respectively. Experimentally, the height of the activation energy barrier, E_a, is measured by determining the rate constant as function of temperature:

$$k = A' e^{-E_a/RT} \qquad (6)$$

In the usual case of an ion-molecule reaction with no activation energy barrier for the exothermic channel, the height of the energy barrier for the endothermic channel must be equal to the endothermicity of reaction, and one should be able to determine that energy barrier height through an Arrhenius treatment of k. However, certain unique features of such ion-molecule reactions dictate caution in such an approach.

It is well known that the experimentally-determined activation energy barrier, E_a in equation 6 , is equal to the average total energy (relative translational plus internal) of all reacting pairs of reactants minus the average total energy of all pairs of reactants. This phenomenological quantity, however, does not necessarily correspond to the actual barrier height, but is related to it by the expression:

$$\Delta H^{\circ} = E_a + (s-3/2)RT \qquad (7)$$

where ΔH° is the actual barrier height, and s is half the number of squared terms into which the critical energy can be factored [14]. Thus, one must determine whether ΔH° and E_a differ significanty for endothermic ion-molecule reactions. This question obviously has wider implications for systems in which reaction proceed without the existence of a "transition state" of higher energy and lower entropy than either set of separated reactants. That is, it is only because of the central barrier that the thermodynamic quantities characteristic of the overall reaction usually emerge in the equilibrium constant in an unequivocal fashion from the ratio of the rate constants:

$$K_{eq} = \frac{k_f}{k_r} = e^{-[\Delta H_f^{\dagger}/RT - \Delta H_r^{\dagger}/RT]} e^{[\Delta S_f^{\dagger}/R - \Delta S_r^{\dagger}/R]} = e^{-\Delta H_{rn}/RT} e^{\Delta S_{rn}/R} \tag{8}$$

If $\Delta H_f^{\dagger}$ in equation 8 is zero, then $\Delta H_r^{\dagger}$ must equal ΔH_{rn} if the observed equilibrium constant is a true thermodynamic equilibrium constant. Thus, under these conditions, the implications of equation 7 are disturbing.

Table 4 shows the rate constants for some corresponding exothermic and endothermic reactions measured as a function of temperature [15] in systems for which the value of ΔH_{rn} is well-established through independent measurements (not involving ion-molecule equilibrium constant determinations). These reactions are all charge transfer reactions, so the enthalpy changes of the reactions are just the differences in the ionization potentials of the two reactant molecules after correction to 300 - 400 K [13].

$$A^+ + B \rightleftarrows B^+ + A \qquad \Delta H_{rn} = IP(B)_{300\ K} - IP(A)_{300\ K} \tag{9}$$

The rate constants given in Table 4 are shown as Arrhenius plots in Figure 1. The values of ΔH_{rn} derived from well-established ionization potentials are given in the Table, and are compared with the values of E_a derived from the Arrhenius treatments of the temperature dependence of the endothermic rate constants. For these reactions, the values of E_a are within ±0.3 kcal/mol of the values of $\Delta H°$. That is, $\Delta H°$ and E_a are equal, within the error limits of the two values. The corresponding equilibrium constants were measured in all the systems, and the values of $\Delta H°$ derived from a van't Hoff treatment of the temperature dependence of the equilibrium constants are also given. Since for these reactions, the rate constants for the exothermic channels exhibit no changes with temperature, it is not surprising that in every case the energy barriers derived from the equilibrium constant measurements are within experimental error equal to those derived from the rate constants of the endothermic reaction.

The Arrhenius treatments of these rate constants also allow a determination of the pre-exponential factors for the endothermic reactions. Because of the limited temperature range of the experiments and the logarithmic dependence of the pre-exponential factor, these values, also shown in the Table are not accurately known. However, they do show an approximate correspondence to the collision rate constants of the endothermic reactions. This suggests that for these efficient reactions:

$$k_r \cong Z_r e^{-E_a/RT} \cong Z_r e^{-\Delta H_r^\circ/RT} \tag{10}$$

Table 4. Rate Constants of Corresponding Exothermic and Endothermic Reactions as a Function of Temperature.[a]

Reaction	T K	k_f (cm^3/molecule-s × 10^{10})	^bZ_f	k_r	^bZ_r	ΔH (kcal/mol)	$E_{a(Exp)}$	$RT\ln[k_r/Z_r]$
$C_6H_5CH_3^+ + C_6H_5C_2H_5 \underset{r}{\overset{f}{\rightleftarrows}} C_6H_5C_2H_5^+ + C_6H_5CH_3$								
	312	13.1±0.3	12.6-14.1	2.3±0.2	11.8-13.0	1.2±0.4[d] (1.4)[f]	1.3±0.2[e]	1.0-1.3
	384	12.5±0.5	12.6-14.1	3.2±0.4	11.8-13.0			1.0-1.3
	399	12.5±0.5	12.6-14.1	3.8±0.3	11.8-13.0			0.9-1.0
				[c]A∼18.				
$c\text{-}C_4H_4O^+ + C_6H_5CH_3 \underset{r}{\overset{f}{\rightleftarrows}} C_6H_5CH_3^+ + c\text{-}C_4H_4O$								
	338	12.7±0.2	13.2-14.6	1.5±0.2	11.0-12.5	1.4±0.3[d] (1.1)[f]	1.1±0.2[e]	1.4-1.5
	379	-	-	1.8±0.2	10.8-12.3			1.4-1.5
	409	13.0±0.3	13.2-14.6	2.0±0.2	10.8-12.3			1.4-1.5
				[c]A∼ 8.				
$C_2H_5I^+ + NO \underset{r}{\overset{f}{\rightleftarrows}} NO^+ + C_2H_5I$								
	316	1.2±0.3	6.1-7.1	1.0±0.1	20.7-30.0	2.0±0.2[d] (2.4)[f]	2.3±0.2[e]	1.9-2.1
	358	0.9±0.4	6.1-7.1	1.4±0.1	20.1-29.1			1.9-2.1
	386	-	-	2.0±0.1	19.7-28.9			1.8-2.0
	417	1.0±0.4	6.1-7.1	2.4±0.2	19.4-28.1			1.8-2.0
				[c]A∼40.				

(Table 4 cont.)

Reaction	T							
$C_6H_6^+ + C_6H_5F \underset{r}{\overset{f}{\rightleftarrows}} C_6H_5F^+ + C_6H_6$								
	311	-	-	3.2±0.2	11.6	1.3±0.2[d] (1.2)[f]	1.0±0.2[e]	0.8
	313	-	-	3.0±0.2	11.6			
	328	6.2±0.5	14.7-22.8	-	-			
	333	-	-	3.7±0.2	11.6			0.8
	340	-	-	3.8±0.1	11.6			0.8
	351	-	-	4.1±0.2	11.6			0.7
	379	5.9±0.4	14.3-22.2	4.2±0.4	11.6			0.8
	381	-	-	4.2±0.4	11.6			0.8
	397	5.6±0.6	14.1-21.8	4.3±0.6	11.6			0.8
				[c] A∿13.				

a. Results from reference 15. Error limits represent standard deviation of measurements.

b. Rate constants for complex-forming collision calculated according to formulations listed in reference 7.

c. Pre-exponential factor (equation 6) derived from Arrhenius treatment of rate constants (Figure 1).

d. ΔH_{rn} derived from differences in 300 K ionization energies of reactant species (reference 13 and references cited therein).

e. E_a derived from Arrhenius treatment of rate constants (Figure 1).

f. ΔH derived from Van't Hoff treatment of equilibrium constants.

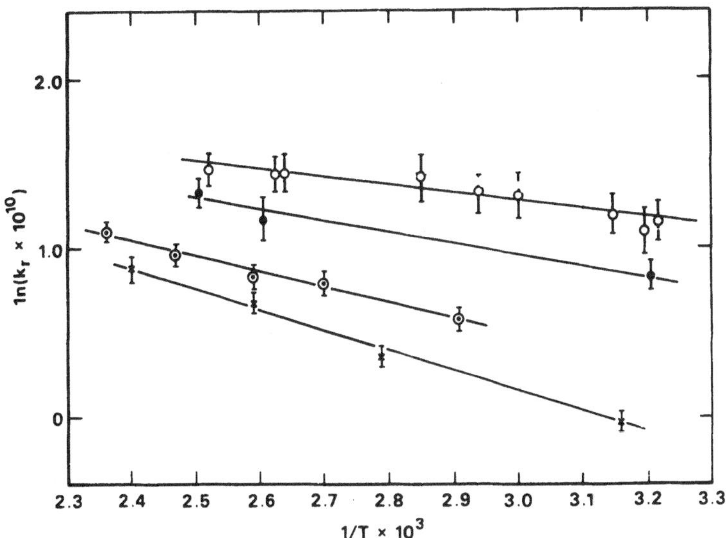

Fig. 1. The logarithm of $k_r \times 10^{10}$ as a function of the reciprocal of the temperature in Kelvin for the reactions: $C_6H_5CH_3^+ + C_6H_5C_2H_5$ (●); $C_6H_5CH_3^+ + c\text{-}C_4H_4O$ (⊙); $NO^+ + C_6H_5I$ (x); $C_6H_6^+ + C_6H_5F$ (o).

(For a discussion of the role of the reaction entropy change, see below. At this juncture, note that for every endothermic reaction under consideration here, ΔS° is either zero or positive.) To the extent that equation 10 is a valid approximation, one should be able to estimate values of E_a from the reaction efficiency at a given temperature:

$$RT\ln k_r/Z_r \cong -E_a \cong -\Delta H_r^\circ/RT \qquad (11)$$

It is shown in the Table that at every temperature in every system given here, values of E_a derived in this way are within the experimental error limits of the values obtained from the temperature dependence of the endothermic reaction rate constants.

The first two reactions given in the Table:

$$C_6H_5CH_3^+ + C_6H_5C_2H_5 \rightleftarrows C_6H_5C_2H_5^+ + C_6H_5CH_3 \qquad (12)$$

and

$$c\text{-}C_4H_4O^+ + C_6H_5CH_3 \rightleftarrows C_6H_5CH_3^+ + c\text{-}C_4H_4O \qquad (13)$$

are predicted to have no overall change in the vibrational, rotational, or electronic entropy. In both cases, the rate constants for the exothermic channels are effectively equal to the corresponding collision rate constants.

In the case of the reaction:

$$C_2H_5I^+ + NO \rightleftarrows NO^+ + C_2H_5I \qquad (14)$$

there is an entropy change associated with the changes in electronic degeneracy; the overall entropy change of ±3.7 cal/deg-mol is negative for the exothermic channel. Since the results given in Table 4 show that equation 10 adequately describes the endothermic rate constant in this system, if the ratio of the rate constants is related to the thermodynamic equilibrium constant, the $e^{\Delta S°/R}$ must influence the rate constant for the exothermic reaction. The results given in Table 4 show that this rate constant is considerably lower than the collision rate constant. The average reaction efficiency is 0.16 independent of temperature. The value of $e^{\Delta S°/R}$(taking $\Delta S°$= -3.7 cal/deg-mol) is 0.16. Thus, for reaction (14)

$$k_f \cong Z_f e^{\Delta S°/R} \qquad (15)$$

For the reaction:

$$C_6H_5^+ + C_6H_5F \rightleftarrows C_6H_5F^+ + C_6H_6 \qquad (16)$$

there is a predicted rotational, vibrational, and electronic entropy change of -2.7 cal/deg-mol, assuming a D_{2h} symmetry for the benzene ion [13]. Thus $e^{\Delta S°/R}$ is 0.26; the observed efficiency of the exothermic reaction is 0.27-0.41.

Other examples of highly efficient exothermic reactions slowed from the collision rate by a factor of $e^{\Delta S°/R}$ (where $\Delta S°$ represents a negative total rotational, vibrational, and electronic entropy change) have been reported. For example, rate constants for proton transfer from the tert-$C_4H_9^+$ ion to a series of polar bases [2] indicate that the maximum efficiency for the reaction:

$$\text{tert-}C_4H_9^+ + M \rightleftarrows MH^+ + \text{iso-}C_4H_8 \qquad (17)$$

is 0.70±0.03. The rotational entropy change for the half-reaction:

$$\text{tert-}C_4H_9^+ \rightarrow \text{iso-}C_4H_8^+ \qquad (18)$$

is calculated to be -0.6 cal/deg-mol at 300 K (taking into account changes in moments of inertia, external symmetry numbers, and the loss of a rotation for one methyl group upon formation of the double bond), in substantial agreement with experimental results [16]. Thus $e^{\Delta S°/R}$ for reaction 17 is 0.74 when there is no entropy change associated with the protonation of M, and equation 15 can be said to describe the rate constants for these reactions. Also, rate constants have been measured for proton transfer to PH_3 form various ions [2,17]:

$$MH^+ + PH_3 \rightleftarrows PH_4^+ + M \qquad (19)$$

The overall rotational entropy change associated with the half-reaction:

$$PH_3 \rightarrow PH_4^+ \quad (20)$$

is -1.8 cal/deg-mol, $e^{\Delta S°/R}$ is 0.40, and the maximum efficiency which has been observed for reactions 19 is 0.43. Table 3 shows rate constants for proton transfer from $C_6H_5CH_3^+$ (reaction 4), which appear to exhibit a maximum efficiency of ∿0.5. The expected change in rotational symmetry numbers occasioned by the loss of a proton from $C_6H_5CH_3^+$ would lead to a value of 0.5 for $e^{\Delta S°/R}$ for the half-reaction:

$$C_6H_5CH_3^+ \rightarrow C_6H_5CH_2 \quad (21)$$

All the results given in Table 4 pertain to reactions which can be described as "highly efficient". That is, these are reactions in which the only factors leading to a lowering of the reaction efficiency (probability of a reactive collision) below unity for the exothermic and endothermic channels are energy barriers or entropic barriers which correspond at least approximately to the overall ΔS° and/or ΔH° of reaction. By constrast, Table 5 shows the temperature dependence of the rate constants of an "inefficient" reaction:

$$AsH_4^+ + CH_3CH{=}CH_2 \rightleftarrows sec\text{-}C_3H_7^+ + AsH_3 \quad (22)$$

Table 5. Temperature Dependence of the Rate Constants of Exothermic and Endothermic Channels of the Reaction:
$[AsH_4^+ + CH_3CH{=}CH_2 \rightleftarrows \textit{sec}\text{-}C_3H_7^+ + AsH_3]^a$

	cm³/molecule-s x 10¹⁰				kcal/mol
T	k_f	Z_f	k_r	Z_r	ΔG°[b]
317	3.9±0.2	11.1	1.1±0.2	9.6	-0.80
350	3.5±0.2	11.1	0.9±0.2	9.6	-0.87
380	3.2±0.2	11.1	0.7±0.2	9.6	-0.93
					ΔH°=-0.15

a. 317 and 380 K values: S. G. Lias, unpublished results (pulsed ICR). 350 K values: Reference 2.

b. From equilibrium constant measurements; S. G. Lias, unpublished results.

The results discussed above and presented in Tables 1 and 2 have provided us with an estimate that ΔH^{o} for this reaction is -0.2 kcal/mol. As expected for a near-thermo-neutral reaction, it is seen that both the exothermic and endothermic channels exhibit efficiencies much lower than those which would be predicted from equations 10 or 15 . Furthermore, both rate constants exhibit a slight negative temperature dependence. These results illustrate an important, if obvious, point: An Arrhenius treatment of the rate constants for endothermic ion-molecule reactions can yield a valid estimate for the height of the energy barrier only in the event that the rate constant for the exothermic channel exhibits no temperature dependence.

It has recently been suggested [9,18] that the fall-off in reaction efficiencies usually observed for near-thermoneutral ion-molecule reactions can be explained by the existence of a free energy of activation, G_a, which is itself related to the ΔG^{o} of reaction:

$$k = Ze^{-G_a/RT} = Ze^{-[n_G \Delta G^{\circ} + G_a^{\circ} M(n_G)/\ln 2]} \qquad (23)$$

where G_a° is the free energy barrier when ΔG is zero, n_G is the bond order and $M(n_G)$ is the entropy of mixing of the reactants. While the existence of such barriers would certainly explain the generally-observed fall-off behavior, the predictions of equation 23 would be that the rate constants of both channels of such inefficient near-thermoneutral reactions should exhibit a positive temperature dependence. Although there are few data available on the temperature dependences of such reactions, those which have been measured usually show a negative temperature dependence [19]. In general, the decrease in reaction efficiency is most pronounced for non-polar reactants, an observation which suggests that a contributing factor may be a competition between reaction and dissociation of the ion-molecule complex [8].

5. Primary Standards for the Proton Affinity Scale

Through determinations of equilibrium constants for proton transfer reactions in the gas phase, there is available an extensive scale of relative gas phase proton affinities [1c,2,20]. The proton affinity is defined as the enthalpy change of the reaction:

$$AH^{+} \rightarrow H^{+} + A \qquad (24)$$

at 300 K. The position of isobutene in the proton affinity scale has been well established for some time [2,16,20], and its absolute proton affinity, calculated from $\Delta H_f(\text{iso-}C_4H_8)$ and available values of $\Delta H_f(\text{tert-}C_4H_9^{+})$, has often been used as a primary standard for establishing absolute values for the entire proton affinity scale. However, the existence in the literature of disparate values of $\Delta H_f(\text{tert-}C_4H_9)$ [21] as well as two values for the ionization potential of this radical [22], has led to some confusion and to wide variations in values ascribed to the proton affinity scale.

In the discussion above, experiments have been described establishing the position of propylene and of the benzyl radical in the proton affinity scale. As described before, fairly reliable values are now available for the absolute heats of formation of the corresponding protanated ions, $\text{sec-}C_3H_7^+$ and $C_6H_5CH_3^+$, so that absolute proton affinities for these two species can be calculated. In addition, the heat of formation of $\text{tert-}C_4H_9^+$ at 300 K has recently been re-evaluated from the appearance potentials of this ion in alkanes and alkyl halides [3]. Table 2 shows absolute values of the proton affinities of several selected bases from the proton affinity scale, using each of these three species (propylene, benzyl radical, and isobutene) as primary standard. While there is good agreement between the relative proton affinities of the benzyl radical and of propylene as related through the proton affinity scale or calculated from relevant heats of formation, the agreement with values based on the proton affinity of isobutene is less good, although within error limits.

6. Ionization Energies of Alkanes

In a recent study [23], equilibrium constants for charge transfer reactions of alkanes were studied as a function of temperature in a high pressure mass spectrometer. It was found that there were large negative entropy changes associated with processes such as:

$$\text{c-}C_6H_{12}^+ + \text{n-Alkane} \rightarrow \text{n-Alkane}^+ + \text{c-}C_6H_{12} \tag{25}$$

for alkanes having seven or more carbon atoms, and it was suggested that this was evidence that the n-alkane parent ions have a constrained configuration. These equilibria were also examined in a pulsed ion cyclotron resonance instrument with startling results; apparent entropy changes 1.5 to 2 times larger than those seen in the high pressure instrument were measured. However, it was observed that as the temperature was increased, the product alkane ions were undergoing dissociation to an increased extent (as verified by double resonance ejection), and thus, the observed "equilibrium constants" were increasingly too low at higher temperatures, and the "entropy changes" were not real. In a related study carried out on the high pressure mass spectrometer [24], charge transfer equilibria were examined in binary cycloalkane systems, where no significant entropy changes were seen. The scale of relative ionization energies of alkanes and cycloalkanes derived from these two studies [23,24] is shown in Table 6. Of all the alkanes and cycloalkanes included in those studies, only cyclohexane has a well-established adiabatic ionization energy (9.88 eV) sc cyclohexane was used as a primary standard for the thermochemical ladder. Because many aromatic molecules have well known ionization potentials, the authors of the latter study[24], as well as the present author, have attempted to provide additional primary standards for the relative ionization energy scale by examining charge transfer equilibria in aromatic-alkane systems. In the results of these experiments some apparent contradictions arise.

For example, an examination of the temperature dependence of the equilibrium constants for the reaction:

$$C_6H_5CF_3^+ + n\text{-}C_9H_{20} \rightleftarrows C_9H_{20}^+ + C_6H_5CF_3 \tag{26}$$

in the ICR led to an estimate that the ionization energy of n-C_9H_{20} lay about 3 kcal/mol below that of $C_6H_5CF_3$. Since the latter ionization potential is well-established as 9.685±0.005 eV [13,25], this gave an estimate of the ionization energy of n-nonane as approximately 9.55 eV, in fair agreement with the value of 9.63 eV derived indicated an entropy change of about -10 cal/deg-mol associated with reaction 26. In the earlier study, a rotational entropy change of -9.2 cal/deg-mol had been ascribed to the half-reaction [23]:

$$n\text{-}C_9H_{20} \rightarrow n\text{-}C_9H_{20}^+ \tag{27}$$

However, it was found that in the ICR experiments, a large fraction (70% at 400 K) of the n-$C_9H_{20}^+$) ions formed in reaction 26 undergo dissociation under the conditions of the ICR experiments. If one examines the rate constants for the forward and reverse channels of reaction 26 , however, it is seen (Table 7) that the charge transfer from n-$C_9H_{20}^+$ to $C_6H_5CF_3$ occurs at about 90% of the collision rate and exhibits little temperature dependence, behavior usually characterisitic of an exothermic reaction. The charge transfer from $C_6H_5CF_3^+$ to n-C_9H_{20}, on the other hand, has an efficiency which is much lower than unity, and a rate constant which increases with increasing temperature. From the efficiency and the temperature dependence of this rate constant, an estimate can be made that the 300-400 K ionization energy of nonane lies about 0.6-0.9 kcal/mol above that of $C_6H_5CF_3$, i.e. $\Delta H_{ionization}(n\text{-}C_9H_{20})$ = 9.71 eV, rather than 9.63 eV, as derived from the equilibrium constant measurements [23].

Accepting this, the difference between the ionization energies of cyclohexane and n-nonane would be 3.9 kcal/mol, but the difference derived from the thermochemical ladder resulting from the equilibrium constant measurements (Table 6) is 5.8 kcal/mol, a serious discrepancy. However, if one were to assume that for cases in which the rate constant of the exothermic channel is temperature independent, the expressions:

$$k_f \cong Z_f e^{\Delta S°/R} \qquad (\Delta S° < 0) \tag{15'}$$

$$k_r \cong Z_r e^{-\Delta H°/RT} e^{\Delta S°/R} \qquad (\Delta S° < 0) \tag{10'}$$

Table 6. Thermochemical Ladders for Charge Transfer Reactions of Alkanes, Cycloalkanes, and Aromatic Compounds.

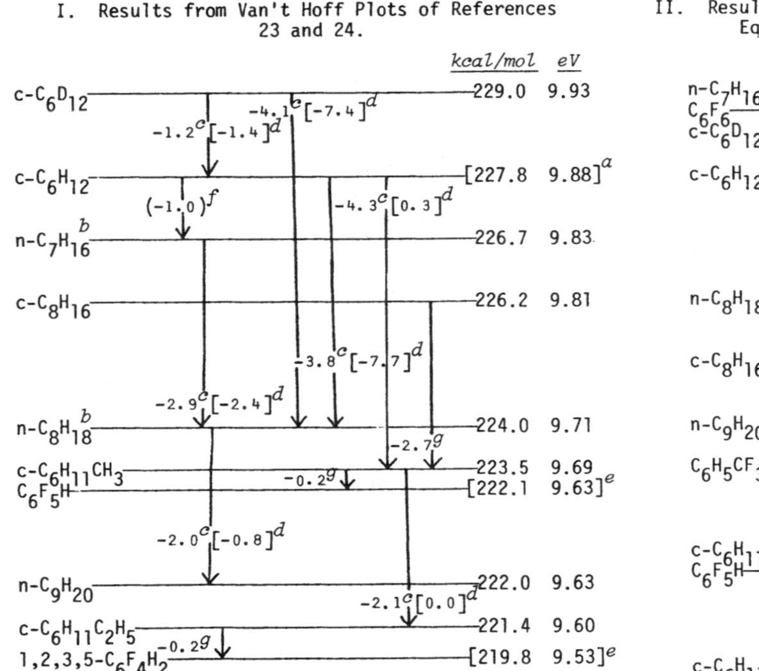

a. Primary standard for ionization energy scale for alkanes and cycloalkanes in reference 23. *b*. In reference 23 it was seen that there is little isotope effect on the ionization energy for these compounds; the rate constants in Table 7 can be used to derive information about the ionization energies of undeuterated analogues. *c*. Ionization energy at 300 K (in kcal/mol) from Van't Hoff plots of reference 23. *d*. Entropy change (in cal/deg-mol) from Van't Hoff plots of reference 23. *e*. Ionization energy from reference 13 and references cited therein. *f*. From temperature dependence of equilibrium constant in high pressure mass spectrometer; L. W. Sieck, private communication. *g*. From Van't Hoff plots of reference 24. *h*. Literature value, 9.90±0.01 eV (D. R. Smith and J. W. Raymonda, Chem. Phys. Letters 12, 269 (1971)). *i*. 300 K ionization energy based on data shown in this thermochemical ladder; primary standards, cyclohexane and aromatic molecules. *j*. From rate constant data in Table 7. *k*. From rate constant data in Table 7. *l*. From $\Delta G_{330\ K}$ assuming no ΔS. *m*. From rate constant data, Table 7. *n*. From rate constant data, Table 7.

Table 7. Rate Constants of Corresponding Exothermic and Endothermic Charge Transfer Reactions.

Reaction	T	k_f ($cm^3/molecule\text{-}s \times 10^{10}$)	k_f/Z_f	k_r ($cm^3/molecule\text{-}s \times 10^{10}$)	k_r/Z_r	ΔH_r kcal/mol	
$n\text{-}C_9H_{20}^+ + C_6H_5CF_3 \underset{r}{\overset{f}{\rightleftarrows}} C_6H_5CF_3^+ + n\text{-}C_9H_{20}$							
	317	~13[i]	0.82	4.5±1.4[i]	0.38	0.6	*a*
	327	14.7±1.3[i]	0.86	4.4±0.8[i]	0.37	0.6	*a*
	390	12.5±1.5[i]	0.84	6.5±1.1[i]	0.55	0.5	*a*
	407	-	-	6.3±1.1[i]	0.53	0.5	*a*
						0.9 (-1.3)	*b* *h*
$c\text{-}C_6H_{12}^+ + n\text{-}C_8D_{18} \underset{r}{\overset{f}{\rightleftarrows}} C_8D_{18}^+ + c\text{-}C_6H_{12}$							
	318	4.9±0.6	0.42[d]	0.9±0.2[d]	0.069	1.7	*a,c*
	357	5.0±0.8	0.42[d]	1.2±0.3[d]	0.092	1.7	*a,c*
						1.7 (3.8)	*b,e* *h*
$C_6F_6^+ + n\text{-}C_7H_{16} \underset{r}{\overset{f}{\rightleftarrows}} n\text{-}C_7H_{16}^+ + C_6F_6$	324	2.4±0.5	0.26[i]	5.7±0.6[i]	0.61		*e*
	358	2.0±0.5	0.19[i]	6.0±0.5[i]	0.64		*e*
$C_6F_6^+ + c\text{-}C_6H_{12} \underset{r}{\overset{f}{\rightleftarrows}} c\text{-}C_6H_{12}^+ + C_6F_6$	335	9.0±0.7	0.94[i]	3.1±0.5[i]	0.31	0.8	*a,f,c*
$C_6F_6^+ + c\text{-}C_6D_{12} \underset{r}{\overset{f}{\rightleftarrows}} c\text{-}C_6D_{12}^+ + C_6F_6$	335	3.1±0.5	0.31[i]	2.8±0.4[i]	0.30	~0	*a,f,c*
$n\text{-}C_7H_{16}^+ + c\text{-}C_6H_{12} \underset{r}{\overset{f}{\rightleftarrows}} c\text{-}C_6H_{12}^+ + n\text{-}C_7H_{16}$	335	5.3±0.3	0.46[g]	1.1±0.3[g]	0.08	1.7 (-1.0)	*a* *h*
$c\text{-}C_6H_{11}CH_3^+ + c\text{-}C_6H_{11}C_2H_5 \underset{r}{\overset{f}{\rightleftarrows}} C_6H_{11}C_2H_5^+ + C_6H_{11}CH_3$	335	12.1[j]	1.0	0.32[j]	0.027	2.4 2.1	*a* *h,c*
$n\text{-}C_9H_{20}^+ + c\text{-}C_6H_{11}CH_3 \underset{r}{\overset{f}{\rightleftarrows}} c\text{-}C_6H_{11}CH_3^+ + n\text{-}C_9H_{20}$	335	6.1[j]	0.54	0.7[j]	0.05	1.9 1.7 (+1.5)	*a* *c* *h*
$n\text{-}C_8D_{18}^+ + c\text{-}C_8H_{16} \underset{r}{\overset{f}{\rightleftarrows}} c\text{-}C_8H_{16}^+ + n\text{-}C_8D_{18}$	335	9.4[j]	0.82	1.4[j]	0.12	1.4 1.1 (-2.2)	*a* *c* *h*
$c\text{-}C_8H_{16}^+ + n\text{-}C_9H_{20} \underset{r}{\overset{f}{\rightleftarrows}} n\text{-}C_9H_{20}^+ + c\text{-}C_8H_{16}$	335	7.3[j]	0.58	3.2[j]	0.28	0.9 1.0 (4.2)	*a* *c* *h*

Footnotes to Table 7.

a. $\Delta H_r = RT \ln k_r/Z_r$

b. ΔH_r from Arrhenius treatment of temperature-dependence of k_r.

c. Value of ΔH given in Table 6, Part II.

d. M. Mautner, private communication.

e. Because of negative temperature dependence of rate constant designated here as k_f, thermochemical information can not be derived from these rate constants. However, the data are consistent with a near-thermoneutral reaction, for which the channel designated "f" has a negative entropy change (associated with the rotational component of the half-reaction: $C_6F_6^+ \rightarrow C_6F_6$.)

f. A Van't Hoff treatment of the equilibrium constant shows that $\Delta S° \sim 0$. It would be expected that the rotational entropy changes for the half reactions $[C_6F_6^+ \rightarrow C_6F_6]$ and $[c\text{-}C_6(H,D)_{12}^+ \rightarrow c\text{-}C_6(H,D)_{12}]$ would be, respectively, -2.2 and +2.2 cal/deg-mol. (Both ions undergo a Jahn-Teller distortion to a lower symmetry configuration.)

g. Rate constants reported in S. G. Lias, P. Ausloos, and Z. Horvath, Int. J. Chem. Kinetics VIII, 725 (1976).

h. Value of ΔH derived from Van't Hoff plots of references 23 and 24 (Table 6, Part I.)

i. Rate constants determined by S. G. Lias, NBS pulsed ICR.

j. M. Mautner, reference 26.

can give a rough approximation of the rate constants for the exothermic (subscript f) and endothermic (subscript r) reactions, then an examination of the rate constants associated with the charge transfer reactions of some of the steps in the ladder (Table 6) suggests that the values for $\Delta H°$ and $\Delta S°$ have been overestimated in the equilibrium constant measurements. For example, for the reaction:

$$c\text{-}C_6H_{12}^+ + n\text{-}C_8D_{18} \rightleftarrows n\text{-}C_8D_{18}^+ + c\text{-}C_6H_{12} \qquad (28)$$

an analysis of the efficiency and temperature dependence of the rate constant for the endothermic channel leads to an estimate that $\Delta H° \cong 1.7$ kcal/mol (compared to 3.7 kcal/mol measured earlier [23]). The temperature independent efficiency of the exothermic reaction would correspond to an entropy change of approximately -1.7 cal/deg-mol for the overall reaction. Since one would expect a rotational entropy change of -2.2 cal/deg-mol to be associated with the half-reaction:

$$c\text{-}C_6H_{12}^+ \rightarrow c\text{-}C_6H_{12} \qquad (29)$$

due to the Jahn-Teller distortion in the $c\text{-}C_6H_{12}^+$ ion, this value would imply that there is little or no rotational entropy change associated with the ionization of n-octane.

An amended thermochemical ladder based on these considerations is given in Table 6; corollary data are given in Table 7. On this basis, there is good internal consistency between the ionization energies of cyclohexane and the aromatic molecules. Nevertheless, at this writing there is no adequate explanation for the reasons for the discrepancy between the equilibrium constant measurements, and the interpretation of the thermochemistry based on the rate constants given in Table 7. It has been suggested [26] that expressions 10' and 15' do not describe the rate constant for charge transfer reactions in alkanes, but that these reactions follow a different model, in which the probability of the dissociation of the ion-molecule complex to (A^+ + B) or (C^+ + D) is completely independent of the origin of the complex (from A^+ + B or C^+ + D reactants). The suggestion, then, is that the probability that the complex dissociates to the right or the left is governed only by the $\Delta G°$ of reaction:

$$k_A = Z_A e^{-\Delta G°/RT}/(1 + e^{-\Delta G°/RT}) \qquad (30)$$

$$k_B = Z_B\{1 - e^{-\Delta G°/RT}/(1 + e^{-\Delta G°/RT})\} \qquad (31)$$

where k_A and k_B are the rate constants for corresponding exothermic and endothermic channels of the same reaction (either A or B may be exothermic), and $\Delta G°$ is the free energy change for the reaction, which is positive for the channel A. This model also assumes that:

$$k_A/Z_B + k_B/Z_B \cong 1 \qquad (32)$$

It is suggested that a positive $\Delta S°$ of reaction can compensate for the presence of an energy barrier [26], so that a sufficiently exoergic (negative $\Delta G°$) reaction could occur effectively at the collision rate even if it were highly endothermic (positive $\Delta H°$). Of course, in this event, the rate constant of the corresponding exothermic reaction would have to exhibit a negative temperature dependence which was quantitatively tied to the positive $\Delta G°$ of reaction. If one accepts the thermochemical results from the equilibrium constant measurements there are a number of rate constants for these systems which appear to follow this model [26]. However, the necessary elucidation of the temperature dependences of the rate constants has not been carried out. Furthermore, at this writing there is no system which appears to follow the predictions of this model for which independent thermochemical information, not based on ion-molecule equilibrium constant measurements, is available.

The only other reasonable explanation of the discrepancy between the rate constant results given in Table 7 and the thermochemical ladder based on equilibrium constant measurements, Table 6, is that the equilibrium constants for some reason do not reflect the actual thermochemistry of the systems. This might be the case, for instance, if the product linear alkane ions formed in the charge transfer process rearranged to another structure or configuration after departure from the ion-molecule complex:

$$\begin{array}{l} c\text{-}C_6H_{12}^+ + \text{n-Alkane} \underset{2}{\overset{1}{\rightleftharpoons}} \left[\text{n-Alkane}^+\right]_1 + c\text{-}C_6H_{12} \\ \quad\uparrow 3 \qquad\qquad\qquad\qquad \downarrow \\ \qquad\qquad\qquad\qquad\quad \left[\text{Alkane}^+\right]_2 + c\text{-}C_6H_{12} \end{array} \tag{33}$$

where $k_2 \neq k_3$. On the assumption that the rearrangement $[\text{n-Alkane}^+]_1 \rightarrow [\text{Alkane}^+]_2$ involves a higher energy structure going to a lower energy structure or configuration, the observed temperature dependences suggest that the process diminishes in importance at higher temperatures. Evidence in favor of this interpretation is the observation that the values of $\Delta G°$ observed at the highest temperature used in the high pressure mass spectrometric study [23] are just the values which would be predicted from the values of $\Delta H°$ and $\Delta S°$ deduced here from rate constant measurements; values at lower temperatures are generally in serious disagreement.

7. References

1. a) M.T. Bowers, D.H. Aue, H.M. Webb, and R.T. McIver, Jr., J. Am. Chem. Soc. 93 (1971) 4314; For reviews, see: b) S.G. Lias in "Ion Cyclotron Resonance Spectrometry" (H. Hartmann and K.-P. Wanczek, editors), Springer-Verlag Lecture Notes in Chemistry Series, 7 (1978); c) D.H. Aue and M.T. Bowers, Chapter 9 in "Gas Phase Ion Chemistry, Vol. 2 (M.T. Bowers, editor), Academic Press (1979).

2. S.G. Lias, D.M. Shold, and P. Ausloos, J. Am. Chem. Soc. 102 (1980) 2540.

3. J.C. Traeger and R.G. McLoughlin, J. Am. Chem. Soc., 103 (1981) 3647.

4. T. Baer, J. Am. Chem. Soc. 102 (1980) 2482.

5. H.M. Rosenstock, R. Buff, M.A.A. Ferreira, S.G. Lias, A.C. Parr, R.L. Stockbauer, and J.L. Holmes, J. Am. Chem. Soc., submitted for publication.

6. All heats of formation of ions cited here are given using the so-called "stationary electron" convention. That is, the integrated heat capacity of the electron is taken to be zero at all temperatures.

7. At this writing, there are several formulations available (listed below) for estimating rate constants for complex-forming collisions of ions with polar molecules. For the results given in Tables 1 and 3, collision rate constants have been calculated using the A.D.O. theory, reference (c) below. In Table 4, we cite the range of values calculated using the different appoaches from references (c), (d), and (e). The rate constants cited or referred to in Tables 5 and 7 are Langevin-Gioumousis-Stevenson values, references (a) and (b). (a) P. Langevin, Ann. Chim. Phys. 5 (1905) 245; (b) G. Gioumousis, and D.P. Stevenson, J. Chem. Phys. 29 (1958) 294; (c) M.T. Bowers and T. Su in "Interactions between Ions and Molecules" (P. Ausloos, editor), Plenum (1975); (d) W.J. Chesnavich, T. Su, and M.T. Bowers, in "Kinetics of Ion-Molecule Reactions" (P. Ausloos, editor), Plenum (1979); (e) D.P. Ridge in "Kinetics of Ion-Molecule Reactions" (P. Ausloos, editor), Plenum (1979).

8. See for example: S.G. Lias, J.R. Eyler, and P. Ausloos, Int. J. Mass Spectrom. Ion Phys. 19 (1976) 219.

9. See for example reference 2 and D.K. Bohme, G.I. Mackay, and H.I. Schiff, J. Chem. Phys. 73 (1980) 4976.

10. See for example S.G. Lias, P. Ausloos, and Z. Horvath, Int. J. Chem. Kinetics, VIII, 725 (1976).

11. M. Mautner, J. Am. Chem. Soc., in press.

12. W. Tsang, Int. J. Chem. Kinetics 10 (1978) 41.

13. S.G. Lias and P. Ausloos, J. Am. Chem. Soc. 100 (1978) 6027.

14. See for instance, E.A. Moelwyn-Highes, Chapter XXII of "Physical Chemistry", Pergamon-Press (1961).

15. S.G. Lias, J. Am. Chem. Soc., submitted for publication.

16. P. Ausloos and S.G. Lias, J. Am. Chem. Soc. 100 (1978) 1953.

17. P. Ausloos and S.G. Lias, J. Am. Chem. Soc. 103 (1981) 3641.

18. N. Agmon and R.D. Levine, Chem. Phys. Letters, 52 (1977) 197.

19. S.G. Lias, unpublished results.

20. a) J.F. Wolf, R.H. Staley, I. Koppel, M. Taagepera, R.J. McIver, Jr., J.L. Beauchamp, and R.W. Taft, J. Am. Chem. Soc. 99 (1977) 5417; b) Y.K. Lau, Ph.D. thesis, University of Alberta, 1979; c) R. Yamdagni and P. Kebarle, J. Am. Chem. Soc. 98 (1976) 1320.

21. a) W. Tsang, Int. J. Chem. Kinetics 10 (1978) 821; b) A.L. Castelhano, P.R. Marriott, and D. Griller, J. Am. Chem. Soc. 103 (1981) 4262; c) C.E. Canosa and R.M. Marshall, Int. J. Chem. Kinetics 13 (1981) 303; d) M. Rossi and D.M. Golden, Int. J. Chem. Kinetics 11 (1979) 969.

22. a) F.A. Houle and J.L. Beauchamp, J. Am. Chem. Soc. 101 (1979) 4067; b) J. Dyke, N. Jonathan, E. Lee, A. Morris, and M. Winter, Phys. Scr. 16 (1977) 197.

23. M. Meot-Ner (Mautner), L.W. Sieck and P. Ausloos, J. Am. Chem. Soc., in press.

24. L.W. Sieck and M. Mautner, to be submitted.

25. V.J. Hammond, W.C. Price, J.P. Teegan, and A.D. Walsh, Disc. Faraday Soc. 9 (1950) 53.

26. M. Mautner, J. Am. Chem. Soc., submitted for publication.

ION-MOLECULE ASSOCIATION REACTIONS

Lewis Bass and Michael T. Bowers
Department of Chemistry
University of California, Santa Barbara, California 93106, USA

1. Introduction

The overall reaction for the general ion-molecule association process is

$$A^+ + B \xrightarrow{k_2} AB^+ \tag{1}$$

where k_2 is the second-order rate constant for formation of the association product AB^+. In this paper we will deal with association reactions which proceed through the formation of a long-lived excited intermediate complex according to the mechanism

$$A^+ + B \underset{k_b}{\overset{k_f}{\rightleftharpoons}} (AB^+)^* \begin{cases} \xrightarrow{k_s\,[M]} AB^+ \\ \xrightarrow{k_r} AB^+ + h\nu \\ \xrightarrow{k_d} C^+ + D \end{cases} \tag{2}$$

where k_f, k_b, k_s, k_r and k_d represent the rate constants for formation of the excited complex, back-dissociation of the complex, stabilization of the complex via collisions with bath gas M, stabilization of the complex via radiative emission of a photon, and dissociation of the complex into some fragment channel other than the reactant channel. Note that the incorporation of more than one such channel into mechanism 2 will not be discussed here, but is straight-forward. For each of the specific systems which will be discussed here either k_r or k_d or both will be zero.

Application of the steady-state hypothesis to $(AB^+)^*$ in eq. 2 yields

$$k_2 = \frac{k_f(k_s[M] + k_r)}{k_b + k_s[M] + k_r + k_d} \tag{3}$$

In order to predict theoretical values of k_2 it is necessary to obtain estimates of k_f, k_s, k_r, k_b and k_d. For ion-molecule systems k_f and k_s are usually estimated from classical collision theory [1] (for k_s an estimated stabilization efficiency may also be included). Radiative rate constants can be estimated from integrated absorption intensities if such data is available [2], otherwise some sort of simplifying assumption needs to be used. In the model discussed in this paper the dissociation rate constants, k_b and k_d, are both calculated on the basis of statistical reaction rate theory [3-6]. All of the assumption inherent in this treatment will be stated explicitly in Section 3 of this paper.

In addition to the overall rate constant, k_2, other rate constants of interest are the three-body association rate constant, k_3,

$$A^+ + B + M \xrightarrow{k_3} AB^+ + M \tag{4a}$$

$$k_3 = \frac{k_f k_s}{k_b + k_s[M] + k_r + k_d} \tag{4b}$$

the bimolecular rate constant for radiative association, k_{RA},

$$A^+ + B \xrightarrow{k_{RA}} AB^+ + h\nu \tag{5a}$$

$$k_{RA} = \frac{k_f k_r}{k_b + k_s[M] + k_r + k_d} \tag{5b}$$

and the bimolecular rate constant for reactive fragmentation, k_2',

$$A^+ + B \xrightarrow{k_2'} C^+ + D \tag{6a}$$

$$k_2' = \frac{k_f k_d}{k_b + k_s[M] + k_r + k_d} \tag{6b}$$

In Section 2 of this paper a brief outline of the methods used to analyze experimental data in order to estimate k_2, k_3, k_{RA}, and k_2' will be discussed. In Section 3 the theoretical model will be outlined, stating explicitly all the assumptions which are utilized. In Section 4 three specific examples of ion-molecule association will be discussed. These are: (a) the formation of the proton-bound dimers of ammonia and the methylamines, which have been studied experimentally by meot-ner and Field [7] and Neilson et al. [8], and have been modeled theoretically by Olmstead et al. [9] and Bass,Chesnavitch and Bowers [10]; (b) the radiative association reaction of CH_3^+ with HCN, which has been studied by Schiff and Bohme [11], McEwan et al.[12] and Bass et al. [13]; and, (c) the clustering reactions of methanol with protonated methanol, which are currently being studied by Cates, Bass and Bowers [14].

2. Experimental Analysis

The experimental work discussed here consists of measurements of the biomolecular rate constants, k_2 and k_2', through the use of mass spectrometry [4], ICR (ion cyclotron resonance) spectrometry [5,9,11], and the flowing afterglow technique [8]. The rate constants k_f and k_s are usually estimated independently of the experimental results) from classical collision theory [1]. Thus, it is implicity assumed that all A^+-B collisions result in the formation of a long-lived $(AB^+)^*$ complex, and that all $(AB^+)^*$-M collisions result in the stabilization of this intermediate. The latter assumption is often modified to state that only a fraction of $(AB^+)^*$-M collisions result in stabilization. In this case k_s is given by

$$k_s = \beta k_s' \tag{7}$$

where k_s' is the collision rate constant and β is the stabilization efficiency.

The mechanism in eq 2 provides the basis for estimating values of k_b, k_r, k_d, k_{RA} and k_s from the experimental values of k_2 and k_2' and the collision theory values of k_f and k_s. It is implicitly assumed when invoking mechanism 2 as written that all the rate constants therein are independent of pressure. The fact that these rate constants may, in fact, be functions of pressure will be seen in Section 3 , after a microscopic theoretical model for these reactions is presented. In the remainder of this section the estimation of k_b, k_d, k_r, k_{RA} and k_3 from the experimental data will be outlined.

The experimental values of k_2 as a function of bath gas pressure can be used to determine k_3 and k_{RA} from the relations

$$k_3 = \left[\frac{\partial k_2}{\partial [M]}\right]_{[M]\to 0} \tag{8}$$

$$k_{RA} = (k_2)_{[M]\to 0} \tag{9}$$

Substitution of eq 3 into 8 and 9 leads directly to eqs 4b and 5b, respectively. In the limit of low [M] the k_s[M] term is negligible in the denominator of eq 3 and a plot of k_2 vs. [M] should be linear with the slope equal to k_3 and the intercept equal to k_{RA}. Thus, k_3 and k_{RA} can both be determined from measurements in the linear, third-order pressure regime.

Expressions for k_b, k_d and k_r in the low-pressure limit can be otained by combining eqs. 4b, 5b and 6b to yield

$$k_r^o = \frac{k_s k_{RA}}{k_3} \tag{10}$$

$$k_d^o = \frac{k_s k'_2}{k_3} \tag{11}$$

$$k_b^o = \frac{(k_f - k_{RA} - k'_2)k_s}{k_3} \tag{12}$$

where the zero superscript indicates that these are low pressure values.

Estimates of k_r, k_d and k_b as functions of pressure are also desireable. However, this type of analysis is not always possible, depending upon the particular details of the system being studied. In this paper the following three cases are considered:

Case (a): Pure three-body association. In this case $k_d = 0$ and $k_r = 0$. The behavior of k_2 vs [M] predicted by eq 3 is illustrated in Figure 1. Equation 12 reduces to

$$k_b^o = \frac{k_f k_s}{k_3} \tag{13}$$

and k_b^o can be obtained directly from the slope of k_2 vs [M] in the linear, low-pressure regime. At higher pressures eq 3 can be rearranged to yield

$$k_b = k_s[M]\left[\frac{k_f - k_2}{k_2}\right] \tag{14}$$

which may be used to calculate k_b as a function of pressure by using experimental k_2 values measured at specific pressures. It should also be noted here that the use of linear plots of k_2^{-1} vs $[M]^{-1}$ to determine k_b and k_f [5] has been shown to be invalid and to lead to large errors in the estimated value of k_f [7]. It will be shown in Section 4 that comparison between theoretical and experimental values of k_b is most meaningful when both the temperature and the pressure are specified explicitly.

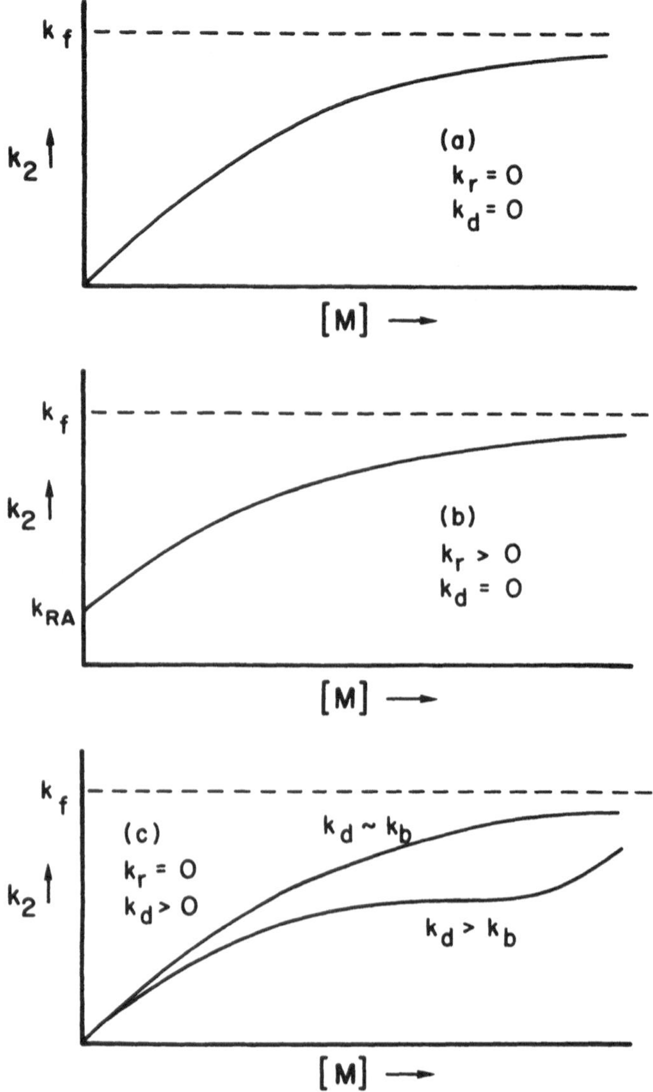

Fig. 1. Schematic diagram of k_2 vs M as predicted by eq 3 : (a) Pure three-body association; (b) three-body and radiative association; (c) three-body association with an exothermic fragmentation channel available. See text for discussion.

Case (b): Three-body and radiative association. In this case $k_d = 0$ and $k_r > 0$. The expected behavior of k_2 vs [M] is illustrated in Figure 1b. Note that the intercept at [M] = 0 is finite (and equal to k_{RA}) in contrast to the pure three-body case. At higher pressures collisional stabilization dominates the radiative step. Thus, in this regime cases (a) and (b) appear similar. In both cases $k_s \rightarrow k_f$ as $[M] \rightarrow \infty$.

The low pressure values of k_r and k_b can be obtained from eq 10 and eq 12, respectively. At high pressures, eq 3 can be rearranged to yield

$$k_b = (k_s M + k_r) \left[\frac{k_f - k_2}{k_2} \right] \tag{15}$$

If some specific pressure dependence can be assumed for k_r, then k_b may be calculated as a function of pressure. If this is not the case, then k_b can only be estimated in the low-pressure regime.

Case (c): Three-body association with a product fragment channel available. In this case $k_d > 0$ and $k_r = 0$. The behavior expected for k_2 vs [M] is illustrated in Figure 1c. Note that the exact shape of the curve depends on the relative values of k_b and k_d over the pressure range studied. If there is a substantial difference between the values of k_b and k_d then the competition between stabilization and dissociation may occur in two distinct phases. For instance, if $k_d \gg k_b$ (which is likely for an exothermic fragment channel without a barrier) then there may be some intermediate pressure range where k_2 begins to level off since stabilization essentially dominates back-dissociation, but has not yet begun to compete effectively with forward dissociation into the $C^+ + D$ channel. At some higher pressure stabilization would begin to compete with forward-dissociation and k_2 would begin to rise again, eventually reaching the collision limit. If k_b and k_d are close in magnitude, then the competition between stabilization and dissociation would occur in a single pressure regime.

Formally, eqs 3 and 6b can be solved simultaneously for k_b and k_d to yield (for $k_r = 0$)

$$k_d = \frac{k_2' k_s [M]}{k_2} \tag{16}$$

$$k_b = \frac{(k_f - k_2 - k_2') k_s [M]}{k_2} \tag{17}$$

These equations may not always prove useful. For instance, for an exothermic product channel with no barrier it is likely that k_2' will be close in value to k_f. Hence, eq 17 would not provide a very accurate estimate of k_b. The utility of these equations should be discussed with regard to specific systems, rather than in general terms.

3 Theoretical Analysis

The basic assumptions of the theoretical model discussed in this work can be stated as follows:

(1) Ion-molecule collision rates are determined by passage across the centrifugal barrier in the long-range part of the effective potential. Thus, thermal ion-molecule collision rate constants can be calculated from classical Langevin theory [1a] for non-polar neutrals and from ADO theory [1b] for polar neutrals.

(2) All A^+/B collisions result in the formation of an $(AB^+)^*$ complex. Thus, the rate constant k_f can be calculated according to assumption (1).

(3) A fraction, β, of all $(AB^+)^*/M$ collisions results in stabilization of the $(AB^+)^*$ complex. Thus, the stabilization rate constant can be written as $k_s = k_s'\beta$, where k_s' is the collision rate constant calculated according to assumption (1) and β is the stabilization efficiency.

(4) The rate constant for radiative stabilization, k_r, is independent of the internal state of the $(AB^+)^*$ complex. This is intended to be only a first approximation. Dunbar [2] and Woodin and Beauchamp [15] have used an approach in which the radiative rate is calculated as a function of the internal energy of the complex. This will be discussed further in section 4.

(5) The $(AB^+)^*$ complex reaches a state of quasi-equilibrium (energy randomization) before dissociating. Hence, subject only to restrictions imposed by conservation of energy and angular momentum all states of the complex are populated uniformly.

(6) The rate of dissociation of the $(AB^+)^*$ complex into any available fragment channel is determined by the corresponding rate of passage across the appropriate transition state in that channel. In terms of the system dynamics the transition state is defined as a dividing surface in the system phase space such that all reactive trajectories (those which proceed from $(AB^+)^*$ to separated fragments) cross this surface once and only once; all non-reactive trajectories do not cross the surface at all [6].

In order to satisfy the principle of microscopic reversibility, the back-dissociation of the complex into the reactant channel must be described by the orbiting transition state, located at the centrifugal barrier. Hence, both the formation of the complex (according to assumptions (1) and (2) above) and its back-dissociation are governed by the same transition state. Dissociation into other fragment channels may be described by orbiting or tight transition states, depending upon the nature of the process.

Assumptions (5) and (6) provide the basis for statistical reaction rate theory, most commonly applied in the form of RRKM theory [3] or QET [4]. When the orbiting transition state is utilized and angular momentum conservation rigorously adhered to (in the limit of classical, spherical top species) the formalism is usually referred to as "phase space" theory [5]. Several reviews of statistical theory have recently been presented [6]. The application of the phase space model to ion-molecule association has previously been discussed [10,13], so only the main points will be outlined here.

Mechanism 2 can be rewritten to show explicitly the dependence of the rate constants on the internal state of the complex:

$$A^+ + B \underset{k_b'(E,J)}{\overset{k_f F(E,J)}{\rightleftharpoons}} AB^+(E,J)^* \quad (18a)$$

$$AB^+(E,J)^* + M \xrightarrow{k_s' \beta} AB^+ + M \quad (18b)$$

$$AB^+(E,J)^* \xrightarrow{k_r} AB^+ + h\nu \quad (18c)$$

$$AB^+(E,J)^* \xrightarrow{k_d'(E,J)} C^+ + D \quad (18d)$$

where F(E,J) is the distribution function for the activation process and the primes on $k_f'(E,J)$ and $k_d'(E,J)$ help the distinguish these microscopic values from the phenomenological rates k_b and k_d. The calculation of these microscopic quantities, and of F(E,J), will be discussed shortly. Note that if the collisional and radiative steps are treated on a microscopic level, then steps 18b and 18c must be written in terms of transition between the initial state (E,J) and some final state (E',J'), where the final state may or may not be stabilized with regard to dissociation. Thus, stepwise deactivation taking place at each level. Application of the steady-state hypothesis to $AB^+(E,J)^*$ at each (E,J) in a model such as this leads to a set of simultaneous equations which must be solved iteratively [16].

In the present treatment, transitions between states are not considered explicitly. The stabilization steps are treated only in terms of average rates, as indicated in mechanism 18 . Application of the steady-state hypothesis to $AB^+(E,J)^*$ in this mechanism leads to the steady-state distribution function

$$P_{ss}(E,J,[M]) = \frac{F(E,J)}{k_t'(E,J,[M])} \Bigg/ \int_{E_0}^{\infty} dE \int_0^{J_{max}} dJ \; \frac{F(E,J)}{k_t'(E,J,[M])} \qquad (19)$$

where

$$k_t'(E,J,[M]) \equiv k_b'(E,J) + k_d'(E,J) + k_s'\beta[M] + k_r \qquad (20)$$

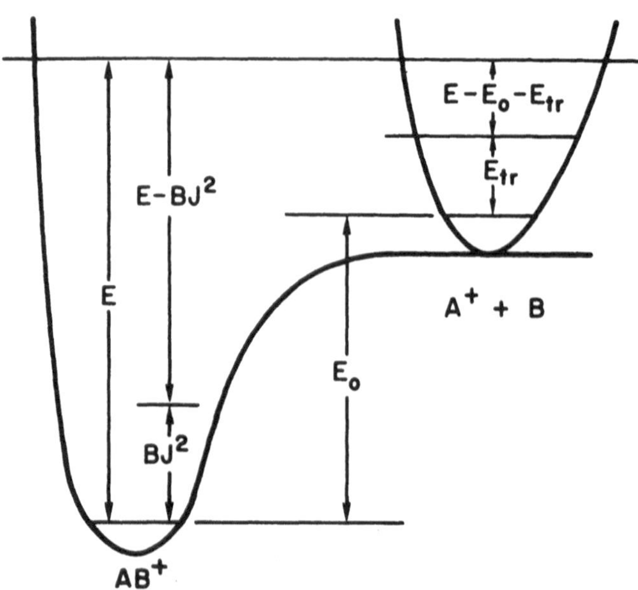

Fig. 2. Schematic potential energy surface for unimolecular dissociation.

The phenomenological dissociation rates are calculated by convoluting the microscopic rate constants with the steady-state distribution function:

$$k_b = \int_{k_o}^{\infty} dE \int_{0}^{J_{max}} dJ\ k_b'(E,J)P_{ss}(E,J,[M]), \tag{21}$$

$$k_d = \int_{E_o}^{\infty} dE \int_{0}^{J_{max}} dJ\ k_d'(E,J)P_{ss}(E,J[M]). \tag{22}$$

Note that k_b and k_d calculated from eqs 21 and 22 respectively, are pressure dependent due to the pressure dependence of the steady-state distribution in eq 19. By substituting k_b and k_d into eqs 3 through 6 theoretical values of k_2, k_3, k_{RA} and k_2' can be obtained for comparison with experiments.

The lower limit on the energy integrals in eqs 19, 21 and 22 is the zero-point energy difference between the complex and the separated reactants, as shown in Figure 2. The upper limit on the J integrals in these equations corresponds to the maximum value of J for which the A^+ + B reactant pair can overcome the centrifugal barrier at a given total energy E [10].

The distribution function for the activation process is determined by thermally weighting the collision flux of the A^+ + B pair as a function of E and J:

$$F(E,J) = \frac{e^{-E/k_BT}\,\Phi(E,J)}{\int_{E_o}^{\infty} dE\ e^{-E/k_BT} \int_{0}^{J_{max}} dJ\Phi(E,J)} \tag{23}$$

where $\Phi(E,J)$ is the flux of A^+ + B collision pairs through the centrifugal barrier, or orbiting transition state, to form a complex with energy E and angular momentum J. This flux can be written as [5]

$$\phi(E,J) = \frac{2J}{h} \int_{E_{tr}^{\neq}}^{E} dE_{tr}\,(E_{tr},J)\rho_v(E - E_o - E_{tr}) \qquad (24)$$

where h is Planck's constant, $\Gamma(E_{tr},J)$ is the rotational-orbital sum of states for the $A^+ + B$ collision pair at translational-rotational energy E_{tr} and angular momentum J, $\rho_v(E - E_o - E_{tr})$ is the vibrational density of states at vibrational energy $E - E_o - E_{tr}$, and $E_{tr}^{\neq}$ is the minimum translational-rotational energy for which the pair can overcome the centrifugal barrier at angular momentum J. The 2J classical degeneracy corresponds to the range of values possible for J_z, the projection of J on a space-fixed axis. The actual evaluation of the integral in eq 13 has been discussed elsewhere [5b].

According to statistical theory, as outlined in assumption (5) and (6) the rate constant for unimolecular dissociation into a particular fragment channel is written as [3-6]

$$k(E,J) = \frac{\phi(E,J)}{\rho(E,J)} \qquad (25)$$

where $\phi(E,J)$ is the flux at the transition state and $\rho(E,J)$ is the density of states of the unimolecular reactant (in this case the excited complex). For the orbiting transition state this expression becomes [5]

$$k(E,J) = \frac{S \int_{E_{tr}^{\neq}}^{E} dE_{tr}\Gamma(E_{tr},J)\rho_v(E - E_o - E_{tr})}{2J\ h\ \rho_v'\ (E - BJ^2)} \qquad (26)$$

where S is the reaction path degeneracy, $\rho_v'(E-BJ^2)$ is the vibrational energy $E - BJ^2$, B is the geometric-mean rotational constant of the complex, and BJ^2 represents classically the rotational energy of the complex (approximated as a spherical top) at angular momentum J. For tight transition states eq 25 is written as [3].

$$k^{\neq}(E,J) = \frac{G_v^{\neq}(E - E^{\neq} - B^{\neq}J^2)S}{h\ \rho_v'(E - BJ^2)} \qquad (27)$$

where $G_v^{\neq}(E - E^{\neq} - B^{\neq}J^2)$ is the vibrational sum of states at the transition state with vibrational energy $E - E^{\neq} - B^{\neq}J^2$, $E^{\neq}$ is the zero-point energy difference between the complex and the tight transition state, and $B^{\neq}$ is the geometric-mean rotational constant of the transition state.

4. Examples

4.1. The Proton-Bound Dimers of Ammonia and the Methylamines

The reactions of ammonia and the methylamines with their corresponding protonated analogs lead to the formation of the proton-bound dimers according to the mechanism

$$AH^+ + A \underset{k_b}{\overset{k_f}{\rightleftharpoons}} (A_2H^+)^* \xrightarrow{k_s[M]} A_2H^+ \tag{28}$$

where $A = NH_3$, CH_3NH_2, $(CH_3)_2NH$ and $(CH_3)_3N$. These reactions have been studied by Neilson et al. [8] using ICR (ion cyclotron resonance) spectrometry and by Meot-Ner and Field [7] using high pressure mass spectrometry. The ICR study was performed at pressures of A ranging from 10^{-4} to 10^{-3} torr with the neutral parent molecule serving as the third-body stabilizer. The mass spectrometry study was performed with mixtures of 1% NH_3 in CH_4 bath gas and 1% CH_3NH_2 or $(CH_3)_2NH$ in i-C_4H_{10} bath gas (no results were reported by Meot-Ner and Field for the $(CH_3)_3N$ system). In both of these studies the data were analyzed assuming that the stabilization step proceeds with unit efficiency, i.e., the stabilization efficiency, β, defined in eq 1 and in assumption (3), was taken to be unity. This assumption is reasonable because only a small amount of energy needs to be removed from the $(A_2H^+)^*$ to stabilize it with regard to back-dissociation (the only fragmentation channel available), and because the bath gases are polyatomic species similar (or identical) to the neutral parent and therefore have a significant number of internal modes capable of absorbing energy from the excited complex.

Statistical theory has been applied to these system by Olmstead et al. [9], who used the RRKM formalism [3] to model the results of Meot-Ner and Field, and by Bass et al. [10], who used the phase space formalism [5] to model both sets of experimental data. Both theoretical treatments are based on the general approach outlined in Section 3 , with the stabilization efficiency, β, set equal to unity and the collision rates k_f and k_s

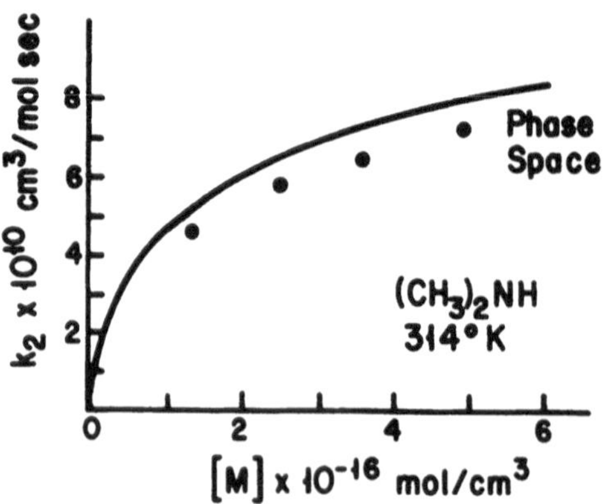

Fig. 3. Comparison of theoretical and experimental pressure dependence of k_2 for the reaction

$$(CH_3)_2NH_2^+ + (CH_3)_2NH \xrightarrow{k_2} ((CH_3)_2NH)_2H^+$$

in the mass spectrometry pressure regime. Phase space calculation from Ref. 10; experimental points from Ref. 7.

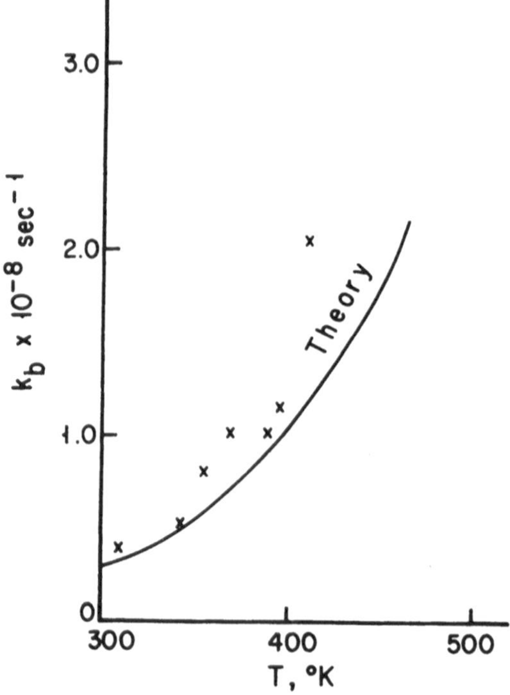

Fig. 4. Comparison of theoretical and experimental temperature dependence of k_b for the reaction

$$[((CH_3)_2NH)_2H^+]^* \xrightarrow{k_b} (CH_3)_2NH_2{}^+ + (CH_3)_2NH$$

in the mass spectrometry pressure regime. Phase space theory from Ref. 10; experimental points from Ref. 7.

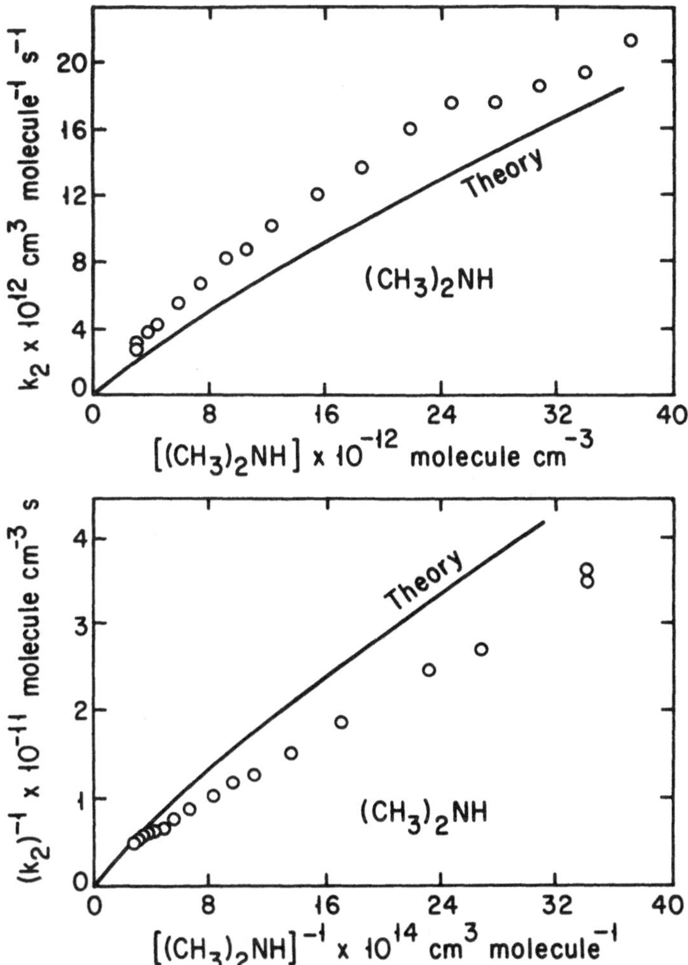

Fig. 5. Comparison of theoretical and experimental pressure dependence of k_2 for the reaction

$$(CH_3)_2NH_2^+ + (CH_3)_2NH \xrightarrow{k_2} ((CH_3)_2NH)_2H^+$$

in the ICR pressure regime. Phase space theory from Ref. 10; experimental points from Ref. 8. Also shown is the inverse plot used in Ref. 9 to estimate k_b.

calculated from Langevin [1a] or ADO theory [1b]. The main difference between the two treatments is that RRKM theory does not account rigorously for angular momentum conservation while phase space theory does. A comparison between the two approaches was made by Bass et al. [10]. Only the phase space results will be considered here.

In the mass spectrometry study of Meot-Ner and Field values were reported for k_2 as a function of pressure for the CH_3NH_2 and $(CH_3)_2NH$ systems, and for k_b as a function of temperature for the NH_3, CH_3NH_2 and $(CH_3)_2NH$ systems. The k_2 vs pressure plots were reported at specific temperatures, as illustrated in Figure 3 for the $(CH_3)_2NH$ system. Experimental values of k_b were determined by substituting measured values of k_2 and the corresponding concentration [M], into eq 14. Figure 4 shows a plot of k_b vs T as reported by Meot-Ner and Field for the $(CH_3)_2NH$ system. Since these authors did not specify the particular pressure at which these k_b values were obtained, it was assumed in the theoretical studies [9,10] that these data represent pressures intermediate in the pressure range of the experiments (roughly 0.3 to 1.8 Torr).

In the ICR study of Neilson et al. k_2 values were reported as a function of pressure, as illustrated in Figure 5 for the $(CH_3)_2NH$ system. Also shown in this figure is an inverse pressure plot, i.e. a plot of k_2^{-1} vs $[M]^{-1}$. These authors noted that the experimental inverse plot is essentially linear and that eq 3 can be invereted to yield (for $k_r = k_d = 0$)

$$\frac{1}{k_2} = \frac{1}{k_f} + \frac{k_b}{k_f k_s [M]} \tag{29}$$

Thus, they assumed k_b to be constant in the experimental pressure range and estimated k_b from the slope of the inverse plot. Bass et al. [10] pointed out that the inverse plot may be linear not only for constant k_b, but also if k_b varies linearly with pressure. Hence, it is not clear if k_b obtained on the basis of eq 29 corresponds to the low pressure limit, k_b^o, or to some average value throughout the pressure range of the experiments. Bass et al. [10] applied eq 14 to the data of Neilson et al. and estimated k_b as a function of pressure, as illustrated in Figure 6 for the $(CH_3)_2NH$ system. The arrow on the ordinate in this figure represents the value of k_b estimated from the slope on the inverse plot, and the points represent the values obtained from eq 14. Although the inverse plot estimate of k_b appears to provide a reasonable estimate of k_b^o, there has been no formal proof that

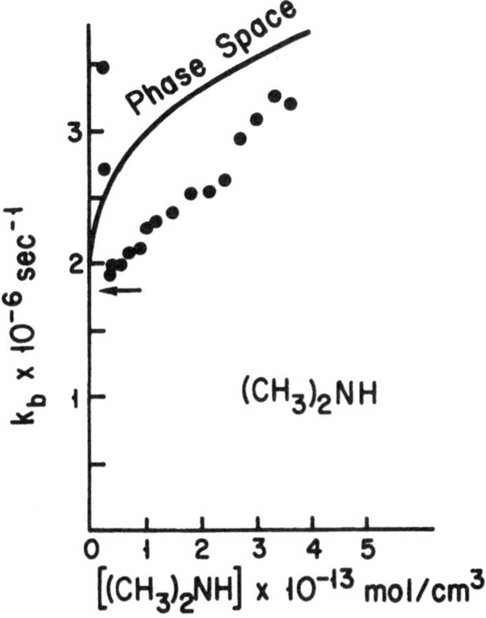

Fig. 6. Comparison of theoretical and experimental values of the pressure dependence of k_b for the reaction

$$[((CH_3)_2NH)_2H^+]^* \xrightarrow{k_b} (CH_3)_2NH_2^+ + (CH_3)_2NH$$

in the ICR pressure regime. Phase space theory from Ref. 10; experimental points as calculated in Ref. 10 from the data in Ref. 8.

this will always be the case. Hence, it has been suggested that the most meaningful comparison between theoretical and experimental values of k_b will be obtained when both the pressure and temperature are specified explicitly [10].

It should also be noted that Neilson et al. estimated values for k_f from eq 29 by extrapolating the experimental inverse plot to $[M]^{-1} \rightarrow 0$. However, Bass et al. pointed out that this method involves extrapolating low-pressure data to the region of infinite pressure and therefore is not valid. They also demonstrated that such an approach can lead to errors of several orders of magnitude in the estimated value of k_f. For instance, consider the theoretical curve in the inverse plot in Figure 5. The curve is essentially linear throughout the range $[(CH_3)_2NH]^{-1} > 8 \times 10^{-14}$ cm^3 $molecule^{-1}$. If this portion of the curve were extrapolated to the y axis an intercept of roughly 0.5×10^{11} molecule·s/cm^{-3} would result. This yields $k_f = 2 \times 10^{-11}$ cm^3 $molecule^{-1}$ s^{-1}. However, this curve was calculated using the ADO value $k_f = 1.5 \times 10^{-9}$ cm^{-3} $molecule^{-1}s^{-1}$. Hence, even if the theoretical model described the experiments exactly, the extrapolation of the linear portion of the inverse plot would lead to an estimate of k_f approximately two orders of magnitude too small.

In general, Figures 3 through 6 show reasonably good agreement between theory and experiment for the $(CH_3)_2NH$ system. Similar agreement was obtained for the CH_3NH_2 and $(CH_3)_3N$ systems. For the ammonia system the discrepancy between theory and experiment was somewhat more substantial. This was discussed in terms of the possible breakdown of assumptions (2) and (5) for this system [10]. Since the ammonia dimer has only a few low frequency modes to absorb the collision energy it is possible that this energy is not transferred out of the collision coordinate rapidly enough to prevent immediate back-dissociation. This may be interpreted from two perspectives: First, the process may be viewed as a "hard" collision in which the two species rebound from each other without forming a complex.
Second, the process may be viewed as the formation of a complex which back-dissociates before the excess collision energy is randomized throughout all the modes. Either perspective is valid.

In summary, the results presented by Bass et al. [10] tend to support the energy transfer mechanism in eq 28 for the formation of the proton-bound dimers of the methylamines, but there is some room for doubt as far as the ammonia system is concerned. These authors also demonstrated that the use of inverse plots to obtain experimental values of k_f and k_b may yield misleading results. Hence, it is best to use eq 14, to calculate k_b directly at specific pressures and temperatures.

4.2. The Radiative and Three-Body Association Reaction of CH_3^+ with HCN

The association reaction of CH_3^+ with HCN has been studied using ICR spectrometry by McEwan et al [12] and by Bass et al. [13] and has been interpreted according to the mechanism

$$CH_3^+ + HCN \underset{k_b}{\overset{k_f}{\rightleftarrows}} (CH_3 \cdot HCN^+)^* \begin{cases} \xrightarrow{k_s[M]} CH_3 \cdot HCN^+ \\ \xrightarrow[k_r]{} CH_3 \cdot HCN^+ + h\nu \end{cases} \tag{30}$$

In both cases helium was used as the third-body and the overall association rate constant, k_2 was reproted as a function of helium pressure in the range 10^{-6} to 2×10^{-2} Torr at room temperature, and as a function of temperature in the limit of zero helium pressure. A single experimental point at 0.5 Torr helium pressure was reported by Schiff and Bohme using the flowing afterglow technique [11].

In the theoretical model there are two factors which make this system more involved than the amine dimerization in the proceeding example. First is the fact that radiative step is present and therefore must be accounted for. Second is the fact that the bath gas is monatomic helium, for which the stabilization efficiency is expected to be significantly less than unity [17]. In a realistic model the collisional stabilization process should be considered as taking place in a series of steps, with each step removing only a fraction of the energy which must be removed in order to stabilize the $(CH_3HCN^+)^*$ complex with respect to back-dissociation. This type of treatment leads to an overall stabilization efficiency which is pressure-dependent [16]. The radiative rate constant also is expected to be a function of the internal energy of the complex and therefore the phenomenological value, k_r, will be a function of pressure.

Statistical theory has been applied to ion-molecule radiative association reactions by Herbst [18], by Woodin and Beauchamp [15], and by Bass et al. [13]. Herbst has utilized several different formulations of the theory, and has recently [18b] incorporated some modifications suggested by Bates [19]. Woodin and Beauchamp treated the radiative rate constant explicitly as a function of the internal energy of the complex. Unfortuneately, there is not sufficient information available on the CH_3HCN^+ complex to allow this approach to be used for this species. None of the theoretical models to date have accounted explicitly for the energy transfer processes involved in the collisional stabilization step, although this step has been considered in detail for non-radiative systems by a number of authors [20].

In the theoretical model applied by Bass et al. [13] the radiative and collisional steps were treated only in an the approximate manner described in Section 3. No attempt was made to model these steps on a truly microscopic level. Note that most of the complexes formed via the CH_3^+ + HCN collisions have energies in the range from about 4.2 to 4.5 eV above the zero-point level of CH_3HCN^+. Data presented by Woodin and Beauchamp [15] for other compounds indicate that k_r is not likely to vary by more than about 25% in this energy range. Hence, the use of a single value of k_r throughout the range appears reasonable for this system. Although the stabilization efficiency, β, was treated as being independent of internal state, and therefore, independent of pressure, calculations were performed for a number of values of β between 0.1 and 1.0. These results provide some indication of the results which would be expected from a more sophisticated treatment which did account explicitly for the pressure dependence of β.

The experimental data obtained by Bass et al. [13] at the lower end of their experimental pressure regime are summarized in Table 1, along with the theoretical results for several values of k_r and β. The experimental values of k_3 and k_{RA} were obtained by applying eqs. 8 and 9 to the linear least-squares fit of the experimental data, shown in Figure 7. The low pressure values of k_r and k_b can then be obtained from eqs 10 and 12. However, k_s in these equations must be replaced by k_s' from eq 7 since $\beta < 1$. The rate constants can then be reported as the ratios k_r/β and k_b/β. The values obtained were $k_r/\beta = 2.1 \times 10^5$ sec^{-1} and $k_b/\beta = 7.1 \times 10^6$ sec^{-1}. Table 1 shows the dependence of the theoretical results on k_r and β.

It is clear that good agreement between theory and experiment is obtained only for $k_r \sim 10^4$ sec^{-1}. Interpolation based on k_2 indicates that k_4 = 1.4×10^4 sec^{-1} will yield the best agreement with experiment.

The theoretical curves in Figure 7 were calculated using $k_r = 1.4 \times 10^4$ sec^{-1} and four values of β as indicated. The same results shown in Figure 7 are shown again in Figure 8 but on a pressure scale which allows the high pressure flow tube data point of Schiff and Bohme [11] to be shown also. It is interesting to note that the theory predicts that saturation is not reached in the region of the flow tube measurement at 0.5 Torr, even for the case β = 1.0. Hence, "high pressure" experiments do not necessarily correspond to saturation. Saturation occurs when the condition

$$k_s[M] \gg k_b'(E,J) \tag{31}$$

holds over the entire manifold of (E,J) states produced by the activation process. When this happens then stabilization overwhelms dissociation at each (E,J) state, and essentially all complexes formed will be stabilized. Thus, the saturation pressure depends upon the magnitude of the microscopic dissociation rate constants, not on the high pressure limit of a particular type of experiment.

The data in Figure 7 show good agreement between theory and experiment if the stabilization efficiency in this regime is about 0.1. The data in Figure 8 indicate that if β = 0.1 at 0.5 Torr then theory and experiment will not agree as well at this pressure as in the low pressure ICR regime. Agreement between theory and experiment will be best if β is an increasing function of pressure, going from 0.1 at low pressure to about 0.25 at 0.5 Torr.

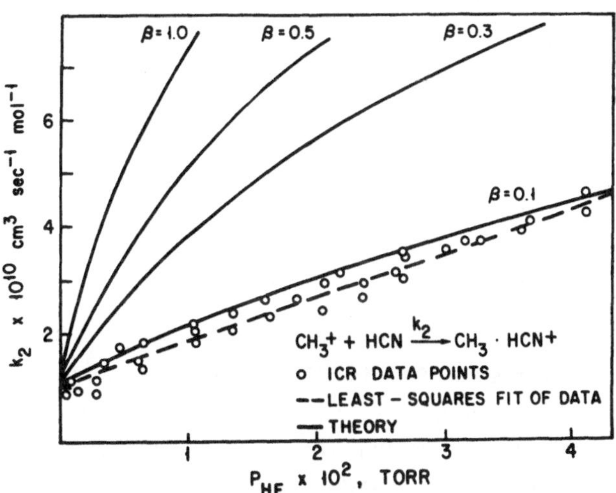

Fig. 7. Comparison of theoretical and experimental pressure dependence of k_2 for the reaction

$$CH_3^+ + HCN \xrightarrow{k_2} CH_3 \cdot HCN^+$$

in the ICR pressure regime. Dashed line represents linear least-squares fit of experimental data. Taken from Ref. 13.

Table 1: Theoretical and Experimental Rate Data in the ICR Pressure Regime at 300 K.

	Exp	Theory using k_r sec^{-1}			
		10^2	10^3	10^5	10^5
$k_{RA} \times 10^{10}$ $cm^3 mol^{-1} sec^{-1}$	1.0 ± 0.1^a	0.014^b	0.11^b	0.78^b	4.7^b
$k_3 \times 10^{25}$ $cm^6 mol^{-2} sec^{-1}$	2.7 ± 0.1^c	5.8^d 23^e	5.4^d 22^e	4.4^d 20^e	2.7^d 13^e
$k_b \times 10^{-5}$ sec^{-1}	7.1^d 36^e	3.3^e 4.2^e	3.5^d 4.2^e	4.3^d 4.6^e	6.2^d 6.3^e

[a] Ref. 13; corresponds to k_2 in the limit of zero He pressure.
[b] Calculated using He pressure = 0 (thus, results are independent of β).
[c] Ref. 13; assumed to correspond to the lower end of the experimental pressure regime shown in Fig. 2 (about 10^{-3} Torr).
[d] Calculated using He pressure = 10^{-3} Torr and β = 0.1.
[e] Calculated using He pressure = 10^{-3} Torr and β = 0.5.

The temperature dependence of k_2 is illustrated in Figure 9. The theoretical curves were calculated using $k_r = 1.4 \times 10^4$ sec^{-1} and the values of β indicated. Note that at zero helium pressure β does not enter into the calculation. The data here is not substantial enough to allow any detailed comparison between theory and experiment, although it is clear that the theory does predict k_2 values of the correct magnitude in the experimental temperature range. One interesting point is that at temperature above 300 K the theory predicts approximately a T^{-3} temperature dependence. However, at lower temperatures the theory deviates substantially from this behavior.

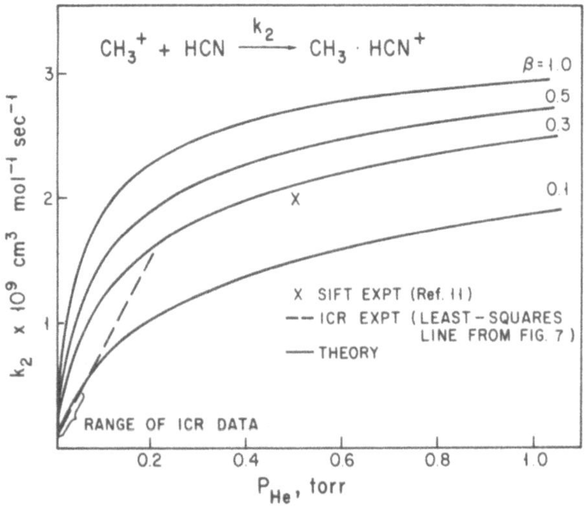

Fig. 8. Comparison of theoretical and experimental values of k_2 for the reaction

$$CH_3^+ + HCN \xrightarrow{k_2} CH_3 \cdot HCN^+$$

at pressures up to 1 Torr. Taken from Ref. 13.

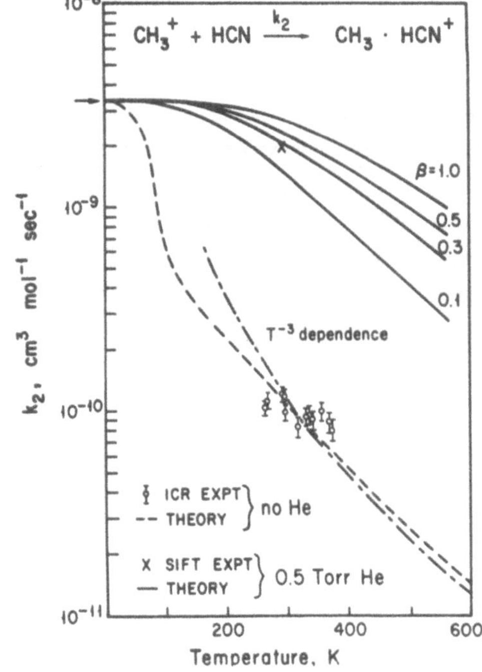

Fig. 9.
Comparison of theoretical and experimental temperature dependence of k_2 for the reaction

$$CH_3^+ + HCN \xrightarrow{k_2} CH_3 \cdot HCN^+$$

in two different pressure regimes. Taken from Ref. 13. Also shown is the behavior predicted by the equation $k_2(T) = k_2(300\ k) \times (T/300)^{-3}$.

This is significant because on the basis of simpler formulations of statistically theory [18b,21] a T^{-n} dependence is predicted for k_2. These formulations utilize a classical or semi-classical approximation to the number of vibrational and rotational states available to both the complex and the transition state. The corresponding sums and densities of states are then calculated using analytical expressions [22] rather than the direct count procedure [23] used by Bass et al. As temperature is decreased the complex is formed with less energy in excess of the dissociation limit. The classical approximation thus becomes worse at lower temperature. The direct count, however, is valid at all temperatures. The curves in Figure 9 indicate that extrapolation of experimental data to very low temperatures, as has been done by some authors to estimate association rates in interstellar clouds [24], should not be based on the T^{-n} dependence observed at higher temperatures. Note that the T^{-n} dependence must break down at low enough temperatures because it predicts that the association rate becomes infinite as $T \rightarrow 0$.

In summary, the basic model presented in Section 3 provides a reasonable representation of the CH_3^+ + HCN reaction over the temperature and pressure range of the ICR and flow tube experiments. Further experimental data in the flow tube pressure regime may provide a more stringent test of the model. A more sophisticated treatment in which both the collisional and radiative steps are dealt with on a microscopic level may prove useful in this case. The model discussed here provides an order of magnitude estimate of $k_r \sim 10^4$ sec^{-1} and indicates that β is in the range from 0.1 to 0.3. Although approximate formulations of statistical theory indicate that k_2 should exhibit a T^{-n} temperature dependence, these approximations must break down at low enough temperatures. The more rigorous treatment discussed here yields a T^{-3} temperature dependence for $T \gtrsim 300$ K, but shows substantial deviation from this behavior at lower temperatures. Hence, extrapolation based on the T^{-n} dependence observed over some experimental temperature range do not necessarily provide reliable estimates of k_2 at lower temperatures.

4.3. The Clustering Reactions of Methanol With Its Protonated Ion

Some preliminary results have recently been reported on the system of reactions [14]

$$CH_3OH_2^+ + CH_3OH \xrightarrow{k_2} (CH_3OH)_2H^+ \qquad (32a)$$

$$CH_3OH_2^+ + CH_3OH \xrightarrow{k_2'} (CH_3)_2OH^+ + H_2O \qquad (32b)$$

$$CH_3OH + (CH_3OH)_2H^+ \xrightarrow{k_t} (CH_3OH)_3H^+ \qquad (32c)$$

$$CH_3OH + (CH_3)_2OH + \xrightarrow{k_m} ((CH_3)_2O)(CH_3OH)H^+ \qquad (32d)$$

where all the rate constants are second-order. The subscripts t and m represent trimer and mixed dimer, respectively, and indicate the nature of the product of the corresponding reactions. Note that the first two reactions can be treated independently of the last two in terms of the theoretical analysis. Experimentally, however, the system must be treated as a whole. The ICR power absorption equations [25] must be solved for each of the ion intensities as a function of the rate constants in reactions 32. An iterative procedure is then used determine the rate constants from the experimental ion intensities.

The reaction sequence in 32 is assumed to proceed via the mechanism

$$CH_3OH_2^+ + CH_3OH \underset{k_b}{\overset{k_f}{\rightleftharpoons}} \left[(CH_3OH)_2H^+\right]^* \begin{array}{l} \xrightarrow{k_s[A]} (CH_3OH)_2H^+ \\ \xrightarrow[k_d]{} (CH_3)_2OH^+ + H_2O \end{array} \qquad (33a)$$

$$(CH_3OH)_2H^+ + CH_3OH \underset{k_{b3}}{\overset{k_{f2}}{\rightleftharpoons}} \left[(CH_3OH)_3H^+\right]^* \xrightarrow{k_{s2}[A]} (CH_3OH)_3H^+ \qquad (33b)$$

$$(CH_3)_2OH^+ + CH_3OH \underset{k_{b3}}{\overset{k_{f3}}{\rightleftharpoons}} \left[((CH_3)_2O)(CH_3OH)H^+\right]^* \xrightarrow{k_{s3}[A]} ((CH_3)_2O)(CH_3OH)H^+ \qquad (33c)$$

Step 33a in the above mechanism corresponds to case (c) discussed in Section 2. Steps 33b and 33a correspond to case (a) with one additional complication - the reactant ions in these steps are formed in 33a and therefore are not necessarily thermalized.

Hence the method of calculating the distribution function for the complex outlined in eqs 23, 24 and 25 must be modified to account for the internal excitation of the reactant ion. Both the theoretical and experimental results on this system are still in the preliminary stage. Hence, any conclusions presented here are still somewhat tentative.

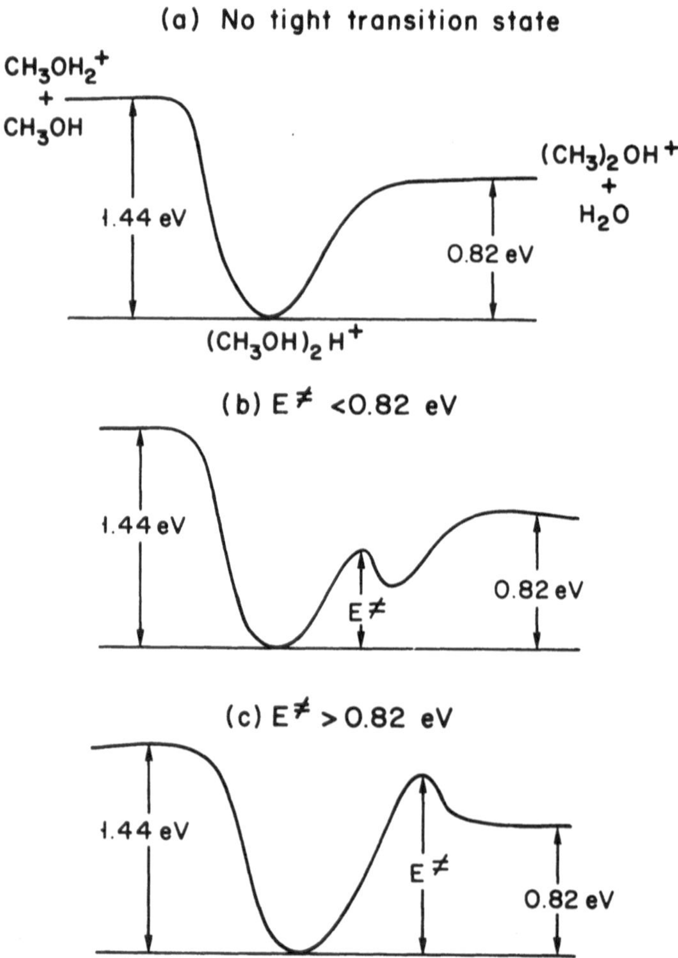

Fig. 10. Three possible potential energy surfaces for the system
$CH_3OH_2^+ + CH_3OH \rightleftharpoons [(CH_3OH)_2H^+]^* \longrightarrow (CH_3)_2OH^+ + H_2O$.

In order to model step 33a it is necessary to make some assumptions about the nature of the potential energy surface and transition state in the water loss channel. A schematic diagram of some possible potential energy surfaces is shown in Figure 10. Calculations performed using surfaces (a) and (b) predict that essentially 100% of all $CH_3OH_2^+$ + CH_3OH collisions lead to reactive dissociation into the water loss channel, i.e., $k_2' = k_f$. Since the experimental results show that k_2' is only about 10% of k_f, a potential energy surface which includes a barrier, such as that in Figure 10c, must be used in order for the theoretical model to yield good agreement with experiment.

The transition state located at the maximum on the potential energy surface is likely to correspond to a cyclic intermediate from which water loss may proceed. The following two structures were considered:

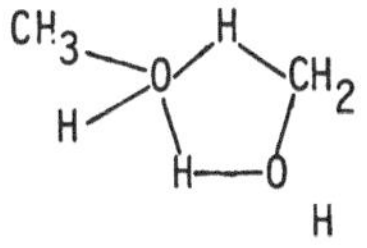

```
CH3      H                    CH3
    >O/   \CH2                    >O——CH3
H /   \    /                  H /  |    |
       H——O                        H——O
           H                           \H
      (I)                         (II)
```

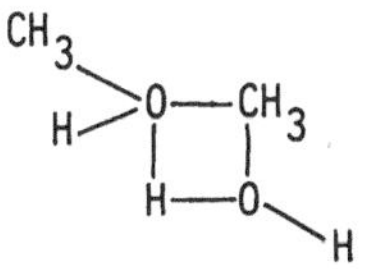

Reasonable limits were placed on the "tightness" and "looseness" of such structures (in other words upper and lower limits were placed on the set of vibrational frequencies and rotational constants in order to set limits on the sum of states at the transition state). A series of calculations was then performed for various values of $E^{\neq}$ using the tightest and loosest transition states reasonable. The results indicate that theory can be brought into agreement with experiment by choosing $E^{\neq}$ in the range from 1.0 to 1.1 eV, depending upon the transition state parameters used. No attempt has been made up to this point to optimize the agreement between theory and experiment by adjusting the transition state parameters and $E^{\neq}$.

Typical results for k_2 vs CH_3OH pressure are shown in Figure 11 for T = 270 K. The two curves shown were calculated for $E^{\neq}$ = 1.0 eV and 1.1 eV, and using transition state parameters intermediate between the tightest and loosest limits. It is not clear at this point if the discrepancy in the experimental data at 267 K and 271 K represents experimental scatter, or if some experimental artifact is responsible for one set of data being invalid. Additional experiments in this temperature range will help clarify this point. In any case, agreement between theory and experiment for k_2 is fairly

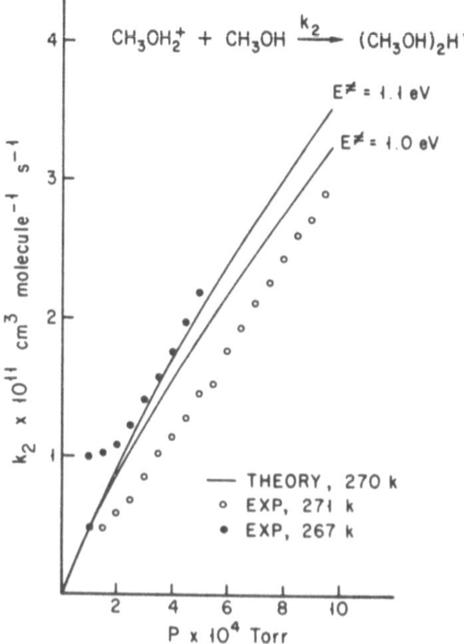

Fig. 11. Comparison of theoretical and experimental pressure dependence of k_2 for the reaction

$$CH_3OH_2^+ + CH_3OH \xrightarrow{k_2} (CH_3OH)_2H^+.$$

Theoretical results based on the potential surface in Fig. 10c using $E^{\neq}$ = 1.0 and 1.1 eV; experimental points correspond to preliminary data from Ref. 14. See text for discussion.

good, and is relatively insensitive to changes in $E^{\neq}$ in the range 1.0 to 1.1 eV. Additional calculations show that these results are also fairly insensitive to changes in the other parameters for the transition state in the water loss channel. This indicates that for any reasonable model the rate of water loss is dominant over the rate of stabilization throughout the range of (E,J) states produced in the activation process. Thus, even if the water loss rate changes by a factor of two or three this rate is still dominant over the stabilization rate. Thus, the competition between stabilization and water loss is affected only slightly by changes in the transition state parameters for the water loss channel.

The predicted values of k_2' at 270 K vary from about 0.5 to 2.0 x 10^{-10} cm^3 $molecule^{-1}$ s^{-1} depending upon the transition state parameters used. The experimental value at this temperature is about 1 x 10^{-10} cm^3 $molecule^{-1}s^{-1}$. Thus, the theoretical and experimental values of k_2' agree to within about a factor of two for any reasonable transition state model.

Results similar to those just discussed for 270 K have also been obtained at 294 K. However, since these are the only two temperatures for which theoretical and experimental results have been compared thus far, no definite conclusions about the temperature dependence can be drawn at this time. The results indicate that it will be possible to find a reasonable set of transition state parameters which will yield good agreement between theoretical and experimental values of k_2 and k_2' over the temperature and pressure range of the experiments.

Step 33b can be modeled similarly to reaction 28, except that the AH^+ reactant ion in 28 is thermalized whereas the $(CH_3OH)_2H^+$ reactant ion in 33b is not necessarily thermalized. As a first approximation the $(CH_3OH)_2H^+$ species is assumed to be thermalized translationally and rotationally, but excited vibrationally by an amount of energy E_x in excess of the thermal vibrational energy. Since the translational and rotational states have thermal populations, the $(CH_3OH)_2H^+/CH_3OH$ collision can be treated as described in Section 3 . But eqs 23 and 24 must be modified to account for the fact that the vibrational energy distribution in the reactant has been shifted upwards by an amount E_x. Hence, the right hand side of eq 24 now yields the flux corresponding to the formation of the complex with total energy E + E_x rather than total energy E. The modified flux at total energy E and angular momentum J, $\phi'(E,J)$, is then given by

$$\phi'(E,J) = \phi(E - E_x,J) \quad (34)$$

where $\phi(E - E_x,J)$ is given by eq 24 with E replaced by E - E_x. The distribution function for the excited trimer $[(CH_3OH)_3H^+]^*$, is then given by eq 23 with $\phi(E,J)$ replaced by $\phi'(E,J)$.

Preliminary calculations indicate that this approach will yield reasonably good agreement with experiment if E_x is on the order of a few hundredths of an eV. However, it has not yet been determined if a single value of E_x will yield good agreement over the whole experimental temperature range, or if some temperature dependent function E_x must be used. The fact that E_x is on the order of only a few hundredths of an eV indicates that the $(CH_3OH)_2H^+$ species is close to thermalized before trimerization occurs.

No attempt has been made thus far to model the reaction or eq 33c leading to the formation of the mixed dimer. In this case the major complication is that the $(CH_3)_2OH^+$ ion is formed via fragmentation of $[((CH_3)_2O)(CH_3OH)H^+]^*$. Thus, this species may have excess translational, as well as excess rotational and vibrational energy. If the excess translatonal energy is substantial then this must be taken into account in eq 23. As first approximation the ion may be treated as thermalized, and comparison between theory and experiment may give some indication of the validity of this assumption.

In summary, the results on the methanol system indicate that the dimerization and water loss processes can be modeled adequately using statistical theory. The water loss channel can be described with a tight transition state located at an energy maximum about 1.0 to 1.1 eV above the $(CH_3OH)_2H^+$ ground level. Results on the trimerization reaction are not very conclusive at this point, but the most likely interpretation at present is that the dimer ion is close to thermalized (with $\lesssim$.02 eV excess vibrational energy) before the trimerization reaction occurs. The theory has not yet been applied to the formation of the mixed dimer.

5. Summary

A generalized model for the application of statistical reaction rate theory to ion-molecule association reactions has been outlined. The model is based on a mechanism which involves the formation of a long-lived intermediate complex which either dissociates back to reactants, is stabilized, or dissociates into some product fragment channel. The stabilization processes considered here are collisional stabilization and radiative stabilization. The stabilization steps are treated in an average sense, while the dissociation steps are considered on a microscopic level. In the three examples discussed here the model was seen to yield reasonably good agreement with experimental data. The use of inverse plots to obtain experimental estimates of k_f and k_b was shown to lead to inaccurate results. It was also pointed out that the temperature dependence of the bimolecular association rate constant dues not necessarily follow a T^{-n} dependence down to low temperatures. Thus, the T^{-n} dependence which may be observed in the laboratory temperature range does not necessarily provide a valid basis for extrapolation to lower temperatures.

Application of the model in sequential clustering reactions is complicated by the fact that for all but the first step of the sequence the reactant ions are not initially thermalized. This aspect of the problem was not dealt with in any detail in this work. Preliminary work on the clustering reaction discussed here indicates that for this system deviations from the thermal distribution are probably small. Hence, a relatively simple approach for modeling sequential clustering and eventual nucleation may be satisfactory. However, a more complete model with wider applicability can be obtained by incorporation of a detailed treatment of the energy transfer processes, both collisional and radiative, into the present model. This may be necessary in modeling some clustering reactions, since it is possible the energy and angular momentum distributions will become further removed from thermalization for each successive step in the clustering process. If this is the case, it will be necessary to calculate these distributions by considering the details of the collisional energy transfer process.

6. Acknowledgement

We gratefully acknowledge the National Science Foundation, Grant No. CHE80-20464, and the donors of the Petroleum Research Fund, administered by the American Chemical Society, for support of this research.

7. References

1. P. Langevin, Ann. Chim. Phys. 5(1905)245; G. Gioumousis and D.P. Stevenson, J. Chem. Phys. 29(1958)294; E. Vogt and G.H. Wannier, Phys. Rev. 95(1954)1190.

2. T. Su and M.T. Bowers, J. Chem. Phys. 58(1973)3027; Int. J. Mass Spectrom. Ion Physics, 17(1975)309; L. Bass, T. Su, W.J. Chesnavich and M.T. Bowers, Chem. Phys. Lett. 34(1975)119.

3. (a) R.A. Marcus and O.K. Rice, J. Phys. Colloid Chem. 55(1951)894; G.M. Wieder and R.A. Marcus, J. Chem. Phys. 37(1962)1835; R.A. Marcus, ibid.,

20(1952)359; (b) P.J. Robinson and K.A. Holbrook, "Unimolecular Reactions", Wiley-Interscience, New York, 1972; W. Forst, "Theory of Unimolecular Reactions", Academic Press, New York, 1973.

4. H.B. Rosenstock, M.B. Wallenstein, A.L. Wahrhaftig and H. Eyring, Proc. Natl. Acad. Sci., USA, 38(1952)667.

5. (a) P. Pechukas and J.C. Light, J. Chem. Phys. 42(1965)3281; J. Lin and J.C. Light, ibid., 43(1965)3209; J.C. Light, Disc. Faraday Soc., 44(1967) 14; E. Niktin, Teor. Eksp. Khim., 1(1965)135,144,428 [Theor. Exp. Chem. 1(1965)83,90,275]; (b) W.J. Chesnavich and M.T. Bowers, J. Chem. Phys. 66(1977)2306; J. Am. Chem. Soc. 98(1976)8301; W.J. Chesnavich, Ph.D. Thesis, University of California, Santa Barbara, CA (1976).

6. W.J. Chesnavich and M.T. Bowers in "Gas Phase Ion Chemistry", Vol.,I, M.T. Bowers (Ed.), Academic Press (1979), pp. 119-153; P. Pechukas in "Dynamics of Molecular Collisions", Part B, W.H. Miller (Ed.), Plenum Press (1976).

7. M. Meot-Ner (Mautner) and F.H. Field, J. Am. Chem. Soc., 97(1978)5339.

8. P.V. Neilson, M.T. Bowers, M. Chan, W.R. Davidson and D.H. Aue, J. Chem. Soc. 100(1978)3649.

9. W.N. Olmstead, M. Lev-On, D.M. Golden and J.I. Brauman, J. Am. Chem. Soc. 99(1977) 992.

10. L.M. Bass, W.J. Chesnavich, and M.T. Bowers, J. Am. Chem. Soc. 101(1979) 5493.

11. H.I. Schiff and D.K. Bohme, Ap. J. 232(1979)740.

12. M.J. McEvan, V.G. Anicich, W.J. Huntress, P.R. Kemper, and M.T. Bowers, Chem. Phys. Lett., 75(1980)278.

13. L.M. Bass, P.R. Kemper, V.G. Anicich and M.T. Bowers, J. Am. Chem. Soc. in press.

14. M.T. Bowers, presented at the 2nd International ICR Conference, Mainz, West Germany, March 1981; R.D. Gates, L.M. Bass and M.T. Bowers, to be submitted.

15. R.L. Woodin and J.L. Beauchamp, Chem. Phys. 41(1979)1.

16. D.C. Tardy and B.S. Rabinovitch, Chem. Rev. 77(1977)369 and references therein.

17. R.D. Cates and M.T. Bowers, J. Am. Chem. Soc., 102(1980)3994.

18. (a) E. Herbst, Ap. J. 205(1976)95; (b) E. Herbst, J. Chem. Phys. 70(1979) 2201; Ap. J. 237(1980)462; J. Chem. Phys., 72(1980)5284; Ap. J. Oct. 15, 1980.

19. D.R. Bates, J. Phys. B: Atom. Molec. Phys., 12(1979)4135.

20. See references 3b and 16 and references therein.

21. D. Smith and N.G. Adams, Chem. Phys. Lett. 54(1978)535; M. Meot-Ner (Mautner) and F.H. Field, J. Chem. Phys. 61(1974)3742.

22. For a discussion of some of the more common analytical approximations for sums and densities of states see Ref. 3b.

24. D. Smith and N.G. Adams, Ap. J. 220(1978)L87.

25. P.R. Kemper, Ph.D. Thesis, University of California, Santa Barbara, CA (1977).

THEORY FOR PULSED AND RAPID SCAN ION CYCLOTRON RESONANCE SIGNALS

Richard L. Hunter and Robert T. McIver, Jr.
Department of Chemistry, University of California
Irvine, California 92717

1. Introduction

Several recent developments have stimulated renewed interest in use of the cyclotron resonance principle for high performance mass spectrometry. A number of laboratories have constructed Fourier transform mass spectrometers (FT-MS) which are capable of ultrahigh mass resolution and rapid scanning [1-4]. These instruments function by storing ions in a one region ion cyclotron resonance (ICR) cell and detecting them by exciting their cyclotron motion in a homogeneous magnetic field [5,6]. The cyclotron resonance principle has also been used in conjunction with ion trapping techniques to study the energetics and dynamics of gaseous ion-molecule reactions [7-10] and to study laser photodetachment and laser photodissociation of ions [11-13]. The most advanced new instruments utilize high-field superconducting magnets to store the ions efficiently in the analyzer cell for long periods of time (up to several minutes).

The basis for use of the cyclotron resonance principle in mass spectrometry is that ions of a particular mass-to-charge ratio can be accelerated by an alternating electric field which is perpendicular to the magnetic field. In the presence of a magnetic field, gaseous ions are constrained to move in small circular orbits, and the frequency of the cyclotron motion is given approximately by $\omega = qB/m$, where m/q is the mass-to-charge ratio of an ion and B is the magnetic field strength. A number of ion detection methods are based on this principle. In the omegatron mass spectrometer an electrometer is used to measure the current of resonant ions impinging on a collector electrode [14]. Cyclotron resonance has also been detected with an audiomodulated bridge circuit developed by Wobschall in 1965 [15]. At the present time, however, marginal oscillators are the most widely used method [16-19]. With a marginal oscillator, the ICR cell is incorporated as a capacitance element in a parallel resonant circuit. When the circuit oscillates at a frequency which is the same as the cyclotron frequency of an ion, energy absorbed by the resonant ions is detected as a slight drop in the oscillation level of the tank circuit. Marginal oscillators are widely used because of their high sensitivity and simplicity of design. However, a major drawback is that the marginal oscillator is a narrowband detector which is difficult to scan in frequency. A mass spectrum is normally obtained by slowly scanning the magnetic field strength while the marginal oscillator is set at a fixed observing frequency. Of course this mode of operation is impractical with superconducting magnets which operate best at fixed field strength.

A capacitance bridge detector (CDB) has been described as an alternate circuit for the detection of ion cyclotron resonances [20]. One of the most important features of the detector is that studies of gaseous ion-molecule reactions can be performed at constant magnetic field to minimize problems caused by loss of ions from the analyzer cell. We have also used it in rapid scan and FT-MS experiments for increased sensitivity and mass resolution.

In this paper, we describe the theory of operation and performance features of the capacitance bridge detector for the pulsed ICR and rapid scan ICR detection modes.

2. Theory

A theory for the full transient response of the capacitance bridge circuit in detecting ion cyclotron resonance signals has been previously published [5]. Unfortunately the solutions are quite complex. In this section a simplified analysis is presented to illustrate more clearly how the circuit functions and what parameters are most important in its performance.

2.1 Basic Concepts

Figure 1 shows a block diagram of a capacitance bridge detector (CBD) circuit. The upper and lower plates of the ICR analyzer cell function as a capacitor which is connected in series with a small balance capacitor to form a capacitance bridge. Gaseous ions are generated by electron impact and stored between the plates of the ICR cell by crossed electric and magnetic fields. They can be accelerated at their cyclotron frequency by an alternating electric field which is established between the two plates. There are two oscillators, a frequency synthesizer and a voltage controlled oscillator (VCO), connected to the upper plate of the ICR cell through two

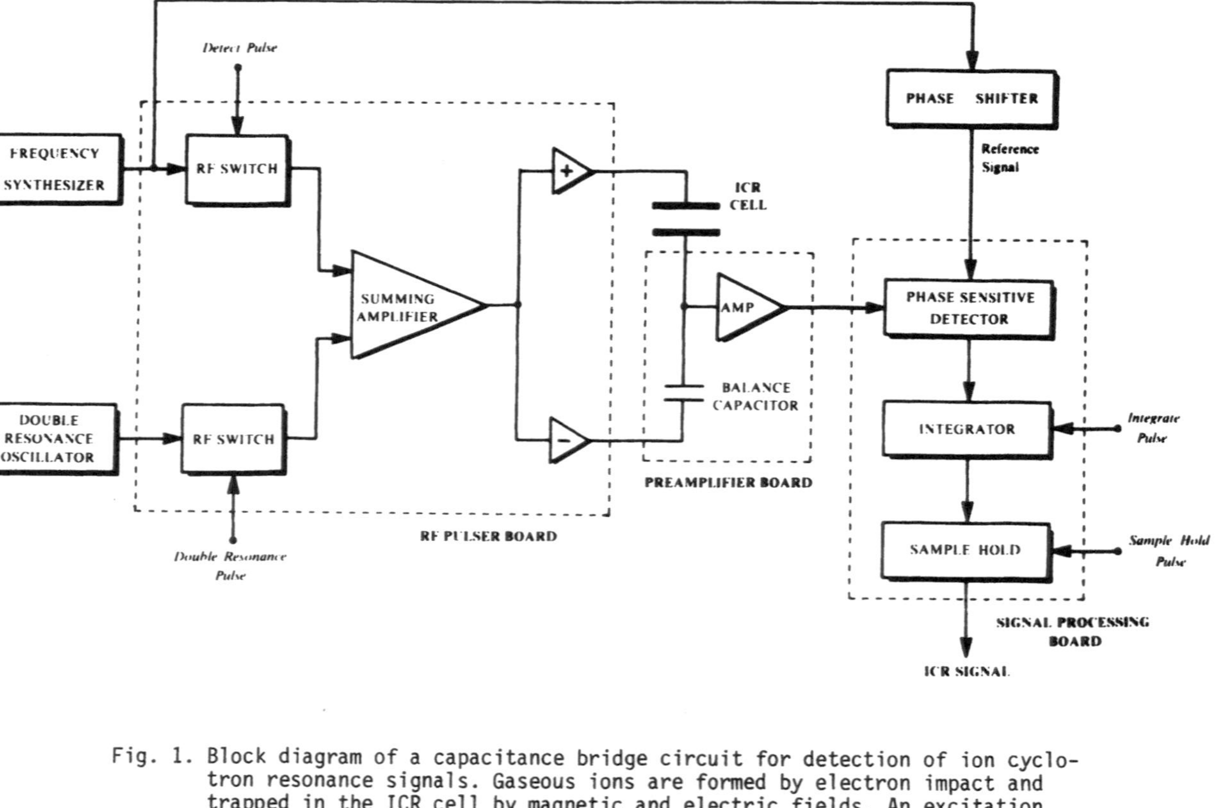

Fig. 1. Block diagram of a capacitance bridge circuit for detection of ion cyclotron resonance signals. Gaseous ions are formed by electron impact and trapped in the ICR cell by magnetic and electric fields. An excitation rf voltage signal applied to the upper plate of the cell accelerates the ions and causes an imbalance signal to appear at the input of the preamplifier.

analog switches and a summing amplifier. The frequency synthesizer provides a reference signal for the phase sensitive detector (PSD) and serves as the source of ω_1, the ion detection frequency. Ion cyclotron double resonance experiments [21-23] are performed using the VCO. The composite rf voltage from the two oscillators is passed to a pair of buffer amplifiers to produce two rf signals 180^0 out of phase. One of the signals is applied to the upper plate of the ICR cell and the other is applied to the balance capacitor. The amplitude of the rf voltage applied to the balance capacitor can be independently adjusted to balance the bridge and produce a virtual ground at the input of the preamplifier. When ions in the ICR cell come into resonance with the applied rf electric field, the bridge is thrown out of balance and rf voltage at the resonance frequency of the ions appears at the input of the preamplifier. This signal, which is proportional to the number of resonant ions, is amplified by the preamplifier and passed to a phase sensitive detector referenced to a phase-shifted version of the ω_1 rf signal. The PSD functions as a synchronous demodulator to produce a low frequency transient signal. The low frequency signal is then integrated and acquired by a sample/hold unit to make it suitable for display on an X-Y recorder or for acquisition by a computer data system.

With most types of mass spectrometers ions are detected by having them collide with the surface of a device such as a continuous dynode electron multiplier. A fundamentally different method is used, however, with the capacitance bridge detector: the circuitry detects the <u>image current</u> induced by coherent cyclotron motion of the ions. Since this phenomenon is not widely understood, we felt further explanation of it would be useful.

Figure 2a shows a packet of positive ions of charge Nq and velocity v moving between two electrodes. Electrode 1 is at ground potential and electrode 2 is connected through a resistor to ground. As the positive ions move closer to electrode 2, electrons are attracted to it by the electric field of the ions. A small negative charge accumulates on electrode 2 as the electron current (called image current) flows through the resistor. Of course, once the ions strike the surface of electrode 2 they are neutralized by the accumulated electrons, but the main point is that the electron current begins to flow even before the ions have impacted on the surface of the electrode.

The phenomenon of ion image current was analyzed in 1938 in connection with the motion of electrons in high frequency vacuum tubes [24]. For the case of the parallel electrodes shown in Figure 2a, the image current I is given by

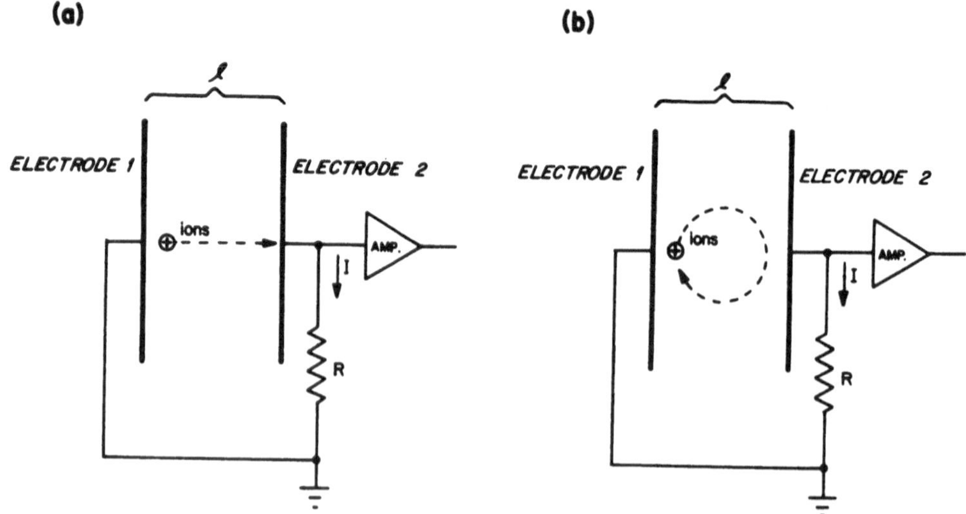

Fig. 2. Ion image current is caused to flow from electrode 1 to electrode 2 by positive ions moving from left to right. (a) In the absence of a magnetic field, a large current is generated, but only for a few microseconds. (b) Wit a strong magnetic field and low sample pressure, the image current can be maintained for up to several seconds since the ions are prevented from striking the electrodes.

$$I = Nqv/\ell \tag{1}$$

where q is the charge of the ions, v is their velocity toward electrode 2 and ℓ is the separation of the two electrodes. For 1000 ions (total charge 1.6×10^{-16} C) moving at 10,000 m/s between electrodes 0.02 m apart, the image current is 8×10^{-11} A. This is surprisingly large; it appears that the 1000 ions can be detected easily before they have even collided with the electrode.

Unfortunately, the above example is misleading for the following reason: the image current flows only for the short period of time while the ions are in transit between the two electrodes. At a velocity of 10,000 m/s the ions move from one electrode to the other in just 2 μs, a time far too short for the method to be practical. However, if the ions and the electrodes are placed in a magnetic field (as illustrated in Figure 2b), coherent motion of the ions will induce alternating image currents having the same frequency as the cyclotron motion of the ions. The magnetic field also insures that the image current will persist far longer because the ions are prevented from colliding with the electrodes. This is precisely the principle utilized by the capacitance bridge circuit to detect ions stored in an ICR cell [20,25]. In the next section a quantitative theory for how the circuit works is presented.

2.2 Transient Response of the CBD - A Simplified Approach

In our previous analysis of the transient response of the capacitance bridge detector [5], ions stored in the ICR analyzer cell were represented by a generalized two-port device connected in parallel with the upper and lower plates of the cell. Power absorbed by the ions from the rf electric field was calculated as the product of the potential difference across the cell plates and the ion image current I. Then, application of Kirchhoff's current equation to the bridge network gave the following equation for the balanced capacitance bridge:

$$(C_1 + C_2 + C_3)\ dV_0/dt + V_0/R = I \qquad (2)$$

In this equation, V_0 is the rf imbalance voltage signal at the input of the preamplifier, C_1 is the capacitance between the upper and lower plates of the ICR cell, C_2 is the capacitance of the balance capacitor, C_3 is the imput capacitance of the preamplifier and the shielded cable leading to the lower plate, and R is the input resistance of the preamplifier. The most important feature of Eq. 2 is the connection it makes between the detected signal V_0 and ion image current I. Substitution of Eq. 1 for the ion image current gives

$$(C_1 + C_2 + C_3)\ dV_0/dt + V_0/R = Nqv/\ell \qquad (3)$$

This equation is of central importance; it relates motion of the ions at velocity v between the upper and lower plates of the ICR cell to the voltage signal V_0 induced at the input of the preamplifier.

General solutions to eq. 3 are given in reference 5. Unfortunately, the general solutions are so complex that it is difficult to develop an intuitive understanding for how the capacitance bridge detector functions. We have developed, therefore, the following simplified approach which focuses on the special case of power absorption at resonance in the zero-pressure limit.

The model follows the earlier theory but utilizes several approximates to simplify the results. The first approximation is to assume that the second term in eq 3 is negligible. For this to be valid, most of the image current must flow through the capacitance elements rather than the resistance R. This condition is met by the circuits we have designed because the impedance of the capacitance elements is on the order of one-thousand times lower than the input resistance of the preamplifier [20]. With this approximation, eq 3 reduces to

$$(C_1 + C_2 + C_3)\ dV_o/dt = Nqv/\ell \tag{4}$$

In order to solve for the ICR signal V_o, one must have an expression for v, the velocity of the ions. The first step is to model the excitation rf voltage signal. For pulsed ICR, a rf signal at a fixed frequency is turned on for a short time. In this case the approximate form for the rf electric field in the ICR cell is

$$E = (V_p/\ell)\ \sin \omega_1 t \tag{5}$$

where ω_1 is the frequency of the rf and V_p is the base-to-peak amplitude of the excitation rf voltage applied to the upper plate of the ICR cell. Combining this with the Lorentz equation and neglecting power lost to ion-neutral collisions gives [5]

$$dv_x/dt = \omega v_y + \omega_T^2 x + (qV_p/m\ )\ \sin \omega_1 t \tag{6}$$

$$dv_y/dt = -\omega v_x \tag{7}$$

where the x and y coordinates are perpendicular to the magnetic field with v_x being the speed of ion motion between the upper and lower plates. Three frequencies appear in eqs 6 and 7:

(1) $\omega = qB/m$ is the classical cyclotron frequency;

(2) ω_1 is the frequency of the alternating electric field; and

(3) ω_T is the frequency of oscillation of an ion in the electrostatic potential well created by the dc voltages on the plates of the ICR cell [26].

Simultaneous solution of eqs 6 and 7 is achieved by diffentiating eq 6 with respect to time and then substituting eq 7 for dv_y/dt. Rearrangement of terms gives

$$d^2v_x/dt^2 + (\omega^2 - \omega_T^2)v_x = (qV_{p\ 1}/m\ell)\cos\omega_1 t \tag{8}$$

This equation describes simple harmonic motion driven at frequency ω_1. At resonance $\omega_1^2 = \omega^2 - \omega_T^2$, the solution of eq 8 with the initial condition $v_x = 0$ at $t = 0$ is

$$v_x = \frac{q\,V_p}{2\,m\,\ell}(t \sin\omega_1 t) \tag{9}$$

where t is the time the ions have been exposed to the excitation rf signal. Substitution of eq 9 for velocity v in eq 4, followed by integration gives

$$V_o = \frac{N\,q^2\,V_p}{2\,m\,\ell^2\,(C_1 + C_2 + C_3)\,\omega_1^2}(\sin\omega_1 t - \omega_1 t\cos\omega_1 t) \tag{10}$$

After phase sensitive detection and low-pass filtering, the transient ICR signal observed on an oscilloscope is

$$V_{ICR} = \frac{N\,q^2\,V_p\,G\,t}{2\,m\,\ell^2\,(C_1 + C_2 + C_3)\,\omega_1} \tag{11}$$

where G is the overall gain of the amplifiers. This equation predicts how the observed transient signals depend on important experimental parameters such as number of ions, size of the cell, circuit capacitance and detection frequency. Linearity between the CBD signal (V_{ICR}) and number of resonant ions N is a particularly important property of the circuit for quantitative studies of ion-molecule reactions.

Further insight into the operation of the CBD circuit is provided by combining the terms m and ω_1 in the denominator of eq 11. Normally ω_T, the frequency of oscillation parallel to the magnetic field, is 5 to 10 times smaller than the cyclotron frequency. Thus, to within a few percent, the resonance condition can be restated as $\omega_1 = \omega = qB/m$. Applying this approximation to eq 11 gives

$$V_{ICR} = \frac{N\,q\,V_p\,G\,t}{2\,\ell^2\,(C_1 + C_2 + C_3)\,B} \tag{12}$$

This shows the important result that at constant magnetic field strength the sensitivity of the detector is constant for ions of different mass-to-charge ratio. Experimental confirmation of the linear dependence of the signals on excitation rf level V_p, irradiation time t, number of ions N, and magnetic field strength B is presented in the Performance Tests section of this paper.

This signal from the capacitance bridge detector can also be related in a simple manner to the radius of gyration of the cyclotron orbits. Radius of gyration r is found using $\omega = v/r$, and $v = (v_x^2 + v_y^2)^{1/2}$, where v_x and v_y are the solutions to eqs. 6 and 7. For cyclotron energies much greater than random thermal translational energy, the radius of gyration is given by

$$r = \frac{V_p\ t}{2\ \ell\ B} \tag{13}$$

Substituting this experession for r into eq 12 gives the result

$$V_{ICR} = \frac{N\ q\ G\ r}{\ell\ (C_1 + C_2 + C_3)} \tag{14}$$

Thus, in the final analysis, the capacitance bridge detector provides a response signal that is proportional to the number of resonant ions and their radius of gyration.

2.3 Transient Response of the CBD to Rapid Scan Excitation

The theory of the capacitance bridge detector can be extended to include other excitation methods by substitution of an alternate expression of the rf electric field in eq 5. The rapid scan ICR detection scheme uses a fast linear-frequency sweep of the excitation rf voltage, [2] and the rf electric field that can be approximately expressed as

$$E = (V_p/\ell)\ \sin\ (\omega_0 t + bt^2) \tag{15}$$

where ω_0 is the initial frequency of the excitation rf voltage and b is the frequency sweep rate in radians/sec^2. When this expression for the excitation rf signal is substituted for eq 5 in the theory, the following differential equation is obtained:

$$d^3V_0/dt^3 + (\omega^2 - \omega_T^2)dV_0/dt = \gamma\beta V_p(\omega_0 + 2bt)\ \cos\ (\omega_0 t + bt^2) \tag{16}$$

where $\gamma\beta$ is $q^2N/m^2(C_1 + C_2 + C_3)$, and V_o is the transient rapid scan signal. Analytical solutions for this are quite complex because it is a third-order, nonolinear differential equation. We have employed a numerical integration method which is based on a Runge-Kutta algorithm [27]. The program calculates V_o using values for the amplitude of the excitation rf signal, resonance frequency, starting frequency and scan rate. The results we have obtained so far predict that the rapid scan signals are linearly proportional to the amplitude of the excitation rf voltage and the number of ions in the cell. The dependence on scan rate is more complicated. Comparison of the calculated rapid scan signals with experimental results is given in the Performance Tests section below.

3. Performance Tests

3.1 Pulsed ICR

Predictable response and high sensitivity are two of the most important requirements for the capacitance bridge detector since it is intended for quantitative measurements of ion-molecule reaction rate constants and equilibrium constants. In this section the results of varios performance tests are compared with the predictions of eqs 11 and 12 to establish a range of useful operating parameters.

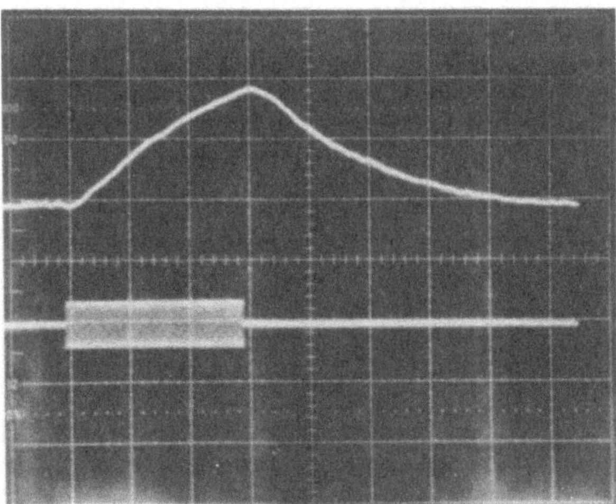

Fig. 3. Oscilloscope traces of V_{ICR}, the transient output signal from the capacitance bridge detector (upper trace), and an 160 mV (p-p) pulsed excitation rf signal (lower trace) which is applied to the upper plate of the ICR cell. The timebase is 2 ms/div. and the transient signal is produced by approximately 4×10^5 ions of m/z 78 at a pressure of 1.2×10^{-6} torr.

A transient signal from the capacitance bridge detector is shown in the upper part of Figure 3. This signal was produced by approximately 400,000 ions of m/z 78 in benzene at a pressure of 1.2×10^{-6} Torr. The ions were produced by electron impact at 20 eV with an emission current of 0.10 μA which was gated on for 11.35 ms [28]. The pulsed excitation rf signal shown in the lower part of Figure 3 has a frequency of 222 kHz, an amplitude of 160 mV (p-p), and a duration of 5.9 ms. There are two features of importance to note from Figure 3:

(1) the transient signal rises linearly with time when the excitation rf is turned on, and

(2) it decays once the excitation rf is turned off.

The linear rise is predicted by eq. 12, but the decay is not explained by the simplified theory. To understand all the features of the signal, reference must be made to the more complete theory [5]. The results may be summarized as follows. While the excitation rf is on, the ions are accelerated and produce transient signal of the form

<u>Excitation Period:</u>

$$V_{ICR} \propto \frac{(1 - e^{-ct})}{c} \tag{16}$$

where c, the reduced collision frequency for momentum transfer, is proportional to pressure and dependent on factors such as the reduced mass of the ion-neutral collision pair and the polarizability of the neutral. In the zero-pressure limit ($c \to o$) the exponential term in eq 16 can be expanded to first-order to get simply $V_{ICR} \propto t$, just as predicted by eq 12. In the high-pressure limit c is so large that the exponential term is effectively zero and the transient signal rises quickly to a constant level proportional to 1/c. At intermediate pressures, such as in Figure 3, the transient rises linearly and then begins to level out as the ions collide with the neutral gas molecules in the ICR cell. Some departure from linearity is apparent in Figure 3 after about 4 ms.

The complete theory also provides an explanation for the transient signal observed after the excitation rf has been turned off. This period is called the relaxation period and the transient signal has the form

<u>Relaxation Period:</u>

$$V_{ICR} \propto \frac{e^{-ct}}{c} \tag{17}$$

The rate of exponential decay decreases as pressure is decreased, and at pressures on the order of 10^{-8} torr several seconds are required for the signal to decay. Under these conditions ultra-high mass resolution signals can be obtained with the capacitance bridge detector. Figure 3 shows an example of the damping which is observed at pressures around 10^{-6} torr. Under the conditions used for Figure 3 one excitation rf pulse 5 ms in duration can detect approximately 7000 ions with a signal-to-noise ratio of 1:1.

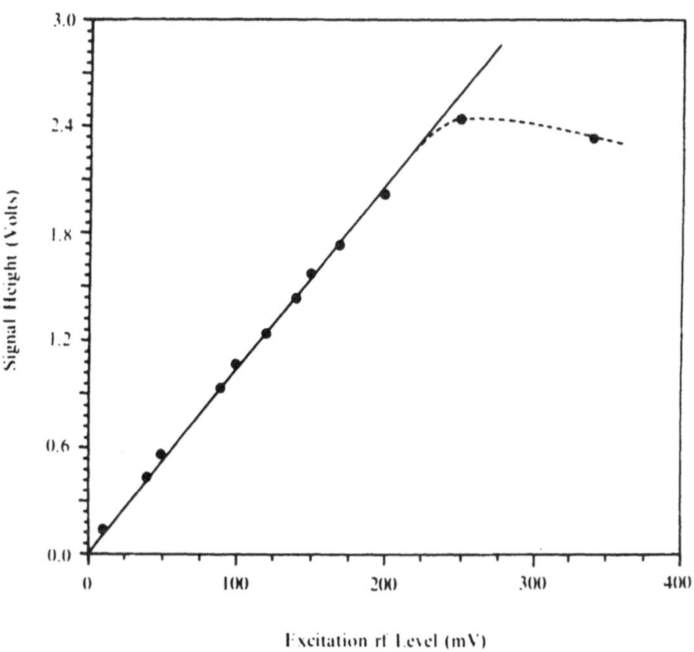

Fig. 4. A linear dependence of signal height on V_p, the level of the excitation rf signal, is predicted by eqs 11 and 12. Above about 220 mV (p-p) this relation no longer holds because the ions are ejected from the ICR cell.

Figure 4 shows the dependence of the transient signal on the level of the excitation rf signal. The experimental conditions were essentially the same as in Figure 3. From eq 11 one expects a linear dependence on V_p, and this is exactly what is observed up to about 225 mV (p-p). Beyond 225 mV there is

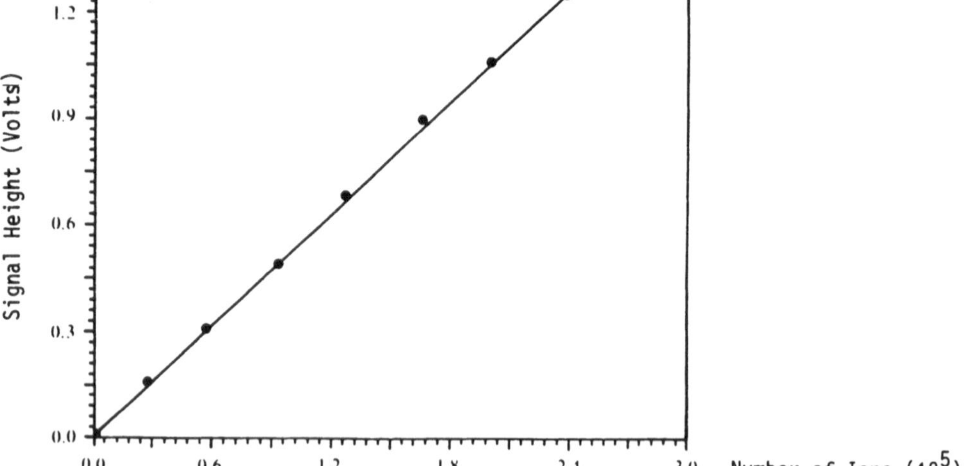

Fig. 5. Excellent linearity is observed between the signal height and the number of resonant ions in the ICR cell. Too many ions, however, creates an excessive space charge condition which distorts the signals and causes a departure from linearity.

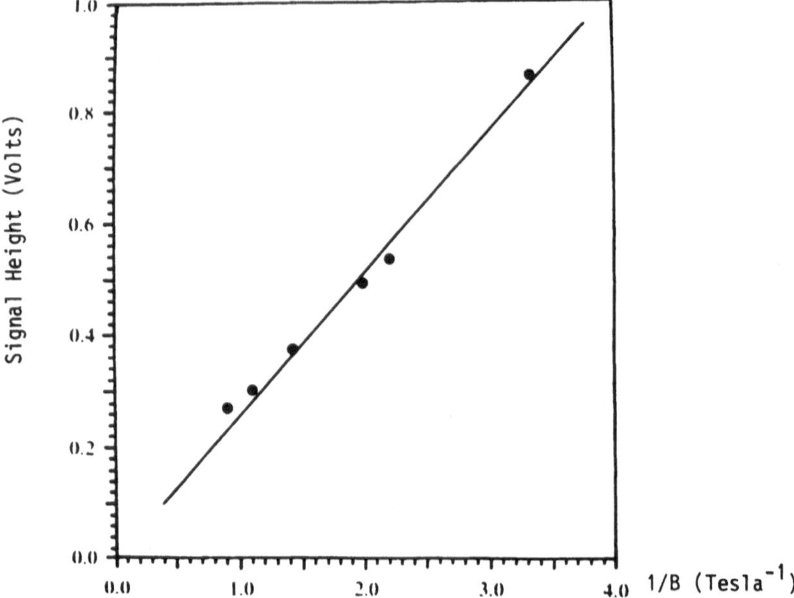

Fig. 6. Signals from the capacitance bridge detector are inversely proportional to the magnetic field strength B, just as is predicted by eq. 12. This relationship was observed to break down at magnetic field strengths lower than about 0.2 T because ions were then easily ejected from the cell.

a rapid drop off which is caused by ions being accelerated so strongly that they strike the upper and lower plates of the ICR cell. Using eq 13 one can calculate that a radius of gyration of 1.1 cm (half the height of the ICR cell) is reached after 5 ms with an excitation rf level, V_p = 108 mV (216 mV_{p-p}). This is in good agreement with the turn over point seen in Figure 4. At a radius of 1.1 cm the translational energy of the ions ($T = m(\omega r)^2/2$) is about 95 eV.

Figure 5 shows a linear relationsship between the amplitude of the CBD signal and the number of ions N in the cell, just as is expected from eq 11. For quantitative work care must be taken to limit the number of ions in the cell. When there are too many ions, space charge effects cause the resonance frequency to decrease and the CBD signals to become unstable. To achieve stable operation with our current ICR cell, which has dimensions of 2.2 x 2.2 x 8.9 cm, the number of ions should be kept lower than about 350,000. Larger ICR cells recently tested in our laboratory are capable of storing more ions and providing a corresponding increase in dynamic range.

The final performance test for the capacitance bridge detector is shown in Figure 6 where signal height is plotted as a function of the magnetic field strength. These data were taken for $C_6H_6^+$ ions with a 30 mV (p-p) excitation rf level; the other parameters were the same as for Figure 3. As suggested by eq 12 the signal height is inversely proportional to magnetic field strength. This was a difficult experiment to perform because at low magnetic field strength the ions are easily ejected from the cell. In addition, the pulsed electron emission current decreased by as much as 25 % at high magnetic field, and the raw data had to be normalized for this effect. These problems emphasize the desirability of working at constant magnetic field with as high a field strength as is feasible.

3.2 Rapid Scan ICR

The simplified rapid scan ICR theory was tested in a manner similar to that employed for the performance tests of the pulsed ICR theory. It should be noted that the conditions under which the rapid scan transients are acquired differ from those of the pulsed ICR experiments. For the results reported below, the rapid scan experiments are performed with a continuous electron beam and no quench pulse [29].

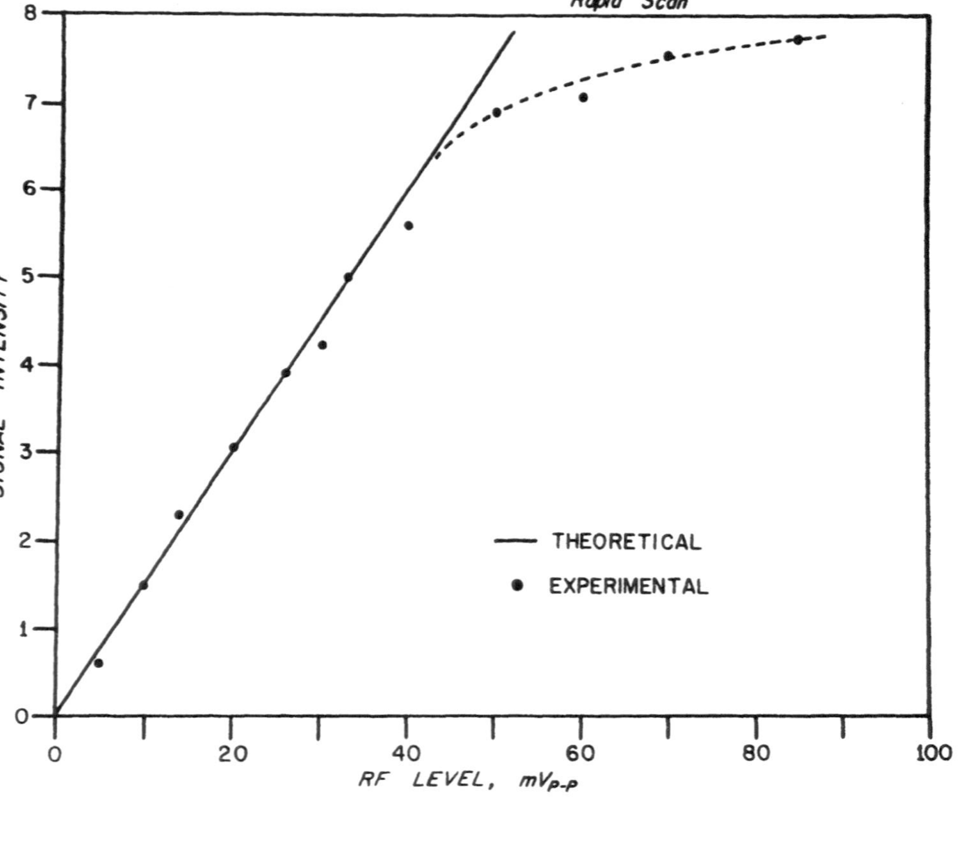

Fig. 7. The amplitude of rapid scan ICR signal is proportional to the excitation rf level, as is predicted by the theory. Above about 45 mV (p-p), the experimental results deviated from linear behavior due to ejection of the ions from the ICR cell.

Figure 7 shows the dependence of the rapid scan ICR signal on the amplitude of the excitation rf voltage. These data are for m/z 78 ions in benzene vapor at 4.5 x 10^{-7} torr and a continuous electron emission of 9 nA at 16 eV. Note that the predicted linear relationship is observed up to excitation rf voltages of approximately 40 mV(p-p). We believe that the nonlinear response at higher rf voltages is primarily due to ejection of the ions from the cell by collison with the upper and lower plates.

Figure 8 shows the relationship between the rapid scan signals and the electron beam current. These data are for benzene at 1.0 x 10^{-6} torr at 20 eV with an rf level of 20 mV(p-p). Under the steady-state conditions by which these experiments were performed, the total number of ions in the cell is proportional to the electron emission current used [29]. This has been demonstrated in other experiments using an electrometer to measure the total ion current and the electron emission current. Figure 8 shows the predicted linear behavior is observed up to approximately 2.5 nA. At higher emission currents space charge interactions cause the signals to become unstable.

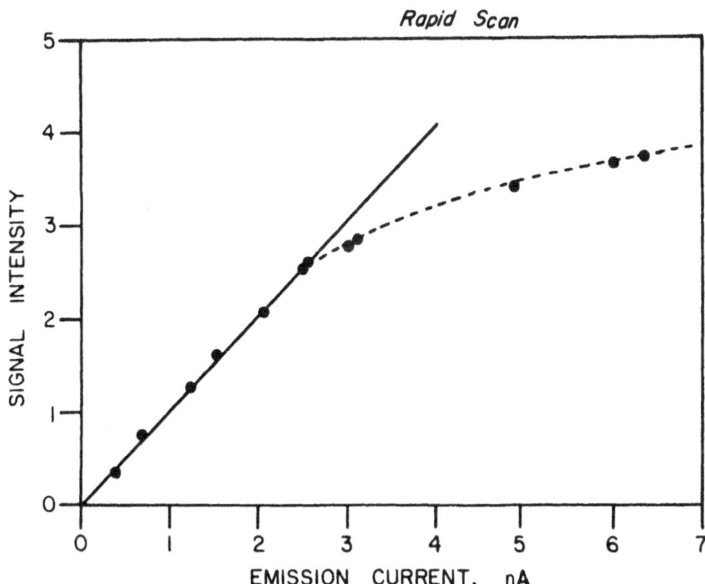

Fig. 8. Linear dependence of the rapid scan ICR signals is observed with electron emission currents below about 3 nA (when the pressure is about 1 x 10^{-6} torr). With higher emission currents, too many ions are created, and the excessive space-charge causes the signals to become distorted.

Experiments were also performed to test how rapid scan ICR signals depend on the scan rate. The solid line in Figure 9 is the theoretical prediction and the points are experimental data for the rapid scan signal obtained at various scan rates. There is very good agreement with the theory for the faster scan rates, but at slow scan rates, the ICR signals are smaller than predicted by the theory. At the slower scan rates, more power is delivered to the ions, and this causes them to be ejected from the cell. As the scan rate is made faster and faster, the amplitude of the rapid scan signals becomes constant. In fact, the fastest scan rates shown in Figure 9 are comparable to the 10^6 to 10^7 Hz/sec scan rates used in FT-ICR experiments.

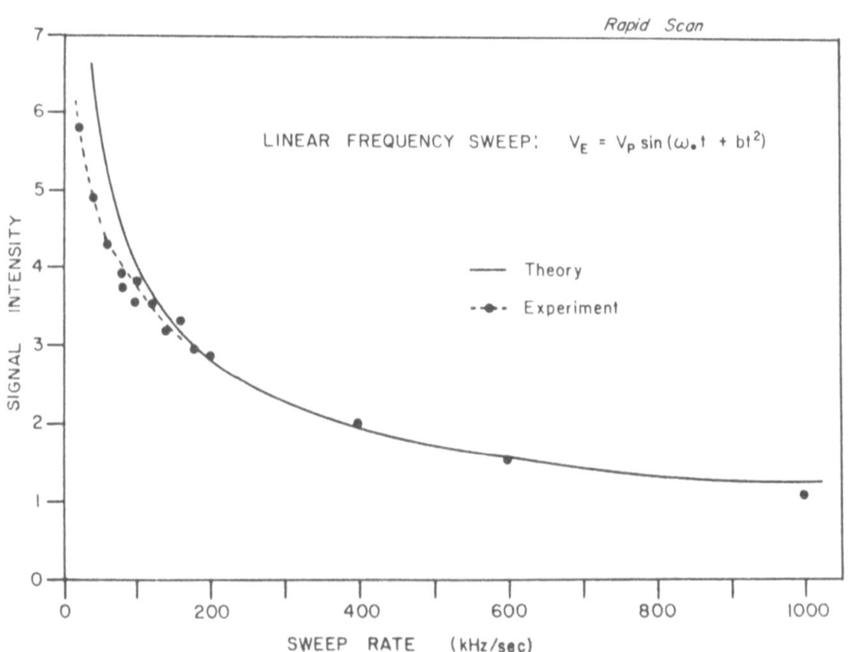

Fig. 9. At constant excitation rf level, V_p, the rapid scan ICR signals decrease in amplitude as the scan rate is increased because less power is delivered to the ions. For very fast scan rates, the amplitude of the rapid scan signals approaches a constant value corresponding to the impulse excitation response. Deviation of the experimental data from the predicted curve at very slow scan rates is probably due to ion ejection.

4. Conclusions

The capacitance bridge circuit is a sensitive and versatile means for detecting gaseous ions. It can be used both with ICR drift cells and trapped ion cells as a direct replacement for the marginal oscillator and can be used in pulsed, rapid scan, and Fourier transform ICR experiments. Its main features for ion chemistry studies are (1) sensitivity which is dependent of the mass and the composition of the detected ions, (2) high detection efficiency and negligible sampling errors since all ions in the cell are detected, and (3) variable frequency operation at constant magnetic field strength. Experiments are typically done with a few hundred thousand ions in the ICR cell, and the linear dynamic range is about 50:1. The simplified theory for pulsed ICR is an adequate model for the performance of the CBD circuit under the conditions normally used for studying ion chemistry.

The rapid scan ICR theory is not as easily applied since it does not have a simple analytical solution. However, it has proved to be valuable in correlating and understanding the performance of the CBD in the rapid scan excitation.

5. Acknowledgements

We gratefully acknowledge grant support from the National Science Foundation (CHE77-10024), the National Institutes of Health (GM 27039) and the donors of the Petroleum Research Fund administered by the American Chemical Society.

6. References

1. (a) M.B. Comisarow and A.G. Marshall, Chem. Phys. Lett., 25 (1974) 282,
   (b) M.B. Comisarow in "Transform Techniques in Chemistry", (P. Griffiths, Ed.), Plenum Press, New York, 1978, p. 257

2. (a) R.L. Hunter and R.T. McIver, Jr., Chem. Phys. Lett., 49 (1977) 577;
   (b) R.L. Hunter and R.T. McIver, Jr., Am. Lab. 9 (1977) 13.

3. E.B. Ledford, Jr., S. Ghaderi, R.L. White, R.B. Spencer, P.S. Kulkarni, C.L. Wilkins, and M.L. Gross, Anal. Chem., 52 (1980) 463.

4. G. Parisod and T. Gaumann, Chimica, 34 (1980) 271.

5. R.T. McIver, Jr., E.B. Ledford, Jr. and R.L. Hunter, J. Chem. Phys., 72 (1980) 2535.

6. (a) M.B. Comisarow and A.G. Marshall, J. Chem. Phys. 64 (1976) 110;
(b) A.G. Marshall, M.B. Comisarow and G. Parisod, J. Chem. Phys., 71 (1979) 4434.

7. J.E. Bartmess and R.T. McIver, Jr. in "Gas Phase Ion Chemistry," Vol. 2 (M.T. Bowers, Ed.) Academic Press, New York, 1979, pp 88-119.

8. J.E. Bartmess, J.A. Scott and R.T. McIver, Jr., J. Am. Chem. Soc., 101 (1979) 6046.

9. R.T. McIver, Jr., Rev. Sci. Instrum., 49 (1978) 111.

10. J.F. Wolf, R.H. Staley, I. Kopple, M. Taagepera, R.T. McIver, Jr., J. L. Beauchamp and R.W. Taft, J. Am. Chem. Soc., 99 (1977) 5417.

11. B.K. Janousek and J.I. Brauman in "Gas Phase Ion Chemistry", Vol. 2, (M.T. Bowers, Ed.) Academic Press, New York, 1979, pp. 53 - 83.

12. R.C. Dunbar in "Kinetics of Ion-Molecule Reactions", (P. Ausloos, Ed.) Plenum Press, New York, 1979, pp 53 - 83.

13. R.L. Woodin, D.S. Bomse and J.L. Beauchamp, J. Am. Chem. Soc., 100 (1978) 3248.

14. (a) J.A. Hipple, H. Sommer and H.A. Thomas, Phys. Rev., 76 (1949) 1877;
(b) H. Sommer, H.A. Thomas and J.A. Hipple, Phys. Rev., 82 (1951) 697;
(c) R.T. McIver, Jr., E.B. Ledford, Jr., and J.S. Miller, Anal. Chem., 47 (1975) 692.

15. D. Wobschall, Rev. Sci. Instrum., 36 (1965) 466.

16. H. Sommer and H.A. Hipple, Phys. Rev., 78 (1950) 806.

17. (a) R.T. McIver, Jr., Rev. Sci. Instrum., 44 (1973) 1071;
(b) R.T. McIver, Jr. and A.D. Baranyi, Int. J. Mass Spectrom. Ion Phys., 14 (1974) 449.

18. A. Warnick, L.R. Anders, T.E. Sharp, Rev. Sci. Instrum., 45 (1974) 929.

19. C. Amano, Y. Goto and M. Inoue, Int. J. Mass Spectrom. Ion Phys., 32 (1979) 67.

20. R.T. McIver, Jr., R.L. Hunter, E.B. Ledford, Jr., M.L. Locke and T.J. Francl, Int. J. Mass Spectrom. Ion Phys., 39 (1981) 65.

21. L.R. Anders, J.L. Beauchamp, R.C. Dunbar and J. Baldeschwieler, J. Chem. Phys., 45 (1966) 1062.

22. R.T. McIver, Jr. and R.C. Dunbar, Int. J. Mass Spectrom. Ion Phys., 7 (1971) 471.

23. D.J. DeFrees, W.J. Hehre, R.T. McIver, Jr. and D.H. McDaniel, J. Phys. Chem., 83 (1979) 232.

24. (a) W. Shockley, J. Appl. Phys., 9 (1938) 635;
(b) M.D. Skirkis and N. Holonyah, Jr., Am. J. Phys., 34 (1966) 943.

25. M.B. Comisarow, J. Chem. Phys., 69 (1978) 4097.

26. Motion of ions in a one-region trapped ICR cell has been discussed by T.E. Sharp, J.R. Eyler and E. Li, Int. J. Mass Spectrom. Ion Phys., 9 (1972) 421.

27. J.A. Zonneveld, "Automatic Numerical Integration", Mathematisch Centrum Amsterdam, 1964

28. We estimate that the ionization efficiency for benzene at 20 eV is 20.5 positive ions per electron per cm per torr of pressure. This estimate was derived from the absolute value of 6.4 ions per electron per cm per torr for C_2H_2 (J.T. Tate and P.T. Smith, Phys. Rev., 39 (1932) 270) and the value of 3.2 for the relative ionization efficiency of C_6H_6 relative to C_2H_2 (J.W. Otvos and D.P. Stevenson, J. Am. Chem. Soc., 78 (1956) 546.

29. R.L. Hunter and R.T. McIver, Jr., Anal. Chem., 51 (1979) 699.

SIGNALS, NOISE, SENSITIVITY AND RESOLUTION IN ION CYCLOTRON RESONANCE SPECTROSCOPY

Melvin B. Comisarow
Chemistry Department
University of British Columbia
Vancouver, B. C., Canada V6T 1Y6

1.Introduction

This chapter discusses ion cyclotron resonance (ICR) signal generation, noise generation in ICR spectrometers, and ICR mass resolution. The quotient of the ICR signal strength and the ICR noise level is the ICR sensitivity. Many important aspects of the above topics have been separately discussed in the literature. This chapter is an attempt to discuss the topics in a coherent manner and to describe additional aspects of the topics which have heretofor been ignored. In the following sections, the origin of ICR signals is discussed in terms of a signal model which accounts for the properties of the signal. The time dependence of the ICR signal is then shown to be related to the ICR resolution. Many factors which affect the time dependence of the ICR signal are discussed and the relative importance of the factors is assessed. The origin of electronic noise in ICR spectrometers is also discussed.

In the discussion to follow ICR, phenomena will usually be discussed in the context of the Fourier transform ion cyclotron resonance (FT-ICR) experiment although the principles being discussed will generally apply to all ICR experiments.

2.ICR Signal Generation

In all ICR spectrometers the "cyclotron motion" of ions is excited by subjecting ions to an alternating electric field whose frequency equals the natural cyclotron frequency of the ion. This cyclotron frequency is given by

$$\omega = \frac{qB}{m} \qquad \text{(mks)} \qquad (1a)$$

or alternatively,

$$f = 1.535 \times 10^7 \frac{q B}{m} \qquad (1b)$$

Equation 1b gives the cyclotron frequency in Hz, for an ion of mass m amu and charge q units of the elementary charge in a magnetic field of B Tesla. Equation 1a is the corresponding equation in mks units. The interaction of the electric field with the ion causes the radius of the ion orbit to increase. All ions of a particular mass are accelerated together and after an excitation period of t seconds the orbital radius of the ion motion is given by

$$r = \frac{E t}{2B} \quad \text{(mks)} \qquad (2a)$$

or alternatively,

$$r = 10^{-2} \frac{E t}{2B} \qquad (2b)$$

Equation 2b gives the orbital radius in cm for an ion in a magnetic field of B Tesla which was irradiated by an electric field of E millivolts/cm for t milliseconds at the cyclotron frequency.

Equation 2a is the corresponding equation in mks units. The "excited cyclotron motion" consisting of a "rotating electric monopole" is shown in Figure 1. This rotating electric monopole is the signal generating physical model which accounts for the signal generating properties of the excited cyclotron motion [1].

The physical basis of the ICR signal generation is as follows: The plates of the ICR cell form a parallel plate capacitor on which the charge of the ion is to be neutralized. For the instantaneous position of the rotating electric monopole of Figure 1, let the neutralizing charge on the top plate be Q. Then, as the monopole rotates the charge on the top plate will change. If the plates of the capacitor are connected by an external conductor then the current in the conductor will just be the time derivative of the charge on the top plate of the capacitor and it can be shown [1] that the root mean square value of the ICR signal current is given by

$$I_S = \frac{N q^2 r B}{\sqrt{2}\, m\, d} \quad \text{(mks)} \qquad (3a)$$

or equivalently,

$$I_S = 1.093 \times 10^{-4} \frac{N q^2 r B}{m d} \qquad (3b)$$

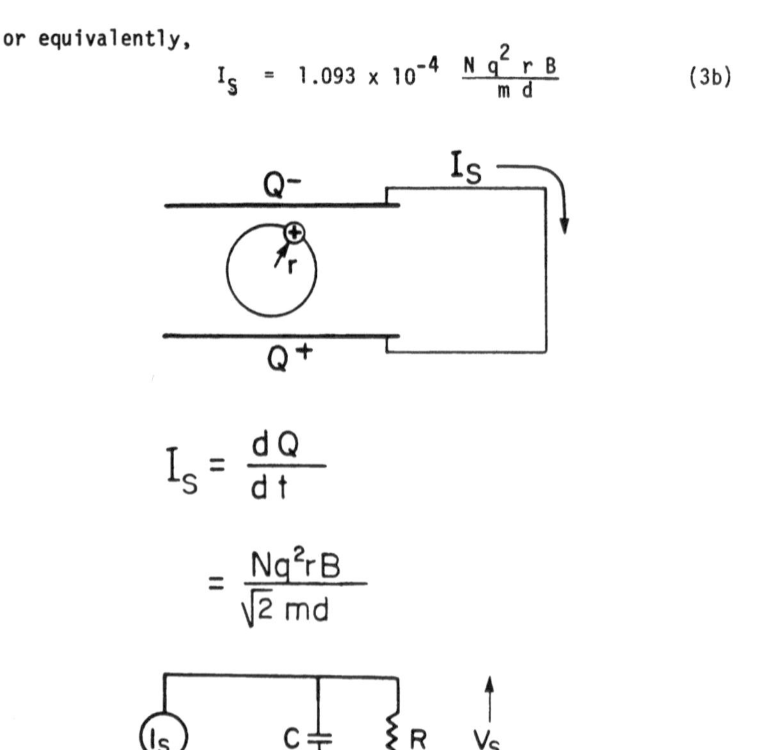

Figure 1. Excited cyclotron motion generates an ICR current signal, I_S.

Equation 3b gives the root mean square ICR signal current in micro-amperes as a function of N, the number of ions, q, the ion charge in units of the elementary charge, r, the orbital radius in cm, B the magnetic field strength in Tesla, m, the ion mass in amu and d, the ICR cell spacing in cm. Equation 3a is the corresponding equation in mks units.

The ICR current signal, I_S, is usually converted to a voltage signal by passing the signal current through an impedence [1] as indicated in Figure 2. The magnitude of the ICR signal voltage is given by the product of the current and the resistance. In actual experimental practice however, the introduction of the resistor and the connection to the electronics will also introduce some parallel capacitance which should

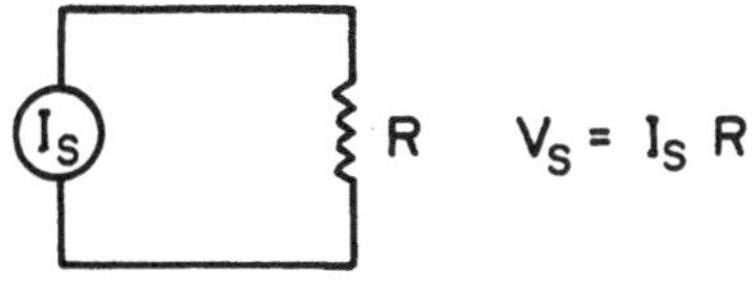

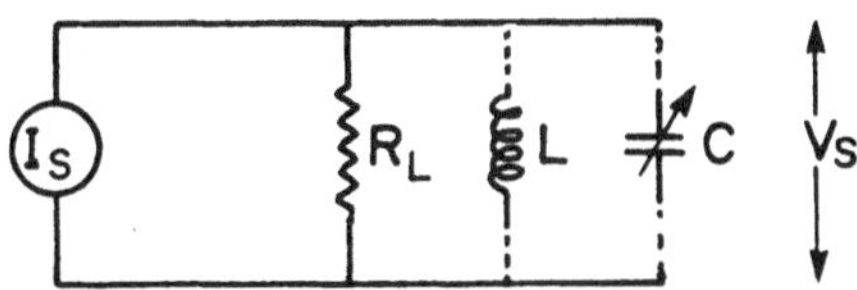

Figure 2. Generation of an ICR signal voltage, V_S, by passing the ICR signal current through a resistor.

also be accounted for in the signal model [1]. Figure 1 shows the resultant circuit. It can be shown [1] that for the circuit of Figure 1 the ICR voltage signal is given by Equation 13 of ref. 1 and in the limit

$$1/R \ll \omega C$$

this voltage is given by [1]

$$V_S = \frac{N\,q\,r}{\sqrt{2}\,dC} \tag{4a}$$

$$V_S = 1.1329 \times 10^{-7}\,\frac{N\,q\,r}{d\,C} \tag{4b}$$

Equation 4b gives the rms ICR voltage signal in volts as a function of N, the number of ions, q, the ion charge in units of the elementary charge, r, the orbital radius in cm, d, the ICR cell spacing in cm and C the shunt capacitance in picofarads, Equation 4a is the corresponding equation in mks units. In the opposite limit

$$1/R \gg \omega C$$

that the circuit be predominantly resistive, the ICR voltage signal strength is given by

$$V_S = \frac{N q^2 r B R}{\sqrt{2}\ m\ d} \qquad (5a)$$

or equivalently

$$V = 1.0931 \times 10^{-1} \frac{N q^2 r B R}{m d} \qquad (5b)$$

Equation 5b gives the rms ICR voltage strength in volts for a predominantly resistive circuit as a function of N, the number of ions, q, the ionic charge in units of the elementary charge, r, the orbital radius in cm, B the magnetic field strength in Tesla, R the shunt resistance in megohms, m the ion mass in amu and d, the ICR cell spacing in cm. Equation 5a is the corresponding equation in mks units.

A predominantly resistive circuit is formed not by using a resistor, but rather by using an inductor and a capacitor in parallel, a circuit known as a tank circuit. At the specific frequency, ω given by

$$\omega = (LC)^{-1/2}$$

the tank circuit behaves as a resistor R_L, whose value is given by

$$R_L = \frac{\omega^2 L^2}{R_{DC}} \qquad (6)$$

where R_{DC} is the series resistance of the inductor. Such tank circuits are used in marginal oscillator ICR spectrometers [1]. (cf Figure 2)

3.ICR Linewidth and ICR Mass Resolution

The relationship between the frequency scale linewidth of an ICR spectrum and the mass scale linewidth has been previously derived [2,3] to be

$$\Delta m = \frac{m^2}{qB} \Delta\omega \qquad \text{(mks)} \qquad (7a)$$

or equivalently,

$$\Delta m = 1.037 \times 10^{-6} \frac{m^2}{qB} \Delta\omega \qquad (7b)$$

Equation 7b gives the mass linewidth of a peak in an ICR spectrum in amu as a function m, the ion mass in amu, q, the ion charge in units of the electronic charge, B the magnetic field strength in Tesla and $\Delta\omega$ the frequency scale linewidth in rad/sec. Equation 7a is the corresponding equation in mks units. The ICR mass resolution $m/\Delta m$, is found by dividing Equation 7b into m:

$$\frac{m}{\Delta m} = 9.645 \times 10^5 \frac{qB}{m\ \Delta\omega} \qquad (8)$$

Equation 8 gives the ICR mass resolution as a function of m, the ion mass in amu, q, the ion charge in units of the elementary charge, B, the magnetic field strength in Tesla and $\Delta\omega$, the linewidth in rad/sec. It follows from Equation 8 that the ICR mass resolution will be inversely proportional to the ion mass and inversely proportional to the frequency scale linewidth. A discussion of ICR mass resolution then becomes principally a discussion of the factors which affect the ICR frequency scale linewidth.

It should be noted that Equations 7 and 8 are very general equations which apply to any ICR experiment.

In the following sections many factors which affect the ICR frequency linewidth will be discussed. The reason for this approach is that for many factors an exact quantitative model can be derived which allows theoretical calculation of $\Delta\omega$. Knowledge of $\Delta\omega$ then gives the ICR mass resolution via Equation 8.

4. Spectroscopic Relaxation and Spectroscopic Linewidths

Spectroscopy is based upon measurement of the frequency and the intensity of wave phenomena. The measurement of the frequency of any wave is ultimately limited by the observation time during which the wave is monitored. A simple demonstration of this phenomenon is the ±1 count uncertainty which is observed when a frequency is monitored with a "zero-crossing" type electronic counter. If a frequency of 1 kilohertz is monitored for 1 second, the uncertainty of ±1 count gives a frequency uncertainty of ±1 Hertz and a fractional error of 0.1%. A monitoring time of 0.1 seconds gives an uncertainty of ±10 Hertz and a monitoring time of 10 seconds gives a frequency uncertainty of ±0.1 Hertz, etc. In spectroscopy, the uncertainty with which a frequency can be measured is manifested as a non-zero linewidth in the frequency spectrum. Thus, as a general summary statement we can say that the linewidth of a spectral peak will be limited by the observation time of a waveform and if the waveform itself has a finite lifetime then this lifetime will also contribute to the spectral linewidth.

Figure 3 graphically demonstrates the relationship between the lifetime of a waveform, the observation time of a waveform and the linewidth of the frequency spectrum of the waveform. For a waveform of the type

$$F(t) = \cos(\omega t)\exp(-t/\tau) \qquad (0 < t < T) \qquad (9)$$

which is characterized by a frequency ω, a relaxation time τ, and an observation time, T, the absorption mode frequency spectrum is given by Equation 23 of ref. 3 and the magnitude mode frequency spectrum is given by Equation 24 of the same reference.

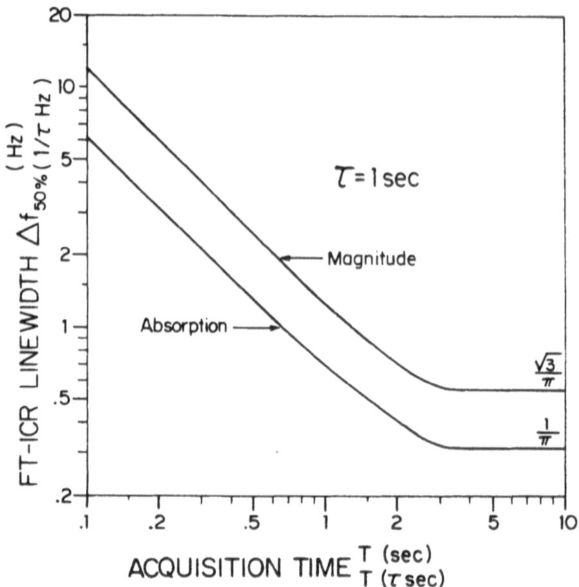

Figure 3.

The full linewidths at 50% of the peak height of the absorption mode and magnitude lineshapes are functions of the observation time, T, and the relaxation time τ. The linewidths are graphically displayed in Figure 3 for the specific value of the relaxation time, τ = 1.0 sec. Note that for a specific value of the relaxation time, the linewidth progressively decreases as the observation time increases. However a limiting minimum linewidth is reached and further increases in the observation time have no effect. This minimum limiting linewidth is controlled by the relaxation time and for τ = 1 second is given by the indicated values in Figure 3. Note that while the curves of Figure 3 were derived for the specific relaxation time value of τ = 1 second, the curves can also be used to determine the linewidth for any value of τ and T. For example, for a value of τ = 0.01 seconds and T = 0.02 seconds, the absorption mode linewidth is 40 Hz.

Figure 4 shows a series of FT-ICR spectra which demonstrate the dependence of the ICR mass resolution and ICR linewidth upon the acquisition time T. These were taken from ref. 2.

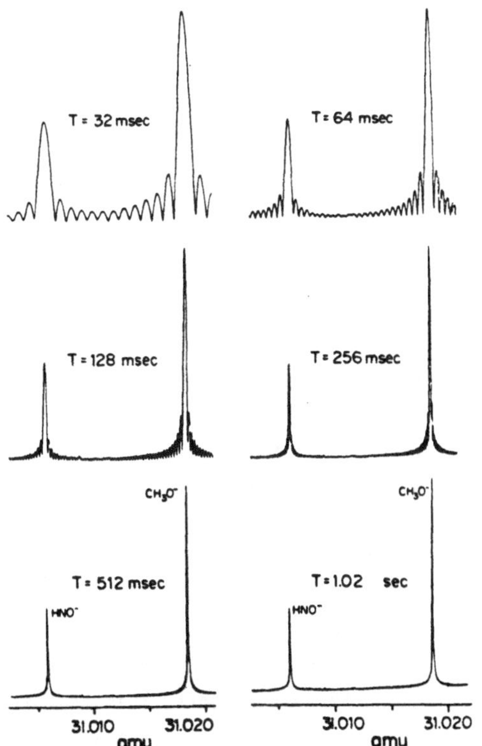

Figure 4. Ultra-High Resolution FT-ICR Spectra as a Function of Acquisition Time.

5. Homogeneous and Inhomogeneous Relaxation and Line Broadening

The treatment of the preceding section leaves unanswered the question: "What determines the relaxation time, τ?" In a very general spectroscopic context there are two distinct types of relaxation mechanisms which cause the amplitude of a waveform of the type Equation 9 to eventually decay to zero. These relaxation mechanisms are called homogeneous relaxation and inhomogeneous relaxation [4-6]. Homogeneous relaxation mechanisms reduce the internal energy content of an oscillating system, whereas inhomogeneous relaxation mechanisms cause the summed amplitude of oscillations of individual oscillators to decrease due to a dephasing of individual oscillators. It should be noted that the terms "homogeneous relaxation" and "inhomogeneous relaxation" refer to time domain processes. The corresponding processes in the frequency domain are "homogeneous broadening" and "inhomogeneous broadening" of spectral lines.

The concept of homogeneous relaxation, where some process removes energy from an oscillating system is a fairly simple concept which is easily grasped. Inhomogeneous relaxation, a process which does not affect the energy content of an oscillating system, but nevertheless causes the oscillation to decay is a more subtle concept but can be easily illustrated by Figure 5. Let us assume that an oscillating system consists of two individual oscillators which nominally have the same oscillation frequency. However, if some process shifts the frequency of one of the oscillators from the nominal value, the summed amplitude of the two oscillations will decay with time, even if the oscillators were exactly in phase with each other at zero time. This is shown in Figure 5. For a macroscopic ensemble of inhomogeneously relaxed oscillators the summed lineshape will be as shown in the bottom figure of Figure 5.

One characteristic of inhomogeneous relaxation which should be noted is that the damped oscillation will itself oscillate with a beat frequency which is related to the "degree of inhomogeneous relaxation". In certain cases such as spin-echo NMR experiments, this beat frequency can be observed. However if the homogeneous relaxation mechanisms of a particular system are comparable to or greater than the inhomogeneous mechanisms, the inhomogeneous mechanisms merely cause an additional amount of relaxation, with the "inhomogeneous beat pattern" being unobservable.

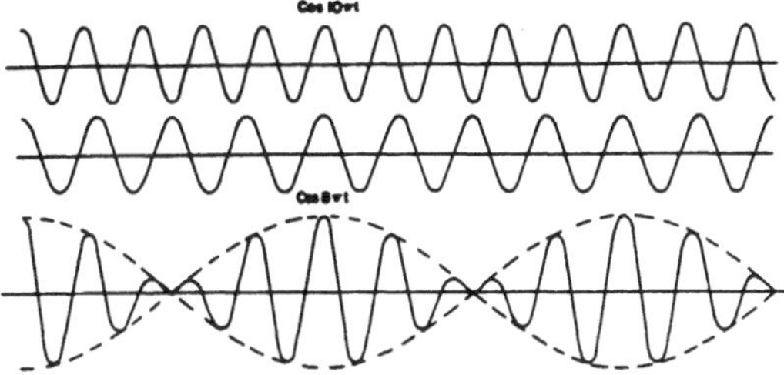

Figure 5.

The meaning of homogeneous relaxation and inhomogeneous relaxation in the context of the ICR experiment is illustrated by the diagrams in Figure 6. The top diagram in Figure 6 shows a two-ion ICR system with average position of the ions being the amplitude of the rotating electric monopole. For a homogeneous relaxation mechanism (which removes energy from the ICR system) the amplitude of the monopole decreases with time because the cyclotron radii of the individual ions becomes less. This is indicated in the middle diagram. For an inhomogeneous relaxation mechanism, the amplitude of the rotating electric monopole decreases because the ions get out of phase with each other. This is indicated in the bottom diagram of Figure 6. Formulae which relate individual ion positions to the magnitude and position of the rotating monopole are given in ref. 1.

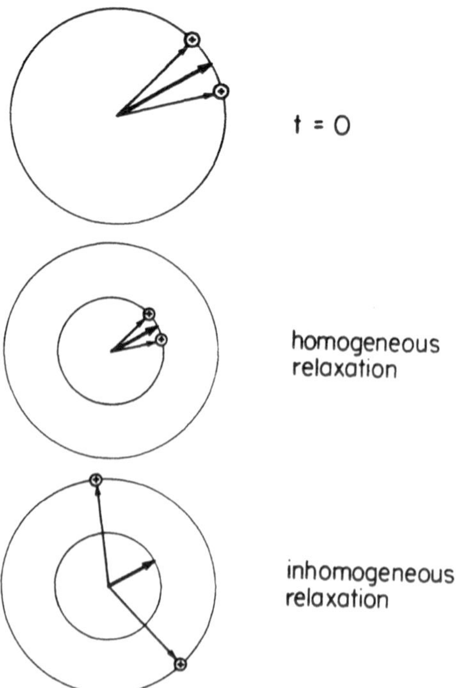

Figure 6. Homogeneous and Inhomogeneous Relaxation in ICR.

In the discussion to follow, various relaxation mechanisms which apply in ICR spectroscopy will be discussed. The individual mechanisms will be classified according to their homogeneous or inhomogeneous character and the quantitative importance of the mechanism will be assessed.

6. Radiation Damping of Excited Cyclotron Motion

Any oscillating charge will emit electromagnetic radiation at the oscillation frequency and conservation of energy requires this radiation to come from the oscillation energy of the oscillator. This phenomenon is called radiation damping of the oscillation amplitude [7]. Radiation damping occurs in all spectroscopic systems and is the most fundamental form of spectroscopic relaxation. Radiation damping is the classical counterpart of the quantum mechanical process known as spontaneous emission. It is a homogeneous relaxation mechanism.

For an oscillating, charged particle of charge q´, the rate of radiative energy loss is given by [7]

$$P = \frac{q'^2 \omega^2 x^2}{12\pi \varepsilon_o c^3} \quad \text{(mks)} \tag{10}$$

where P is the power loss, ω is the oscillation frequency, x is the oscillation amplitude, ε_o is the permittivity of free space and c is the velocity of light. For any damped harmonic oscillator, the Q of the oscillator, the number of radians required for the <u>energy</u> of the oscillator to decay to 1/e of the initial value, is related to the power loss by

$$Q = \frac{\omega U}{P} \tag{11}$$

where U is the instantaneous energy content of the oscillator. Moreover, the quality factor Q, is related to τ, the relaxation time for decay of the <u>amplitude</u> of oscillation to 1/e of its initial value, by

$$Q = \frac{\tau \omega}{2} \tag{12}$$

For excited cyclotron motion the oscillation amplitude is the cyclotron radius and the oscillation energy is given by

$$U = \tfrac{1}{2} m' \omega^2 r^2 \tag{13}$$

where m´ is the mass of the rotating monopole and ω is the cyclotron frequency. Combination of Equations 1, 10, 11 and 12 gives

$$\tau = \frac{12\pi m'^3 \varepsilon_o c^3}{q'^4 B^2} \quad \text{(mks)} \tag{14a}$$

or equivalently

$$\tau = 4.3713 \times 10^{14} \frac{m^3}{N q^4 B^2} \tag{14b}$$

Equation 14b gives τ, the relaxation time in seconds for ICR radiation damping, as a function of m, the ions mass in amu, q, the ion charge in units of the elementary charge, B, the magnetic field strength in kilogauss and N, the number of ions. Equation 14a is the corresponding equation in mks units as a function of the monopole mass, m´ and the monopole charge q´.

For the typical ICR values of m = 100 amu, N = 10^5 ions, q = 1 and B = 10 kilogauss, τ = 4.3 x 10^{13} seconds = 1.4 million years. Radiation damping is an insignificant line-broadening mechanism in ICR spectroscopy.

7. Resistive Damping of Excited Cyclotron Motion

As indicated in Figure 1, ions whose cyclotron motion has been excited can be modelled as a rotating electric monopole which generates the ICR signal. The rotating electric monopole induces an alternating signal current in and an alternating signal voltage across a resistor which connects the plates of the ICR cell [1]. The magnitude of the signal current is proportional to the number of ions and is given by Equation 3. When the current is passed through the resistor a voltage signal is developed but in addition there is an Ohmic power dissipation in the resistor. Conservation of energy requires this power loss to come from the kinetic energy of the rotating electric monopole [1]. As a consequence, the energy of the rotating monopole decays with time [1]. This "resistive damping" of excited ICR motion is a homogeneous damping mechanism.

The time constant for decay of excited ICR motion depends upon the nature of the circuit which connects the ICR plates [1]. For circuits which have only a resistor connecting the plates this time constant is given by [1]

$$\tau = \frac{2\ m\ d}{N\ q^2\ R} \qquad \text{(mks)} \qquad (15a)$$

or equivalently

$$\tau = 12.938\ \frac{m\ d^2}{N\ q^2\ R} \qquad (15b)$$

Equation 15b gives τ, the time constant in seconds for resistive damping of excited ICR motion for a predominantly resistive circuit, as a function of m, the ion mass in amu, d, the ICR cell spacing in cm, N, the number of ions, q, the ion charge in units of the elementary charge and R the shunt resistance in Megohms, Equation 15a is the corresponding equation in mks units.

For a typical case of N = 10^5 ions, m = 100 amu, d = 2 cm, q = 1 and R = 2.2 Megohm, Equation 15b predicts a resistive time constant of 0.023 sec. Resistive damping of ICR motion can be a significant factor in limiting the resolution of ICR spectrometers which use purely resistive circuits for signal detection. This damping is proportional to the number of ions and can be minimized for any particular spectrometer by working with a smaller number of ions.

For the signal detection circuit used in FT-ICR spectrometers and some bridge circuit spectrometers, the circuit across which the signal voltage is developed comprises a resistor and capacitor in parallel. For this circuit the capacitive impedance is much less than the resistive impedance and the time and the constant for resistive decay of excited cyclotron motion is given by [1]

$$\tau = \frac{2\ R\ B^2\ C^2\ d^2}{N\ m} \quad \text{(mks)} \qquad (16a)$$

or equivalently

$$\tau = 1.2044 \times 10^5\ \frac{R\ B^2\ C^2\ d^2}{N\ m} \qquad (16b)$$

Equation 16b gives the time constant in seconds for resistive damping in ICR as a function of R, the shunt resistance in Megohms, B the magnetic field strength in kilogauss, C the shunt capacitance in picofarads d, the cell spacing in cm, m, the ion mass in amu and N, the number of ions, Equation 16a is the corresponding equation in mks units.

For the typical case of R = 10 Megohms, B = 10 kilogauss, C = 10 picofarads, d = 2 cm, $N = 10^5$ ions and m = 100 amu, τ is calculated (Eq. 16b) to be 4,818 sec. Resistive damping is an insignificant relaxation mechanism for the predominantly capacitive ICR detection circuits used in broad band ICR spectrometers such as FT-ICR spectrometers.

8. Collisional Damping of Excited ICR Motion

An ICR system whose cyclotron motion has been excited may collide with neutral molecules. If the collision is chemically reactive then the number of excited ions, N(t), will decay with time according to

$$N(t) = N_0\ e^{-kt} \qquad (17)$$

where N_0 is the number of ions at t = 0 and k is the first order rate constant for the chemical reaction. If the collision is nonreactive then the forward momentum of the rotating monopole will damped [8-11]. The damping due to nonreactive collisions lowers the value of the monopole radius, r, (Figure 5) and may be represented in Equation 18 by a damping coefficient, ξ, called the reduced collision frequency [8-11).

$$r(t) = r_0 e^{-\xi t} \qquad (18)$$

The ICR current signal then becomes

$$I(t)_{rms} = I_0 e^{-kt} e^{-\xi t} \qquad (19)$$

where I_0 is the rms value of the ICR signal current (Equation 3) in the absence of collisional relaxation. Equation 19 may be rewritten in the form

$$I(t)_{rms} = I_0 \exp(-t/\tau_{collisions}) \qquad (20)$$

where $\tau_{collisions}$, the overall relaxation time for all types of collisions is given by Equation 21

$$\tau_{collisions} = 1/(k + \xi) \qquad (21)$$

Collisional damping of coherent ICR motion is both a homogeneous damping mechanism and an inhomogeneous damping mechanism.

Estimation of a numerical value for $\tau_{collision}$ requires presumption of a particular model for the ion-neutral collision. For systems in which the only significant potential is the ion-induced dipole potential (the Langevin potential), collisions are naturally separated into orbiting collisions and glancing collisions [12]. A reasonable model for the collision then is to assume that chemical reaction results from only a certain fraction of the orbiting collisions; the rest of the orbiting collisions and the glancing collisions giving only momentum relaxation [11]. For the Langevin potential the orbiting collision frequency, ν, is given by

$$\nu = \frac{n\,q}{2\,\varepsilon_0^{1/2}}\left(\frac{\alpha}{\mu}\right)^{1/2} \qquad \text{(mks)} \qquad (22a)$$

or equivalently

$$\nu = 2.2615 \times 10^{10}\ \frac{P\,q}{K}\left(\frac{(m + M)}{m\,M}\right)^{1/2} \qquad (22b)$$

where n is the neutral number density, α is the atomic polarizability of the neutral, μ is the reduced mass of the ion-neutral pair, ε_0 is the permittivity of free space, and where the terms in Equation 22b are as defined below.

The first order rate constant, k, for chemical reaction between the ion and the neutral is related to the orbiting collision frequency, ν, by

$$k = \nu f \tag{23}$$

where f is the fraction of reactive orbiting collisions. The corresponding reactive relaxation time, $\tau_{reactive}$, for ICR signal damping due to reactive collisions is just the inverse of the rate constant, and is given by Equation 24.

$$\tau_{reactive} = \frac{1}{k} = 4.422 \times 10^{-11} \frac{K}{f\ P\ q} \left(\frac{m\ M}{\alpha(m+M)}\right)^{\frac{1}{2}} \tag{24}$$

Equation 24 gives $\tau_{reactive}$, the ICR signal relaxation time in seconds for reactive ion-molecule collisions as a function of K, the absolute temperature f, the fraction of orbiting collision which are chemically reactive, P, the neutral pressure in torr, q, the ion charge in units of the elementary charge, α, the Gaussian atomic polarizability of the neutral in $Å^3$, m, the ion mass in amu and M, the neutral mass in amu. Equation 22b is the corresponding equation in the same units for the orbiting collision frequency, ν. Equation 22a is the corresponding equation in mks units for the orbiting collision frequency.

For a Langevin system which undergoes only nonreactive collisions ξ, the reduced collision frequency, is given by

$$\xi = 1.1052 \left(\frac{M}{m + M}\right) \tag{25}$$

where the numerical term in Equation 25 includes momentum relaxation due to glancing collisions. For a Langevin system which undergoes both reactive and nonreactive collisions ξ is given by [11]

$$\xi = (1.1052 - f)\ \nu \left(\frac{M}{m + M}\right) \tag{26}$$

and the overall collisional relaxation time, $\tau_{collision}$, follows from combination of Equations 21-26.

$$\tau_{collision} = \frac{1}{\nu[(1.1052 - f)\frac{m}{m + M} + f]} \tag{27}$$

For the typical case of m = M = 100 amu, K = 300°, α = 100 $Å^3$, q = 1, and f = 1, Equations 22b and 27 give a collisional relaxation time of 0.94 seconds for P = 10^{-8} torr and 0.00094 sec. for P = 10^{-5} torr. For the same case but with only nonreactive collisions (i.e. f = 0),Equation 27 gives relaxation times of 1.70 seconds for P = 10^{-8} torr and 0.0017 seconds for P = 10^{-5} torr. It is clear that at pressures above 10^{-8} torr collisional relaxation is a significant factor which limits ICR resolution.

For cases of a heavy ion colliding with a light neutral, the collisional relaxation time may be somewhat longer than that for the above "typical case". For an ion of mass m = 100 amu colliding non-reactively with 10^{-8} torr of He (M = 4 amu, $\alpha = 0.2$ Å^3), Equation 27 gives a collisional relaxation time of 5.5 seconds which is 6 times longer than 0.94 seconds characteristic of the M = m = 100 amu "typical case" given above.

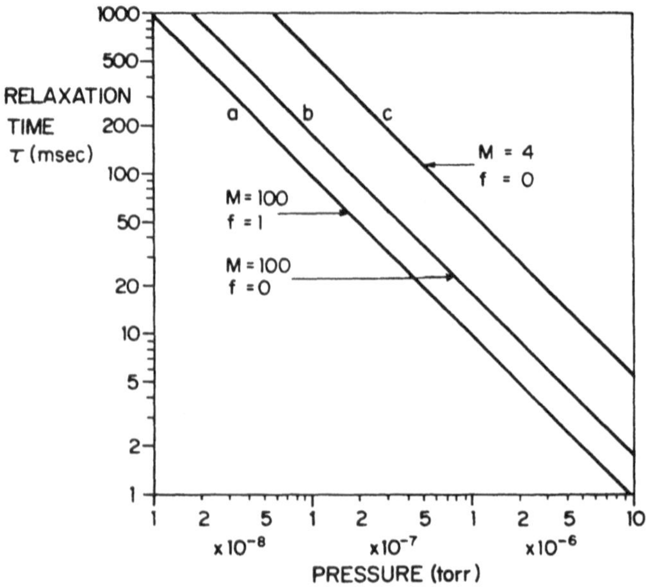

Figure 7.

Figure 7 shows the dependence of the collisional relaxation time, $\tau_{collisions}$, upon neutral pressure for a Langevin system. The three cases illustrated are an ion of mass 100 amu colliding reactively with a neutral of mass 100, an ion of mass 100 colliding non-reactively with a neutral of mass 100 and an ion of 100 colliding non-reactively with a neutral of mass 4. The longer relaxation time resulting from collisions with a lighter neutral results in greater mass resolution. A practical application of these considerations is that of Ridge and coworkers [13] and Gross and Wilkins and coworkers [14] who have noted that using Helium as a carrier gas in GC-ICR experiments will give greater ICR mass resolution than using heavier carrier gases, due to lesser collisional broadening.

The relaxation times calculated in the previous paragraph were specific to the chosen parameters and other choices for a "typical case" would give different relaxation times. The overall relaxation time is a function of the collision frequency (Equation 22) to which intermolecular potential terms other than the ion-induced dipole term may contribute [15]. For example, the presence of permanent dipole in the neutral can increase the collision frequency (and shorten the relaxation time) by up to a factor of 3. Also, charge exchange reactions may occur with a rate which may be in excess of the Langevin rate [12].

9. Doppler Relaxation of Excited ICR Motion

In most forms of spectroscopy, the sample molecules are moving and this movement relative to the radiation detector causes an additional broadening of the spectral lines. For emission spectroscopy, those molecules which are moving towards the detector when they emit will be perceived by the detector as having emitted a photon whose frequency is higher than the normal frequency. Molecules which are moving away from the detector are perceived as having emitted a lower frequency photon. For a random distribution of molecular velocities the spectral line can be broadened considerably by this phenomenon. This broadening of spectral lines due to the random velocities of the sample molecules is known as Doppler broadening. Doppler broadening is the classic form of inhomogeneous broadening [4-6].

For purposes of discussion in this communication, ICR relaxation due to the random motion of ions will be termed "Doppler relaxation". As discussed below, there is a direct effect due to random ion velocities which will be termed "first order Doppler relaxation" and an indirect effect which will be termed "second order Doppler relaxation".

The direct effect of random ion velocities on ICR relaxation can be determined with the model illustrated in Figure 8. Figure 8a shows ensemble of four ions in an ICR experiment which are instantaneously formed at the same location in space but have different velocities. For simplicity the velocities are taken to be of the same magnitude but in four different directions as indicated by the arrows in Figure 8a. A short interval of time after ionization the ion positions change to the positions indicated in Figure 8b. After ICR excitation to a monopole radius r the centers of the individual ion orbits are as shown in Figure 8c. Note that the relative positions of the individual ions in Figure 8c are the same as the relative positions of the centers of the ion orbits in Figure 8b. In the absence of relaxation phenomena these relative positions will remain the positions of the individual ions relative to one another, and implicitly their phase relationship to each other will remain constant with time. Thus the fact that the ions were moving relative to one another at their instant of formation will not result in <u>any</u> dephasing at a later time. First order Doppler relaxation, the relaxation due to the random motion of ions, is identicaly zero in ICR spectroscopy.

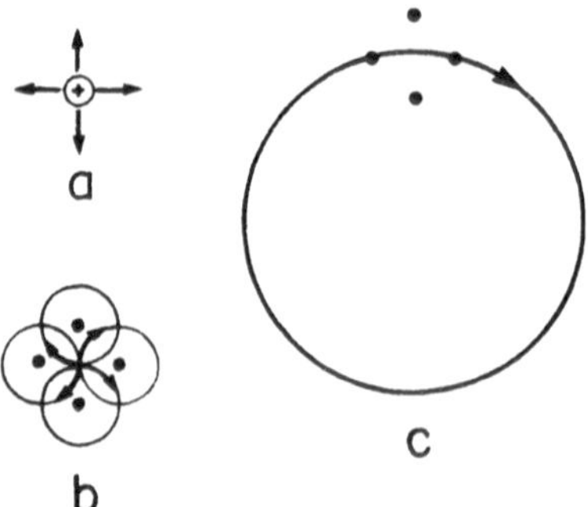

Figure 8

According to the model of Figure 8 the ensemble of four ions was formed in an infinitesimal volume of space. However due to differing velocities the ensemble of ions will occupy a finite volume of space which would be zero if the ions had no velocity. If for some reason the ICR frequency varied throughout space (say because of magnetic field inhomogeneities) then there would be an <u>additional</u> inhomogeneous relaxation due to the random motions of ions at their instant of formation. This additional relaxation will be termed second order Doppler relaxation.

According to kinetic theory, the root mean square velocity of molecules is

$$v_{rms} = \left(\frac{3\ k\ K}{m}\right)^{\frac{1}{2}} \qquad \text{(mks)} \qquad (28)$$

where v_{rms} is the velocity, K is the absolute temperature, k is Boltzmann's constant, and m is the ion mass. Now the relationship between ion velocity and ion cyclotron radius is

$$v = \omega r = \frac{q\ B}{m}\ r \qquad \text{(mks)} \qquad (29)$$

Solving Equation 29 for r and substituting Equation 28 yields

$$r = \frac{(3\ k\ K\ m)^{\frac{1}{2}}}{q\ B} \qquad \text{(mks)} \qquad (30a)$$

or equivalently

$$r = 2.835 \times 10^{-1}\ \frac{m^{\frac{1}{2}}}{q\ B} \qquad (30b)$$

Equation 30b gives the rms ion radius in cm for a thermal ion at 300°K as a function of m, the ion mass in amu, q the ion charge in elementary units and B the magnetic field strength in Tesla. Equation 30a is the corresponding equation in mks units. For the typical case of m = 100 amu, q = 1 and B = 1 Tesla, r (Equation 30b) is 0.023 cm. The volume occupied by thermal ions would have a diameter of twice this figure (Figure 8b) and unless the cyclotron frequency varied significantly over a distance of 0.05 cm, second order ICR Doppler relaxation would be insignificant for thermal ions.

The second order Doppler relaxation phenomenon is obviously more significant if the ions have a suprathermal velocity at their instant of formation. Such may be the case for fragment ions which are formed with a large kinetic energy release and for ions which are formed by laser irradiation.

10. Relaxation Due to Magnetic Field Inhomogenieties

The derivation of Equation 1 presumed that the magnetic field strength was constant throughout space and therefore that all ions of a given mass m, would have the exact same cyclotron frequency, ω. In any experimental situation this presumption will be only approximately true and ions in different regions of space will have slightly different resonant frequencies. In a conventional scanning ICR experiment ions in different regions of the magnetic field will resonate at slightly different times as the spectrum is scanned and the spectral peak for a particular ion mass will consist of a sum of individual peaks corresponding to the slightly different spectral positions. The overall line shape of the peak for a particular mass will be broadened accordingly. In transient ICR experiments such as the FT-ICR experiment, all ions of a given mass are excited together but as the resonant frequencies vary slightly, the signals from different ions of a particular mass will get out of phase with each other. The macroscopic transient signal, which is just a sum of the individual signals, will decay due to this dephasing. ICR relaxation due to magnetic field inhomogeneities is an inhomogeneous line broadening process.

Differentiating Equation 1 gives [2]

$$d\omega = \frac{q}{m} dB \tag{31}$$

and equating $\Delta\omega$ and ΔB, a small change in ω or B to the corresponding differentials gives

$$\Delta\omega = \frac{q}{m} \Delta B \tag{32}$$

Dividing Equation 32 by Equation 1 yields Equation 33

$$\frac{\Delta\omega}{\omega} = \frac{\Delta B}{B} \tag{33}$$

which shows that the fractional variation, $\Delta\omega/\omega$, in the cyclotron frequency which results from a fractional variation, $\Delta B/B$, in the magnetic field is numerically identical to $\Delta B/B$. This identity is independent of ion mass, is valid for all ICR experiments, and implies that the ICR broadening from magnetic inhomogenity should be independent of the monomer in which the ICR experiment is performed.

For the ICR experiments of this laboratory the maximum magnetic field variation, ΔB, over the volume of the cubic trapped ion cell [16], at a magnetic field strength of B = 20 kilogauss, is 500 milligauss which gives a fractional variation in the magnetic field strength of 1/40,000. According to Equation 33 the ICR frequency resolution should be limited to this fraction. Two inconsistencies are immediately apparent. First, the maximum mass resolution obtained in this laboratory is 550,000, an order of magnitude greater than the magnetic field homogeniety of 40,000. Second, <u>conventional</u> trapped-ion cell experiments in this laboratory, carried out with the same magnet and same ICR cell as was used for the early FT-ICR experiments [17,18] never gave a mass resolution of in excess of 5,000 at m/e = 31. These two inconsistencies are discussed in the following paragraphs.

Figure 9 shows the path of the rotating electric monopole during the excitation period of any ICR experiment. For conventional, scanning ICR experiments this is also the path of the monopole during the detection period. Figure 9 shows the path of the monopole during the detection period of the Ft-ICR experiment. As argued below, this path difference during the detection period makes the FT-ICR experiment less sensitive to magnetic field inhomogeneities than is implied by Equation 33.

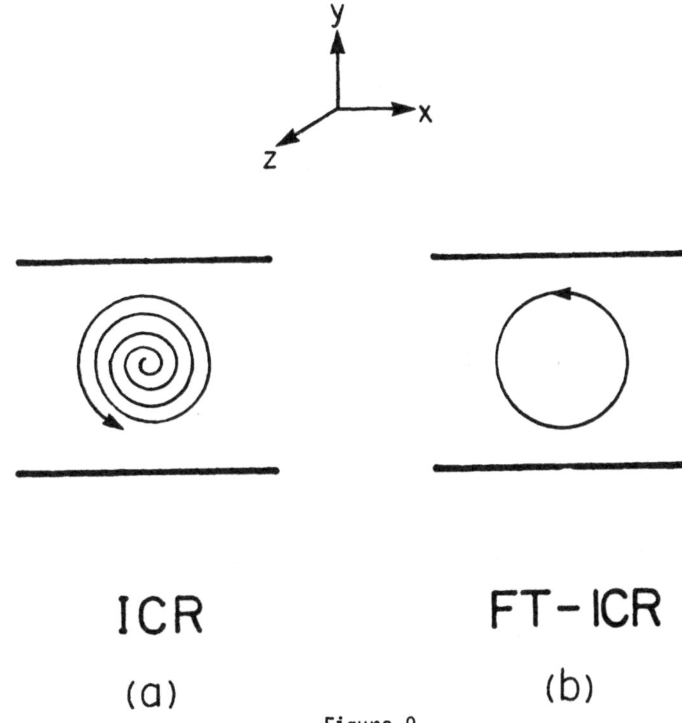

Figure 9

Consider the case where the magnetic field increased slightly in the radial direction. For this case the ion in the conventional ICR experiment (Figure 9a) would experience a changing resonant frequency as the monopole was detected whereas the monopole in the FT-ICR experiment (Figure 9b) would experience a constant magnetic field as it was detected. Even in these cases where there are other (non-radial) magnetic inhomogeneities the motion of Figure 9b will tend to average the magnetic field inhomogeneities seen by the rotating monopole to zero. This averaging will not occur for the motion of Figure 9a since the monopole does not retrace the same path on each revolution.

The averaging process is similar in concept to the averaging of magnetic field inhomogeneities in high resolution NMR spectroscopy by the technique of "spinning the sample" [19-22]. In NMR spectroscopy, the magnetic nuclei will experience only the average value of an inhomogeneous magnetic field if the time, t', required for a particular nucleus to experience the range of inhomogeneities, satisfies the inequality

$$t' < \frac{2\pi}{\gamma\ \Delta B} = \frac{2\pi}{\Delta\omega} \qquad (34)$$

where γ is the magnetogyric ratio of the nucleus, ΔB is the magnitude of the magnetic inhomogeneity, and $\Delta\omega$ is the range of NMR frequencies which results from the magnetic inhomogeneity. For modern high resolution NMR magnets, $2\pi/\Delta\omega$, the NMR frequency linewidth in Hz arising from magnetic field inhomogeneities, is of the order of 2 Hz. In current practice, these inhomogeneities can be averaged to zero by a spinning frequency of about 30 Hz; $t' = 1/30$ sec, which easily satisfies Inequality (34).

For ICR, we can write (by analogy to the NMR case)

$$t' < \frac{2\pi}{\Delta\omega} \qquad (35)$$

where t' is the time for an ion to experience the range of magnetic inhomogeneities and $\Delta\omega$ is given by Equation 32. For ICR, t' is the period of one cycle so that the criterion for "motional averaging of magnetic inhomogeneities in ICR" is given by

$$\frac{\Delta\omega}{\omega} < 1 \qquad (36)$$

or alternatively, from Equation 33,

$$\frac{\Delta B}{B} < 1 \qquad (37)$$

Inequality 37 implies that ICR motional averaging of magnetic field inhomogeneities will be effective even for magnetic inhomogeneities of the order of the magnetic field strength!

The derivation of the NMR Inequality 34 requires that the magnetic nucleus retrace the same path on each "averaging" cycle. In NMR this is a realistic presumption given that the sample spinning is mechanically imposed by an external "sample spinner". If we assume that an analogous process occurs in ICR, then Inequality 35 naturally follows. The enormous inhomogeneities which can be averaged (Inequality 37) result because the "averaging frequency" is the cyclotron frequency and is therefore very high compared to a typical NMR "averaging frequency". That Inequality 37 is an inadequate criterion for ICR motational averaging of magnetic inhomogeneities is discussed in the next paragraph.

There is reason to believe that Inequality 37 is not a stringent enough criterion to guarantee the effect of ICR motational averaging. For example, it has been assumed that the ion exactly retraces its path on each cycle and that by analogy to the NMR case, Inequality 35 is valid. However, it is known that magnetic field gradients will cause a slow drift of the centers of cyclotron orbits in a direction perpendicular to the direction of the gradient [23]. For a magnetic gradient in the y direction (Figure 9) the drift velocity in the x direction of the centers of the cyclotron orbits is given by

$$v_x = \frac{q\, r^2\, (\partial B/\partial y)}{2\, m} \qquad \text{(mks)} \qquad (38)$$

where q is the ion charge, r is the ion cyclotron radius, $\partial B/\partial y$ is the magnetic field gradient and m is the ion mass. From the field plots for the magnet of this laboratory the maximum gradient is 0.0005 T/ 0.025m = 4 x 10^{-3} T/m which for an ion of radius 1 cm and mass 100 amu gives a drift velocity (Equation 38) of 0.1 m/sec. Since the "retracing of the same path" required for validity of Inequality 35 requires that the center of the ion orbit be fixed for a period of 1 second or so it would appear the magnetic gradients as described above would invalidate the "motational averaging" arguments given above. That this is not entirely so is argued in the next paragraph.

In the presence of crossed static electric and magnetic fields the centers of the cyclotron orbits will drift in a direction perpendicular to the electric field, E, with a velocity, v, given by

$$v = \frac{E}{B} \qquad \text{(mks)} \qquad (39)$$

If this drift velocity is to cancel a "magnetic gradient drift" of 0.1m/sec., the magnitude of the required electric field for B = 2T is only 0.24 v/m = 2mv/cm. This small value for the applied static electric field is easily achieved by adjustment of the ICR cell voltages. Complete cancellation of "magnetic-gradient-ion-drift" everywhere in the ICR cell would require an electric field which could be adjusted for every point in the cell. Since this is an experimental impossibility, it is only possible to have an optimal setting of the electric gradient for minimizing magnetic gradient drift. For this reason only modest magnetic gradients can be nulled by the above procedure.

In summary, it seems that magnetic field gradients will inhomogeneously broaden ICR spectra lines and that this inhomogeneous broadening is less severe for FT-ICR experiments than for conventional ICR experiments because the volume of the magnetic field occupied by the signal-generating rotating monopole during the ICR detection period is smaller for the FT-ICR experiment than for the conventional ICR experiment. In addition, it seems that there is a motational averaging of modest magnetic inhomogeneities in the FT-ICR experiment which is absent in the conventional ICR experiment. This motational averaging is possible because ion drift in crossed electric and magnetic fields can be used to cancel ion drift due to magnetic field gradients.

11. Relaxation due to Electric Field Gradients

In all ICR spectrometers ion motion transverse to the magnetic field is constrained by the magnetic field and ion motion parallel to the magnetic field is constrained by an electric field which is imposed by the trapping plates [10]. This electric field, E_z, given by

$$E_z = -\frac{\partial \Phi}{\partial z} \qquad (40)$$

where is the electric potential inside the ICR cell, forces ions towards the central plane, z = 0, of the cell. However, there will always be components, E_x and E_y, of the electric field, given by

$$E_x = -\frac{\partial \Phi}{\partial x} \qquad E_y = -\frac{\partial \Phi}{\partial y} \qquad (41)$$

in the x and y directions. These fields are radial to the ion motion and will add to the magnetic force which causes cyclotron motion.

$$\underset{\sim}{F} = q(\underset{\sim}{v} \times \underset{\sim}{B}) \qquad (42)$$

Since the ion velocity, v, increases linearly with ion radius according to

$$\underset{\sim}{v} = \omega \underset{\sim}{r} \qquad (43)$$

the magnetic force F (Equation 42) will also increase linearly, keeping ω constant. If the electric field, Φ, is quadrupolar

$$\Phi = \text{constant } (z^2 + r^2)$$

Then the radial electric field, E_r, will increase linearly with distance from the center of the cell and the total radial force on the ions will increase linearly with ion radius.

$$F_{radial} = q(\omega rB - cr) \qquad \text{(c, constant)} \qquad (44)$$

$$= qr(\omega B - c) \qquad (45)$$

The net effect of the radial electric field is to lower the experimental cyclotron frequency from that calculated from Equation 1 [10]. However, if the electric field is not quadrupolar then the radial electric gradient is not a linear function of ion radius and the experimental cyclotron frequency will be a function ion radius. Since the experimental electric potential in rectangular ICR cells such as the cubic trapped ion cell [16] is only approximately quadrupolar, a variation of cyclotron frequency with ion radius can be expected to occur. The resulting dephasing if all ions are not excited to exactly the same ion radius is an inhomogeneous relaxation mechanism.

The effects of non quadrupolar electric fields on ICR resolution are similar to those of magnetic field inhomogeneities as discussed in the preceeding section. These effects can be expected to be more severe for conventional ICR than for FT-ICR experiments because the volume of space occupied by the rotating monopole during the detection period is less for FT-ICR than for conventional ICR and because motational averaging of the electric gradient will occur in FT-ICR but not conventional ICR (cf. preceeding section and Figure 9).

12. Noise and Sensitivity in ICR Spectrometers

The principal sources of noise in ICR spectrometers arise from the electronic components which are used to monitor the signal from the rotating electric monopole. The first source of noise is the Johnson noise of the resistor which is used to convert the ICR current signal into a voltage signal. The magnitude of the noise current is given by

$$I_N = \left(\frac{4k\ K\ \Delta f}{R}\right)^{-\frac{1}{2}} \qquad \text{(mks)} \qquad (46)$$

where k is Boltzmann's constant, K is the absolute temperature, Δf is the detection bandwidth and R is the resistance (Figure 1). If this were the only source of noise then it can be shown that the ICR sensitivity, the ratio of the ICR signal to the ICR noise, is proportional to $R^{\frac{1}{2}}$ [1].

Other forms of noise however are introduced by the electronic components which are used to amplify the voltage across the resistor, R. The additional noise coming from the electronic components can be modelled by a current noise source, i_N and a voltage noise source, e_n. The total voltage noise is then given by

$$V_N = I_n Z_{RC} + i_N Z_{RC} + e_N \qquad (47)$$

where the circuit model for the generation of the total noise, V_N, is shown in Figure 10. The total voltage noise appears across the terminals on the right hand side of Figure 10. The ICR signal, V_S, appears across the same terminals.

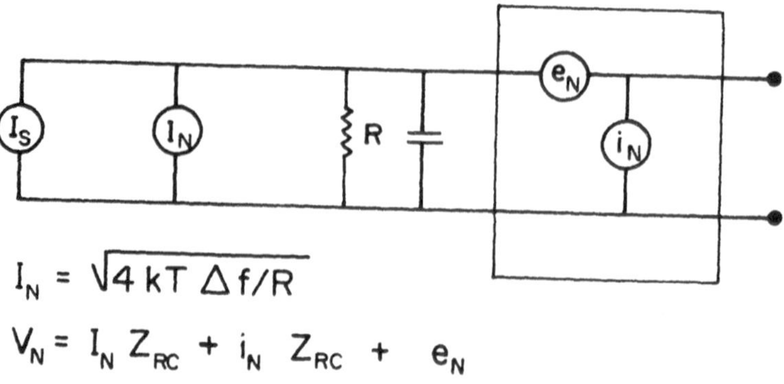

$I_N = \sqrt{4\,kT\,\Delta f/R}$

$V_N = I_N\, Z_{RC} + i_N\, Z_{RC} + e_N$

Figure 10.

The noise voltage e_N and the noise current, i_N, which are frequency dependent, are available in the data sheets of electronic devices. Generally, the noise due to the first amplification stage of the electronics will determine the overall signal to noise ratio. For low noise field effect transistors a typical value of e_N is 10^{-9} volts/($Hz^{\frac{1}{2}}$) over the frequency band 10kHz - 10MHz. Over the same band, the current noise, i_N, increases approximately linearly from 1 fA to 1 pA. For simplicity, these values can be taken as the typical values for calculations involving Equation 47. Figure 11 shows the results for a tuned ICR circuit with an effective resistance value (Equation 6) of 1 Megohm. The individual terms in Equation 47 are also shown in Figure 11. Note that from 10 KHz to 1 MHz, the total noise is dominated by the Johnson noise (Equation 46) but that above this frequency the current noise of the electronics is the dominant source of noise. Also shown on the Figure is the ICR signal calculated from Equation 5 for r = 1cm, m = 100 amu, N = 10 ions.

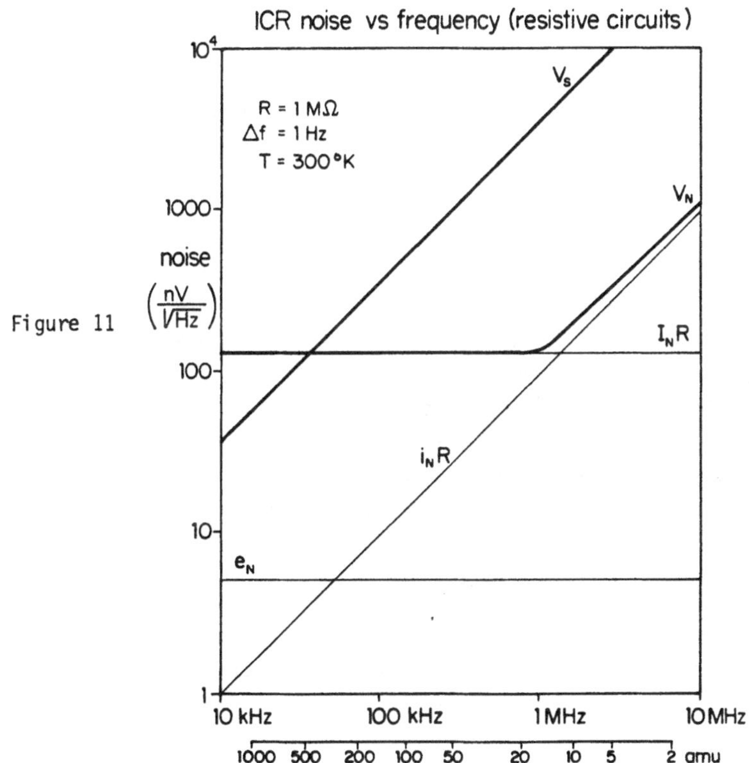

Figure 11

Figure 12 shows the same relationships as in Figure 11, but for a predominantly capacitive circuit.

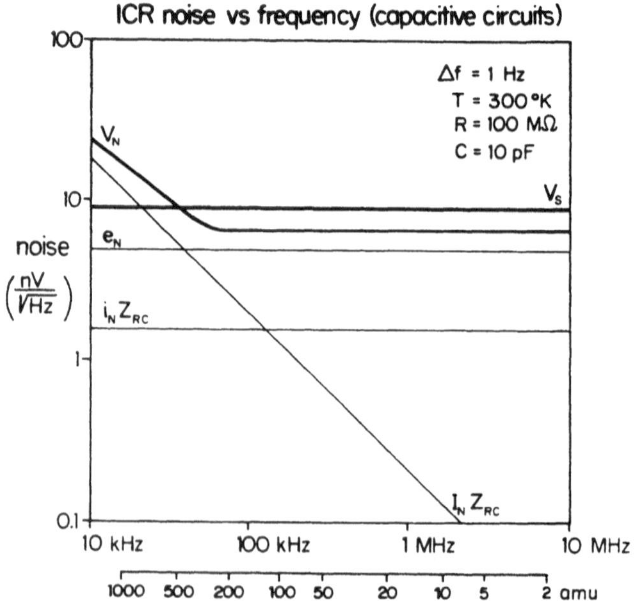

Figure 12.

Comparison of Figures 11 and 12 shows that for the "typical values" chosen, the sensitivity, V_S / V_N, of tuned circuits is slightly greater than the sensitivity of the "broad band" capacitive circuits used for FT-ICR. These considerations however are only important if only a single peak in the spectrum is being (single ion monitoring). The fact that the FT-ICR spectrometer "gives the whole spectrum at once" permits time averaging of the whole spectrum and when this factor is taken into consideration broadband detection circuitry is more sensitive than is the tuned circuitry used for scanning ICR.

13. Acknowledgement

This research was supported by the Natural Sciences and Engineering Research Council of Canada.

14.References

1. M. B. Comisarow, J. Chem. Phys. 69(1978)4097.
2. M. B. Comisarow and A. G. Marshall, J. Chem. Phys. 64(1976)110.
3. A. G. Marshall, M. B. Comisarow and G. Parisod, J. Chem. Phys. 71 (1979)4434.
4. J. I. Steinfeld, "Molecules and Radiation", Harper and Row, New York (1974).
5. M. Sargent, M. O. Scully, W. I. Lamb, Jr., "Laser Physics", Addison Wesley, Reading, Mass., (1974).
6. A. M. Portis, Phys. Rev. 91(1953)1071.
7. R. P. Feynman, "The Feynman Lectures on Physics", Addison-Wesley Publishing Company, Reading, Mass., (1963).
8. D. Wobschall, J. R. Graham and D. P. Malone, Phys. Rev 131(1963)1565.
9. J. L. Beauchamp, J. Chem. Phys. 46(1967)1231.
10. T. A. Lehman and M. M. Bursey, "Ion Cyclotron Resonance Spectrometry", John Wiley and Sons, New York (1976), p.6.
11. M. B. Comisarow, J. Chem. Phys. 55(1971)205.
12. S. G. Lias and P. Ausloos, "Ion-Molecule Reactions", American Chemical Society, Washington, D.C. (1975) Chapter 4.
13. M. T. Nguyen, J. Wronka and D. P. Ridge, paper presented at ASMS meeting, Seattle, June 3-8, 1979.
14. E. B. Ledford, R. L. White, S. Ghaderi, M. L. Gross and C. L. Wilkins, Anal. Chem. 52(1980)1090.
15. T. Su and M. T. Bowers, Chapter 3 in "Gas Phase Ion Chemistry", M. T. Bowers, Editor, Academic Press, New York, 1979.
16. M. B. Comisarow, Int. J. Mass Spec. Ion Phys. 37(1981)247.
17. M. Comisarow, Adv. Mass Spec. 7(1978)1042.
18. M. Comisarow, Chapter 10 in "Transform Techniques in Chemistry", P. Griffiths, Editor, Plenum (1978).
19. Halbach, Helv. Phys. Acta 29(1956)37.
20. G. A. Williams and H. S. Gutowsky, Phys. Rev. 104(1956)
21. F. Bloch, Phys. Rev. (1954)496.
22. W. A. Anderson and J. T. Arnold, Phys. Rev. (1954)497.
23. J. D. Jackson, "Classical Electrolynomics" Wiley (1962).

THEORETICAL TOOLS FOR THE DESCRIPTION OF ION MOTION IN ICR SPECTROMETRY

Dieter Schuch, Kyu-Myung Chung, and Hermann Hartmann
Institut für Physikalische und Theoretische Chemie der Johann Wolfgang Goethe-Universität, Robert-Mayer-Str. 11, 6000 Frankfurt (Main), FRG

1. Methods for the Description of Ion Motion

In ICR spectrometry the motion of ions under the influence of magnetic and electric field is considered. This motion can be described by different methods which may be principally divided into two groups. One group includes descriptions based on classical particle mechanics, the other group utilizes field theoretical methods, where especially the quantum theoretical description has to be mentioned which will be discussed in more detail later on. In the following, we want to consider examplarily a single ion with thermal energy. Additional statistical and thermodynamical effects may be taken into account subsequently.

1.1 Classical Particle Mechanics

In classical theory the essential physics involved in the mechanics of a particle is contained in the equation of motion which can be formulated in different ways. Here we want to restrict ourselves to the two most common formulations, Newton's equation of motion, for a particle of mass m in electric and magnetic fields given by

$$\vec{F} = m\dot{\vec{v}} = q\,(\vec{E} + \frac{1}{c}\,[\vec{v} \times \vec{B}]),\ \vec{F}\text{: force}$$
$$\text{with } \vec{E} = -\nabla\Phi,\ \vec{E}\text{: electric field} \qquad (1)$$
$$\Phi\text{: electric potential},$$

where q is the particle's charge and $\vec{v}$ its velocity normal to $\vec{B}$, the magnetic induction which is connected with the macroscopic quantity magnetic field, $\vec{H}$, via the constant μ, the so-called permeability, according to $\vec{B} = \mu\vec{H}$ [1], and Hamilton's equations of motion

$$\dot{q}_i = \frac{\partial H}{\partial p_i},\ \dot{p}_i = -\frac{\partial H}{\partial q_i}, \qquad (2)$$

(For the following discussion we will accept $\mu = 1$, i.e. $\vec{B} = \vec{H}$) where q_i are generalized coordinates, p_i generalized momenta, and the Hamiltonian function H for the particular case of electromagnetic forces has the form

$$H = \frac{1}{2m}\,(\vec{p} - \frac{q}{c}\,\vec{A})^2 + V$$
$$\text{with } \vec{p}\text{: canonical momentum} \qquad (3)$$
$$\vec{A}\text{: magnetic vector potential}$$

$V = q\Phi$: potential energy.

Only in a few cases these classical equations of motion can be solved exactly. Usually, the influence of interfering fields, field inhomogeneities, collisions and similar effects entail complicated, often coupled or even nonlinear differential equations. In most cases, these equations cannot be solved analytically and even a numerical evaluation is often quite cumbersome - if possible at all - and the results are only numerical values.

1.2. Quantum Theoretical Description

Another method to describe a dynamical system is quantum theory, where normally the Hamiltonian function is taken as basis. The vectors of position and canonical momentum and the energy of the system are replaced by properly defined operators. Acting with the Hamiltonian operator obtained in this way on wave functions, the so-called state vectors, we arrive at the quantum mechanical equation of motion, the Schrödinger equation, in our particular case given by

$$i\hbar\dot{\Psi}(\vec{r},t) = H\Psi(\vec{r},t) = \{\frac{1}{2m}(\frac{\hbar}{i}\nabla - \frac{q}{c}\vec{A})^2 + q\Phi\}\,\Psi(\vec{r},t), \qquad (4)$$

where the dot denotes the time derivative.

Here we can see that completely different mathematical objects are used to describe the state of a dynamical system in classical particle theory and in quantum theory [2]. In classical mechanics the state of the system is fully described by giving the trajectory. In quantum theory we describe the state of the system by the state vector, but the information contained in the state vector has to be extracted with the help of various operators.

In order to compare directly the classical and the quantum description, we must first formulate both theories in terms of mathematical objects of the same type.

The proper objects in quantum theory - which become in the classical limit the classical solutions of the equation of motion - are expectation values of quantum operators evaluated using coherent states, $\langle\ldots\rangle_{wp}$.

The quantum mechanical pure states are quite different from the classical states. However, from the infinite superposition of pure states, minimized according to the uncertainty principle, one can construct wave packets, the so-called coherent states, which are necessary for our description of quasi-classically behaving ions [3-10].

The quantum theoretical results obtained in this way can be converted to the classical results by making use of the relations between classical energy and angular momentum, and the corresponding quantum theoretical expressions [9-13,15] here shown for the cyclotron motion:

$$\frac{m}{2}\omega_c^2 < R^2>_{wp} = \frac{m}{2}\omega_c^2\, a^2 = (\bar{N}+1)\hbar\omega_c$$
$$<[\vec{R}\times\vec{p}]_z>_{wp} = [\vec{a}\times\vec{p}_{kl}]_z = -\hbar\bar{N} \qquad (5)$$

where:

$\bar{N} = \frac{a^2 b}{2}$ with $b = \frac{qB}{\hbar c}$, a: classical orbit radius (cf. Fig. 1)

$(\bar{N}+\frac{1}{2})\,\hbar\omega_c = k_B T$ with ω_c: $\frac{qB}{mc}$ = cyclotron frequency

k_B : Boltzmann's constant

T: temperature in Kelvin.

Now, we want to add a few words elucidating the advantages of a quantum theoretical description. Besides some essential advantages of quantum theory already pointed out in earlier works [9,10], we would like to draw attention to the difficulty mentioned above, that besides a few exceptions only numerical integration of the complicated differential equations of classical particle mechanics is possible yielding only numbers.

On the other hand, if it is possible to solve the Schrödinger equation of a rather simple problem exactly, more complicated smaller effects may be regarded as perturbations and treated accordingly by the quite well developed quantum theoretical methods of perturbation theory if an exact solution is not possible. In this way problems can also be investigated which are not attainable with classical calculations. Furthermore, as result of perturbation calculation relations between physical quantities and experimental parameters can be obtained instead of numerical values only, which makes basical relationships more transparent.

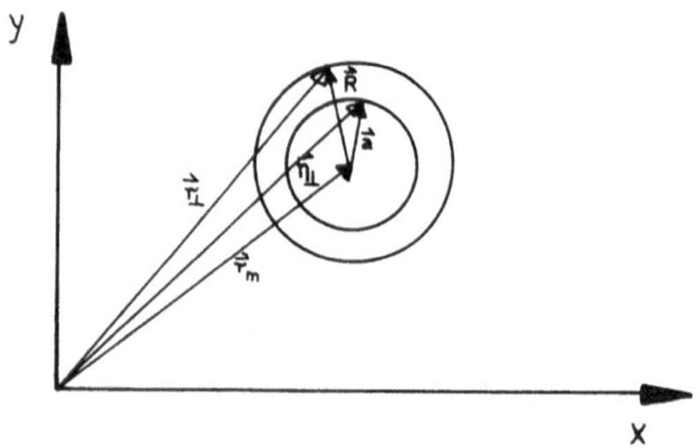

Fig. 1. Projection of ion motion onto the xy plane perpendicular to the magnetic field direction.

$\vec{r}$: vector of position in the xy plane, $\vec{r}_\perp$: classical trajectory in the xy plane, $\vec{r}_m$: centre of circular motion, $\vec{R} = \vec{r}_\perp - \vec{r}_m$: with $R = |\vec{R}|$ = radius of circular motion, $\vec{a} = \vec{r}_\perp - \vec{r}_m$ with $a = |\vec{a}|$ = classical radius of circular motion.

Finally, for all so-called quantum effects naturally a quantum theoretical description is the only adequate.

So, the scheme in Figure 2 shows how to proceed with a given problem:
- One may try to solve the classical equations of motion or - if this is not convenient or not possible, the equations of classical particle mechanics have to be quantized in order to obtain the corresponding field equation, the Schrödinger equation.

Solving this equation, or if this is not possible a simplified version of it, we get the pure states. A superposition of these pure states provides us with the coherent states which are used to evaluate quantum theoretical expectation values.

Finally, the quantum theoretical results have to be converted to the equivalent classical expressions.

In order to construct coherent states, however, we actually need the pure states, i.e. the eigenfunctions of the corresponding Schrödinger equation. In the following we will therefore draw attention to the Schrödinger equations of some selected exactly solvable problems which are of importance in ICR spectrometry and can be used as a basis for the treatment of more complicated problems.

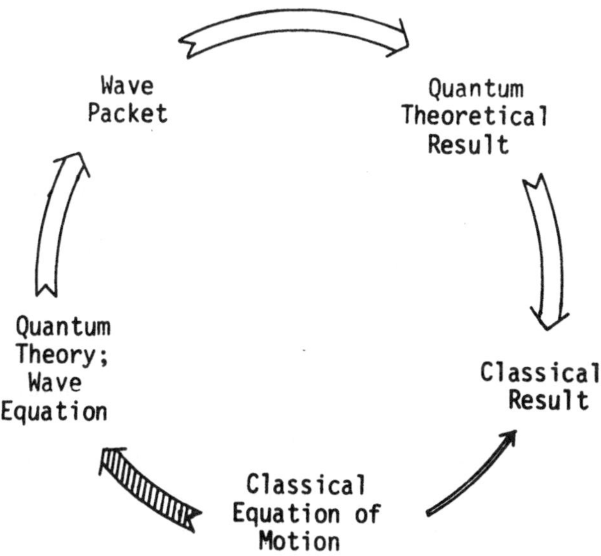

Fig. 2. Different ways for the description of a nearly classical behaving system as it is represented by an ion in the ICR cell, for example.

For our calculation we choose the coordinate frame in a way that for an ICR drift cell the drift direction is identical with the x direction, the direction of RF oscillation is the y direction, and the trapping and usual magnetic field direction is the z direction.

First we will regard the case of homogeneous magnetic field in z direction, including electric fields of increasing complexity. After the discussion of a change of magnetic field orientation, finally the influence of velocity dependent frictional forces will be investigated.

2. Homogeneous Magnetic Field $\vec{B} = (0,0,B)$, and Electric Fields of Increasing Complexity

2.1. Schrödinger Equation of a Charged Particle Exclusively Influenced by a Homogeneous Magnetic Field

For application to ICR spectometry, the most simple problem is that of an ion exclusively under the influence of a homogeneous magnetic field oriented in z direction. The corresponding Schrödinger equation, given by

$$i\hbar\dot{\Psi}(\vec{r},t) = \frac{1}{2m}\left(\frac{\hbar}{i}\nabla - \frac{q}{c}\vec{A}(\vec{r})\right)^2 \Psi(\vec{r},t) \qquad (6)$$

can be separated using the product ansatz

$$\Psi(\vec{r},t) = \Psi_{\perp}(\vec{r}_{\perp},t)\cdot\Psi_{\parallel}(z,t), \quad \text{with } \vec{r}_{\perp} = (x,y,0).$$

The solution of the Schrödinger equation

$$i\hbar\dot{\Psi}_{\parallel}(z,t) = -\frac{\hbar^2}{2m}\frac{\partial^2}{\partial z^2}\Psi_{\parallel}(z,t) \qquad (7)$$

describing the motion in z direction consists of normalized plane waves,

$$\Psi_{\parallel}(z,t) = \psi_{\parallel}(z)\cdot\chi_{\parallel}(t) = (L_z)^{-1/2}\, e^{ik_z z}\cdot e^{-\frac{i}{\hbar}E_{\parallel}t} \qquad (8)$$

$$E_{\parallel} = \frac{\hbar^2}{2m}k_z^2 = \frac{p_z^2}{2m} \quad ,$$

with the energy $E_{\parallel}$ (in z direction, parallel to the magnetic field) appearing in the time dependent factor $\chi_{\parallel}(t)$.

The solution functions describing the motion in the plane perpendicular to magnetic field direction, $\Psi_{\perp,N}(\vec{r}_{\perp},t)$, also consist of an only time dependent factor $\chi_{\perp,N}(t)$ containing the energy $E_{\perp,N}$ with the cyclotron frequency ω_c,

$$\Psi_{\perp,N}(\vec{r}_{\perp},t) = \psi_{\perp,N}(\vec{r}_{\perp})\cdot\chi_{\perp,N}(t) = \psi_{\perp,N}(\vec{r}_{\perp})\cdot e^{-\frac{i}{\hbar}E_{\perp,N}t} \qquad (9)$$

$$E_{\perp,N} = (N+\tfrac{1}{2})\hbar\omega_c$$

and of the so-called screw-functions $\psi_{\perp,N}(\vec{r}_\perp)$ which are solutions of the stationary Schrödinger equation

$$H_\perp \psi_{\perp,N}(\vec{r}_\perp) = \frac{1}{2m}\left(\frac{\hbar}{i}\nabla_\perp - \frac{q}{c}\vec{A}\right)^2 \psi_{\perp,N}(\vec{r}_\perp) = E_{\perp,N}\,\psi_{\perp,N}(\vec{r}_\perp) \tag{10}$$

for the symmetric gauge $\vec{A}(\vec{r}) = \frac{1}{2}[\vec{B} \times \vec{r}_\perp]$ of the vector potential [14]. The explicit screw-functions,

$$\psi_{\perp,N}(\vec{r}_\perp) = \left(\frac{b}{2\pi}\left(\frac{b}{2}\right)^N \frac{1}{N!}\right)^{1/2} \exp\left\{\frac{iq}{\hbar c}\vec{A}(\vec{r}_m)\cdot\vec{r}_\perp - \frac{b}{4}(\vec{r}_\perp - \vec{r}_m)^2\right\} |\vec{r}_\perp - \vec{r}_m|^N e^{-iN\phi} \tag{11}$$

with $\phi = \mathrm{arctg}\,(y - y_m/x - x_m)$, $\vec{r}_\perp - \vec{r}_m = \vec{R}$ (cf. Fig. 2),

show , that they are characterized by the quantum number N and the position of the centre of cyclotron orbit, $\vec{r}_m$ entering as parameter.

These functions have qualities which simplify explicit calculations and are similar to the properties of the harmonic oscillator eigenfunctions. Combining the operators of velocity components linearly, the operators v_+ and v_- are obtained:

$$\begin{aligned} v_x &= \frac{\hbar}{im}\left(\frac{\partial}{\partial x} + i\frac{b}{2}y\right) \qquad v_y = \frac{\hbar}{im}\left(\frac{\partial}{\partial y} - i\frac{b}{2}x\right) \\ v_+ &= v_x - iv_y \qquad\qquad\qquad v_- = v_x + iv_y \\ v_\mp &= \left\{\frac{\hbar}{im}\;\frac{\partial}{\partial x} \pm i\frac{\partial}{\partial y} \pm \frac{b}{2}(x \pm iy)\right\} \end{aligned} \tag{12}$$

which are acting like the raising and lowering operators of the harmonic oscillator:

$$\begin{aligned} v_+ \psi_{\perp,N} &= -\frac{\hbar}{im}(2b(N+1))^{1/2}\;\psi_{\perp,N+1} \\ v_- \psi_{\perp,N} &= \frac{\hbar}{im}(2bN)^{1/2}\;\psi_{\perp,N-1}\,. \end{aligned} \tag{13}$$

By means of these operators, the Hamiltonian and the operators of velocity components may be rewritten in the form

$$\begin{aligned} H_\perp &= \frac{m}{4}(v_+v_- + v_-v_+) \\ v_x &= \frac{1}{2}(v_+ + v_-) \\ v_y &= \frac{i}{2}(v_+ - v_-). \end{aligned} \tag{14}$$

Taking care of the commutation relation

$$[v_-, v_+] = \frac{2\hbar^2 b}{m^2} = \frac{2}{m}\hbar\omega_c \tag{15}$$

time-derivatives of expectation values can evaluated quite easily. As an example the first and second time-derivatives of the x and y component of velocity are given by:

$$\begin{aligned}
\frac{d}{dt}\langle v_x\rangle &= \frac{1}{i\hbar}\langle[v_x,H_\perp]\rangle = \omega_c\langle v_y\rangle = \langle\dot{v}_x\rangle \\
\frac{d^2}{dt^2}\langle v_x\rangle &= \omega_c\frac{1}{i\hbar}\langle[v_y,H_\perp]\rangle = -\omega_c^2\langle v_x\rangle = \langle\ddot{v}_x\rangle \\
\frac{d}{dt}\langle v_y\rangle &= \frac{1}{i\hbar}\langle[v_y,H_\perp]\rangle = -\omega_c\langle v_x\rangle = \langle\dot{v}_y\rangle \\
\frac{d^2}{dt^2}\langle v_y\rangle &= -\omega_c\frac{1}{i\hbar}\langle[v_x,H_\perp]\rangle = -\omega_c^2\langle v_y\rangle = \langle\ddot{v}_y\rangle .
\end{aligned} \tag{16}$$

2.1.1. Wave Packet Solution

Examplarily, the wave packet solution describing the motion in xy plane shall be given together with a short discussion of its properties. The wave packet $\Psi_{\perp,wp}(\vec{r}_\perp,t)$ can be expanded in terms of stationary wave functions [11]

$$\Psi_{\perp,wp}(\vec{r}_\perp,t) = e^{-i(\omega_c/2)t}\sum_{N=0}^{\infty} A_N\ \psi_{\perp,N}(\vec{r}_\perp)\ e^{-iN\omega_c t} \tag{17}$$

with the arbitrary constants A_N, which should be determined for the minimum wave packet. We assume that at t = 0, the general solution (17) has the form of the normalized minimum packet

$$\begin{aligned}
\Psi_{\perp,wp}(\vec{r}_\perp,0) &= \left(\frac{b}{2\pi}\right)^{1/2}\exp\left\{\frac{i}{\hbar}(m\dot{\vec{\eta}}_\perp + \frac{q}{c}\vec{A}(\vec{\eta}_\perp))(\vec{R}-\vec{a})\right\}\cdot\exp\left\{-\frac{b}{4}(\vec{R}-\vec{a})^2\right\} \\
&= \sum_{N=0}^{\infty} A_N\psi_{\perp,N}(\vec{r}_\perp)
\end{aligned} \tag{18}$$

where the quantities $\vec{\eta}_\perp$, $\vec{R}$, $\vec{a}$ are defined in Figure 1. Making use of the orthogonality of the wave functions [11], a particular coefficient A_N can be determined and we get, choosing appropriate initial conditions, the simple form

$$A_N = \left(\left(\frac{b}{2}\right)^N\frac{1}{N!}\right)^{1/2}\ |\vec{a}|^N\ \exp\left\{-\frac{b}{4}a^2 - \frac{iq}{\hbar c}\vec{A}(\vec{r}_m)\cdot\vec{a}\right\}. \tag{19}$$

Inserting this weight function 19 into 17, we obtain the time dependent wave packet solution.

The coefficients A_N fulfil the relation $\sum_{N=0}^{\infty} A_N A_N^* = 1.$ (20)

The absolute square of this wave packet gives a position probability density

$$|\Psi_{\perp,wp}(\vec{r}_\perp,t)|^2 = \frac{b}{2\pi}\exp\left\{-\frac{b}{2}(\vec{R}-\vec{a})^2\right\}, \tag{21}$$

which attains its maximum at the classical cyclotron orbit, $\vec{R} = \vec{a}$.

In the limit $|\vec{a}|\to 0$, the wave packet $\Psi_{\perp,wp}(\vec{r}_\perp,t)$ approaches the ground state eigenfunction $\Psi_{\perp,0}(\vec{r}_\perp,t)$. The larger $|\vec{a}|$ becomes, the larger the number of stationary states that contribute significantly to the packet, and the larger the quantum number N_{max} for which A_N of Eq. 19 has a maximum. As in our case $N \gg 1$, we can use Stirling's formula to maximize $\ln A_N$; neglecting terms of order $\ln N$ and lower, finally we obtain

$$N_{max} \approx \frac{b}{2} a^2 = \frac{m\,\omega_c^2\,a^2}{2\,\hbar\,\omega_c} . \tag{22}$$

Thus, the energy level $E_{\perp,N_{max}} = (N_{max}+1/2)\hbar\omega_c$, from whose neighbourhood most of the contribution to $\Psi_{\perp,wp}$ comes, is approximately equal to the energy $\frac{m}{2}\omega_c^2 a^2$ of the classical motion in the same magnetic field. Taking usual ICR conditions as a basis, N_{max} is in the region of 10^6.

Inserting 22 into 19 we get

$$|A_N^* A_N| = (\frac{b}{2} a^2)^N \frac{1}{N!} e^{-\frac{b}{2}a^2} \approx \frac{(N_{max})^N}{N!} e^{-N_{max}} , \tag{23}$$

i.e. the probability to find the system in the state N obeys a Poisson distribution with maximum at N_{max}.

2.2. Consideration of Drift Motion

Still remaining in the xy-plane, we take now into account an additional electric field in y direction causing a drift motion in x direction with a drift velocity given by

$$\vec{v}_D = \frac{c[\vec{E}_D \times \vec{B}]}{B^2}$$
$$\text{with } \vec{E}_D = (0,E_D,0) = \text{constant} \tag{24}$$
$$\Phi = -\vec{E}_D \cdot \vec{r}.$$

Changing to a moving coordinate system by the Galilean transformation $\vec{r}' = \vec{r} - \vec{v}_D t$ accompanied by a gauge transformation of the potentials according to

$$\begin{aligned} \hat{\Phi}(\vec{r}',t) &= \Phi'(\vec{r}' + \vec{v}_D t, t) - \frac{1}{c}\frac{\partial \hat{F}}{\partial t} \\ \hat{\vec{A}}(\vec{r}',t) &= \vec{A}(\vec{r},t) + \text{grad}\,\hat{F}(\vec{r}',t) \\ \Phi'(\vec{r}' + \vec{v}_D t) &= \Phi(\vec{r}) - \frac{1}{c}(\vec{v}_D \cdot \vec{A}(\vec{r})) \\ \hat{F}(\vec{r}',t) &= -(\vec{r}' \cdot [\vec{B}/2 \times \vec{v}_D t]) \end{aligned} \tag{25}$$

we can also find an exact solution to this problem [16].

The solution functions in the laboratory system

$$\Psi_{D,N}(\vec{r},t) = \left(\frac{b}{2\pi}\left(\frac{b}{2}\right)^N \frac{1}{N!}\frac{1}{L_z}\right)^{1/2} \exp\left\{-\frac{b}{4}(\vec{r}-\vec{r}_m)^2 + i\left[\frac{q}{\hbar c}\vec{A}(\vec{r}_m)\cdot\vec{r} + \frac{m}{\hbar}\vec{v}_D\cdot\vec{r} + k_z z\right]\right\}$$

$$\cdot\ |\vec{r}-\vec{r}_m|^N e^{-iN\phi} \cdot \exp\left\{-\frac{i}{\hbar}\left[(N+\frac{1}{2})\hbar\omega_c + \frac{p_z^2}{2m} - \frac{q}{2}E_D y_m + \frac{m}{2}v_D^2\right]\cdot t\right\} \qquad (26)$$

show three additional terms: an additional momentum in drift direction and additional kinetic and potential energies due to the drift motion in the electric field.

2.3. Consideration of Trapping Motion

In a next step, the trapping field in z direction should be taken into account. The most simple approximation is the assumption of a harmonic oscillation in z direction.

The Schrödinger equation of the resulting three-dimensional problem

$$i\hbar\dot{\Psi}(\vec{r},t) = \left\{\frac{1}{2m}\left(\frac{\hbar}{i}\nabla - \frac{q}{c}\vec{A}\right)^2 + \frac{m}{2}\omega_t^2 z^2\right\}\Psi(\vec{r},t) \qquad (27)$$

with ω_t: trapping frequency,

also can be separated by a product ansatz $\Psi(\vec{r},t) = \Psi_\perp(\vec{r}_\perp,t)\ \ \Psi_\parallel(z,t)$.
The functions $\Psi_\perp(\vec{r}_\perp,t)$ are already known from 9 and 11. The functions $\Psi_\parallel(z,t)$ are the well-known eigenfunctions of the harmonic-oscillator wave equation. Wave packet solutions for the motion in z direction can be constructed in analogy to $\Psi_{\perp,wp}(\vec{r}_\perp,t)$ (see e.g. Schiff [17], Landau and Lifschitz [18]).

2.4. Consideration of the Actual Electric Potential

The most general and most complex problem is attained if we take into consideration the actual electric potential according to the geometry of the cell and the applied voltages (see Figure 3).

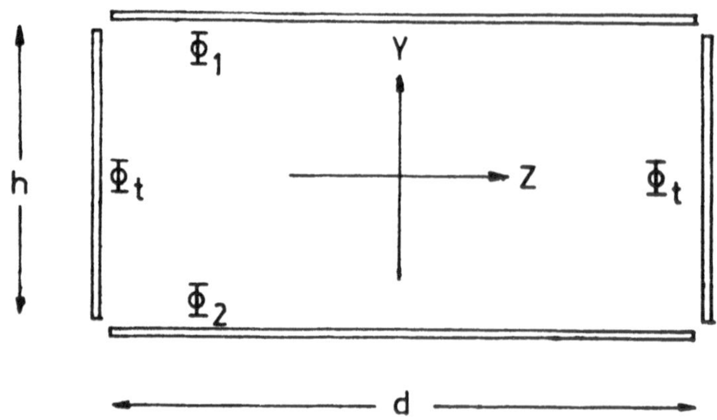

Fig. 3. Cross section of an ICR drift cell.
Φ_1 and Φ_2: drift voltages, Φ_t: trapping voltage.

This potential is quite complicated as one can see for instance regarding the potential of a drift cell given by

$$\Phi(y,z) = \Phi_t - \frac{2}{\pi}\sum_{k=0}^{\infty}\frac{(-1)^k}{(2k+1)}\cdot\{[2\Phi_t - (\Phi_1+\Phi_2)]\,\frac{\cosh[(2k+1)\pi y/d]}{\cosh[(2k+1)\pi h/2d]} - (\Phi_1-\Phi_2)\cdot\frac{\sinh[(2k+1)\pi y/d]}{\sinh[(2k+1)\pi h/2d]}\}\cos[(2k+1)^{\pi z/d}]. \tag{28}$$

Because of the strong coupling between y and z components, a separation cannot be achieved - neither classically nor quantum mechanically.

But, as the cyclotron frequency ω_c is higher by about an order of magnitude than the trapping frequency ω_t, it is legitimate to parametrize the z coordinate for the xy motion and vice versa. That means: the variables z or y respectively, in the potential 28 are replaced by the corresponding quantum mechanical expectation value. In this way, the equations for the z direction and the xy plane can be decoupled and treated separately by perturbation theory. For the ICR drift cell, this has been shown by Baykut [15].

3. More Complicated Magnetic Field $\vec{B} = (0,B_{\parallel},B_{\perp})$

After the discussion of problems with electric fields of increasing complexity, we now regard a quite different problem characterized by a different magnetic field orientation

For this purpose, we rotate an ICR drift cell around its longitudinal axis, the x axis, by a certain angle θ. Our coordinate frame shall be fixed in the cell. Now, drift field and magnetic field are no longer perpendicular, but the magnetic field has now a z component and a y component,

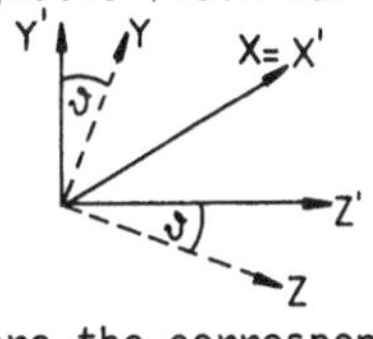

$$\vec{B} = B\vec{e}_{z'} = B_{\parallel}\vec{e}_y + B_{\perp}\vec{e}_z \tag{29}$$

$$B_{\parallel} = B\sin\theta,\quad B_{\perp} = B\cos\theta$$

where the corresponding vector potential is given by

$$\vec{A}(\vec{r}) = \frac{1}{2}[\vec{B}\times\vec{r}] = \frac{1}{2}(-B_{\perp}y + B_{\parallel}z,\ B_{\perp}x,\ -B_{\parallel}x). \tag{30}$$

What consequences for the cyclotron frequency ω_c should this manipulation entail?

Through the rotation, the components of the magnetic field change according to the relation

$$\theta = 0 : \vec{B} = (0,0,B) \rightarrow \theta \neq 0 : \vec{B} = (0,B_{\parallel},B_{\perp}), \tag{31}$$

the z component especially is given by $B_z = B_{\perp} = B\cos\theta$.

As it is just this component perpendicular to the plain of cyclotron motion, that usually determines the cyclotron frequency, one could at first sight expect the resonance frequency to change from ω_c to ω_c^* :

$$\theta = 0: \quad \omega_c = \frac{qB}{mc} \xrightarrow{(?)} \theta \neq 0: \quad \omega_c^* = \frac{qB_\perp}{mc} = \omega_c \cos\theta . \tag{32}$$

But, in consequence of the rotation, an additional coupling of y and z motions arises. So we need a more detailed investigation of this problem.

3.1. Classical Treatment

The classical Hamiltonian function corresponding to this problem is

$$H = \frac{1}{2m}(p_x^2 + p_y^2 + p_z^2) + \frac{\omega_\perp}{2}(yp_x - xp_y) + \frac{\omega_\parallel}{2}(xp_z - zp_x) + \frac{m}{8}(\omega_\perp^2 + \omega_\parallel^2)\,x^2$$

$$+ \frac{m}{8}(\omega_\perp^2 y^2 + \omega_\parallel^2 z^2 - 2\omega_\perp\omega_\parallel yz) + V(\vec{r}) \tag{33}$$

with $\omega_\perp = \omega_c \cos\theta$, $\omega_\parallel = \omega_c \sin\theta$.

For the sake of simplicity, we will discuss only a reduced potential given by

$$\Phi(\vec{r}) = \left(\frac{\Phi_D}{2} + \frac{\Phi_t}{2} - \frac{\Phi_D}{h}\,y - \frac{2\Phi_t}{h^2}\,y^2 + \frac{2\Phi_t}{d^2}\,z^2\right) \tag{34}$$

with Φ_D: drift voltage; Φ_t: trapping voltage; d,h of Figure 3; that means, only a homogeneous drift field in y direction and a quadratic approximation of the potential in yz direction will be utilized [19].

From the classical Hamiltonian we can get the time-derivatives of the velocity components

$$\ddot{v}_x = -\omega_c^2 v_x + \omega_\perp\omega_c v_D + \omega_\perp\omega_t^2 y + \omega_\parallel\omega_t^2 z$$

$$\ddot{v}_y = -(\omega_\perp^2 - \omega_t^2)\,v_y + \omega_\perp\omega_\parallel v_z$$

$$\ddot{v}_z = -(\omega_\parallel^2 + \omega_t^2)v_z + \omega_\perp\omega_\parallel v_y \tag{35}$$

$$\text{with } \omega_t = \left(\frac{4q\Phi_t}{m\,a^2}\right)^{1/2}$$

$$v_D = \frac{\Phi_D\,c}{B\,h} ,$$

where the coupling of the motions in y and z direction becomes quite obvious. It is possible to decouple these equations and to obtain the effective frequency for the oscillation in y direction which fulfils

$$\ddot{v}_y = -\omega_{eff}^2\, v_y$$

$$\omega_{eff}^2 = \omega_c^2\left\{\frac{1}{2} + \frac{1}{2}\left[1 - 4(1 - \sin^2\theta)\left(\frac{\omega_t}{\omega_c}\right)^2 + 4\left(\frac{\omega_t}{\omega_c}\right)^4\right]^{1/2}\right\} . \tag{36}$$

When the angle θ approaches zero, this frequency becomes identical with the well-known expression found by Beauchamp and Armstrong [19].

$$\lim_{\theta\to 0} \omega_{eff}^2 = \omega_{eff,0}^2 = \omega_c^2 \left(1-\left(\frac{\omega_t}{\omega_c}\right)^2\right) . \tag{37}$$

So, a comparison of equivalent approximations for the effective frequency,

$$\begin{aligned} &\text{without coupling: } \hat{\omega}_{eff} \approx \omega_c \cos\theta - \frac{\omega_t^2}{2\omega_c \cos\theta} \\ &\text{with coupling: } \omega_{eff} \approx \omega_c - \frac{\omega_t^2 \cos 2\theta}{2\omega_c} \end{aligned} \tag{38}$$

shows that the frequency shift would be much larger neglecting the coupling contrary to the more detailed investigation.

First preliminary experimental results are in agreement with this more sophisticated calculation.

3.2. Basis for a Quantum Mechanical Treatment

A quantum mechanical treatment of this problem requires the solution of the corresponding Schrödinger equation

$$\begin{aligned} &H^{*}\Psi_N^{*}(\vec{r}) = E_N^{*}\Psi_N^{*}(\vec{r}) \\ &\text{with } H^{*} = \frac{m}{2}\{v_x^{*2} + v_y^{*2} + v_z^{*2}\} + V(\vec{r}) \end{aligned} \tag{39}$$

where the velocity operators are defined by

$$\begin{aligned} v_x^{*} &= \frac{\hbar}{im}\left(\frac{\partial}{\partial x} + \frac{ib_\perp}{2} y - i\frac{b_\parallel}{2} z\right) \\ v_y^{*} &= \frac{\hbar}{im}\left(\frac{\partial}{\partial y} - \frac{ib_\perp}{2} x\right) \\ v_z^{*} &= \frac{\hbar}{im}\left(\frac{\partial}{\partial z} + \frac{ib_\parallel}{2} x\right) \end{aligned} \tag{40}$$

with $b_\perp = b\cos\theta$; $b_\parallel = \sin\theta$.

Starting again with the most simple case, i.e. without electric potential, we are abel to solve this equation exactly.

At first sight, the stationary solutions given by

$$\begin{aligned} \Psi_N^{*}(\vec{r}) = &\left(\frac{b}{2\pi}\left(\frac{1}{2b}\right)^N \frac{1}{N!}\frac{1}{L_z}\right)^{1/2} \exp\left\{\frac{ib_\perp}{2}(x_m y - x y_m) - \frac{ib_\parallel}{2}(x_m z - x z_m)\right. \\ &\left. + i(k_y \sin\theta + k_z\cos\theta)(y\sin\theta + z\cos\theta)\right\} \cdot \exp\left\{-\left[\left(\frac{\sqrt{b}}{2}(x-x_m)\right)^2\right.\right. \\ &\left.\left. + \left(\frac{b_\perp}{2\sqrt{b}}(y-y_m) - \frac{b_\parallel}{2\sqrt{b}}(z-z_m)\right)^2\right]\right\} \cdot \left[b(x-x_m) - ib_\perp(y-y_m) + ib_\parallel(z-z_m)\right]^N \end{aligned} \tag{41}$$

seem to be quite clumsy. Fortunately it is possible also for these functions to define appropriate raising- and lowering operators

$$v_+^* = v_x^* - iv_y^* \cos\theta + iv_z^* \sin\theta$$
$$v_-^* = v_x^* + iv_y^* \cos\theta - iv_z^* \sin\theta \qquad (42)$$

which show again these favourable qualities:

$$v_+^* \Psi_N^* = -\frac{\hbar}{im} (2b(N+1))^{1/2} \Psi_{N+1}^*$$
$$v_-^* \Psi_N^* = \frac{\hbar}{im} (2bN)^{1/2} \Psi_{N-1}^* \ . \qquad (43)$$

Therefore, once more calculations like the determination of time-derivatives of physical quantities reduce to the application of simple algebraic relations.

As an example, the time-derivatives of the expectation values of the velocity components

$$v_x^* = \frac{1}{2} (v_+^* + v_-^*)$$
$$v_y^* = \frac{i}{2} (v_+^* - v_-^*) \cos\theta + \frac{\hbar}{m} (k_y \sin\theta + k_z \cos\theta) + \sin\theta \qquad (44)$$
$$v_z^* = \frac{i}{2} (v_+^* - v_-^*) \sin\theta + \frac{\hbar}{m} (k_y \sin\theta + k_z \cos\theta) \cos\theta$$

can be easily determined, utilizing the commutation relation

$$[v_-^*, v_+^*] = \frac{2}{m} \hbar\omega_c \qquad (45)$$

and the Hamiltonian operator in this formulation

$$H^* = \frac{m}{4} (v_+^* v_-^* + v_-^* v_+^*) + \frac{\hbar^2}{2m} (k_y \sin\theta + k_z \cos\theta)^2 \ . \qquad (46)$$

Comparison of the results

$$\frac{d}{dt} \langle v_x^* \rangle = \langle \dot{v}_x^* \rangle = \omega_\perp \langle v_y^* \rangle - \omega_\parallel \langle v_z^* \rangle$$
$$\frac{d^2}{dt^2} \langle v_x^* \rangle = \langle \ddot{v}_x^* \rangle = -\omega_c^2 \langle v_x^* \rangle$$
$$\frac{d}{dt} \langle v_y^* \rangle = \langle \dot{v}_y^* \rangle = -\omega_\perp \langle v_x^* \rangle \qquad (47)$$
$$\frac{d^2}{dt^2} \langle v_y^* \rangle = \langle \ddot{v}_y^* \rangle = -\omega_\perp^2 \langle v_y^* \rangle + \omega_\perp\omega_\parallel \langle v_z^* \rangle$$
$$\frac{d}{dt} \langle v_z^* \rangle = \langle \dot{v}_z^* \rangle = \omega_\parallel \langle v_x^* \rangle$$
$$\frac{d^2}{dt^2} \langle v_z^* \rangle = \langle \ddot{v}_z^* \rangle = -\omega_\parallel^2 \langle v_z^* \rangle + \omega_\perp\omega_\parallel \langle v_y^* \rangle$$

with Eq. (35) shows that the expectation values obey the classsical equations of motion neglecting $\vec{v}_D$ and ω_t, as we did not regard any electric potential in our quantum mechanical treatment of this problem up till now.

Like shown in the case of usual magnetic field orientation, successively the electric potential may be taken into account up to the actual potential in the cell.

4. Velocity Dependent Frictional Forces

The last point of our discussion shall be the description of dissipative, velocity dependent frictional forces. These forces summarize collisional effects which slow down the ion during its flight in the ICR cell.

In classical particle mechanics Langevin's equation

$$m\dot{\vec{v}} = -\nabla V - m\gamma\vec{v} + F(t) \tag{48}$$

is assumed to give an adequate phenomenological description of this damping process.

- The first term on the right side is a conservative external force which can be derived from a potential V;
- the second term is a dissipative frictional force proportional to velocity, where γ is a friction constant. On this term we will concentrate in the following;
- finally, there is a rapidly fluctuating force $F(t)$ which is purely random and allows only statistical statements. (For a classical statistical treatment of this force cf. e.g. [20].)

For some external potentials - like for example the harmonic oscillator potential - it is possible to solve the classical equation of motion including the frictional force $-m\gamma\vec{v}$. Taking into account also the random force, the problem becomes by far more difficult (see again [20]).

On the other side, a field theoretical description of dissipative systems is not as easy as it might seem to be at first sight.

Especially, one has to proceed with caution if somehow also quantum theoretical aspects shall be taken into account. Actually, there exist several approaches to derive a quantum theoretical description of dissipation, but most of these suffer from serious difficulties.

The two major approaches use either a linear, but time-dependent Hamiltonian, or introduce a nonlinear damping potential into the Schrödinger equation, (e.g. see Refs. cited in [31,32]).

The first group, due to Kanai [21], uses canonical quantization to derive an explicitly time-dependent Hamiltonian operator on the basis of a classical Hamiltonian function which does not represent the energy of the system. The most serious shortcoming of this method is violation of the uncertainty principle [22,23]. Modifications of the Kanai Hamiltonian try to get rid of these difficulties but need concepts like variable mass particles [24] for this purpose.

In order to avoid difficulties with nonphysical Hamiltonians, the representatives of the second approach try to introduce additional "friction potentials" into the linear Schrödinger equation by various other quantization methods like stochastic quantization or only by phenomenological arguments. In consequence of these methods, nonlinear Schrödinger-type equations (NLSE) are obtained, where the first nonlinear approach is due to Kostin [25], but recently several modifications have been published (cf. Refs. cited in [31,32]).

However, all the NLSEs proposed hitherto suffer from the weakness that the solutions of these equations show at least in one respect unphysical behaviour. Especially the fact that the undamped solutions are also solutions of the problem including friction is difficult to interpret. Furthermore, the solutions of the damped harmonic oscillator do not contain the right reduced frequency.

A difficulty all these nonlinear Schrödinger-type equations (NLSE) have in common, is that the superposition principle which is valid in "orthodox" linear quantum mechanics, now no longer applies. But this fact by itself should not exclude that these NLSEs might be interpreted as quantum theoretical equations in the sense of a probabilistic interpretation of the wave function (if we want to restrict the term quantum mechanics to the linear theory), if such an equation is able to describe the dynamics of physical systems without inconsistencies and in accordance with empiricism, also on a microscopic level. In order to avoid terminological confusion, perhaps it might be even advantageous in this case to replace the term quantum theory by some other more appropriate expression.

As it is known, many phenomena which within the scope of a corpuscular consideration cannot be explained until quantization, can be understood within the scope of an undulatory consideration already on classical level. Prominent example is the treatment of the linear Schrödinger equation as classical field equation, which is discussed in detail e.g. by Hund [26]. Interpreting the Schrödinger field as classical field of matter, physical qualities can be associated with this field by the aid of properly defined classical mean values and operators. Although there exists a striking similarity between these operators and mean values and the respective quantum mechanical operators and expectation values, this similarity is only a formal one, as the meaning and interpretation of these quantities is quite different in classical and quantum theories.

So, an undulatory theory for the description of dissipative systems has not necessarily to be a quantum theoretical description. If the interpretation of a field equation (and the constituting quantities) is not compatible with fundamental principles of quantum theory, nevertheless such an undulatory theory may yield interesting and valuable results already on the classical level, as it can be seen taking the linear Schrödinger equation of a material field as example [26-28].

There does not exist a unique and reliable method for the "derivation" of a new field equation. So, for example several different methods can be used to make plausible how the undulatory theory using the linear Schrödinger equation as fundamental field equation can be established on the basis of a few physically resonable assumptions.

As already mentioned, hitherto attempts to obtain a field equation for the description of dissipative systems by the use of methods known from the "derivation" of the linear Schrödinger equation were not yet entirely successful.

Madelung and Mrowka [29,30] established a method different from those used by the authors cited above to derive a field equation, the linear Schrödinger equation, where classical particle mechanics enters only in Newton's formulation. An attempt to incorporate damping effects into the linear Schrödinger equation using the method of Madelung and Mrowka, led to rather cumbersome perturbation theoretical calculations and was therefore not completely satisfactory [11].

Now, we developed a new method similar to that of Madelung and Mrowka to derive a nonlinear field equation corresponding to the Langevin equation, which is capable to describe dissipative systems [31-33]. As there does not yet exist any pendant to our NLSE, a priori we have to take into account different possibilities of interpretation. In Refs. [31-33], this method and the possibility of classical undulatory or probabilistic (e.g. quantum theoretical) interpretations of our field equation are discussed in more detail. As we could not find an absolutely convincing argument favouring one of the interpretations and as we do not know an *experimentum crucis* to decide with certainty which possibility is the only right one, at the moment there only remains to indicate the different possible interpetations, remarking that in each case one has to investigate whether the chosen interpretation leads inconsistencies or not.

Because of the formal similarity between classical undulatory theory and quantum theory, however, calculation can be performed without the necessity to decide previously in favour of one of the possible interpretations. Only the occurring quantities have to be provided with the meaning corresponding to the accepted interpretation. So, e.g. the function Ψ in classical field theory is proportional to the amplitude of a field of matter, in quantum theory, however, it has to be interpreted as probability amplitude, and the symbol <A> denotes the correspondingly defined classical mean value of a quantity A or its quantum theoretical expectation value, respectively. Therefore, undermentioned terms like wave function, operator, mean value etc. shall be used synonymously for all possible interpretations (as long as no particular interpretation is accepted). For formal reasons, namely to facilitate comparison of the results of our nonlinear field theory with results of other authors or results obtained in linear quantum mechanics, the quantity $\hbar$ ($\hbar = h/2\pi$ with h: Planck's constant) appears in our NLSE and the results appertaining to this equation. In the course of the development of our field theory it is possible, as we will show, to ignore the existance of $\hbar$ and to avoid the appearance of quantities like m and q which usually denote corpuscular properties like mass or charge of a particle. Therefore, prefering the classical undulatory interpretation of our NLSE it is easy via the definitions $\nu = m/\hbar$ and $\zeta = q/\hbar$ to introduce appropriate quantities which can be interpreted completely within the framework of classical field theory. In this work, however, we shall neither use this notation (for formal reasons), nor discuss the consequences of the classical undulatory interpretation in more detail.

Comparing the results of our nonlinear field theory with those obtained by the other authors using NLSEs we find that the solutions of our NLSE no longer show the unphysical behaviour mentioned above.

In the following we will give a short outline of the derivation of the NLSE for the damped motion under the influence of magnetic and electric fields which approximately represent the conditions in an ICR cell (for a more detailed discussion of the method cf. [31-33]).

4.1. One-dimensional Damped Harmonic Oscillator

We want to illustrate our method using the one-dimensional damped harmonic oscillator as an example which can be a useful model for the description of trapping motion in the ICR cell.

Axiomatic basis of our method are three general principles taken from empiricism:

1. Uncertainty principle or complementarity, respectively,
2. the fact that interference phenomena show up in experiments with material systems, and
3. correspondence principle or Ehrenfest's theorem, respectively.

The mathematical form of a theory taking into account this empirical knowledge follows from the structure of these principles. Due to uncertainty principle exact initial conditions in the sense of classical particle mechanics cannot be given. Therefore, it is only possible to develop a theory where mean values $\overline{Q}$ of quantities Q are determined with the aid of a distribution function ρ. According to the interpretation accepted, ρ is directly connected with the material density of the field or ρ is a probability density, respectively. Paying regard to the irreversible character of dissipative processes, ρ is assumed to fulfil the Fokker-Planck equation

$$\dot{\rho} + \mathrm{div}\,(\rho\vec{v}) - D\Delta\rho = 0 \quad . \tag{49}$$

Bearing in mind our second axiom, in analogy to optics (where intensity is a quatratic function of the amplitudes) the bilinear ansatz $\rho = \alpha \cdot \beta \geq 0$ with the field amplitudes $\alpha(\vec{r},t)$ and $\beta(\vec{r},t)$ shall be used. With these amplitudes also the flux $\vec{j} = \rho\vec{v}$ can be defined (in the absence of magnetic fields) by a bilinear form,

$$\begin{aligned} \vec{j} &= \rho\vec{v} = C(\beta\nabla\alpha - \alpha\nabla\beta) \\ \text{with } \vec{v} &= C\nabla \ln\frac{\alpha}{\beta} = C\left(\frac{\nabla\alpha}{\alpha} - \frac{\nabla\beta}{\beta}\right) \end{aligned} \tag{50}$$

where C is a constant.

Finally, the third axiom shall be taken into consideration in the form of Ehrenfest's theorem, i.e. at an average the classical equations of motion are valid:

$$\overline{m\frac{d^2}{dt^2}\vec{r}} = \overline{m\frac{d}{dt}\vec{v}} = m\frac{d}{dt}\int \rho\vec{v}\,d\tau = m\int\frac{\partial}{\partial\tau}\vec{j}\,d\tau = \overline{\vec{F}} \tag{51}$$

with $d\tau$ = element of volume.

Inserting the preceding definitions into Eq. (51), we obtain

$$\overline{\vec{F}} = 2mC\int\{\dot{\beta}\nabla\alpha - \dot{\alpha}\nabla\beta\}d\tau. \tag{52}$$

In addition, by the help of these definitions Eq. (49) can be separated into two equations containing only α or β, respectively, if furthermore the condition

$$\begin{aligned} -\frac{D\Delta\rho}{\rho} &= \gamma(\ln\rho + \tilde{Z}) \\ \text{with } \tilde{Z} &= Z + Z^{*} = \text{function independent of } \alpha \text{ and } \beta \\ Z &= Z_R + Z_I \\ \gamma &= \text{constant} \end{aligned} \tag{53}$$

is fulfilled. For the discussion of consequences for the functions α and β due to this condition, see [31,32].

So α has to obey the relation

$$\dot{\alpha} + C\Delta\alpha + \gamma(\ln\alpha + Z)\alpha + f\alpha = 0, \tag{54}$$

f being a function independent of α and β.

The conjugate complex function β obeys the corresponding conjugate complex equation. Inserting these equations into the averaged Eq. (52), after some arithmetics we receive

$$\overline{m\dot{\vec{v}}} = \int \rho\{-\nabla(2mCf) - m\gamma\nabla \ln\frac{\alpha}{\beta}\}d\tau. \tag{55}$$

Using Def. (50) of $\vec{v}$ and identifying $2mCf = V$, where V is the potential of a conservative force, we obtain

$$\overline{m\dot{\vec{v}}} = \int\rho(-\nabla V)d\tau + \int\rho\,(-m\gamma\vec{v})d\tau = \overline{-\nabla V} - \overline{m\gamma\vec{v}} \tag{56}$$

which is the averaged Langevin equation without stochastic force.
Replacing α by Ψ, Eq. (54) attains the form of a nonlinear Schrödinger-type equation (NLSE)

$$i\hbar\dot{\Psi} = \{-\frac{\hbar^2}{2m}\Delta + V + \gamma\frac{\hbar}{i}(\ln\Psi + Z_R + iZ_I)\}\Psi = H\Psi \tag{57}$$

if $C = \hbar/2mi$ is chosen. This choice is not the only possible one, as the constant C is not determined unequivocally by the three basic assumptions, but has to be taken from experimental experience. Therefore, from the viewpoint of classical field theory also a choice $C = 1/2\,\nu i$, i.e. a constant without $\hbar$ and m, is possible. With $V = q\Phi$, e.g., and the foregoing definitions of ν and ζ, in this case the NLSE would read

$$i\dot{\Psi} = \{-\frac{1}{2\nu}\Delta + \zeta\Phi + \frac{\gamma}{i}(\ln\Psi + Z_R + iZ_I)\}\Psi, \tag{58}$$

but, as already noticed, for formal reasons this form shall not be used in the following.

Based on physical arguments [31,32] Z_R and Z_I can be determined in a way that the functions Ψ obey the NLSE

$$i\hbar\dot{\Psi} = \{-\frac{\hbar^2}{2m}\Delta + V + \gamma\frac{\hbar}{i}(\ln\Psi - \langle\ln\Psi\rangle\}\Psi = H\Psi\ . \tag{59}$$

From this form of our NLSE it can be taken that the dissipative term is connected with deviations from a mean value. Moreover, $\langle H\rangle = \langle T\rangle + \langle V\rangle$, that means the mean value of the Hamiltonian is at any time equal to the sum of kinetic and potential energies as it is in usual linear quantum mechanics or, in metaphorical meaning, in nondissipative classical particle mechanics, respectively.

With $V = \frac{m}{2}\omega_t^2 z^2$ the NLSE corresponding to the Newton-type equation of motion

$$F = m\ddot{z} = -m\omega_t^2 z - m\gamma\dot{z} \tag{60}$$

is given by

$$i\hbar\dot{\Psi}_{\shortparallel} = \{-\frac{\hbar^2}{2m}\frac{d^2}{dz^2} + \frac{m}{2}\omega_t^2 z^2 + \gamma\frac{\hbar}{i}(\ln\Psi_{\shortparallel} - \langle\ln\Psi_{\shortparallel}\rangle)\}\Psi_{\shortparallel}\ . \tag{61}$$

The solution of this equation

$$\Psi_{II}(z,t;\eta(t)) = \left(\frac{m\Omega}{\hbar\pi}\right)^{1/4} \exp\{\frac{i}{\hbar}[m\dot{\eta}\,(z-\eta) - m\frac{\gamma}{4}(z-\eta)^2] - \frac{m}{2\hbar}\Omega(z-\eta)^2\} \cdot \chi(t)$$

$$\text{where } \chi(t) = \exp\{\frac{i}{\hbar}\int_0^t L(\eta,\dot{\eta},t')dt' - \frac{\hbar}{2}\omega_t\left(\frac{\omega_t}{\Omega}\right)t\} \qquad (62)$$

$$\text{with } L(\eta,\dot{\eta},t) = \frac{m}{2}\dot{\eta}^2(t) - \frac{m}{2}\omega_t^2\eta^2(t),$$

shows a form similar to a wave packet in linear quantum mechanics; it is a soliton-like solution, in particular a so-called gausson [34], and contains the correct reduced frequency $\Omega = (\omega_t^2 - \gamma^2/4)^{1/2}$.

The centre of the gausson is following the classical trajectory $\eta(t)$ which obeys the equation

$$m\ddot{\eta} = -m\omega_t^2\eta - m\gamma\dot{\eta}. \qquad (63)$$

This gausson also can be expanded into a series of wave functions, now oscillating with the frequency Ω,

$$\Psi_{II}(z,t;\eta(t)) = \sum_{n=0}^{\infty} C_n v_n \exp(i\{\chi(t) + f(z,t;\eta)\})$$

$$\text{where: } v_n = \frac{\beta^{1/2}}{\pi^{1/4}}\left(\frac{2^{-n}}{n!}\right)^{1/2} e^{-1/2\beta^2 z^2} H_n(\beta z)$$

$$\text{with } \beta^2 = \frac{m\Omega}{\hbar} \qquad (64)$$

$$H_n(\beta z) = n^{th} \text{ Hermite polynomial}$$

$$\chi(t) = \text{see Eq. (60)}$$

$$f(z,t;\eta) = \frac{1}{\hbar}\{m\dot{\eta}\,(z-\eta) - m\frac{\gamma}{4}(z-\eta)^2\}$$

$$C_n = (2^n\, n!)^{-1/2}(\beta\eta)^n e^{-1/4(\beta\eta)^2},$$

but none of these wave functions v_n $(n>0)$ for itself is solution of the NLSE. Especially, the solutions of the undamped problem are no longer solutions of the problem including friction.

As $t\to\infty$ the gausson asymptotically approaches a final state

$$\Psi_{II,\infty}(z,t) = \left(\frac{m\Omega}{\hbar\pi}\right)^{1/4} \exp\{-\frac{i}{\hbar}\frac{m\gamma}{4}z^2 - \frac{m\Omega}{2\hbar}z^2 - \frac{i}{\hbar}\frac{\hbar}{2}\omega_t\left(\frac{\omega_t}{\Omega}\right)t\} \qquad (65)$$

with stationary density function $\rho(z)$.

This final state in the limit $\gamma\to 0$ turns into the ground state wave function of the undamped harmonic oscillator.

Furthermore, the uncertainty principle and Ehrenfest's theorem are fulfilled, as shown in [31,32].

The diffusion constant D of the Fokker-Planck equation can be determined as

$$D = \frac{\hbar}{2m}\frac{\gamma/2}{\Omega} \qquad (66)$$

where $\lim_{\gamma\to 0} D = 0$.

4.2. Two-dimensional Damped Motion in a Magnetic Field

As under the influence of a magnetic field $\vec{v} = \vec{p}/m - \frac{q}{mc}\vec{A}$ is valid, Def. (50) of $\vec{v}$ has to be replaced by

$$\vec{v} = C\left(\frac{\nabla\alpha}{\alpha} - \frac{\nabla\beta}{\beta}\right) - \frac{q}{mc}\vec{A}. \qquad (67)$$

Inserting this Def. (67) into the Fokker-Planck equation, again it is possible to derive a NLSE,

$$i\hbar\dot{\Psi}_\perp = \left\{\frac{1}{2m}\left(\frac{\hbar}{i}\nabla_\perp - \frac{q}{c}\vec{A}(\vec{r}_\perp)\right)^2 + \gamma\frac{\hbar}{i}\left(\ln\Psi_\perp - \langle\ln\Psi_\perp\rangle\right) - \gamma\frac{q}{c}\vec{A}(\vec{\eta}_\perp)\cdot\vec{r}_\perp\right\}\Psi_\perp \qquad (68)$$

corresponding to the classical Newton-type equation

$$\vec{F} = \frac{q}{c}[\vec{v}\times\vec{B}] - m\gamma\vec{v} \quad . \qquad (69)$$

The solution of this NLSE is also a gausson

$$\Psi_\perp(\vec{r}_\perp,t;\vec{\eta}_\perp(t)) = \left(\frac{\hat{b}}{2\pi}\right)^{1/2}\exp\left\{\frac{i}{\hbar}\left(m\dot{\vec{\eta}}_\perp + \frac{q}{c}\vec{A}(\vec{\eta}_\perp)\right)\cdot(\vec{R}-\vec{a}) - \frac{i}{\hbar}\frac{m\gamma}{4}(\vec{R}-\vec{a})^2 + iK_\perp^\circ(t)\right\} \qquad (70)$$

$$\cdot\exp\left\{-\frac{\hat{b}}{4}(\vec{R}-\vec{a})^2\right\}$$

with $\hat{b} = \frac{m}{\hbar}\hat{\Omega}$, $\quad \hat{\Omega} = (\omega_c^2 - \gamma^2)^{1/2}$

and $K_\perp^\circ(t) = \frac{1}{\hbar}\int_0^t \frac{m}{2}\dot{\vec{\eta}}_\perp^2(t') + \frac{q}{c}\vec{A}(\vec{\eta}_\perp)\,\dot{\vec{\eta}}_\perp(t')\,dt' - \frac{1}{2}\frac{\omega_c}{\hat{\Omega}}\,t$

where $m\ddot{\vec{\eta}}_\perp = \frac{q}{c}[\dot{\vec{\eta}}_\perp\times\vec{B}] - m\gamma\dot{\vec{\eta}}_\perp$ with $\vec{\eta}_\perp = \vec{r}_m + \vec{a}$.

Here, $\vec{a}\to\vec{0}$ as $t\to\infty$, so that

$$\lim_{\vec{a}\to\vec{0}}\Psi_{\perp,\infty}(\vec{r}_\perp,t;\vec{\eta}_\perp) = \Psi_{\perp,\infty}(\vec{r}_\perp,t) = \left(\frac{\hat{b}}{2\pi}\right)^{1/2}\exp\left\{\frac{iq}{\hbar c}\vec{A}(\vec{r}_m)\cdot\vec{R} - \frac{i}{\hbar}\frac{m\gamma}{4}R^2 - \frac{\hat{b}}{4}R^2 - \frac{i}{2}\frac{\omega_c^2}{\hat{\Omega}}t\right\}. \qquad (71)$$

In the limit $\gamma\to 0$ this wave function turns into the (time dependent) screw-function describing the ground state of the undamped motion

$$\lim_{\gamma\to 0}\Psi_{\perp,\infty}(\vec{r}_\perp,t) = \left(\frac{b}{2\pi}\right)^{1/2}\exp\left\{\frac{iq}{\hbar c}\vec{A}(\vec{r}_m)\cdot\vec{R} - \frac{b}{4}R^2 - \frac{i}{2}\omega_c t\right\} = \Psi_{\perp,0}(\vec{r}_\perp,t). \qquad (72)$$

In analogy to the undamped problem investigated in 2.2., an additional drift motion in x-direction with constant velocity $\vec{v}_D$ can be taken into account by Galilean- and gauge transformations, yielding a function similar to (70), but now $\vec{\eta}_\perp$ obeying

$$m\ddot{\vec{\eta}}_\perp = \frac{q}{c}[\dot{\vec{\eta}}_\perp\times\vec{B}] - m\gamma\dot{\vec{\eta}}_\perp - m\gamma\vec{v}_D, \qquad (73)$$

and an additional time-dependent factor

$$\exp\{\frac{i}{\hbar}\int_0^t q\,\vec{E}_D\cdot\dot{\vec{\eta}}_\perp\cdot t'dt'\} \qquad (74)$$

occurs, paying regard to the potential energy due to the applied drift field.

4.3. Three-dimensional Damped Motion in Magnetic and Electric Fields

Considering frictionally damped three-dimensional motion, our method has to be extended to include also anisotropic conditions, as different diffusion constants may be possible for different space directions. Therefore, the diffusion constant D in the Fokker-Planck equation is replaced by a 3 x 3 matrix, the diffusion tensor $\hat{D}$ with components D_{kj}, and $-D\Delta\rho$ by the more general expression

$$\mathrm{div}\,\vec{J}_D = -\sum_{k,j=1}^{3} \nabla_k\nabla_j(D_{kj}\rho) \qquad (75)$$

or, if the elements D_{kj} of $\hat{D}$ are constant,

$$\mathrm{div}\,\vec{J}_D = -\sum_{k,j=1}^{3} D_{kj}\,\nabla_k\nabla_j\,\rho(\vec{r}), \qquad (76)$$

respectively.

From a condition of separability analogue to (53) it follows, that $\hat{D}$ is an anti-symmetric matrix.

As an example, we now regard the motion of ions under the influence of a homogeneous magnetic field in z-direction and an electric field given by the quadratic approximation of Beauchamp and Armstrong [19], neglecting the drift potential and constant contributions of the electric potential.

The NLSE of this problem, correponding to the classical Newton-type equation of motion

$$\vec{F} = q\vec{E} + \frac{q}{c}[\vec{v}\times\vec{B}] - m\gamma\vec{v} \qquad (77)$$

$$\text{with } \vec{E} = (0,\ \frac{m}{q}\omega_t^2 y,\ -\frac{m}{q}\omega_t^2 z) \qquad (78)$$

is given by

$$i\hbar\dot{\Psi} = \{\frac{1}{2m}(\frac{\hbar}{i}\nabla - \frac{q}{c}\vec{A}(\vec{r}))^2 + \gamma\frac{\hbar}{i}(\ln\Psi - \langle\ln\Psi\rangle) - \gamma\frac{q}{c}\vec{A}(\vec{\eta})\cdot\vec{r} + \frac{m}{2}\omega_t^2 z^2 - \frac{m}{2}\omega_t^2 y^2\}\Psi. \qquad (79)$$

The exact solution is again a gausson, similar to those mentioned above, with the density

$$\rho(\vec{r},\vec{\eta}(t)) = \Psi^*\Psi = N_x^2\exp\{-2T_x(x-\eta_x)^2\}\cdot N_y^2\exp\{-2T_y(y-\eta_y)^2\}\cdot N_z^2\exp\{-2T_z(z-\eta_z)^2\}$$

$$= \rho_x(x)\cdot\rho_y(y)\cdot\rho_z(z)\ , \qquad N_{x_i}:\ \text{normalization constant} \qquad (80)$$

showing product form. (For a more detailed discussion see [31,33].)

The diagonal elements of the diffusion tensor, $D_{x_i x_i}$, can be determined from

$$D_{x_i x_i} = \frac{\gamma}{8T_{x_i}} \tag{81}$$

where

$$T_x = \frac{m}{2\hbar}\Omega_x \quad \text{with} \quad \Omega_x = \frac{\omega_c}{2}\left(\frac{4}{(\sqrt{\xi}+\sqrt{\sigma})^2} - 1\right)^{1/2} \cdot \sqrt{\xi}$$

$$T_y = \frac{m}{2\hbar}\Omega_y \quad \text{with} \quad \Omega_y = \frac{\omega_c}{2}\left(\frac{4}{(\sqrt{\xi}+\sqrt{\sigma})^2} - 1\right)^{1/2} \cdot \sqrt{\sigma} \tag{82}$$

$$T_z = \frac{m}{2\hbar}\Omega_z \quad \text{with} \quad \Omega_z = (\omega_t^2 - \gamma^2/4)^{1/2}$$

and

$$\xi = \left(\frac{\gamma}{\omega_c}\right)^2$$

$$\sigma = \frac{\gamma^2 + 4\omega_t^2}{\omega_c^2} \quad .$$

5. Conclusion

We tried to show that a description of ion motion in an ICR cell by means of classical particle mechanics is not the only possible way. An alternative which sometimes can be even advantageous and more powerful is the description within the scope of field theory.

Neglecting dissipative processes, usual quantum mechanics, which is a linear field theory, can be applied for this purpose. In this connection, we discussed some exactly solvable quantum mechanical problems as well as the properties of their solutions. However, to describe the quasi-classical behaviour of ions in ICR spectrometry, quantum mechanical pure states are not adequate, but moreover wave packets, so-called coherent states which can be constructed by superposition of pure states have to be used for this aim.

Regarding also dissipative phenomena like velocity dependent frictional forces, the situation becomes different in a certain respect. In classical particle mechanics a Newton-type equation for the description of such problems exists which can be extended to include also random forces, then known as Langevin equation. Contrary in linear quantum mechanics up till now no simple corresponding Schrödinger equation exists. We showed that it is possible to derive a field equation for the description of dissipative phenomena exhibiting certain similarities with the linear Schrödinger equation.

But our equation is a nonlinear differential equation wherefrom essential differences between our theory and linear quantum mechanics originate. However, because of the formal similarity between classical undulatory and probabilistic theories like quantum theory, calculations can be performed independent of the interpretation of our equation, only the occurring quantities and results have to be interpreted in different ways. It can be shown that the solutions of our equation are stable solitons, in particular so-called gaussons, which are quite similar to the wave packets we had to construct in linear quantum mechanics for application in the theory of ICR spectrometry. These gaussons asymptotically approach a stationary final state as time tends to infinity. Transition from our nonlinear field theory to linear Schrödinger field theory can be accomplished in the limit friction coefficient $\gamma \to 0$.

So, our NLSE might open a new way for the treatment of dissipative phenomena like those occurring in ICR spectrometry within the framework of field theory.

6. References

1. J. D. Jackson, "Classical Electrodynamics" Wiley and Sons, Inc., New York, 1962
2. J. Bialiniki-Birula, Ann. Phys. 67 (1971) 252.
3. E. Schrödinger, Naturwiss. 28 (1926) 664.
4. R. J. Glauber, Phys. Rev. 131 (1963) 2766.
5. L. S. Brown, Am. J. Phys. 41 (1973) 525.
6. J. Mostowski, Phys. Lett. 56A (1976) 369.
7. J. Mostowski, Lett. Math. Phys. 2 (1977) 1.
8. A. O. Barut, Z. Naturforsch. 32a (1977) 369.
9. H. Hartmann and K.-M. Chung, Theoret. Chim. Acta (Berl.) 45 (1977) 137.
10. H. Hartmann and K.-M. Chung, Lecture Notes in Chemistry, Vol. 7. Springer-Verlag, Berlin, 1978.
11. H. Hartmann, K.-M. Chung, D. Schuch and J. Radtke, Theoret. Chim. Acta (Berl.) 53 (1979) 203.
12. H. Hartmann, K.-M. Chung, D. Schuch and K.P. Wanczek, Int. J. Mass Spectrom. Ion Phys. 34 (1980) 303.
13. H. Hartmann, K.-M. Chung, G. Baykut and K.-P. Wanczek, J. Chem. Phys. to be published.
14. A. Janussis, Phys. Status Solidi 6 (1964) 217.

15. G. Baykut, Thesis, University, Frankfurt, 1980.

16. K.-M. Chung and B. Mrowka, Z. Physik 259 (1973) 157.

17. L. I. Schiff, "Quantum Mechanics", McGraw-Hill, Tokyo,1968.

18. L. D. Landau and E. M. Lifschitz,"Lehrbuch der Theoretischen Physik III, Quantenmechanik", Akademie-Verlag, Berlin,1979.

19. J. L. Beauchamp and T. J. Armstrong, Rev. Sci. Instrum. 40 (1969) 123.

20. S. Chandraskhar, Rev. Mod. Phys. 15 (1943) 1.

21. E. Kanai, Progr. Theor. Phys. 3 (1948) 440.

22. W. E. Brittin, Phys. Rev. 77 (1950) 396.

23. J. R. Ray, Nuovo Cim. Lett. 25 (1979) 47.

24. D. M. Greenberger, J. Math. Phys. 20 (1979) 762.

25. M. D. Kostin, J. Chem. Phys. 57 (1972) 3589.

26. F. Hund, "Materie als Feld", Springer-Verlag, Berlin, 1954.

27. E. Fick, "Einführung in die Grundlagen der Quantenmechanik", Akademische Verlagsgesellschaft, Frankfurt, 1974.

28. H. Hartmann,"Die chemische Bindung. Drei Vorlesungen für Chemiker", Springer-Verlag, Berlin, 1971.

29. E. Madelung, "Die mathematischen Hilfsmittel des Physikers", Springer-Verlag, Berlin, 1950.

30. B. Mrowka, Z. Physik, 130 (1951) 164.

31. D.Schuch, Thesis, University, Frankfurt, 1982.

32. D. Schuch, K.-M. Chung and H. Hartmann, submitted for publication to J. Math. Phys.

33. D. Schuch, K.-M. Chung and H. Hartmann, submitted for publication to Int. J. Quant. Chem.

34. J.Bialinicki-Birula and J. Mycielski, Ann. Phys. (N.Y.) 100 (1976) 62.